P9-EJK-588

INDUSTRIAL MOTOR CONTROL

INDUSTRIAL MOTOR CONTROL

Third edition

Stephen L. Herman
Walter N. Alerich

NOTICE TO THE READER

Publisher does not warrant or guarantee any of the products described herein or perform any independent analysis in connection with any of the product information contained herein. Publisher does not assume, and expressly disclaims, any obligation to obtain and include information other than that provided to it by the manufacturer.

The reader is expressly warned to consider and adopt all safety precautions that might be indicated by the activities described herein and to avoid all potential hazards. By following the instructions contained herein, the reader willingly assumes all risks in connection with such instructions.

The publisher makes no representations or warranties of any kind, including but not limited to, the warranties of fitness for particular purpose or merchantability, nor are any such representations implied with respect to the material set forth herein, and the publisher takes no responsibility with respect to such material. The publisher shall not be liable for any special, consequential or exemplary damages resulting, in whole or in part, from the readers' use of, or reliance upon, this material.

Cover Design: design M design W

New Product Acquisitions Editor: Mark W. Huth
Project Editor: Eleanor Isenhart
Senior Design Supervisor: Susan C. Mathews
Senior Production Supervisor: Larry Main

For information, address Delmar Publishers Inc.
3 Columbia Circle, PO Box 15015
Albany, New York 12212-5015

Printed in the United States of America
Published simultaneously in Canada by Nelson Canada,
A Division of The Thomson Corporation

10 9 8 7 XXX 99 98 97

Library of Congress Cataloging in Publication Data

Herman, Stephen L.
Industrial motor control / Stephen L. Herman, Walter N. Alerich — 3rd ed.
p. cm.
Includes index.
ISBN 0-8273-5252-2
1. Electric controllers. 2. Electric motors. I. Alerich, Walter N. II. Title.
TK2851.H47 1992
621.46—dc20 92-15455
CIP

CONTENTS

Section 4 Basic Control Circuits 203

Section 5 Dc Motor Controls 233

Section 6 Ac Motor Control 255

PREFACE

The amount of knowledge an electrician must possess to be able to install and repair control systems in today's industry has increased dramatically in recent years. A constant flow of improved control components allows engineers and electricians to design and install more sophisticated and complex control systems. Programmable logic controllers are replacing the older magnetic relay systems, forcing electricians to have a working knowledge of both types of control systems. The influx of programmable controllers into industrial control systems has provided a bridge that spans the gap between the responsibilities of the industrial electrician and the instrumentation technician. *Industrial Motor Control, Third edition* provides information on analog sensing of pressure, flow, and temperature to help narrow this gap between electricians and instrumentation technicians.

Electronic devices provide dependable, fast service in a wide array of industrial applications. The third edition of *Industrial Motor Control* was written to ensure that electricians have a working knowledge of electronic devices and circuits. This text provides the theory electricians need to form a solid understanding of electronic components and their uses. Devices commonly used in industry are explained from a practical standpoint rather than from a mathematical standpoint. The electronic components covered are the diode, zener diode, junction transistor, unijunction transistor, SCR, diac, and triac. The appendix offers a step-by-step procedure for field testing these components. The text also discusses two important integrated circuits often found in industrial circuits—the 555 timer and the 741 operational amplifier.

Industrial Motor Control covers numerous types of control devices, many of which contain electronic components. For this reason, electronic components are discussed near the beginning of the text. This placement permits different types of control components and devices to be covered in a logical sequence according to their function, where others might divide the text between magnetic devices and electronic devices. Some of the control devices presented in this text are magnetic relays and motor starters, solid state relays, overload relays, timers, (both magnetic and electronic), Hall effect sensors, proximity detectors, (including metal, sonic, and capacitive), and photodetectors. Other types of control components, such as push buttons, float switches, and limit switches, are also presented.

Once the student understands what these devices do, the components are applied in circuits designed for specific functions. Typical industrial control circuits are explained in a step-by-step procedure to help the student understand how to read and interpret schematic diagrams. *Industrial Motor Control* also gives step-by-step instructions for converting ladder diagrams into wiring diagrams. This will help the student understand how to connect actual control circuits in the field.

Industrial Motor Control covers different types of direct current motors, three phase and single phase alternating current motors, and stepping motors. Different methods for starting, accelerating, stopping, and reversing these motors are discussed. A separate section of the text details step-by-step instructions on how to compute the size conductor, overload relay, and short-circuit protective device needed when connecting motors in the field. All calculations are based on the *National Electrical Code.*

The last section of *Industrial Motor Control* deals with basic computer logic and gate circuits. This section also introduces the programmable controller and describes its capabilities. Ladder diagrams are converted into programs for the programmable controller. Two different circuits are used. One circuit assumes the use of a programming terminal and the second illustrates how a program is loaded using *Boolean.* This section also covers analog sensing for programmable controllers and gives common rules for installing programmable controllers.

To complement the text, the *Instructor's Guide* lists the learning objectives of the text for the instructor's convenience, as well as a bank of test questions. The guide also provides answers to the unit review questions.

FEATURES OF THE THIRD EDITION

- Extensive coverage of solid-state control devices in addition to electromagnetic devices.
- Basic electronics is not a prerequisite for studying this text. Sufficient solid-state theory is presented to enable the student to understand and apply the concepts discussed.
- The most commonly used solid-state devices are thoroughly described in terms of both operation and typical application.
- Information on analog devices, which sense pressure, flow, and temperature, is provided to help bridge the gap between the industrial electrician and the instrumentation technician.
- Dc and ac motor theory is included so students will understand the effects of control circuits on motor characteristics.
- The text covers the operating characteristics of stepping motors when connected to either dc or ac voltage.
- Detailed instructions are given for connecting motors in the field, including the size of conductors, overload relays, and fuses or circuit breakers. All calculations are taken from the *National Electrical Code.*
- The principles of digital logic are discribed in sufficient detail for students to understand programmable controllers and prepare basic programs.
- A step-by-step testing procedure for electronic components is provided in the Appendix.

ABOUT THE AUTHORS

Stephen L. Herman has been both a teacher of industrial electricity and an industrial electrician for many years. His formal training was obtained at Stephen F. Austin University in Nacogdoches, Texas and Catawba Valley Technical College in Hickory, North Carolina. Mr. Herman has worked as a maintenance electrician and as a class "A" electrician. He was employed as an electrical maintenance instructor at Randolph Technical College in Asheboro, North Carolina for nine years. Presently, he is teaching industrial electricity at Lee College in Baytown, Texas.

Walter N. Alerich has had an extensive background in the electrical trades and as a teacher. As a journeyman wireman, he has had years of experience in the practical applications of motor control. He has also been a teacher, supervisor, and administrator. A former department head of the Electrical-Mechanical Department at Los Angeles Trade-Technical College, Mr. Alerich has written extensively on the subjects of electricity and motor controls. He presently serves as an international specialist/consultant in the field of electrical trades, developing curricula and designing training facilities.

ACKNOWLEDGMENTS

The following individuals provided detailed critiques of the manuscript and offered valuable suggestions for improvement:

Mr. Richard Cutbirth
Electrical JATC
620 Legion Way
Las Vegas, NV 89110

Mr. Rick Hecklinger
Toledo Electrical JATC
803 Lime City Road
Rossford, OH 43460

Mr. Harry Katz
South Texas Electrical JATC
1223 East Euclid
San Antonio, TX 78212

Mr. Alan Bowden
Central Westmoreland
Area Vocational School
Arona Road
New Stanton, PA 15672

The following companies provided the photographs used in this text:

Allen-Bradley Company
1201 South Second Street
Milwaukee, WI 53204

Automatic Switch Company
50-A Hanover Road
Florham Park, NJ 07932

Eaton Corporation
Cutler-Hammer Products
4201 North 27th Street
Milwaukee, WI 53216

Eagle Signal Controls
A Division of Gulf & Western Manufacturing Company
736 Federal Street
Davenport, IA 52803

Emerson Electric Company
Industrial Controls Division
3300 South Standard Street
Santa Ana, CA 92702

Furnas Electric Company
1007 McKee Street
Batavia, IL 60510

GE Fanuc Automation North America, Inc.
PO Box 8106
Charlottesville, VA 22906

General Electric Company
101 Merritt 7, P.O. Box 5900
Norwalk, CT 06856

Hevi-Duty Electric
A Division of General Signal Corporation
P.O. Box 268, Highway 17 South
Goldsboro, NC 27530

International Rectifier
Semiconductor Division
233 Kansas
El Segundo, CA 90245

McDonnell & Miller, ITT
3500 N. Spaulding Avenue
Chicago, IL 60618

McGraw-Edison Company
Electric Machinery
800 Central Avenue
Minneapolis, MN 55413

Micro Switch
A Honeywell Division
11 West Spring Street
Freeport, IL 61032

RCA
Solid State Division
Route 202
Somerville, NJ 08876

Ramsey Controls, Inc.
335 Route 17
Mahwah, NJ 07430

Reliance Electric
24701 Euclid Avenue
Cleveland, OH 44117

Sparling Instruments, Co. Inc.
4097 North Temple City Boulevard
El Monte, CA 91734

Square D Company
P.O. Box 472
Milwaukee, WI 53201

The Superior Electric Company
Bristol, CT 06010

Struthers-Dunn, Inc.
Systems Division
4140 Utica Ridge Road
P.O. Box 1327
Bettendorf, IA 52722-1327

Tektronix, Inc.
P.O. Box 500
Beaverton, OR 97077

Telemecanique, Inc.
2525 S. Clearbrook Drive
Arlington Heights, IL 60005

Turck Inc.
3000 Campus Drive
Plymouth, MN 55441

U.S. Electrical Motors Division
Emerson Electric Company
125 Old Gate Lane
Milford, CT 06460

Vactec, Inc.
10900 Page Boulevard
St. Louis, MO 63132

Warner Electric Brake & Clutch Company
449 Gardner Street
South Beloit, IL 61080

SECTION 1

Solid-State Devices

UNIT 1

General Principles of Electric Motor Control

Objectives *After studying this unit, the student will be able to:*

- **State the purpose and general principles of electric motor control**
- **State the difference between manual and remote control**
- **List the conditions of starting and stopping, speed control, and protection of electric motors**
- **Explain the difference between compensating and definite time delay action**

There are certain conditions that must be considered when selecting, designing, installing, or maintaining electric motor control equipment. The general principles are discussed to help understanding and to motivate students by simplifying the subject of electric motor control.

Motor control was a simple problem when motors were used to drive a common line shaft to which several machines were connected. It was simply necessary to start and stop the motor a few times a day. However, with individual drive, the motor is now almost an integral part of the machine and it is necessary to design the motor controller to fit the needs of the machine to which it is connected. Large installations and the problems of starting motors in these situations may be observed in figures 1-1 and 1-2.

Motor control is a broad term that means anything from a simple toggle switch to a complex system with components such as relays, timers, and switches. The common function of all controls, however, is to control the operation of an electric motor. As a result, when motor control equipment is selected and installed, many factors must be considered to insure that the control will function properly for the motor and the machine for which it is selected.

MOTOR CONTROL INSTALLATION CONSIDERATIONS

When choosing a specific device for a particular application, it is important to remember that the motor, machine, and motor controller are interrelated and need to be considered as a package. In general, five basic factors influence the selection and installation of a controller.

1. ELECTRICAL SERVICE
   Establish whether the service is direct (dc) or alternating current (ac). If ac, determine the frequency (hertz) and number of phases in addition to the voltage.

2. MOTOR
   The motor should be matched to the electrical service, and correctly sized for the ma-

FIGURE 1-1 Five 2000-hp, 1800-rpm induction motors driving water pumps for a Texas oil/water flood operation. Pumps are used to force water into the ground and "float" oil upward. (Courtesy Electric Machinery Mfg. Co.)

FIGURE 1-2 Horizontal 4000-hp synchronous motor driving a large centrifugal air compressor (Courtesy Electric Machinery Mfg. Co.)

chine load in horsepower rating (hp). Other considerations include motor speed and torque. To select proper protection for the motor, its full load current rating (FLC), service factor (SF), time rating (duty), and other pertinent data—as shown on the motor nameplate—must be used.

3. OPERATING CHARACTERISTICS OF CONTROLLER
   The fundamental tasks of a motor controller are to start and stop the motor, and to protect the motor, machine, product, and operator. The controller may also be called upon to provide supplementary functions such as reversing, jogging or inching, plugging, operating at several speeds or at reduced levels of current and motor torque. (See Glossary.)

4. ENVIRONMENT
   Controller enclosures serve to provide safety protection for operating personnel by pre-

venting accidental contact with live parts. In certain applications, the controller itself must be protected from a variety of environmental conditions which might include:

- Water, rain, snow or sleet
- Dirt or noncombustible dust
- Cutting oils, coolants or lubricants

Both personnel and property require protection in environments made hazardous by the presence of explosive gases or combustible dusts.

5. ELECTRICAL CODES AND STANDARDS

Motor control equipment is designed to meet the provisions of the National Electrical Code® (NEC®). (National Electrical Code® and NEC® are registered trademarks of the National Fire Protection Association Inc., Quincy, MA 02269.) Also, local code requirements must be considered and met when installing motors and control devices. Presently, code sections applying to motors, motor circuits, and controllers and industrial control devices are found in Article 430 on motors and motor controllers, Article 440 on air conditioning and refrigeration equipment, and Article 500 on hazardous locations of the NEC.

The 1970 Occupational Safety and Health Act (OSHA) as amended, requires that each employer furnish employment in an environment free from recognized hazards likely to cause serious harm.

Standards established by the National Electrical Manufacturers Association (NEMA) assist users in the proper selection of control equipment. NEMA standards provide practical information concerning the construction, testing, performance, and manufacture of motor control devices such as starters, relays and contactors.

One of the organizations that actually test for conformity to national codes and standards is Underwriters' Laboratories (UL). Equipment that is tested and approved by UL is listed in an annual publication, which is kept current by means of bimonthly supplements to reflect the latest additions and deletions. A UL listing does not mean that a product is approved by the NEC. It must be acceptable to the local authority having jurisdiction.

PURPOSE OF CONTROLLER

Some of the complicated and precise automatic applications of electrical control are illustrated in figures 1-3 and 1-6. Factors to be considered when selecting and installing motor control components for use with particular machines or systems are described in the following paragraphs.

Starting

The motor may be started by connecting it directly across the source of voltage. Slow and gradual starting may be required, not only to protect the machine, but also to insure that the line current inrush on starting is not too great for the

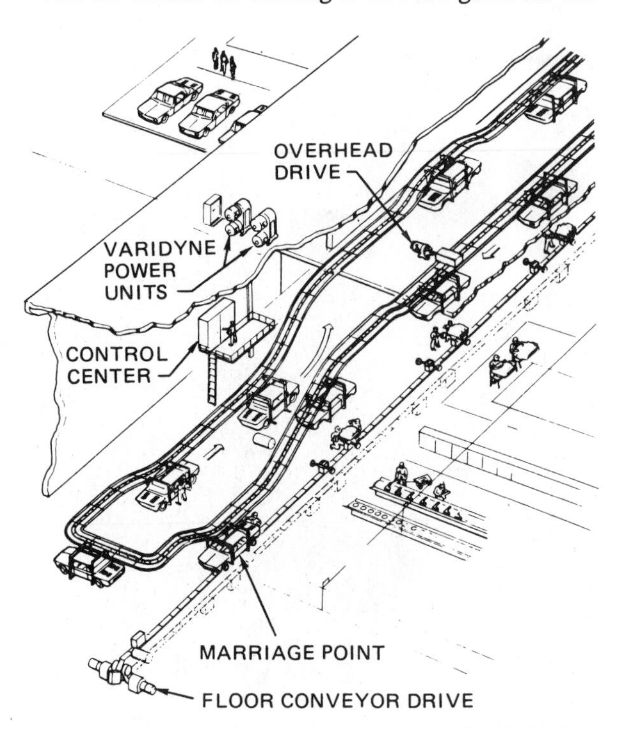

FIGURE 1-3 Synchronizing two automobile assembly systems (Courtesy U.S. Electrical Motors)

power company's system. Some driven machines may be damaged if they are started with a sudden turning effort. The frequency of starting a motor is another factor affecting the controller. A combination motor starter with circuit breaker and control transformer is shown in figure 1-4.

Stopping

Most controllers allow motors to coast to a standstill. Some impose braking action when the machine must stop quickly. Quick stopping is a vital function of the controller for emergency stops. Controllers assist the stopping action by retarding centrifugal motion of machines and lowering operations of crane hoists.

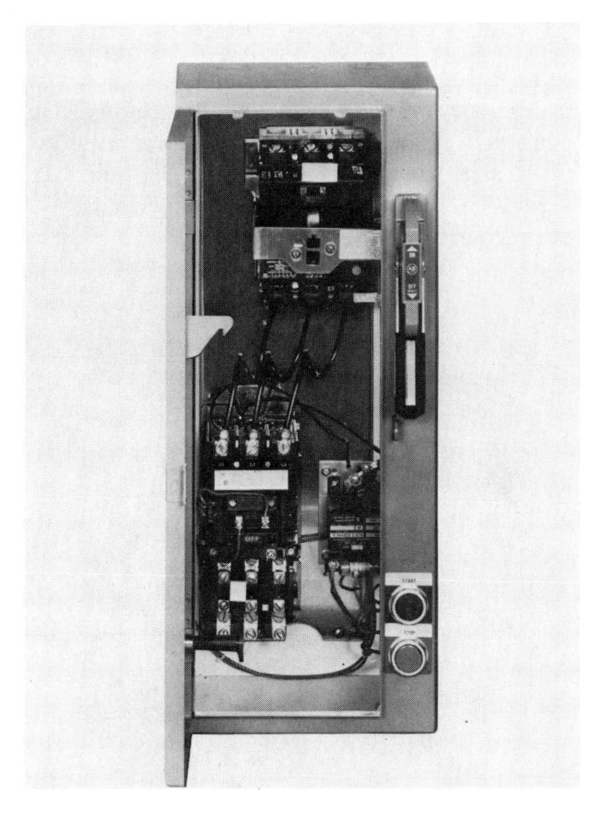

FIGURE 1-4 Combination motor starter with circuit breaker and control transformer. Disconnect switch must be off before the door can be opened. (Courtesy Allen-Bradley Co.)

Reversing

Controllers are required to change the direction of rotation of machines automatically or at the command of an operator at a control station. The reversing action of a controller is a continual process in many industrial applications.

Running

The maintaining of desired operational speeds and characteristics is a prime purpose and function of controllers. They protect motors, operators, machines, and materials while running. There are many different types of safety circuits and devices to protect people, equipment and industrial production and processes against possible injury that may occur while the machines are running.

Speed Control

Some controllers can maintain very precise speeds for industrial processes. Other controllers can change the speeds of motors either in steps or gradually through a continuous range of speeds.

Safety of Operator

Many mechanical safeguards have been replaced or aided by electrical means of protection. Electrical control pilot devices in controllers provide a direct means of protecting machine operators from unsafe conditions.

Protection from Damage

Part of the operation of an automatic machine is to protect the machine itself and the manufactured or processed materials it handles. For example, a certain machine control function may be the prevention of conveyor pileups. A machine control can reverse, stop, slow, or do whatever is necessary to protect the machine or processed materials.

Maintenance of Starting Requirements

Once properly installed and adjusted, motor starters will provide reliable operation of starting time, voltages, current, and torques for the benefit of the driven machine and the power system. The National Electrical Code, supplemented by local codes, governs the selection of the proper sizes of conductors, starting fuses, circuit breakers, and disconnect switches for specific system requirements.

MANUAL CONTROL

A manual control is one whose operation is accomplished by mechanical means. The effort required to actuate the mechanism is almost always provided by a human operator. The motor may be controlled manually using any one of the following devices.

Toggle Switch

A toggle switch is a manually operated electric switch. Many small motors are started with toggle switches. This means the motor may be started directly without the use of magnetic switches or auxiliary equipment. Motors started with toggle switches are protected by the branch circuit fuse or circuit breaker. These motors generally drive fans, blowers, or other light loads.

FIGURE 1-5 Safety disconnect switch (Courtesy EATON Corp., Cutler-Hammer Products)

Safety Switch

In some cases it is permissible to start a motor directly across the full line voltage if an externally-operated safety switch is used, figure 1-5. The motor receives starting and running protection from dual-element, time-delay fuses. The use of a safety switch requires manual operation. A safety switch, therefore, has the same limitations common to most manual starters.

Drum Controller

Drum controllers are rotary, manual switching devices which are often used to reverse motors and to control the speed of ac and dc motors. They are used particularly where frequent start, stop, or reverse operation is required. These controllers may be used without other control components in small motors, generally those with fractional horsepower ratings. Drum controllers are used with magnetic starters in large motors. A drum controller is shown in figure 1-6.

FIGURE 1-6 Drum controller with cover (left) and with cover removed (right) (Courtesy EATON Corp., Cutler-Hammer Products)

FIGURE 1-7 Typical cement mill computer console

REMOTE AND AUTOMATIC CONTROL

The motor may be controlled by remote control using push buttons, figure 1-7. When push-button remote control is used or when automatic devices do not have the electrical capacity to carry the motor starting and running currents, magnetic switches must be included. Magnetic switch control is one whose operation is accomplished by electromagnetic means. The effort required to actuate the electromagnet is supplied by electrical energy rather than by the human operator. If the motor is to be automatically controlled, the following two-wire pilot devices may be used.

Float Switch

The raising or lowering of a float that is mechanically attached to electrical contacts may start motor-driven pumps to empty or fill tanks. Float switches are also used to open or close piping solenoid valves to control fluids, figure 1-8.

Pressure Switch

Pressure switches are used to control the pressure of liquids and gases (including air) within a desired range, figure 1-9. Air compressors, for example, are started directly or indirectly on a call for more air by a pressure switch. Electrical wiring symbols are shown as normally closed and normally open in figure 1-8.

Time Clock

Time clocks can be used when a definite "on and off" period is required and adjustments are not necessary for long periods of time. A typical requirement is a motor that must start every morning at the same time and shut off every night

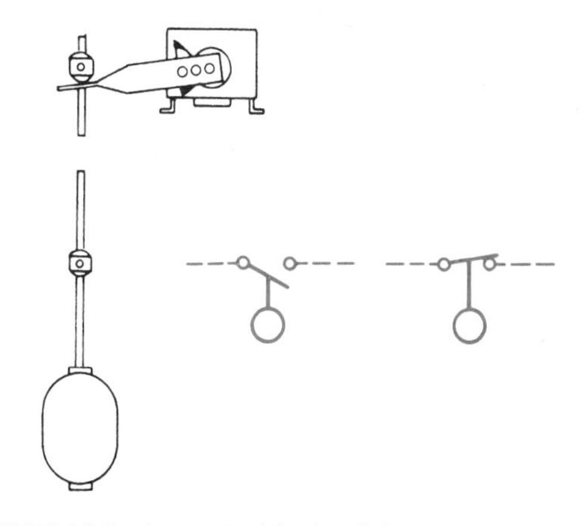

FIGURE 1-8 Rod-operated float switch with electrical wiring symbols

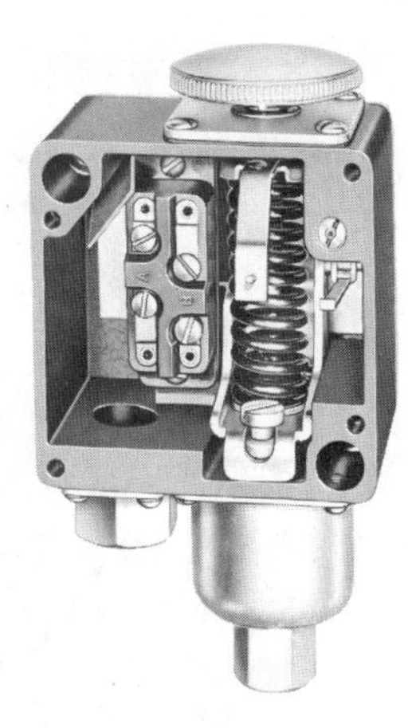

FIGURE 1-9 Pressure switches may be necessary to start the motors shown in figure 1-10 (Courtesy Square D Co.)

FIGURE 1-10 Two 1500-hp vertical induction motors driving pumps (Courtesy Electrical Machinery Mfg. Co.)

at the same time, or that switches the floodlights on and off.

Thermostat

In addition to pilot devices sensitive to liquid levels, gas pressures, and time of day, thermostats sensitive to temperature changes are widely used, figure 1-11. Thermostats indirectly control large motors in air conditioning systems and in many industrial applications to maintain the desired

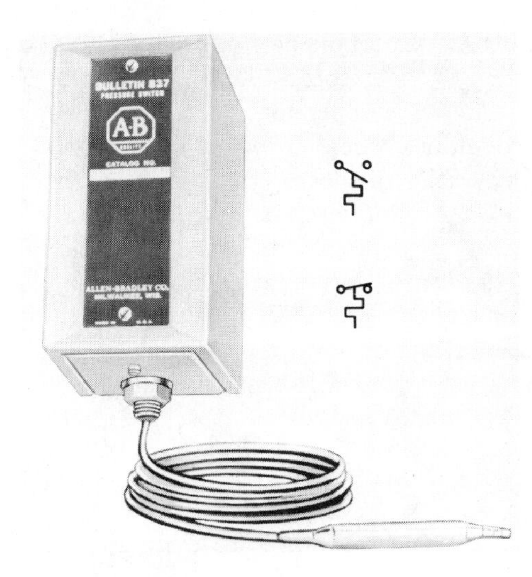

FIGURE 1-11 Industrial temperature switch with extension bulb and electrical wiring symbols (Courtesy Allen-Bradley Co.)

temperature range of air, gases, liquids, or solids. There are many types of thermostats and temperature-actuated switches.

Limit Switch

Limit switches, figure 1-12, are designed to pass an electrical signal only when a predetermined limit is reached. The limit may be a specific

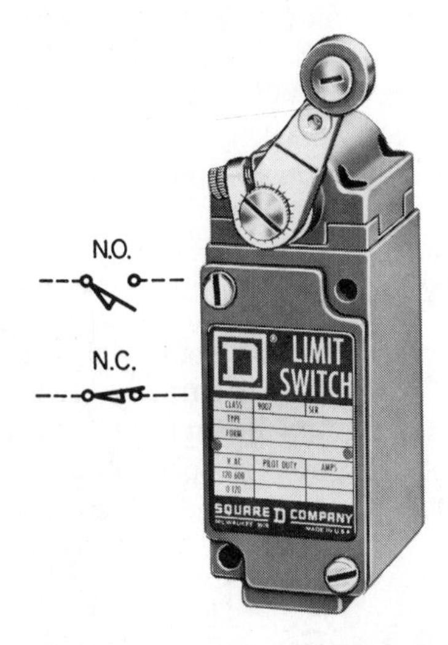

FIGURE 1-12 Limit switch shown with electrical wiring symbols (Courtesy Square D Co.)

position for a machine part or a piece of work, or a certain rotating speed. These devices take the place of a human operator and are often used under conditions where it would be impossible or impractical for the operator to be present or to efficiently direct the machine.

Limit switches are used most frequently as overtravel stops for machines, equipment, and products in process. These devices are used in the control circuits of magnetic starters to govern the starting, stopping, or reversal of electric motors.

Electrical or Mechanical Interlock and Sequence Control

Many of the electrical control devices described in this unit can be connected in an interlocking system so that the final operation of one or more motors depends upon the electrical position of each individual control device. For example, a float switch may call for more liquid but will not be satisfied until the prior approval of a pressure switch or time clock is obtained. To design, install, and maintain electrical controls in any electrical or mechanical interlocking system, the electrical technician must understand the total operational system and the function of the individual components. With practice, it is possible to transfer knowledge of circuits and descriptions for an understanding of additional similar controls. It is impossible—in instructional materials—to show all possible combinations of an interlocking control system. However, by understanding the basic functions of control components and their basic circuitry, and by taking the time to trace and draw circuit diagrams, difficult interlocking control systems can become easier to understand.

STARTING AND STOPPING

In starting and stopping a motor and its associated machinery, there are a number of conditions that may affect the motor. A few of them are discussed here.

Frequency of Starting and Stopping

The starting duty cycle of a controller is an important factor in determining how satisfactorily the controller will perform in a particular application. Magnetic switches, such as motor starters, relays, and contactors, actually beat themselves apart from repeated opening and closing thousands of times. An experienced electrician soon learns to look for this type of component failure when trouble shooting any inoperative control panels. NEMA standards require that the starter size be derated if the frequency of start-stop, jogging, or plugging is more than 5 times per minute. Therefore, when the frequency of starting the controller is great, the use of heavy duty controllers and accessories should be considered. For standard duty controllers, more frequent inspection and maintenance schedules should be followed.

Light or Heavy Duty Starting

Some motors may be started with no loads and others must be started with heavy loads. When motors are started, large feeder line disturbances may be created which can affect the electrical distribution system of the entire industrial plant. The disturbances may even affect the power company's system. As a result, the power companies and electrical inspection agencies place certain limitations on "across-the-line" motor starting.

Fast or Slow Start (Hard or Soft)

To obtain the maximum twisting effort (torque) of the rotor of an ac motor, the best starting condition is to apply full voltage to the motor terminals. The driven machinery, however, may be damaged by the sudden surge of motion. To prevent this type of damage to machines, equipment, and processed materials, some controllers are designed to start slowly and then increase the motor speed gradually in definite steps. This type is often used by power companies and inspection agencies to avoid electrical line surges.

Smooth Starting

Although reduced electrical and mechanical surges can be obtained with a step-by-step motor starting method, very smooth and gradual starting will require different controlling methods. These are discussed in detail later in the text.

Manual or Automatic Starting and Stopping

While the manual starting and stopping of machines by an operator is still a common practice, many machines and industrial processes are started and restarted automatically. These automatic devices result in tremendous savings of time and materials. Automatic stopping devices are used in motor control systems for the same reasons. Automatic stopping devices greatly reduce the safety hazards of operating some types of machinery, both for the operator and the materials being processed. An electrically operated, mechanical brake is shown in figure 1-13. Such a brake may be required to stop a machine's motion in a hurry to protect materials being processed or people in the area.

Quick Stop or Slow Stop

Many motors are allowed to coast to a standstill. However, manufacturing requirements and safety considerations often make it necessary to bring machines to as rapid a stop as possible. Automatic controls can retard and brake the speed of a motor and also apply a torque in the opposite direction of rotation to bring about a rapid stop. This is referred to as plugging. Plugging can only be used if the driven machine and its load will not be damaged by the reversal of the motor torque. The control of deceleration is one of the important functions of a motor control.

Another method of braking electric motors is known as *dynamic braking*. When this method is used to reduce the speed of dc motors, the armature is connected across a load resistor when power is disconnected from the motor. If the field winding of the motor remains energized, the motor becomes a generator and current is supplied to the load resistor by the armature, figure 1-14. The current flowing through the armature winding creates a magnetic field around the armature. This magnetic field causes the armature to be attracted to the magnetic field of the pole pieces. This action in a dc generator is known as *counter torque*. Using counter torque to brake a dc motor is known as dynamic braking.

Ac induction motors can be braked by momentarily connecting dc voltage to the stator

FIGURE 1-13 Typical 30-inch brake (Courtesy EATON Corp., Cutler-Hammer Products)

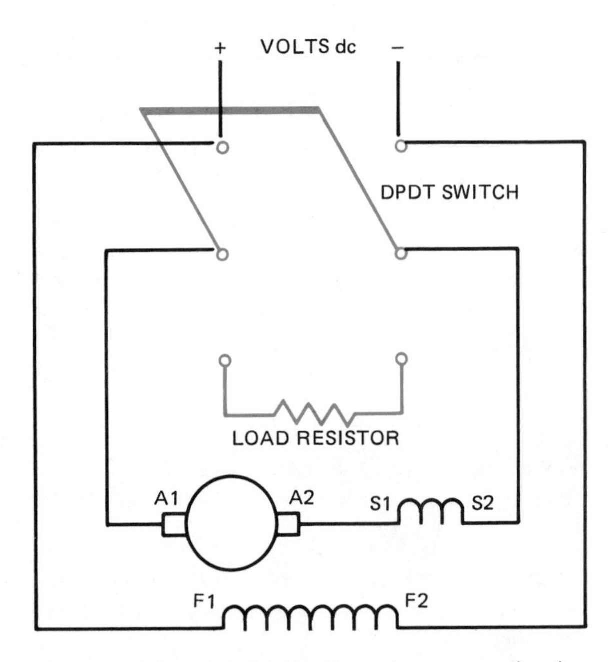

FIGURE 1-14 Dynamic braking for a dc compound motor

winding, figure 1-15. When direct current is applied to the stator winding of an ac motor, the stator poles become electromagnets. Current is induced into the windings of the rotor as the rotor continues to spin through the magnetic field. This induced current produces a magnetic field around the rotor. The magnetic field of the rotor is attracted to the magnetic field produced in the stator. The attraction of these two magnetic fields produces a braking action in the motor.

An advantage of using dynamic braking is that motors can be stopped rapidly without wearing brake linings or drums. It cannot be used to hold a suspended load, however. Mechanical brakes must be employed when a load must be held, such as with a crane or hoist.

Accurate Stops

An elevator must stop at precisely the right location so that it is aligned with the floor level. Such accurate stops are possible with the use of automatic devices interlocked with control systems.

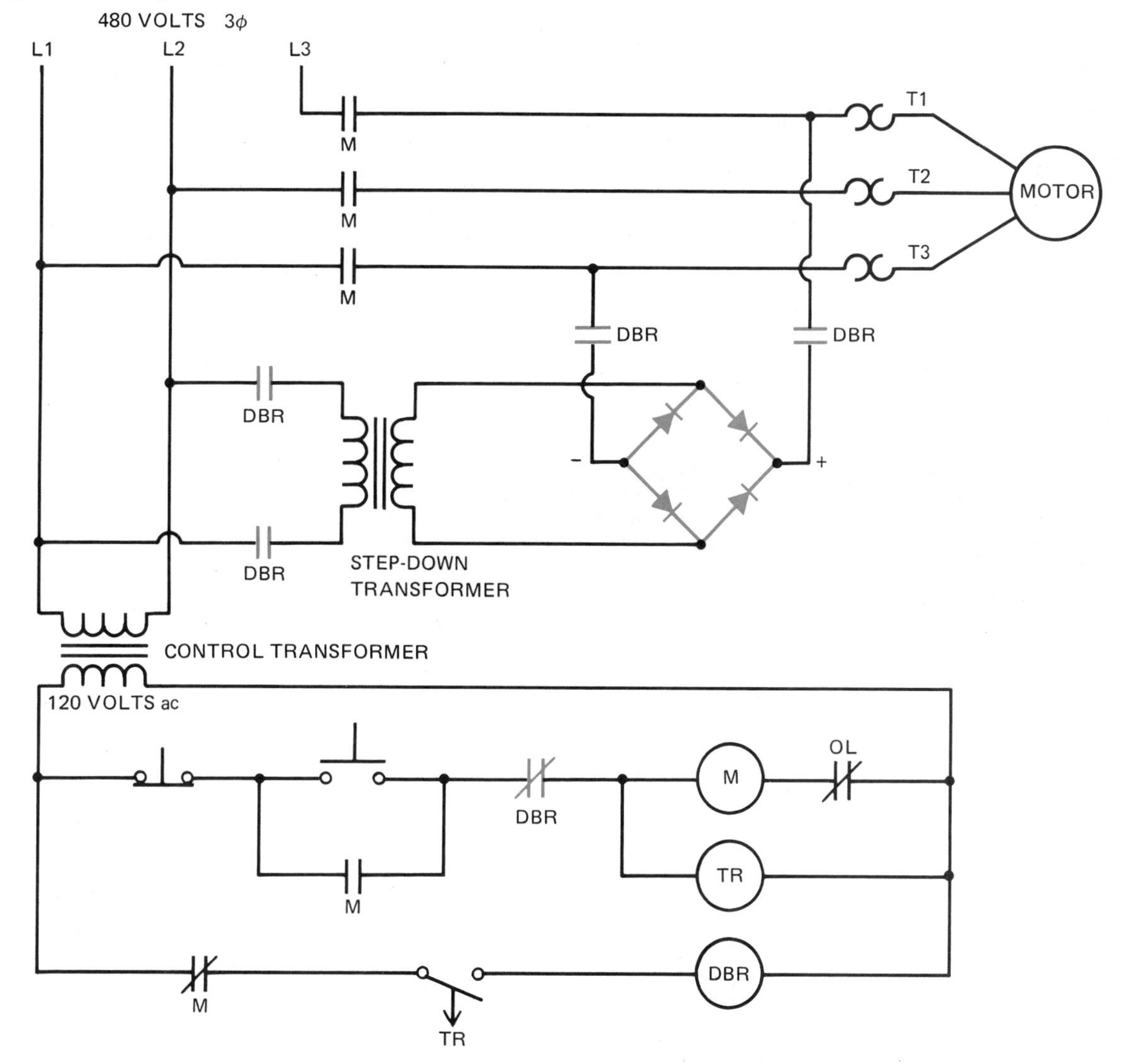

FIGURE 1-15 Dynamic braking for an ac motor

Frequency of Reversals Required

Frequent reversals of the direction of rotation of the motor impose large demands on the controller and the electrical distribution system. Special motors and special starting and running protective devices may be required to meet the conditions of frequent reversals. A heavy duty drum switch-controller is often used for this purpose.

SPEED CONTROL OF MOTORS

The speed control is concerned not only with starting the motor but also with maintaining or controlling the motor speed while it is running. There are a number of conditions to be considered for speed control.

Constant Speed

Constant speed motors are used on water pumps, figures 1-11 and 1-16. Maintenance of constant speed is essential for motor generator sets under all load conditions. Constant speed motors with ratings as low as 80 rpm and horsepower ratings up to 5000 hp are used in direct drive units. The simplest method of changing speeds is by gearing. Using gears, almost any "predetermined" speed may be developed by coupling the input gear to the shaft of a squirrel-cage induction motor. A speed-reducing gear motor is shown in figure 1-17.

Varying Speed

A varying speed is usually preferred for cranes and hoists, figure 1-18. In this type of application, the motor speed slows as the load increases and speeds up as the load decreases.

FIGURE 1-16 Multiple synchronous motors of 3000-hp and 225-rpm driving water pumps (Courtesy Electric Machinery Mfg. Co.)

FIGURE 1-17 Cutaway view of speed-reducing gear motor (Courtesy U.S. Electrical Motors)

Adjustable Speed

With adjustable speed controls, an operator can gradually adjust the speed of a motor over a wide range while the motor is running. The speed may be preset, but once it is adjusted it remains essentially constant at any load within the rating of the motor.

Multispeed

For multispeed motors, such as the type used on turret lathes in a machine shop, the speed can be set at two or more definite rates. Once the motor is set at a definite speed, the speed will remain practically constant regardless of load changes.

PROTECTIVE FEATURES

The particular application of each motor and control installation must be considered to determine what protective features are required to be installed and maintained.

FIGURE 1-18 Large traveling overhead crane (Courtesy Square D Co.)

Overload Protection

Running protection and overload protection refer to the same thing. This protection may be an integral part of the motor or be separate. A controller with electrical overload protection will protect a motor from burning up while allowing the motor to achieve its maximum available power under a range of overload and temperature conditions. An electrical overload on the motor may be caused by mechanical overload on driven machinery, a low line voltage, an open electrical line in a polyphase system resulting in single-phase operation, motor problems such as too badly worn bearings, loose terminal connections, or poor ventilation within the motor.

Open Field Protection

Dc shunt and compound-wound motors can be protected against the loss of field excitation by field loss relays. Other protective arrangements are used with starting equipment for dc and ac synchronous motors. Some sizes of dc motors may race dangerously with the loss of field excitation while other motors may not race due to friction and the fact that they are small.

Open-Phase Protection

Phase failure in a three-phase circuit may be caused by a blown fuse, an open connection, a broken line or other reasons. If phase failure occurs when the motor is at a standstill during attempts to start, the stator currents will rise to a very high value and will remain there, but the motor will remain stationary (not turn). Since the windings are not properly ventilated while the motor is stationary, the heating produced by the high currents may damage them. Dangerous conditions also are possible while the motor is running. When the motor is running and an open-phase condition occurs, the motor may continue to run. The torque will decrease, possibly to the point of motor "stall"; this condition is called *breakdown torque*.

Reversed Phase Protection

If two phases of the supply of a three-phase induction motor are interchanged (phase reversal), the motor will reverse its direction of rotation. In elevator operation and industrial applications, this reversal can result in serious damage. Phase failure and phase reversal relays are safety devices used to protect motors, machines, and personnel from the hazards of open-phase or reversed-phase conditions.

Overtravel Protection

Control devices are used in magnetic starter circuits to govern the starting, stopping, and reversal of electric motors. These devices can be used to control regular machine operation or they can be used as safety emergency switches to prevent the improper functioning of machinery.

Overspeed Protection

Excessive motor speeds can damage a driven machine, materials in the industrial process, or the motor. Overspeed safety protection is provided in control equipment for paper and printing plants, steel mills, processing plants, and the textile industry.

Reversed Current Protection

Accidental reversal of currents in direct-current controllers can have serious effects. Direct-current controllers used with three-phase alternating-current systems that experience phase failures and phase reversals are also subject to damage. Reverse current protection is an important provision for battery charging and electroplating equipment.

Mechanical Protection

An enclosure may increase the life span and contribute to the trouble-free operation of a motor and controller. Enclosures with particular ratings

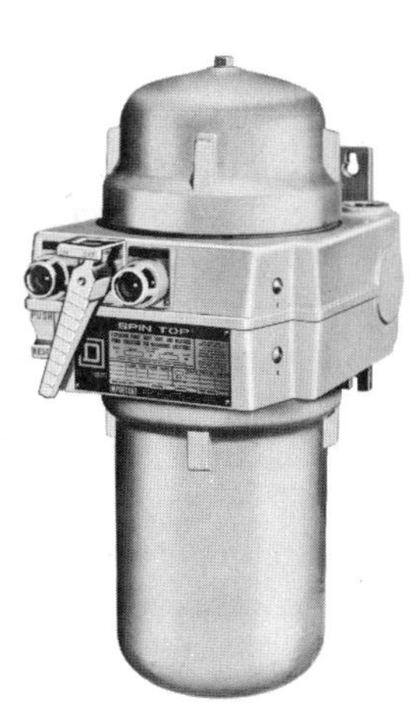

FIGURE 1-19 Spin-on, explosion-proof enclosure for a combination disconnect switch and magnetic motor starter (Courtesy Square D Co.)

such as general purpose, watertight, dustproof, explosionproof, and corrosion resistant are used for specific applications, figure 1-19. All enclosures must meet the requirements of national and local electrical codes and building codes.

Short Circuit Protection

For large motors with greater than fractional horsepower ratings, short circuit and ground fault protection generally is installed in the same enclosure as the motor-disconnecting means. Overcurrent devices (such as fuses and circuit breakers) are used to protect the motor branch circuit conductors, the motor control apparatus, and the motor itself against sustained overcurrent due to short circuits and grounds, and prolonged and excessive starting currents.

CLASSIFICATION OF AUTOMATIC MOTOR STARTING CONTROL SYSTEMS

The numerous types of automatic starting and control systems are grouped into the following classifications: current limiting acceleration and time delay acceleration.

Current Limiting Acceleration

This is also called *compensating time*. It refers to the amount of current or voltage drop required to open and close magnetic switches when used in a motor accelerating controller. The rise and fall of the current or voltage determines a timing period which is used mainly for dc motor control. Examples of types of current limiting acceleration are:

- Counter emf or voltage drop acceleration
- Lockout contactor or series relay acceleration

Time Delay Acceleration

For this classification, *definite time* relays are used to obtain a preset timing period. Once the period is preset, it does not vary regardless of current or voltage changes occurring during motor acceleration. The following timers and timing systems are used for motor acceleration; some are also used in interlocking circuits for automatic control systems.

- Pneumatic timing
- Motor-driven timers
- Capacitor timing
- Electronic timers
- Dashpot timers

TROUBLESHOOTING

One of the primary jobs of an industrial electrician is troubleshooting control circuits. An electrician that is proficient in troubleshooting is sought after by most of industry. The greatest troubleshooting tool an electrician can possess is the ability to read and understand control schematic diagrams. Many of the circuits shown in this text are accompanied by detailed explanations of the operation of the circuit. If the circuit and explanation are studied step by step, the student will have an excellent understanding of control schematics when this text is completed.

Most electricians follow a set procedure when troubleshooting a circuit. If the problem has occurred several times in the past and was caused by the same component each time, most electricians check that component first. If that component proves to be the problem, much time has been saved by not having to trace the entire circuit.

Another method of troubleshooting a circuit is *shot gun* troubleshooting. This method derives its name from the manner in which components are tested. Instead of following the circuit in a logical step-by-step procedure, the electrician quickly checks the major components of the circuit. This approach is used to save time because in many industrial situations an inoperative piece of equipment can cost a company thousands of dollars for each hour it is not working.

When neither of these methods reveals the problem, the electrician must use the control schematic to trace the circuit in a logical step-by-step procedure. The primary tool used to trace a circuit is the volt-ohm-milliammeter (VOM), which measures voltage, current, and resistance. It is often necessary to use jumper leads to bridge open contacts when using the VOM. When a jumper lead is used for this application, it should be provided with short circuit protection. This can be done by connecting a small fuse holder or circuit breaker in series with the jumper lead. In this way if the jumper is accidently shorted, the fuse or circuit breaker will open and protect the rest of the circuit.

When troubleshooting a circuit, most electricians work backward through the circuit. For example, one line of a control schematic is shown in figure 1-20. M relay coil is connected in series with a normally closed overload contact, a normally open limit switch contact, a normally closed pressure switch contact, a normally closed CR relay contact, and a normally open float switch contact. The problem with the circuit is that M relay coil will not energize. The first test should be to measure the voltage at each end of the circuit to confirm the presence of control voltage. The next procedure is to connect the voltmeter across each of the circuit components to determine which one is open and stopping the current flow to the coil. When the voltmeter is connected across a closed contact, there is no voltage drop, and the meter indicates 0 volts. If the voltmeter is connected across an open contact, the meter indicates the full voltage of the circuit.

Assume in this circuit that the full circuit voltage is indicated when the meter is connected across float switch FS. This reading signals that float switch FS is open. The next step is to determine if the switch is bad or if the liquid lever it is sensing has not risen high enough to close the switch. Once that has been determined, the electrician can correct the problem.

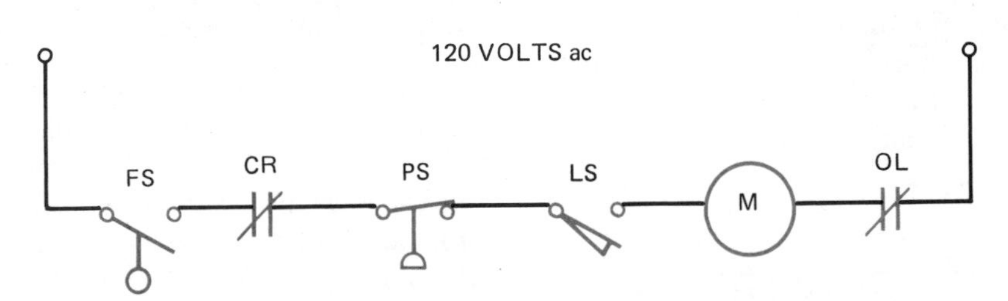

FIGURE 1-20 Troubleshooting a circuit

REVIEW QUESTIONS

1. What is a controller and what is its function? (Use the Glossary and the information from this unit to answer this question.)
2. What is meant by remote control?
3. To what does current limiting, or compensating time, acceleration refer?
4. List some devices that are used to control a motor automatically. Briefly describe the purpose of each device.

Select the *best* answer for each of the following.

5. The general purpose of motor control is
   a. to start the motor
   b. to stop the motor
   c. to reverse the motor
   d. all of the above
6. A motor may be controlled manually by using a
   a. float switch
   b. pressure switch
   c. toggle switch
   d. time clock
7. A motor may be controlled remotely or automatically by using a
   a. drum controller
   b. thermostat
   c. safety switch
   d. faceplate control
8. Conditions that may affect starting and stopping of motor driven machinery are
   a. fast or slow starts
   b. light or heavy duty starting
   c. frequency of starting and stopping
   d. all of the above
9. Which factor is *not* to be considered for motor speed control when the motor is running?
   a. Constant speed
   b. Varying speed
   c. Multispeed
   d. Starting protection
10. Which is not considered a motor controller protective feature?
    a. Overload
    b. Short circuit
    c. Adjustable speed
    d. Mechanical
11. Which function is not a fundamental job of a motor controller?
    a. Start and stop the motor
    b. Protect the motor, machine, and operator
    c. Reverse, inch, jog, speed control
    d. Motor disconnect switch and starting protection

12. What factors are to be considered when selecting and installing a controller?
    a. Electrical service
    b. Motor
    c. Electrical codes and standards
    d. All of the above
13. Dynamic braking for a dc motor is accomplished by
    a. connecting ac voltage to the armature
    b. maintaining dc current flow through the field and connecting the armature to a load resistor
    c. maintaining dc current flow through the armature and connecting a load resistor to the field
    d. disconnecting dc power from the motor and reconnecting the armature to a load resistor
14. Dynamic braking for an ac motor is accomplished by
    a. disconnecting ac power from the motor leads and reconnecting the motor to a load resistor
    b. reversing the direction of rotation of the motor
    c. connecting dc voltage to the stator leads
    d. connecting a load resistor in series with the motor leads

UNIT 2

Semiconductors

Objectives *After studying this unit, the student will be able to:*

- **Discuss the atomic structure of conductors, insulators, and semiconductors**
- **Discuss how a P-type material is produced**
- **Discuss how an N-type material is produced**

Many of the control systems used in today's industry are operated by solid-state devices as well as magnetic and mechanical devices. If an electrician is to install and troubleshoot control systems, he must have an understanding of electronic control devices as well as relays and motor starters.

Solid-state devices, such as diodes and transistors, are often referred to as *semiconductors*. The word, semiconductor, refers to the type of material used to make solid-state devices. To understand how solid-state devices operate, one must first study the atomic structure of conductors, insulators, and semiconductors.

CONDUCTORS

Conductors are materials that provide easy paths for the flow of electrons. Conductors are generally made from materials that have large, heavy atoms. For this reason, most conductors are metals. The best electrical conductors are silver, copper and aluminum.

Conductors are also made from materials that have only one or two valence electrons in their atoms. (*Valence electrons* are the electrons in the outer orbit of an atom, figure 2-1.) An atom that has only one valence electron makes the best electrical conductor because the electron is held loosely in orbit and is easily given up for current flow.

INSULATORS

Insulators are generally made from lightweight materials that have small atoms. The outer orbits of the atoms of insulating materials are filled or almost filled with valence electrons. This means

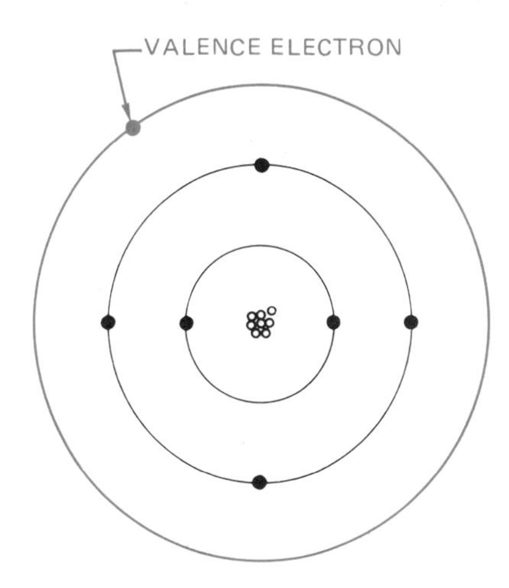

FIGURE 2-1 Atom of a conductor

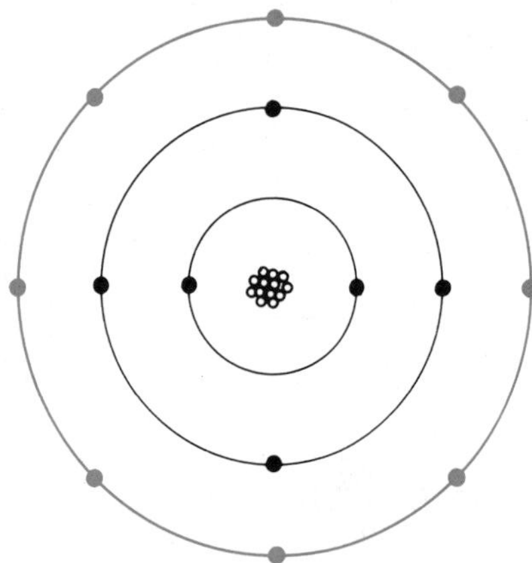

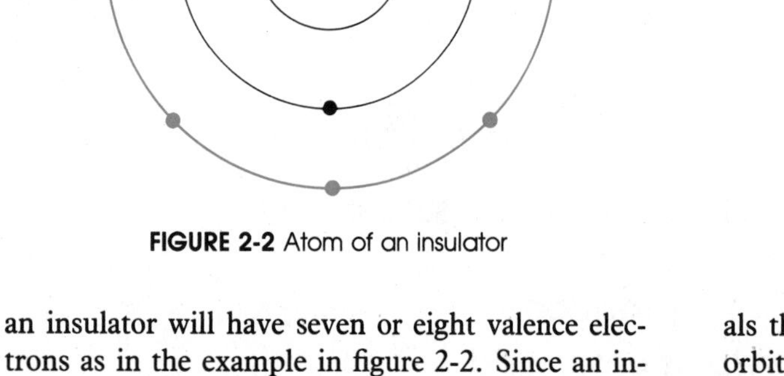

FIGURE 2-2 Atom of an insulator

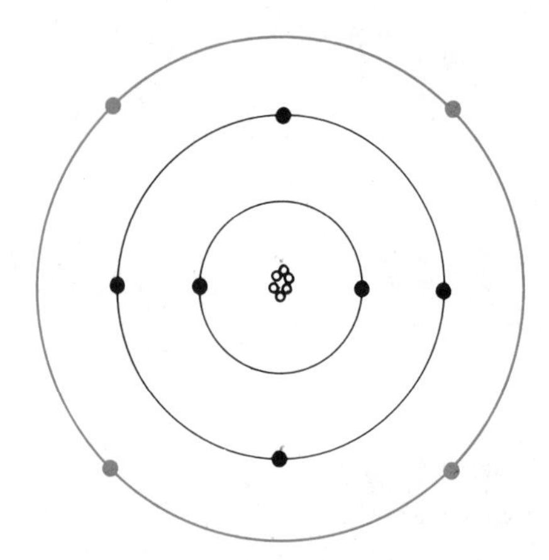

FIGURE 2-3 Atom of a semiconductor

an insulator will have seven or eight valence electrons as in the example in figure 2-2. Since an insulator has its outer orbit filled or almost filled with valence electrons, the electrons are held tightly in orbit and are not easily given up for current flow.

SEMICONDUCTORS

Semiconductors, as the word implies, are materials that are neither good conductors nor good insulators. Semiconductors are made from materials that have four valence electrons in their outer orbits, figure 2-3. Germanium and silicon are the most common semiconductor materials used in the electronics field. Of these materials, silicon is used more often because of its ability to withstand heat.

When semiconductor materials are refined into a pure form, the molecules arrange themselves into a crystal structure which has a definite pattern, figure 2-4. This type of pattern is called a *lattice structure*. A pure semiconductor material such as silicon has no special properties and will do little more than make a poor conductive material. To make semiconductor material useful in the

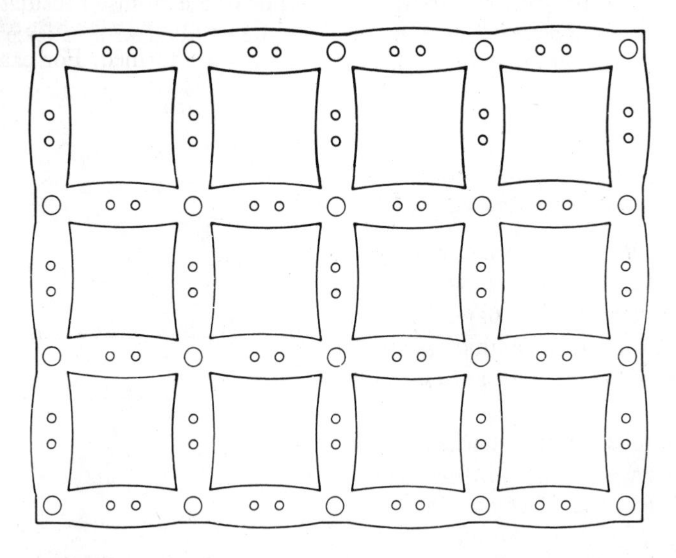

FIGURE 2-4 Lattice structure of a pure semiconductor material

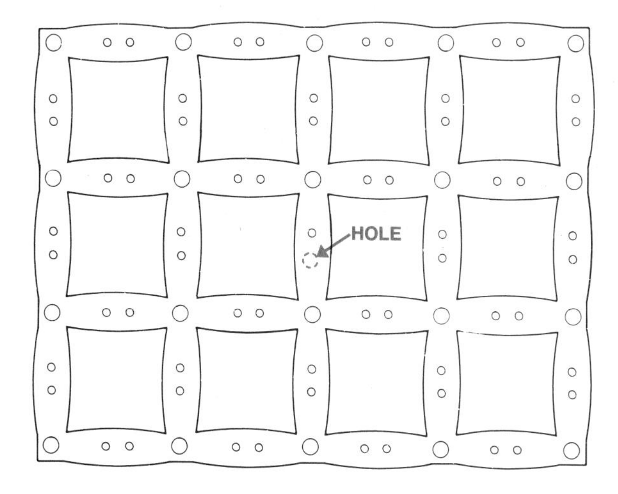

FIGURE 2-5 Lattice structure of a P-type material

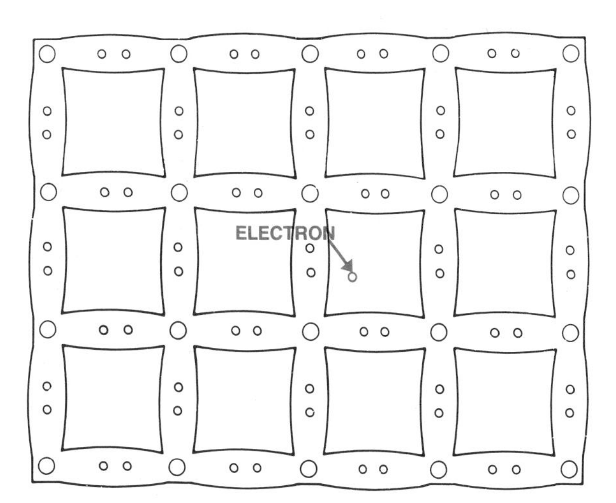

FIGURE 2-6 Lattice structure of an N-type material

production of solid-state components, it is mixed with an impurity. When pure semiconductor material is mixed with an impurity that has only three valence electrons, such as indium or gallium, the lattice structure changes leaving a hole in the material, figure 2-5. This hole is caused by a missing electron. Since the material now lacks an electron, it is no longer electrically neutral. Electrons are negative particles. The hole, which has taken the place of an electron, has a positive charge; therefore, the semiconductor material now has a net positive charge and is called a P-type material.

When a semiconductor material is mixed with an impurity that has five valence electrons, such as arsenic or antimony, the lattice structure has an excess of electrons, figure 2-6. Since electrons are negative particles, and there are more electrons in the material than there should be, the material has a net negative charge. This material is referred to as an N-type material because of its negative charge.

All solid-state devices are made from combinations of P- and N-type materials. The type of device formed is determined by how the P- and N-type materials are connected. The number of layers of material and the thickness of each layer play an important part in determining what type of device is formed. For example, the diode is often called a PN junction because it is made by joining a piece of P-type material and a piece of N-type material, figure 2-7. The transistor, on the other hand, is made by joining three layers of semiconductor materials, figure 2-8.

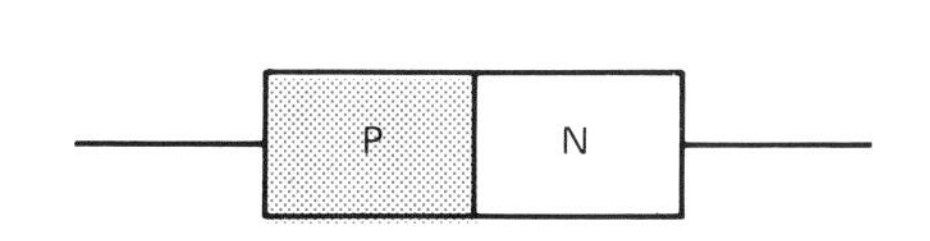

FIGURE 2-7 The PN junction

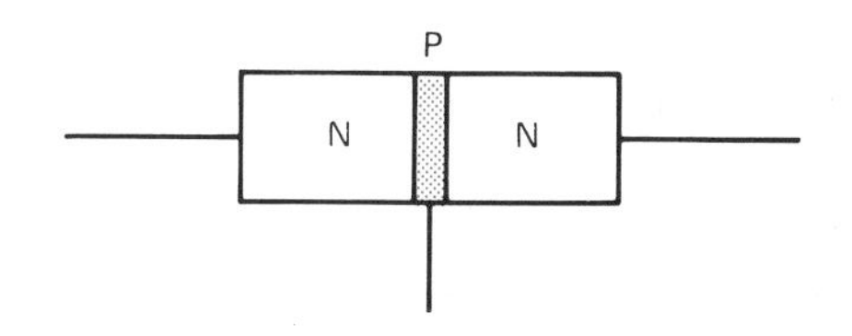

FIGURE 2-8 The transistor

REVIEW QUESTIONS

1. The atoms of a material used as a conductor generally contain __________________ valence electrons.
2. The atoms of a material used as an insulator generally contain __________________ valence electrons.
3. The two materials most often used to produce semiconductor devices are __________________ and __________________.
4. What is a lattice structure?
5. How is a P-type material made?
6. How is an N-type material made?
7. Which type of semiconductor material can withstand the greatest amount of heat?
8. All electronic components are formed from P-type and N-type materials. What factors determine the kind of components formed?

UNIT 3

The PN Junction

Objectives *After studying this unit, the student will be able to:*

- Discuss how the PN junction is produced
- Recognize the schematic symbol for a diode
- Discuss the differences between the conventional current flow theory and the electron flow theory
- Discuss how the diode operates in a circuit
- Identify the anode and cathode leads of a diode
- Properly connect the diode in an electric circuit
- Discuss the differences between a half-wave rectifier and a full-wave rectifier
- Test the diode with an ohmmeter

Hundreds of different electronic devices have been produced since the invention of solid-state components. As stated previously, solid-state devices are made by combining P-type and N-type materials. The device produced is determined by the number of layers of material used, the thickness of the layers of material, and the manner in which the layers are joined.

It is not within the scope of this text to cover even a small portion of these devices. The devices that are covered have been selected because of their frequent use in industry as opposed to communications or computers. These devices are presented in a straightforward, practical manner, and mathematical explanation is used only when necessary.

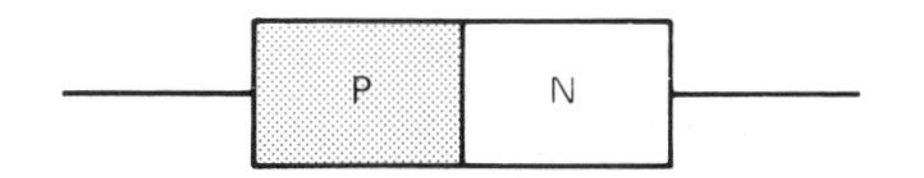

FIGURE 3-1 The PN junction, or diode

The PN junction is often referred to as the *diode*. The diode is the simplest of all electronic devices. It is made by joining a piece of P-type material and a piece of N-type material, figure 3-1. The schematic symbol for a diode is shown in figure 3-2. The diode operates like an electric check valve in that it permits current to flow through it in only one direction. If the diode is to conduct current, it must be forward biased. The diode is forward biased only when a positive voltage is connected to the anode and a negative voltage is connected to the cathode. If the diode is reverse biased, the negative voltage connected to the anode and the positive voltage connected to

FIGURE 3-2 Schematic symbol for a diode

the cathode, it will act like an open switch and no current will flow through the device.

When working with solid-state circuits, it is important to realize that circuits are often explained assuming conventional current flow as opposed to electron flow. *The conventional current flow theory assumes that current flows from positive to negative, while the electron flow theory states that current flows from negative to positive.* Although it has been known for many years that current flows from negative to positive, many electronic circuit explanations assume a positive to negative current flow. There are several reasons for this assumption. One reason is that ground is generally negative and is considered to be 0 volts in an electronic circuit. Any voltage above, or greater, than ground is positive. Most people find it is easier to think of something flowing downhill or from some point above to some point below. Another reason is that all of the arrows in an electronic schematic are pointed in the direction of conventional current flow. The diode shown in figure 3-2 is forward biased only when a positive voltage is applied to the anode and a negative voltage is applied to the cathode. If the conventional current flow theory is used, current will flow in the direction the arrow is pointing. If the electron theory of current flow is used, current must flow against the arrow.

A common example of the use of the conventional current flow theory is the electrical system of an automobile. Most automobiles use a negative ground system, which means that the negative terminal of the battery is grounded. The positive terminal of the battery is considered to be the "hot" terminal, and it is generally assumed that current flows from the "hot" terminal to ground.

The diode can be tested with an ohmmeter (see Procedure 1 in the Appendix). When the leads of an ohmmeter are connected to a diode, the diode should show continuity in only one direction. For example, assume that when the leads of an ohmmeter are connected to a diode, it shows continuity. If the leads are reversed, the ohmmeter should indicate an open circuit. If the diode shows continuity in both directions, it is shorted. If the ohmmeter indicates no continuity in either direction, the diode is open.

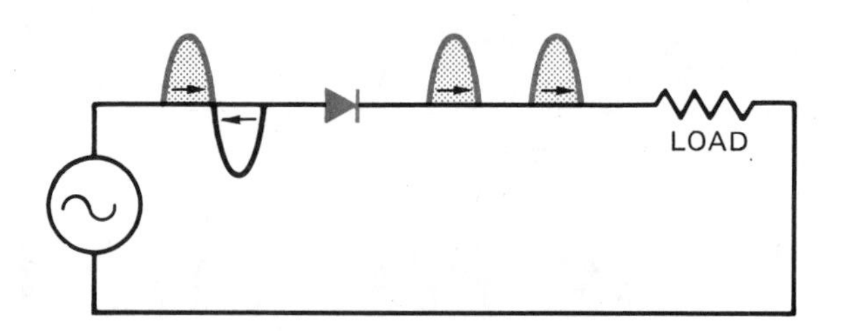

FIGURE 3-3 Half-wave rectifier

The diode can be used to perform many jobs, but it is most commonly used in industry to construct a *rectifier.* A rectifier is a device that changes, or converts, ac voltage into dc voltage. The simplest type of rectifier is the half-wave rectifier, figure 3-3. The half-wave rectifier can be constructed using only one diode. It gets its name from the fact that it will rectify only half of the ac waveform applied to it. When the voltage applied to the anode is positive, the diode is forward biased and current flows through the diode, the load resistor, and back to the power supply. When the voltage applied to the anode is negative, the diode is reverse biased and no current will flow. Since the diode permits current to flow through the load resistor in only one direction, the current is direct current.

Diodes can be connected to produce full-wave rectification, which means that both halves of the ac waveform are made to flow in the same direction. One type of full-wave rectifier is the bridge rectifier, figure 3-4. Notice that four diodes are required to construct the bridge rectifier.

To understand the operation of the bridge rectifier shown in figure 3-4, assume that point X of the ac source is positive and point Y is negative. Current flows to point A of the rectifier. At point A, diode D4 is reverse biased and D1 is forward biased; therefore, the current flows through diode D1 to point B of the rectifier. At point B, diode D2 is reverse biased, so the current must flow through the load resistor to ground. The current returns through ground to point D of the rectifier. At point D, both diodes D3 and D4 are forward biased, but current will not flow from positive to positive. Therefore, the current flows through diode D3 to point C of the bridge, and then to point Y of the ac source, which is negative at this

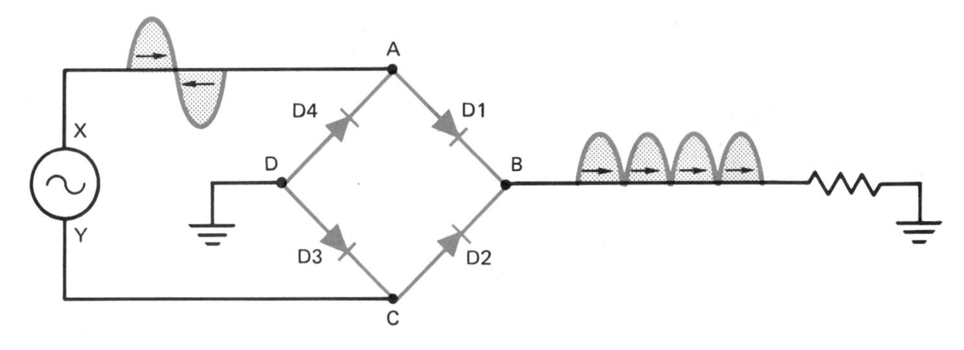

FIGURE 3-4 Bridge rectifier

time. Since current flowed through the load resistor during this half cycle, a voltage developed across the resistor.

Now assume that point Y of the ac source is positive and point X is negative. Current flows from point Y to point C of the rectifier. At point C, diode D3 is reverse biased and diode D2 is forward biased. The current flows through diode D2 to point B of the rectifier. At point B, diode D1 is reverse biased, so the current must flow through the load resistor to ground. The current flows from ground to point D of the bridge. At point D, both diodes D3 and D4 are forward biased. Since current will not flow from positive to positive, the current flows through diode D4 to point A of the bridge and then to point X which is now negative. Current flowed through the load resistor during this half cycle, so a voltage developed across the load resistor. Notice that the current flowed in the same direction through the resistor during both half cycles.

FIGURE 3-5 Bridge rectifiers in a single case (Courtesy International Rectifier)

Bridge rectifiers in single cases are shown in figure 3-5.

In industry three-phase power is used more often than single-phase power. Six diodes can be connected to form a three-phase bridge rectifier that will change three-phase ac voltage into dc voltage, figure 3-6.

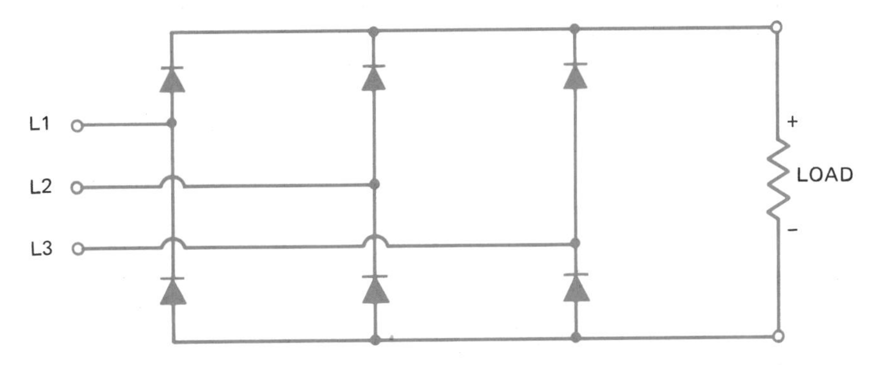

FIGURE 3-6 Three-phase bridge rectifier

When the diode is to be connected in a circuit, there must be some means of identifying the anode and the cathode. Diodes are made in different case styles, as shown in figure 3-7, so there are different methods of identifying the leads. Large stud mounted diodes often have the diode symbol printed on the case to show proper lead identification. Small plastic case diodes often have a line or band around one end of the case, figure 3-8. This line or band represents the line in front of the arrow on the schematic symbol of the diode. An ohmmeter can always be used to determine the proper lead identification if the polarity of the ohmmeter leads is known. The positive lead of the ohmmeter must be connected to the anode to make the diode forward biased.

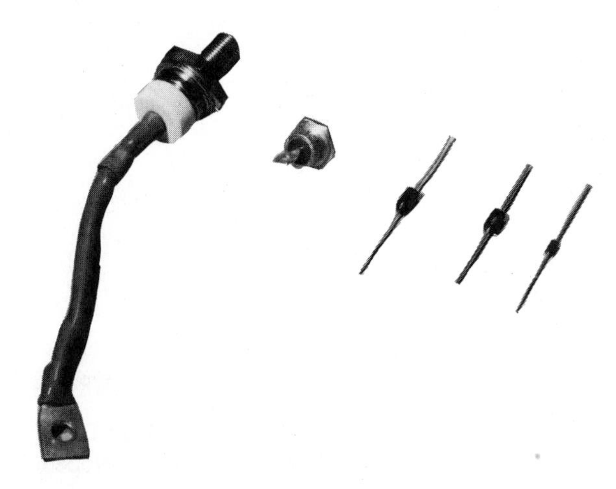

FIGURE 3-7 Diodes shown in various case styles

FIGURE 3-8 Lead identification of a plastic case diode

REVIEW QUESTIONS

1. The PN junction is more commonly known as the ________________.
2. Draw the schematic symbol for a diode.
3. Explain how a diode operates.
4. Explain the difference between the conventional current flow theory and the electron flow theory.
5. Explain the difference between a half-wave rectifier and a full-wave rectifier.
6. Explain how to test a diode with an ohmmeter.

UNIT 4

The Zener Diode

Objectives ***After studying this unit, the student will be able to:***

- **Explain the difference between a junction diode and a zener diode**
- **Discuss common applications of the zener diode**
- **Connect a zener diode in a circuit**

The zener diode is a special device designed to be operated with reverse polarity applied to it. When a diode is broken down in the reverse direction, it enters what is known as the *zener region.* Usually, when a diode is broken down into the zener region, it is destroyed; the zener diode, however, is designed to be operated in this region without harming the device.

When the reverse breakdown voltage of a zener diode is reached, the voltage drop of the device remains almost constant regardless of the amount of current flowing in the reverse direction, figure 4-1. Since the voltage drop of the zener diode is constant, any device connected parallel to the zener will have a constant voltage drop even if the current through the load is changing.

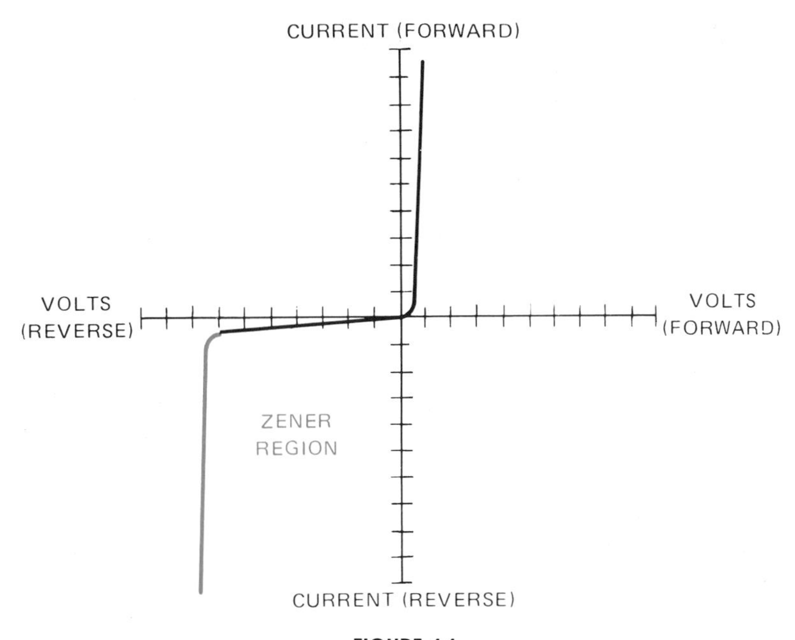

FIGURE 4-1

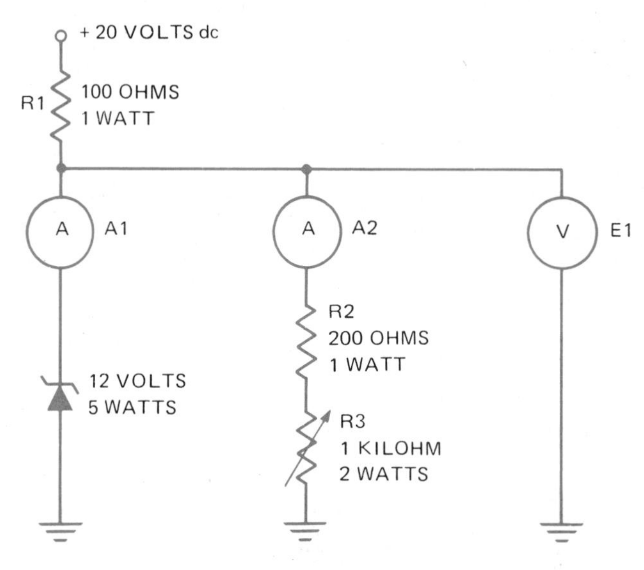

FIGURE 4-2

In figure 4-2, resistor R1 is used to limit the total current of the circuit. Resistor R2 is used to limit the current in the load circuit. Note that the value of R1 is less than the value of R2. This is to insure that the supply can furnish enough current to operate the load. Note also that the supply voltage is greater than the zener voltage. The supply voltage must be greater than the voltage of the zener diode or the circuit cannot operate.

Resistor R1 and the zener diode form a series circuit to ground. Since the zener diode has a voltage drop of 12 volts, resistor R1 has a voltage drop of 8 volts: (20 volts − 12 volts = 8 volts). Therefore, resistor R1 will permit a maximum current flow in the circuit of .08 amperes or 80 milliamps

$$\left(\frac{8}{100} = .08\right)$$

The load circuit, which is a combination of R2 and R3, is connected parallel to the zener diode. Therefore, the voltage applied to the load circuit must be the same as the voltage dropped by the zener. If the zener diode maintains a constant 12-volt drop, a constant voltage of 12 volts must be applied to the load circuit.

The maximum current that can flow through the load circuit is .06 amperes or 60 mA

$$\left(\frac{12 \text{ volts}}{200 \text{ ohms}} = .06 \text{ amps}\right)$$

Notice that the value of R2 (200 ohms) is used to insure that there is enough current available to operate the load.

The maximum current allowed into the circuit by resistor R1 is always equal to the sum of the currents passing through the zener diode and the load. For example, when the load is connected parallel to the zener diode as shown in figure 4-2, and resistor R3 is adjusted to 0 ohms, meter A1 will indicate a current of 20 mA, and meter A2 will indicate a current of 60 mA. Therefore, the maximum current allowed into the circuit by resistor R1 will be 80 mA (20 mA + 60 mA = 80 mA). The voltage value indicated by meter E1 will be the same as the zener voltage value.

If resistor R3 is increased in value to 200 ohms, the resistance of the load will increase to 400 ohms (200 + 200 = 400). Meter A1 will indicate a current of 50 mA and meter A2 will indicate a current of 30 mA. The voltage value indicated by meter E1 will still be the same as the zener voltage value.

The zener diode, therefore, makes a very effective voltage regulator for the load circuit. Although the current through the load circuit changes, the zener diode forces the voltage across the load circuit to remain at a constant value, and conducts the current not used by the load circuit to ground.

The schematic symbol for a zener diode is shown in figure 4-3. The zener diode can be tested with an ohmmeter in the same manner as a common junction diode is tested, provided the zener voltage is greater than the battery voltage of the ohmmeter.

FIGURE 4-3 Schematic symbol for the zener diode

REVIEW QUESTIONS

1. How is a zener diode connected in a circuit as compared to a common junction diode?
2. What is the primary use of a zener diode?
3. A 5.1-volt zener diode is to be connected to an 8-volt power source. The current must be limited to 50 mA. What value of current-limiting resistor must be connected in series with the zener diode?
4. How is a zener diode tested?
5. In a zener diode circuit, the current-limiting resistor limits the total circuit current to 150 mA. If the load circuit is drawing a current of 90 mA, how much current is flowing through the zener diode?

UNIT 5

The Transistor

Objectives

After studying this unit, the student will be able to:

- **Discuss the differences between PNP and NPN transistors**
- **Test transistors with an ohmmeter**
- **Identify the leads of standard, case-style transistors**
- **Discuss the operation of a transistor**
- **Connect a transistor in a circuit**

Transistors are made by connecting three pieces of semiconductor material. There are two basic types of transistors: the NPN and the PNP, figure 5-1. The schematic symbols for these transistors are shown in figure 5-2. These transistors differ in the manner in which they are connected in a circuit. The NPN transistor must have a positive voltage connected to the collector and a negative voltage connected to the emitter. The PNP must have a positive voltage connected to the emitter and a negative voltage connected to the collector. The base must be connected to the same polarity as the collector to forward bias the transistor. Notice that the arrows on the emitters point in the direction of conventional current flow.

An ohmmeter can be used to test a transistor which will appear to the ohmmeter to be two joined diodes, figure 5-3. (For an explanation of how to test a transistor, see Procedure 2 in the Appendix.) If the polarity of the output of the

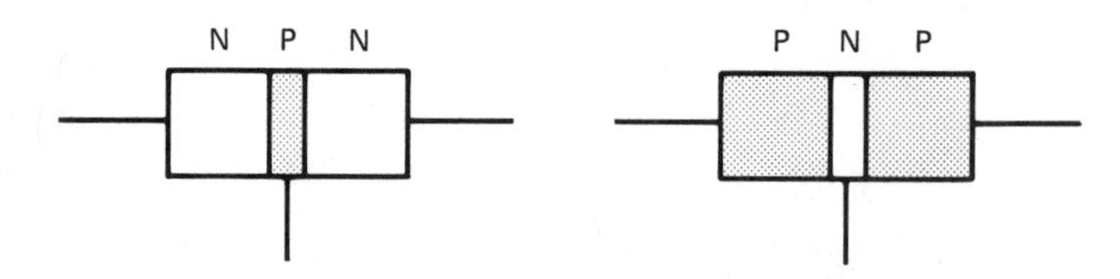

FIGURE 5-1 Two basic types of transistors

+ COLLECTOR
+ BASE
NPN
- EMITTER
- COLLECTOR
- BASE
PNP
+ EMITTER

FIGURE 5-2 Schematic symbols for transistors

FIGURE 5-3 Ohmmeter test for transistors

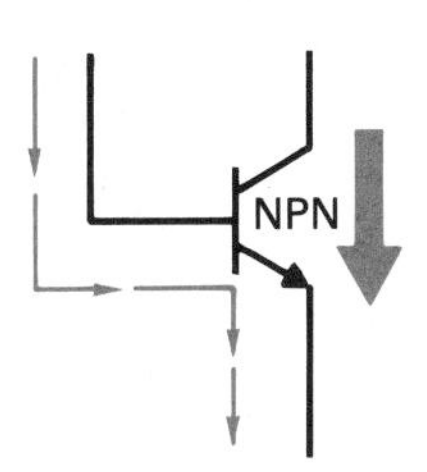

FIGURE 5-4 A small base current controls a large collector current

ohmmeter leads is known, the transistor can be identified as NPN or PNP. An NPN transistor will appear to an ohmmeter to be two diodes with their anodes connected. If the positive lead of the ohmmeter is connected to the base of the transistor, a diode junction should be seen between the base-collector and the base-emitter. If the negative lead of the ohmmeter is connected to the base of an NPN transistor, there should be no continuity between the base-collector and the base-emitter junction.

A PNP transistor will appear to an ohmmeter to be two diodes with their cathodes connected. If the negative lead of the ohmmeter is connected to the base of the transistor, a diode junction should be seen between the base-collector and the base-emitter. If the positive ohmmeter lead is connected to the base, there should be no continuity between the base-collector or the base-emitter.

The simplest way to describe the operation of a transistor is to say that it operates like an electric valve. Current will not flow through the collector-emitter until current flows through the base-emitter. The amount of base-emitter current, however, is small when compared to the collector-emitter current, figure 5-4. For example, assume that when 1 milliamp of current flows through the base-emitter junction, 100 mA of current flow through the collector-emitter junction. If this transistor is a linear device, an increase or decrease of base current will cause a similar increase or decrease of collector current. Therefore, if the base current is increased to 2 mA, the collector current will increase to 200 mA. If the base current is decreased to .5 mA, the collector current will decrease to 50 mA. Notice that a small change in the amount of base current can cause a large change in the amount of collector current. This permits a small amount of signal current to operate a larger device such as the coil of a control relay.

One of the most common applications of the transistor in industry is that of a switch. When used in this manner, the transistor operates like a *digital* device instead of an *analog* device. The term digital refers to a device that has only two states, such as on and off. An analog device can be adjusted to different states. An example of this control can be seen in a simple switch connection. A common wall switch is a digital device. It can be used to turn a light on or off. If the simple toggle switch is replaced with a dimmer control, the light can be turned on, off, or it can be adjusted to any position between on and off. The dimmer is an example of analog control.

If no current flows through the base of the transistor, the transistor acts like an open switch and no current can flow through the collector-emitter junction. If enough base current is applied to the transistor to turn it completely on, it acts like a closed switch and permits current to flow through the collector-emitter junction. This is the same action produced by the closing contacts of a relay or motor starter, but, unlike a transistor, a relay or motor starter cannot turn on and off several thousand times a second.

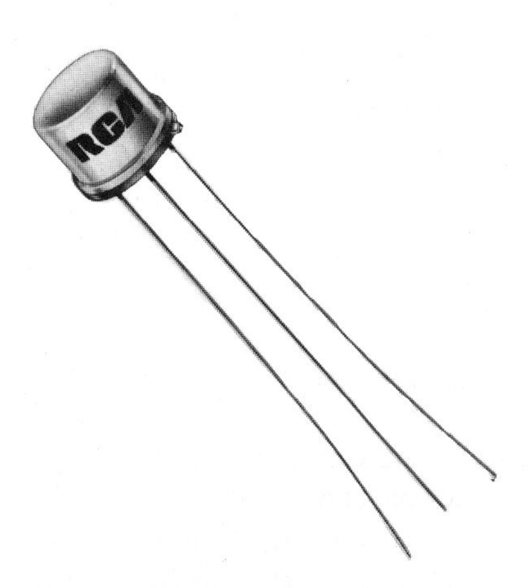

FIGURE 5-5 TO 18 case transistor (Courtesy RCA, Solid State Division)

FIGURE 5-6 TO 220 case transistor (Courtesy RCA, Solid State Division)

FIGURE 5-7 TO 3 case transistor (Courtesy RCA, Solid State Division)

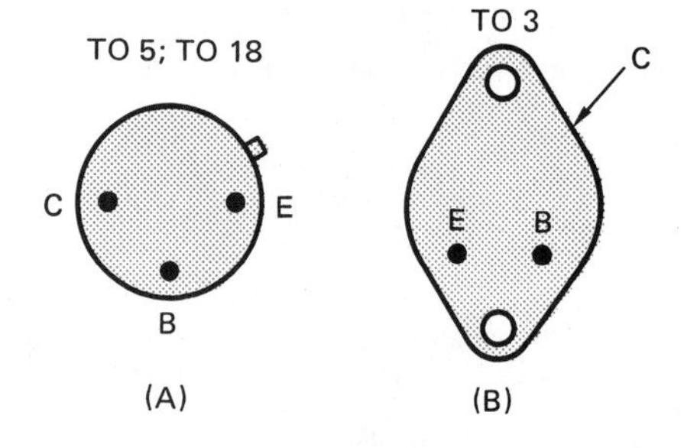

FIGURE 5-8 Lead identification of transistors

Some case styles of transistors permit the leads to be quickly identified, figures 5-5, 5-6 and 5-7. The TO 5 and TO 18 cases, and the TO 3 case are in this category. The leads of the TO 5 and TO 18 case transistors can be identified by holding the case of the transistor with the leads facing you as shown in figure 5-8A. The metal tab on the case of the transistor is closest to the emitter lead. The base and collector leads are positioned as shown.

The leads of a TO 3 case transistor can be identified as shown in figure 5-8B. When the transistor is held with the leads facing you and down, the emitter is the left lead and the base is the right lead. The case of the transistor is the collector.

REVIEW QUESTIONS

1. What are the two basic types of transistors?
2. Explain how to test an NPN transistor with an ohmmeter.
3. Explain how to test a PNP transistor with an ohmmeter.
4. What polarity must be connected to the collector, base, and emitter of an NPN to make it forward biased?
5. What polarity must be connected to the collector, base, and emitter of a PNP transistor to make it forward biased?
6. Explain the difference between an analog device and a digital device.

UNIT 6

The Unijunction Transistor

Objectives

After studying this unit, the student will be able to:

- **Discuss the differences between junction transistors and unijunction transistors**
- **Describe the operation of the unijunction transistor (UJT)**
- **Identify the leads of a UJT**
- **Draw the schematic symbol for a UJT**
- **Test a UJT with an ohmmeter**
- **Connect a UJT in a circuit**

The *unijunction transistor (UJT)* is a special transistor that has two bases and one emitter. The unijunction transistor is a digital device because it has only two states, on and off. It is generally classified with a group of devices known as *thyristors.* Thyristors are devices that are turned completely on or completely off. Thyristors include such devices as the SCR, the triac, the diac and the UJT.

The unijunction transistor is made by combining three layers of semiconductor material as shown in figure 6-1. Figure 6-2 shows the schematic symbol of the UJT with polarity connections and the base diagram.

Current flows in two paths through the UJT. One path is from base #2 to base #1. The other path is through the emitter and base #1. In its

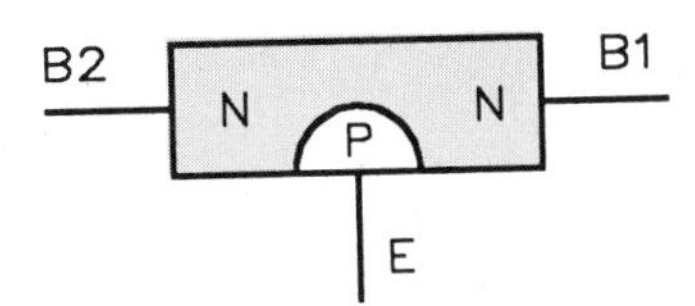

FIGURE 6-1 The unijunction transistor

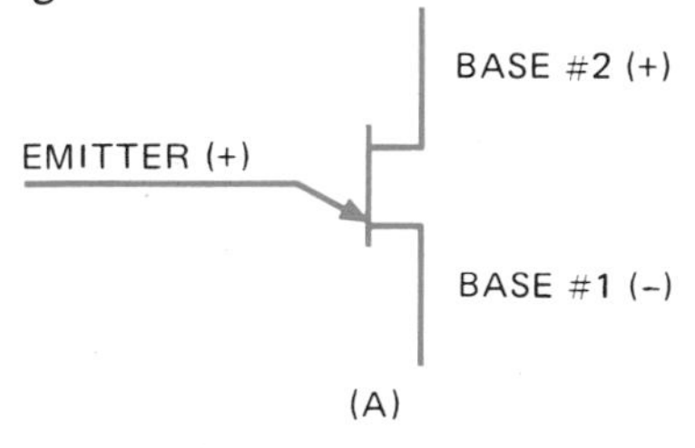

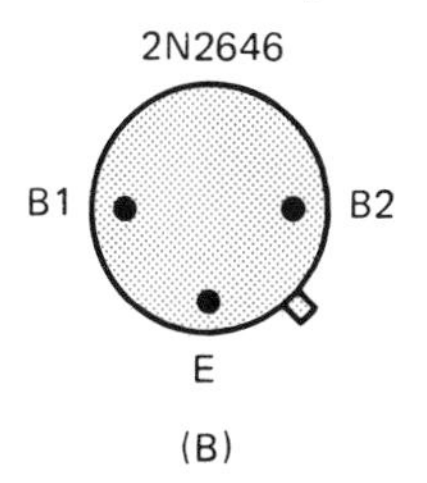

FIGURE 6-2 The schematic symbol for the unijunction transistor with polarity connections and base diagram

normal state, current does not flow through either path until the voltage applied to the emitter is about 10 volts higher than the voltage applied to base #1. When the voltage applied to the emitter is about 10 volts higher than the voltage applied to base #1, the UJT turns on and current flows through the base #1-base #2 path and from the emitter through base #1. Current will continue to flow through the UJT until the voltage applied to the emitter drops to a point that is about 3 volts higher than the voltage applied to base #1. When the emitter voltage drops to this point, the UJT will turn off and will remain off until the voltage applied to the emitter again reaches a level about 10 volts higher than the voltage applied to base #1.

The unijunction transistor is generally connected to a circuit similar to the circuit shown in figure 6-3. The variable resistor controls the capacitor's rate of charge time. When the capacitor has been charged to about 10 volts, the UJT turns on and discharges the capacitor through the emitter and base #1. When the capacitor has been discharged to about 3 volts, the UJT turns off and permits the capacitor to begin charging again. By varying the resistance connected in series with the capacitor, the amount of time needed for charging the capacitor can be changed, thereby controlling the pulse rate of the UJT (T = RC).

The unijunction transistor can furnish a large output pulse because the output pulse is produced by the discharging capacitor, figure 6-4. This large output pulse is generally used for triggering the gate of a silicon-controlled rectifier.

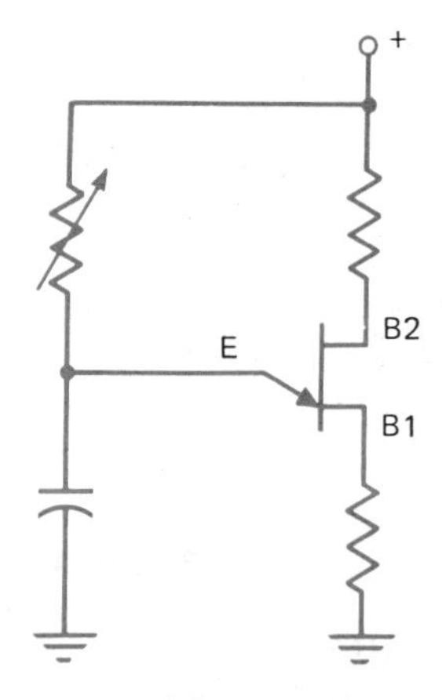

FIGURE 6-3

The pulse rate is determined by the amount of resistance and capacitance connected to the emitter of the UJT. However, the amount of capacitance that can be connected to the UJT is limited. For instance, most UJTs should not be connected to capacitors larger than 10 μF because the UJT may not be able to handle the current spike produced by a larger capacitor, and the UJT could be damaged.

The unijunction transistor can be tested with an ohmmeter in a manner very similar to that used to test a common junction transistor. (For an ex-

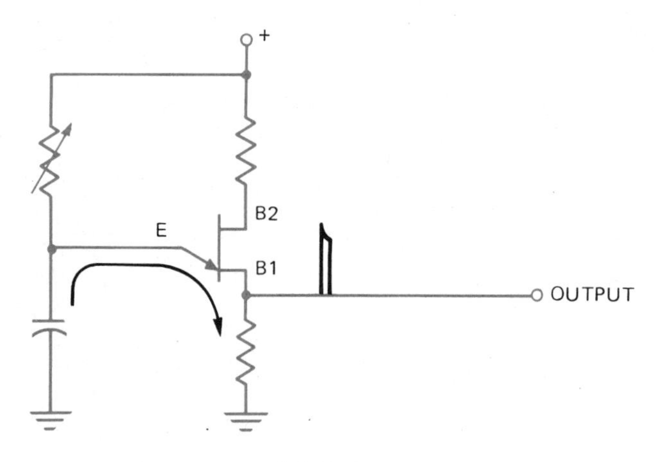

FIGURE 6-4

planation of how to test a unijunction transistor, see Procedure 3 in the Appendix.)

When testing the UJT with an ohmmeter, the UJT will appear as a circuit containing two resistors connected in series with a diode connected to the junction point of the two resistors as shown in figure 6-5. If the positive lead of the ohmmeter is connected to the emitter of the UJT, a circuit should be seen between emitter and base 1 and emitter and base 2. If the negative lead of the ohmmeter is connected to the emitter, no circuit should be seen between the emiter and either base. If the ohmmeter leads are connected to the two bases, continuity will be seen between these two leads provided that the output voltage of the ohmmeter is high enough.

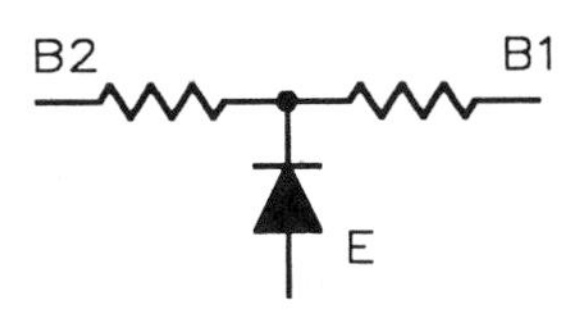

FIGURE 6-5 Testing a UJT

REVIEW QUESTIONS

1. What do the letters UJT stand for?
2. How many layers of semiconductor material are used to construct a UJT?
3. Briefly explain the operation of the UJT.
4. Draw the schematic symbol for the UJT.
5. Briefly explain how to test a UJT with an ohmmeter.

UNIT 7

The SCR

Objectives

After studying this unit, the student will be able to:

- **Discuss the operation of an SCR in a dc circuit**
- **Discuss the operation of an SCR in an ac circuit**
- **Draw the schematic symbol for an SCR**
- **Discuss phase shifting**
- **Test an SCR with an ohmmeter**
- **Connect an SCR in a circuit**

The silicon-controlled rectifier (SCR) is often referred to as a PNPN junction because it is made by joining four layers of semiconductor material, figure 7-1. The schematic symbol for the SCR is shown in figure 7-2. Notice that the symbol for the SCR is the same as the symbol for the diode except that a gate lead has been added. Case styles for SCRs are shown in figure 7-3.

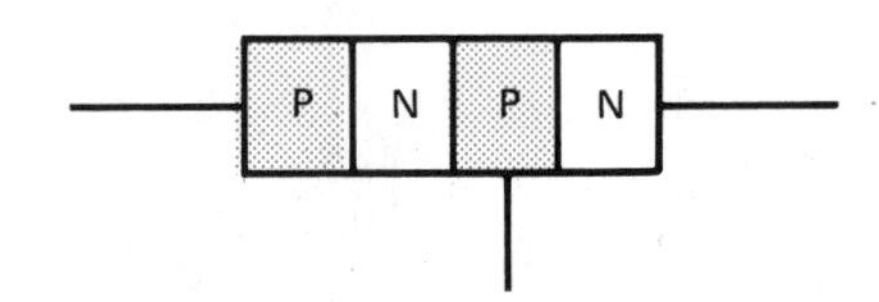

FIGURE 7-1 The PNPN junction

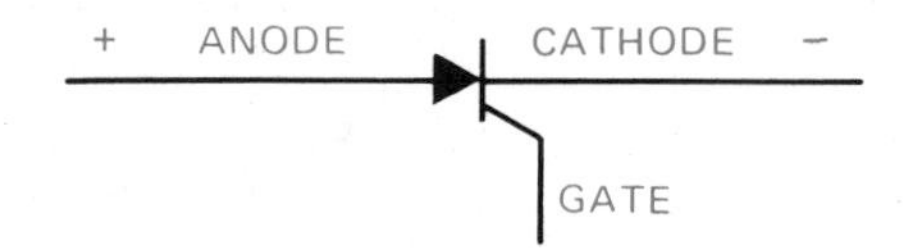

FIGURE 7-2 The schematic symbol for a silicon-controlled rectifier

The SCR is a member of a family of devices known as thyristors. Thyristors are digital devices in that they have only two states, on and off. The SCR is used when it is necessary for an electronic device to control a large amount of power. For example, assume that an SCR has been connected in

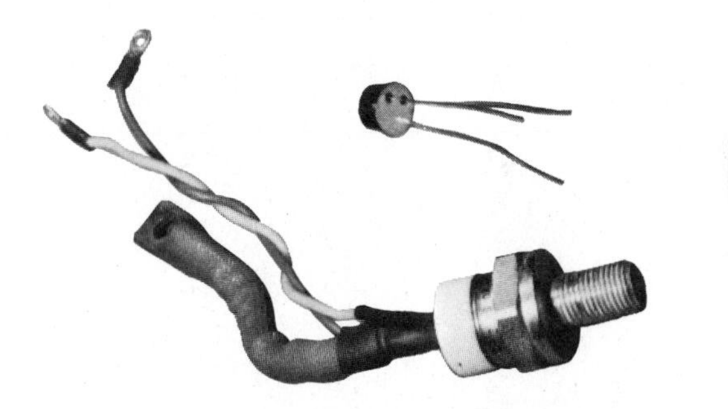

FIGURE 7-3 SCRs shown in different case styles

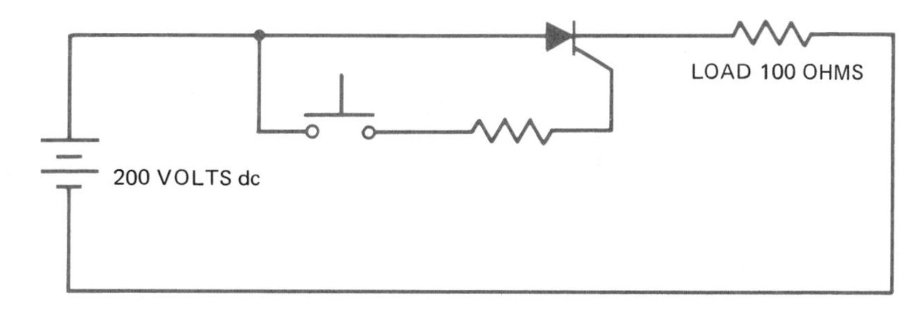

FIGURE 7-4 The SCR is turned on by the gate

a circuit as shown in figure 7-4. When the SCR is turned off, it will drop the full voltage of the circuit and 200 volts will appear across the anode and cathode. Although the SCR has a voltage drop of 200 volts, there is no current flow in the circuit. The SCR does not have to dissipate any power in this condition (200 volts × 0 amperes = 0 watts). When the push button is pressed, the SCR turns on, producing a voltage drop across its anode and cathode of about 1 volt. The load resistor limits the circuit current to 2 amperes

$$\left(\frac{200 \text{ volts}}{100 \text{ ohms}} = 2 \text{ amperes}\right)$$

Since the SCR now has a voltage drop of 1 volt and 2 amperes of current flowing through it, it must dissipate 2 watts of heat (1 volt × 2 amperes = 2 watts). Notice that although the SCR is dissipating only 2 watts of power, it is controlling 400 watts of power.

THE SCR IN A DC CIRCUIT

When an SCR is connected in a dc circuit as shown in figure 7-4, the gate will turn the SCR on, but it will not turn the SCR off. To turn the anode-cathode section of the SCR on, the gate must be connected to the same polarity as the anode. Once the gate has turned the SCR on, the SCR will remain on until the current flowing through the anode-cathode section drops to a low enough level to permit the device to turn off. The amount of current required to keep the SCR turned on is called the *holding current*.

In figure 7-5 assume that resistor R1 has been adjusted to its highest value and resistor R2 has been adjusted to its lowest or 0 value. When switch S1 is closed, no current will flow through the anode-cathode section of the SCR because resistor R1 prevents the amount of current needed to trigger the device from flowing through the gate-cathode section of the SCR. If the value of resistor R1 is slowly decreased, current flow through the gate-cathode section will slowly increase. When the gate reaches a certain level, assume 5 mA for this SCR, the SCR will fire, or turn on. When the SCR fires, current will flow through the anode-cathode section and the voltage drop across the device will be about 1 volt. Once the SCR is turned on, the gate has no control over the device. It could be disconnected from the an-

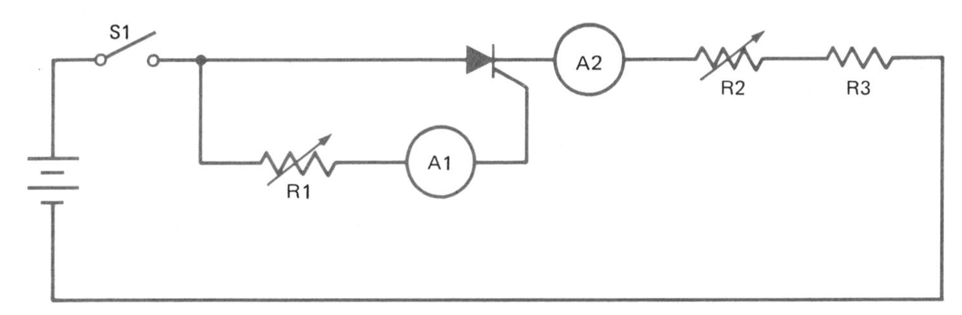

FIGURE 7-5 Operation of an SCR in a dc circuit

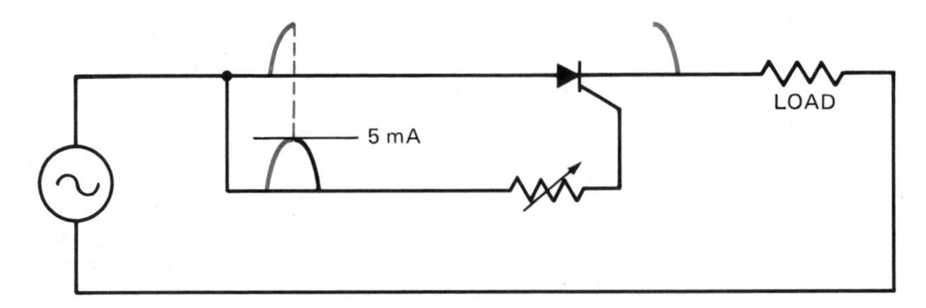

FIGURE 7-6 The SCR fires when the ac waveform reaches peak value

ode without affecting the circuit. When the SCR fires, the anode-cathode section becomes a short circuit and current flow is limited by resistor R3.

Now assume that resistor R2 is slowly increased in value. When the resistance of R2 is slowly increased, the current flow through the anode-cathode section will slowly decrease. Assume that when the current flow through the anode-cathode section drops to 100 mA, the device suddenly turns off and the current flow drops to 0. This SCR requires 5 mA of gate current to turn it on, and has a holding current value of 100 mA.

THE SCR IN AN AC CIRCUIT

The SCR is a rectifier; when it is connected in an ac circuit, the output is direct current. The SCR operates in the same manner in an ac circuit as it does in a dc circuit. The difference in operation is caused by the ac waveform falling back to 0 at the end of each half cycle. When the ac waveform drops to 0 at the end of each half cycle, the SCR turns off. This means that the gate must retrigger the SCR for each cycle it is to conduct, figure 7-6.

Assume that the variable resistor connected to the gate has been adjusted to permit 5 mA of current to flow when the voltage applied to the anode reaches its peak value. When the SCR turns on, current will begin flowing through the load resistor when the ac waveform is at its positive peak. Current will continue to flow through the load until the decreasing voltage of the sine wave causes the current to drop below the holding current level of 100 mA. When the current through the anode-cathode section drops below 100 mA, the SCR turns off and all current flow stops. The SCR will remain turned off when the ac waveform goes into its negative half cycle because during this half cycle the SCR is reverse biased and cannot be fired.

If the resistance connected in series with the gate is reduced, a current of 5 mA will be reached before the ac waveform reaches its peak value, figure 7-7. This will cause the SCR to fire earlier in the cycle. Since the SCR fires earlier in the cycle, current is permitted to flow through the load resistor for a longer period of time, which produces a

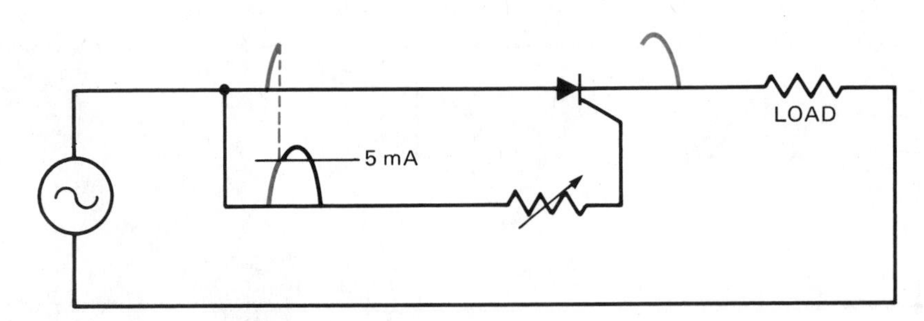

FIGURE 7-7 The SCR fires before the ac waveform reaches peak value

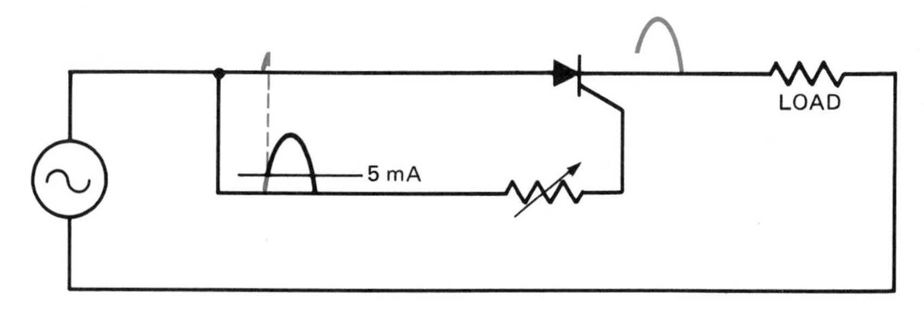

FIGURE 7-8 The SCR fires earlier than in figure 7-7

higher average voltage drop across the load. If the resistance of the gate circuit is reduced again as shown in figure 7-8, the 5 mA of gate current needed to fire the SCR will be reached earlier than in figure 7-7. Current will begin flowing through the load sooner than before, which will permit a higher average voltage to be dropped across the load.

Notice that this circuit enables the SCR to control only half of the positive waveform. The latest the SCR can be fired in the cycle is when the ac waveform is at 90° or peak. If a lamp were used as the load for this circuit, it would burn at half brightness when the SCR first turned on. This control would permit the lamp to be operated from half brightness to full brightness, but it could not be operated at a level less than half brightness.

PHASE SHIFTING THE SCR

The SCR can control all of the positive waveform through the use of *phase shifting*. As the term implies, phase shifting means to shift the phase of one thing in reference to another. In this instance, the voltage applied to the gate must be shifted out of phase with the voltage applied to the anode. Although there are several methods used for phase shifting an SCR, it is beyond the scope of this text to cover all of them. The basic principles are the

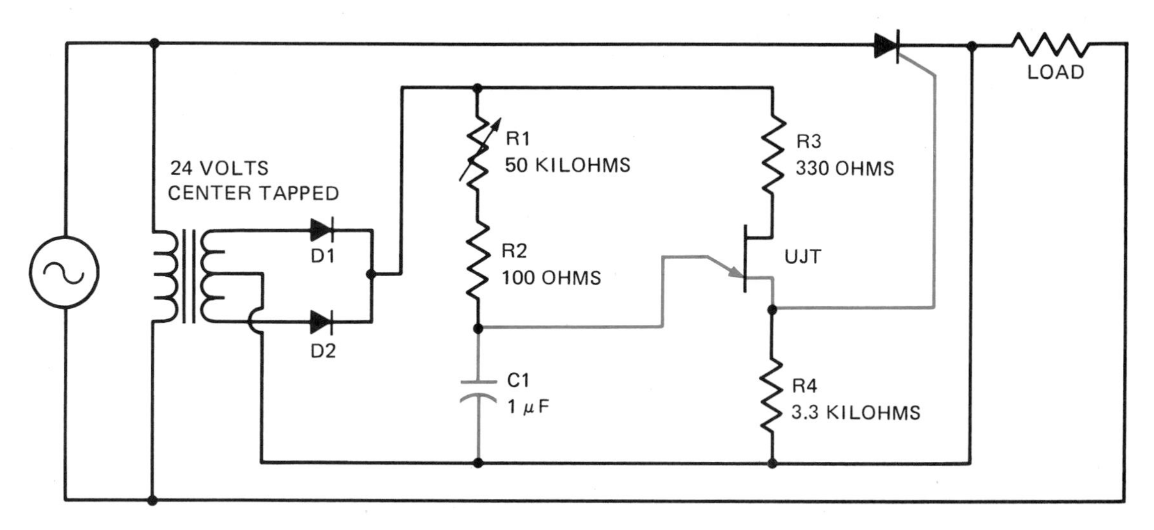

FIGURE 7-9 UJT phase shift for an SCR. SCR gate current is provided by the discharging capacitor when the UJT fires.

same for all of the methods, however, so only one method is covered.

To phase shift an SCR, the gate circuit must be unlocked or separated from the anode circuit. The circuit shown in figure 7-9 will accomplish this. A 24-volt, center-tapped transformer is used to isolate the gate circuit from the anode circuit. Diodes D1 and D2 are used to form a two-diode type of full-wave rectifier to operate the UJT circuit. Resistor R1 is used to determine the pulse rate of the UJT by controlling the charge time of capacitor C1. Resistor R2 is used to limit the current through the emitter of the UJT if resistor R1 is adjusted to 0 ohms. Resistor R3 limits current through the base 1-base 2 section when the UJT turns on. Resistor R4 permits a voltage spike or pulse to be produced across it when the UJT turns on and discharges capacitor C1. The pulse produced by the discharge of capacitor C1 is used to trigger the gate of the SCR.

Since the pulse of the UJT is used to provide a trigger for the gate of the SCR, the SCR can be fired at any time regardless of the voltage applied to the anode. This means that the SCR can now be fired as early or late during the positive half cycle as desired because the gate pulse is determined by the charge rate of capacitor C1. The voltage across the load can now be adjusted from 0 to the full applied voltage.

TESTING THE SCR

The SCR can be tested with an ohmmeter (see Procedure 4 in the Appendix). To test the SCR, connect the positive output lead of the ohmmeter to the anode and the negative lead to the cathode. The ohmmeter should indicate no continuity. Touch the gate of the SCR to the anode. The ohmmeter should indicate continuity through the SCR. When the gate lead is removed from the anode, conduction may stop or continue depending on whether or not the ohmmeter is supplying enough current to keep the device above its holding current level. If the ohmmeter indicates continuity through the SCR before the gate is touched to the anode, the SCR is shorted. If the ohmmeter will not indicate continuity through the SCR after the gate has been touched to the anode, the SCR is open.

REVIEW QUESTIONS

1. What do the letters SCR stand for?
2. If an SCR is connected to an ac circuit, will the output voltage be ac or dc?
3. Briefly explain how an SCR operates when connected to a dc circuit.
4. How many layers of semiconductor material are used to construct an SCR?
5. SCRs are members of a family of devices known as thyristors. What is a thyristor?
6. Briefly explain why thyristors have the ability to control large amounts of power.
7. What is the average voltage drop of an SCR when it is turned on?
8. Explain why an SCR must be phase shifted.

UNIT 8

The Diac

Objectives

After studying this unit, the student will be able to:

- **Draw the schematic symbol for a diac**
- **Discuss the operation of a diac**
- **Connect a diac in a circuit**

The *diac* is a special-purpose, bidirectional diode. The primary function of the diac is to phase shift a triac. The operation of the diac is very similar to that of a unijunction transistor, except that the diac is a two-directional device. The diac has the ability to operate in an ac circuit while the UJT can operate only in a dc circuit.

There are two schematic symbols for the diac, figure 8-1. Both of these symbols are used in electronic schematics to illustrate the use of a diac. Therefore, you should make yourself familiar with both symbols.

The diac is a voltage sensitive switch that can operate on either polarity, figure 8-2. Voltage applied to the diac must reach a predetermined level before the diac will activate. For this example, assume that the predetermined level is 15 volts. When the voltage reaches 15 volts, the diac will turn on, or fire. When the diac fires, it displays a negative resistance, which means that it will con-

FIGURE 8-1 Schematic symbols for the diac

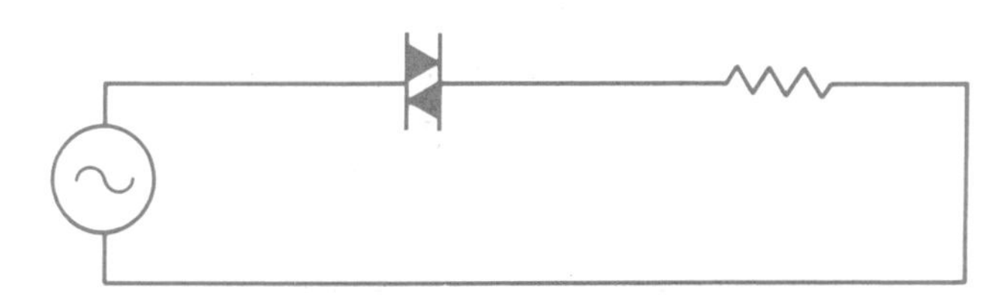

FIGURE 8-2 The diac can operate on either polarity

duct at a lower voltage than the voltage needed to turn it on. In this example, assume that the voltage drops to 5 volts when the diac conducts. The diac will remain on until the applied voltage drops below its conduction level, which is 5 volts, figure 8-3.

Since the diac is a bidirectional device, it will conduct on either half cycle of the alternating current applied to it, figure 8-4. Note that the diac operates in the same manner on both halves of the ac cycle. The simplest way to summarize the operation of the diac is to say that it is a voltage sensitive ac switch.

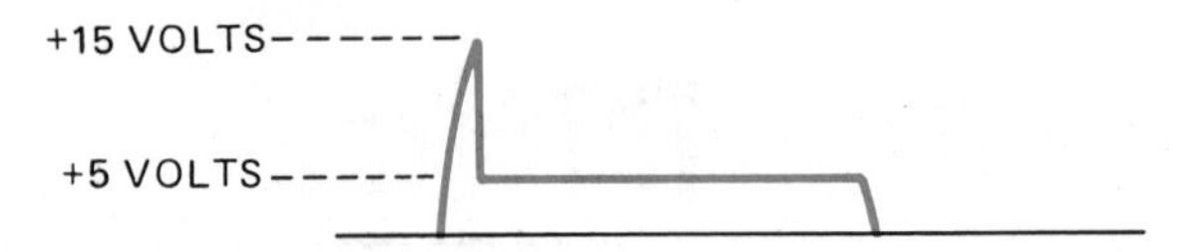

FIGURE 8-3 The diac operates until the applied voltage falls below its conduction level

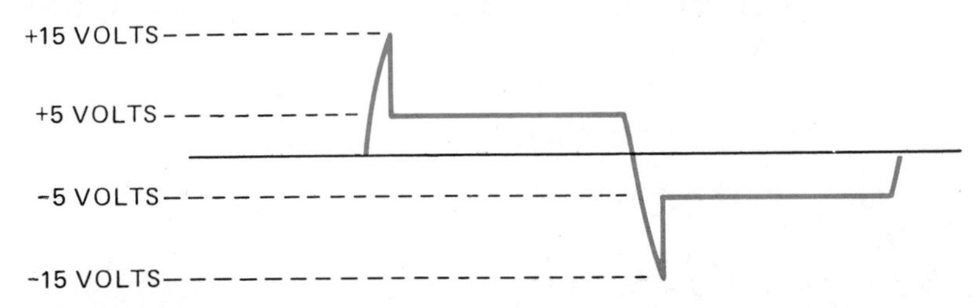

FIGURE 8-4 The diac will conduct on either half of the alternating current

REVIEW QUESTIONS

1. Briefly explain how a diac operates.
2. Draw the two schematic symbols used to represent the diac.
3. What is the major use of the diac in industry?
4. When a diac first turns on, does the voltage drop, remain at the same level, or increase to a higher level?

UNIT 9

The Triac

Objectives *After studying this unit, the student will be able to:*

- **Draw the schematic symbol for a triac**
- **Discuss the similarities and differences between SCRs and triacs**
- **Discuss the operation of a triac in an ac circuit**
- **Discuss phase shifting a triac**
- **Connect a triac in a circuit**
- **Test a triac with an ohmmeter**

The triac is a PNPN junction connected parallel to an NPNP junction. Figure 9-1 illustrates the semiconductor arrangement of a triac. The triac operates in a manner similar to that of two connected SCRs, figure 9-2. The schematic symbol for the triac is shown in figure 9-3.

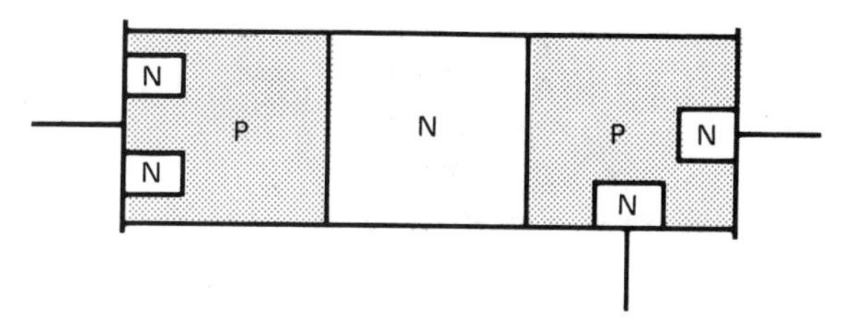

FIGURE 9-1 The semiconductor arrangement of a triac

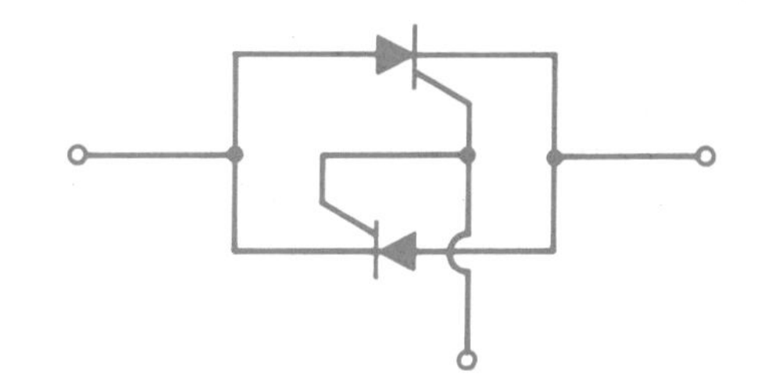

FIGURE 9-2 The triac operates in a manner similar to two SCRs with a common gate

When an SCR is connected in an ac circuit, the output voltage is direct current. When a triac is connected in an ac circuit, the output voltage is alternating current. Since the triac operates like two SCRs that are connected and facing in opposite directions, it will conduct both the positive and negative half cycles of ac current.

When a triac is connected in an ac circuit as shown in figure 9-4, the gate must be connected to the same polarity as MT2. When the ac voltage applied to MT2 is positive, the SCR, which is forward biased, will conduct. When the voltage applied to MT2 is negative, the other SCR is forward biased and will conduct that half of the waveform. Since one of the SCRs is forward biased for each half cycle, the triac will conduct ac current as long as the gate lead is connected to MT2.

The triac, like the SCR, requires a certain amount of gate current to turn it on. Once the

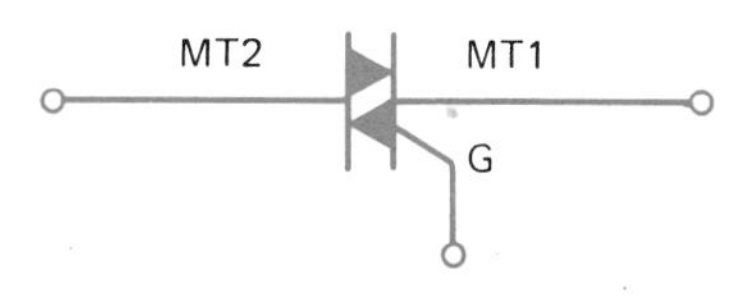

FIGURE 9-3 The schematic symbol for a triac

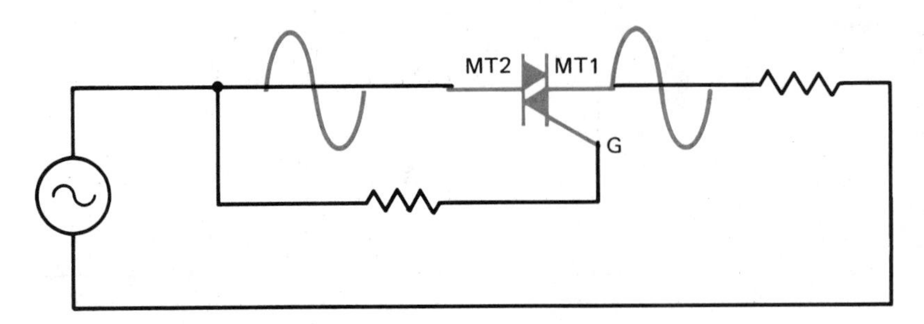

FIGURE 9-4 The triac conducts both halves of the ac waveform

triac has been triggered by the gate, it will continue to conduct until the current flowing through MT2-MT1 drops below the holding current level.

THE TRIAC USED AS AN AC SWITCH

The triac is a member of the thyristor family, which means that it has only two states of operation, on and off. When the triac is turned off, it drops the full applied voltage of the circuit at 0 amps of current flow. When the triac is turned on, it has a voltage drop of about 1 volt and circuit current must be limited by the load connected to the circuit.

FIGURE 9-5 The triac used for low power applications (Courtesy RCA, Solid State Division)

The triac has become very popular in industrial circuits as an ac switch. Since it is a thyristor, it has the ability to control a large amount of voltage and current. There are no contacts to wear out, it is sealed against dirt and moisture, and it can operate thousands of times per second. The triac is used as the output device of many solid-state relays which will be covered later. Two types of triacs are shown in figures 9-5 and 9-6.

THE TRIAC USED FOR AC VOLTAGE CONTROL

The triac can be used to control ac voltage, figure 9-7. If a variable resistor is connected in se-

FIGURE 9-6 The triac shown in a stud mount case (Courtesy RCA, Solid State Division)

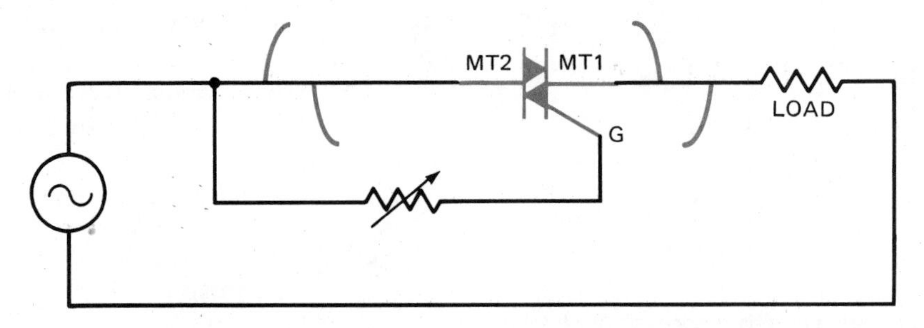

FIGURE 9-7 The triac controls half of the ac applied voltage

ries with the gate, the point at which the gate current is high enough to fire the triac can be adjusted. The resistance can be adjusted to permit the triac to fire when the ac waveform reaches its peak value. This will cause half of the ac voltage to be dropped across the triac and half to be dropped across the load.

If the gate resistance is reduced, the amount of gate current needed to fire the triac will be obtained before the ac waveform reaches its peak value. This means that less voltage will be dropped across the triac and more voltage will be dropped across the load. This circuit permits the triac to control only one half of the ac waveform applied to it. If a lamp is used as the load, it can be controlled from half brightness to full brightness. If an attempt is made to adjust the lamp to operate at less than half brightness, it will turn off.

PHASE SHIFTING THE TRIAC

To obtain complete voltage control, the triac, like the SCR, must be phase shifted. Several methods can be used to phase shift a triac, but only one will be covered in this unit. In figure 9-8, a diac is used to phase shift the triac. Resistors R1 and R2 are connected in series with capacitor C1. Resistor R1 is a variable resistor which is used to control the charge time of capacitor C1. Resistor R2 is used to limit current if resistor R1 is adjusted to 0 ohms. Assume that the diac connected in series with the gate of the triac will turn on when capacitor C1 has been charged to 15 volts. When the diac turns on, capacitor C1 will discharge through the gate of the triac. This permits the triac to fire, or turn on. Since the diac is a bidirectional device, it will permit a positive or negative pulse to trigger the gate of the triac.

When the triac fires, there is a voltage drop of about 1 volt across MT2 and MT1. The triac remains on until the ac voltage drops to a low enough value to permit the triac to turn off. Since the phase shift circuit is connected parallel to the triac, once the triac turns on, capacitor C1 cannot begin charging again until the triac turns off at the end of the ac cycle.

Notice that the pulse applied to the gate is controlled by the charging of capacitor C1, not the amplitude of voltage. If the correct values are chosen, the triac can be fired at any point in the ac cycle applied to it. The triac can now control the ac voltage from 0 to the full voltage of the circuit. A common example of this type of triac circuit is the light dimmer control used in many homes.

TESTING THE TRIAC

The triac can be tested with an ohmmeter (see Procedure 5 in the Appendix). To test the

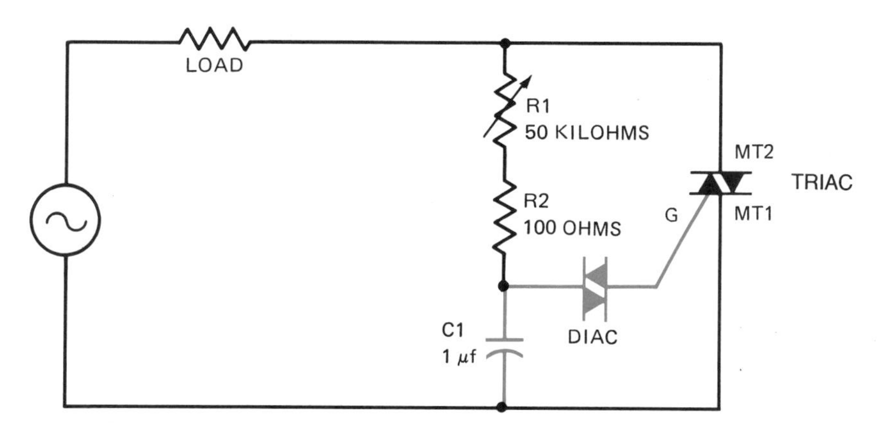

FIGURE 9-8 Phase shift circuit for a triac. When the diac turns on, gate current is supplied to the triac by the discharge of capacitor C1.

triac, connect the ohmmeter leads to MT2 and MT1. The ohmmeter should indicate no continuity. If the gate lead is touched to MT2, the triac should turn on and the ohmmeter should indicate continuity through the triac. When the gate lead is released from MT2, the triac may continue to conduct or it may turn off depending on whether the ohmmeter supplies enough current to keep the device above its holding current level. This tests one half of the triac.

To test the other half of the triac, reverse the connection of the ohmmeter leads. The ohmmeter should indicate no continuity. If the gate is touched again to MT2, the ohmmeter should indicate continuity through the device. The other half of the triac has been tested.

REVIEW QUESTIONS

1. Draw the schematic symbol for a triac.
2. When a triac is connected in an ac circuit, is the output ac or dc?
3. The triac is a member of what family of devices?
4. Briefly explain why a triac must be phase shifted.
5. What electronic component is frequently used to phase shift the triac?
6. When the triac is being tested with an ohmmeter, which other terminal should the gate be connected to if the ohmmeter is to indicate continuity?

UNIT 10

The 555 Timer

Objectives ***After studying this unit, the student will be able to:***

- **Describe the operation of the 555 timer**
- **Discuss the uses of the 555 timer**
- **Connect the timer as an oscillator**
- **Connect the 555 timer as an on-delay timer**

The 555 timer is an eight-pin integrated circuit which has become one of the most popular electronic devices used in industrial electronic circuits. The reason for the 555's popularity is its tremendous versatility. The 555 timer is used in circuits that require a time delay function, and is also used as an oscillator to provide the pulses needed to operate computer circuits.

The 555 timer is most often housed in an eight-pin, in-line integrated circuit (IC), figures 10-1 and 10-2. This package has a notch at one end, or a dot by one pin, which is used to identify pin #1. Once pin #1 has been identified, the other pins are numbered as shown in figure 10-1. The 555 timer operates on voltages that range from about 3 to 16 volts. Following is an explanation of each pin and its function.

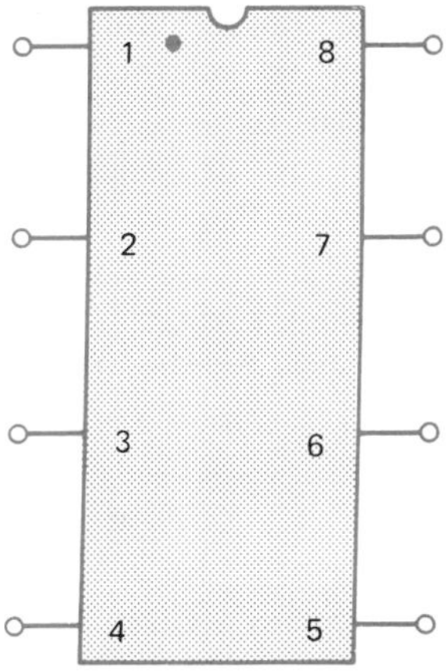

FIGURE 10-1 After pin #1 has been identified, the other pins are numbered as shown

Pin #1 *Ground*—This pin is connected to circuit ground.

Pin #2 *Trigger*—Pin #2 must be connected to a voltage that is less than 1/3 Vcc (the applied voltage) to trigger the unit. This usually is done by connecting pin #2 to ground. The connection to 1/3 Vcc or

FIGURE 10-2 An eight-pin, in-line, integrated circuit (Courtesy RCA, Solid State Division)

ground must be momentary. If pin #2 is not removed from ground, the unit will not operate.

Pin #3 ***Output***—The output turns on when Pin #2 is triggered and turns off when the discharge is turned on.

Pin #4 ***Reset***—When this pin is connected to Vcc, it permits the unit to operate. When it is connected to ground, it activates the discharge and keeps the timer from operating.

Pin #5 ***Control Voltage***—If this pin is connected to Vcc through a variable resistor, the on time is longer, but the off time is not affected. If pin #5 is connected to ground through a variable resistor, the on time is shorter, and the off time is still not affected. If pin #5 is not to be used in the circuit, it is usually taken to ground through a small capacitor. This helps to keep circuit noise from "talking" to pin #5.

Pin #6 ***Threshold***—When the voltage across the capacitor connected to pin #6 reaches 2/3 the value of Vcc, the discharge turns on and the output turns off.

Pin #7 ***Discharge***—When pin #6 turns the discharge on, it discharges the capacitor connected to pin #6. The discharge remains turned on until pin #2 retriggers the timer. The discharge then turns off and the capacitor connected to pin #6 begins charging again.

Pin #8 ***Vcc***—Pin #8 is connected to Vcc.

(For the following explanation, assume that pin #2 is connected to pin #6. This permits the unit to be retriggered by the discharge each time it turns on and discharges the capacitor to 1/3 the value of Vcc.)

The 555 timer operates on a percentage of the applied voltage. This permits the time setting to remain constant even if the applied voltage changes. For example, when the capacitor connected to pin #6 reaches 2/3 of the applied voltage, the discharge turns on and discharges the capacitor until it reaches 1/3 of the applied voltage. If the applied voltage of the timer is connected to 12 volts dc, 2/3 of the applied voltage is 8 volts

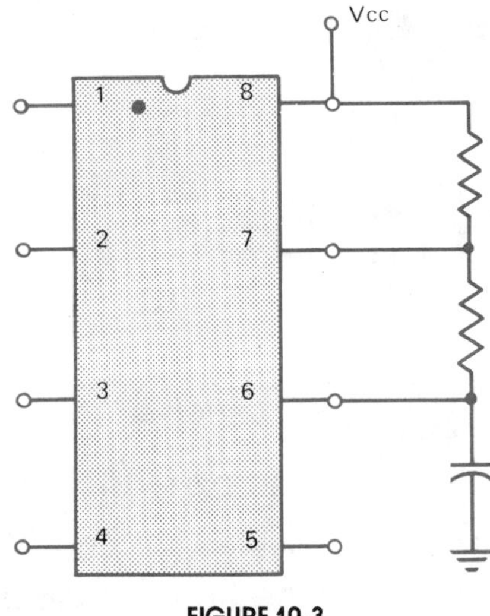

FIGURE 10-3

and 1/3 is 4 volts. This means that when the voltage across the capacitor connected to pin #6 reaches 8 volts, pin #7 will turn on until the capacitor is discharged to 1/3 the value of Vcc, or 4 volts, and will then turn off, figure 10-3.

If the voltage is lowered to 6 volts at Vcc, 2/3 of the applied voltage is 4 volts and 1/3 of the applied voltage is 2 volts. Pin #7 will now turn on when the voltage across the capacitor connected to pin #6 reaches 4 volts and will turn off when the voltage across the capacitor drops to 2 volts.

The formula for a RC time constant is (Time = Resistance × Capacitance). Notice that there is no mention of voltage in the formula. This means that it will take the same amount of time to charge the capacitor regardless of whether the circuit is connected to 12 volts or to 6 volts. If the time it takes for the voltage of the capacitor connected to pin #6 to reach 2/3 of Vcc when the timer has an applied voltage of 12 volts is measured, it will be the same as the amount of time it takes when the applied voltage is only 6 volts. The timing of the circuit remains the same even if the voltage changes.

The circuit shown in figure 10-4 is used to explain the operation of the 555 timer. In figure 10-4, a normally closed switch, S1, is connected between the discharge, pin #7, and the ground, pin #1. A normally open switch, S2, is connected between the output, pin #3, and Vcc, pin #8.

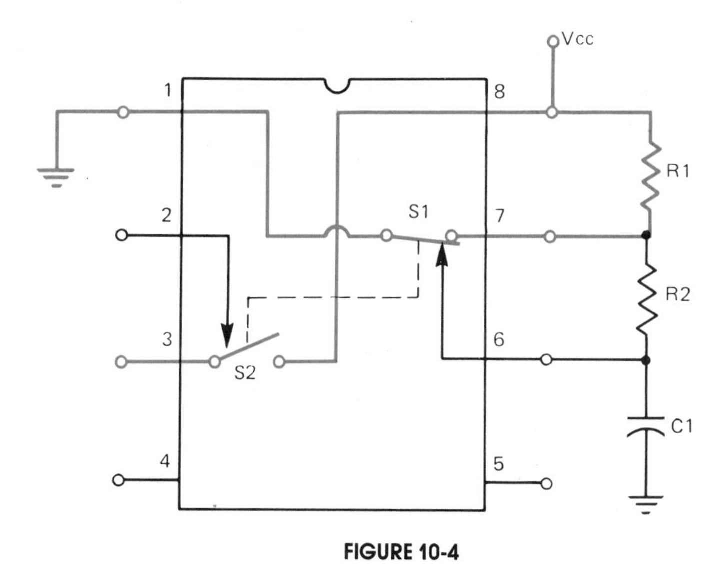

FIGURE 10-4

The dotted line drawn between these two switches shows mechanical connection. This means that these switches operate together. If S1 opens, S2 closes at the same time. If S2 opens, S1 closes. Pin #2, the trigger, and pin #6, the threshold, are used to control these switches. The trigger can close switch S2, and the threshold can close S1.

To begin the analysis of this circuit, assume that switch S1 is closed and switch S2 is open as shown in figure 10-4. When the trigger is connected to a voltage that is less than 1/3 of Vcc, it causes switch S2 to close and switch S1 to open. When switch S2 closes, voltage is supplied to the output at pin #3. When switch S1 opens, the discharge is no longer connected to ground, and capacitor C1 begins to charge through resistors R1 and R2. When the voltage across C1 reaches 2/3 of Vcc, the threshold, pin #6, causes switch S1 to close and switch S2 to open. When switch S2 opens, the output turns off. When switch S1 closes, the discharge, pin #7, is connected to ground. Capacitor C1 then discharges through resistor R2. The timer will remain in this position until the trigger is again connected to a voltage that is less than 1/3 of Vcc.

If the trigger is connected permanently to a voltage less than 1/3 of Vcc, switch S2 will be held closed and switch S1 will be held open. This, of course, will stop the operation of the timer. As stated previously, the trigger must be a momentary pulse, not a continuous connection, in order for the 555 timer to operate.

CIRCUIT APPLICATIONS

The Oscillator

The 555 timer can perform a variety of functions. One of the functions it is commonly used for is that of an oscillator. The 555 timer has become popular for this application because it is so easy to use.

The 555 timer shown in figure 10-5 has pin #2 connected to pin #6. This permits the timer to retrigger itself at the end of each time cycle. When the applied voltage is turned on, capacitor C1 is discharged and has a voltage of 0 volts across it. Since pin #2 is connected to pin #6, and the voltage at that point is less than 1/3 of Vcc, the timer will trigger. When the timer is triggered, two things happen at the same time: the output turns on, and the discharge turns off. When the discharge at pin #7 turns off, capacitor C1 charges through resistors R1 and R2. The amount of time it takes for capacitor C1 to charge is determined by the capacitance of the capacitor and the combined resistance of R1 and R2.

When capacitor C1 is charged to a voltage that is 2/3 of Vcc, the output turns off, and the

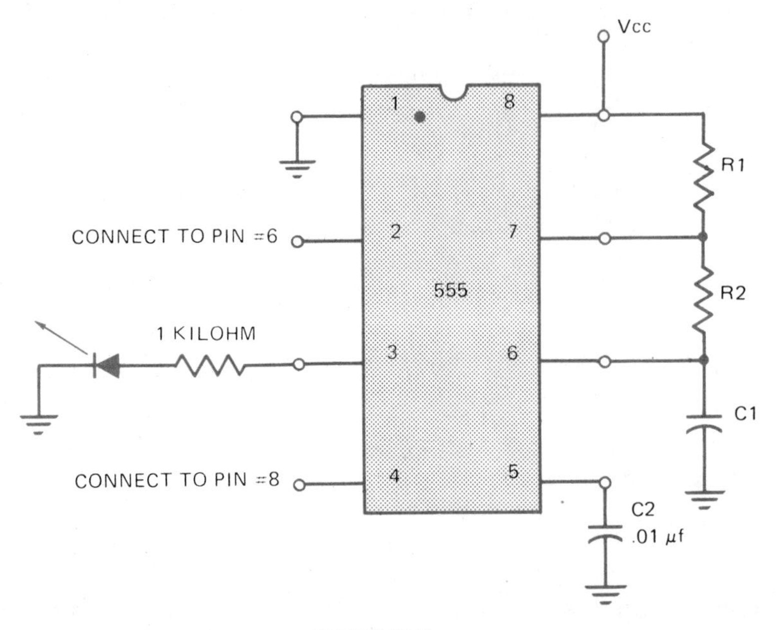

FIGURE 10-5

discharge at pin #7 turns on. When the discharge turns on, capacitor C1 discharges through resistor R2 to ground. The amount of time it takes C1 to discharge is determined by the capacitance of capacitor C1 and the resistance of R2. When capacitor C1 is discharged to a voltage that is 1/3 of Vcc, the timer is retriggered by pin #2 causing the output to turn on and the discharge to turn off. When the discharge turns off, capacitor C1 begins charging again.

The amount of time required to charge capacitor C1 is determined by the combined resistance of R1 and R2. The discharge time, however, is determined by the value of R2, figure 10-6.

Since the timer's output is turned on while capacitor C1 is charging, and turned off while C1 is discharging, the on time of the output is longer than the off time. If the value of resistor R2 is much greater than the value of resistor R1, this condition is not too evident. For example, if resis-

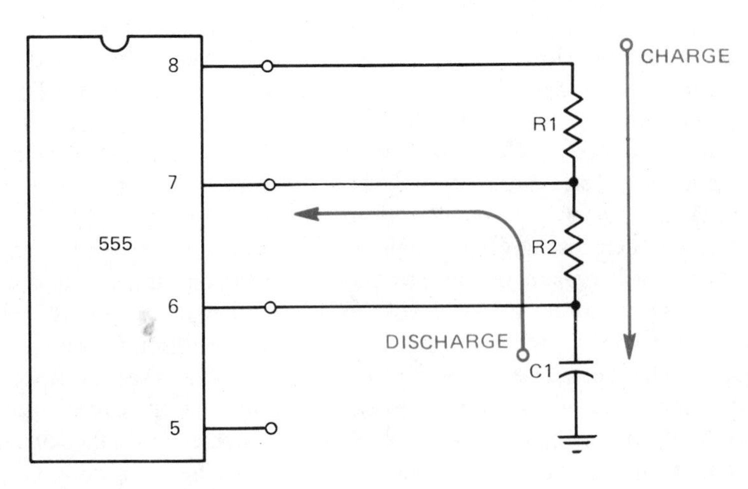

FIGURE 10-6

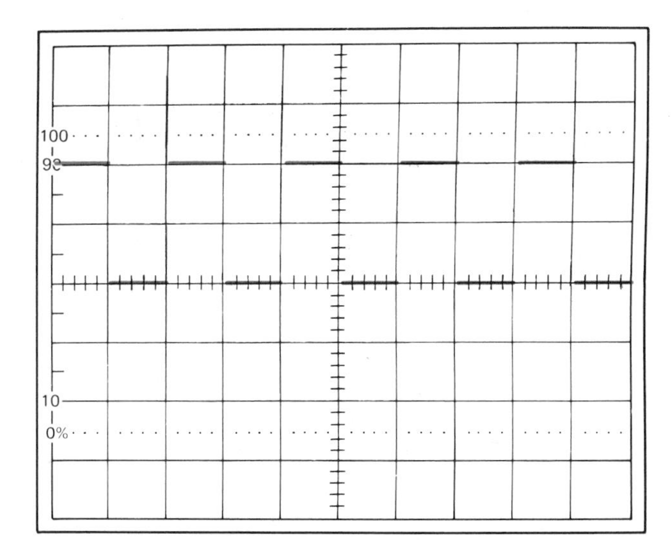

FIGURE 10-7 Waveform produced when an oscilloscope is connected to the output of the 555 timer (Reproduced by permission of Tektronix, Inc., copyright © 1983)

tor R1 has a value of 1 kilohm and R2 has a value of 100 kilohms, the resistance connected in series with the capacitor during charging is 101 kilohms. The resistance connected in series with the capacitor during discharge is 100 kilohms. In this circuit, the difference between the charge time and the discharge time of the capacitor is 1%. If an oscilloscope is connected to the output of the timer, a waveform similar to the waveform shown in figure 10-7 will be seen.

Assume that the value of resistor R1 is changed to 100 kilohms and the value of resistor R2 remains at 100 kilohms. In this circuit, the resistance connected in series with the capacitor during charging is 200 kilohms. The resistance connected in series with the capacitor during discharge, however, is 100 kilohms. Therefore, the discharge time is 50% of the charge time. This means that the output of the timer will be turned on twice as long as it will be turned off. An oscilloscope connected to the output of the timer would display a waveform similar to the one shown in figure 10-8.

Although this condition can exist, the 555 timer has a provision for solving the problem. Pin #5, the control voltage pin, can give complete

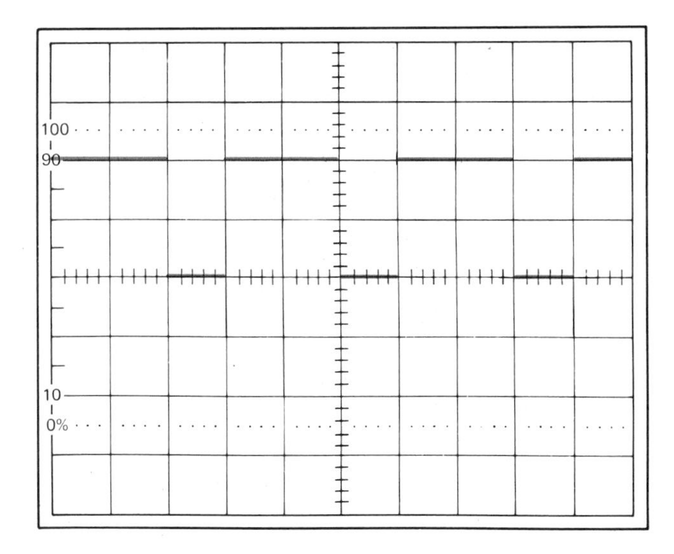

FIGURE 10-8 A different waveform is produced when the value of one of the resistors is changed (Reproduced by permission of Tektronix, Inc., copyright © 1983)

control of the output voltage. If a variable resistor is connected between pin #5 and Vcc, the on time of the output can be lengthened to any value desired. If a variable resistor is connected between pin #5 and ground, the on time of the output can be shortened to any value desired. Since the on time of the timer is adjusted by connecting resistance to pin #5, the off time is set by the values of C1 and R2.

The output frequency of the unit is determined by the values of capacitor C1 and resistors R1 and R2. The 555 timer will operate at almost any frequency desired. It is used in many industrial electronic circuits that require the use of a square wave oscillator.

The On-Delay Timer

In this circuit, the 555 timer is used to construct an on-delay relay. The 555 produces accurate time delays which can range from seconds to hours depending on the values of resistance and capacitance used in the circuit. In figure 10-9, transistor Q1 is used to switch relay coil K1 on or off. A transistor is used to control the relay because the 555 timer may not be able to supply the current needed to operate it.

Transistor Q2 is used as a stealer transistor to steal the base current from transistor Q1. As long as transistor Q2 is turned on by the output of the timer, transistor Q1 is turned off.

Capacitor C3 is connected from the base of transistor Q1 to ground. Capacitor C3 acts as a short time-delay circuit. When Vcc is turned on by switch S1, capacitor C3 is discharged. Before transistor Q1 can be turned on, capacitor C3 must be charged through resistor R3. This charging time is only a fraction of a second, but it insures that transistor Q1 will not turn on before the output of the timer can turn transistor Q2 on. Once transistor Q2 has been turned on, it will hold transistor Q1 off by stealing its base current.

Diode D1 is used as a kickback or freewheeling diode to kill the spike voltage induced into the coil of relay K1 when switch S1 is opened. Resistor R3 limits the base current to transistor Q1 and

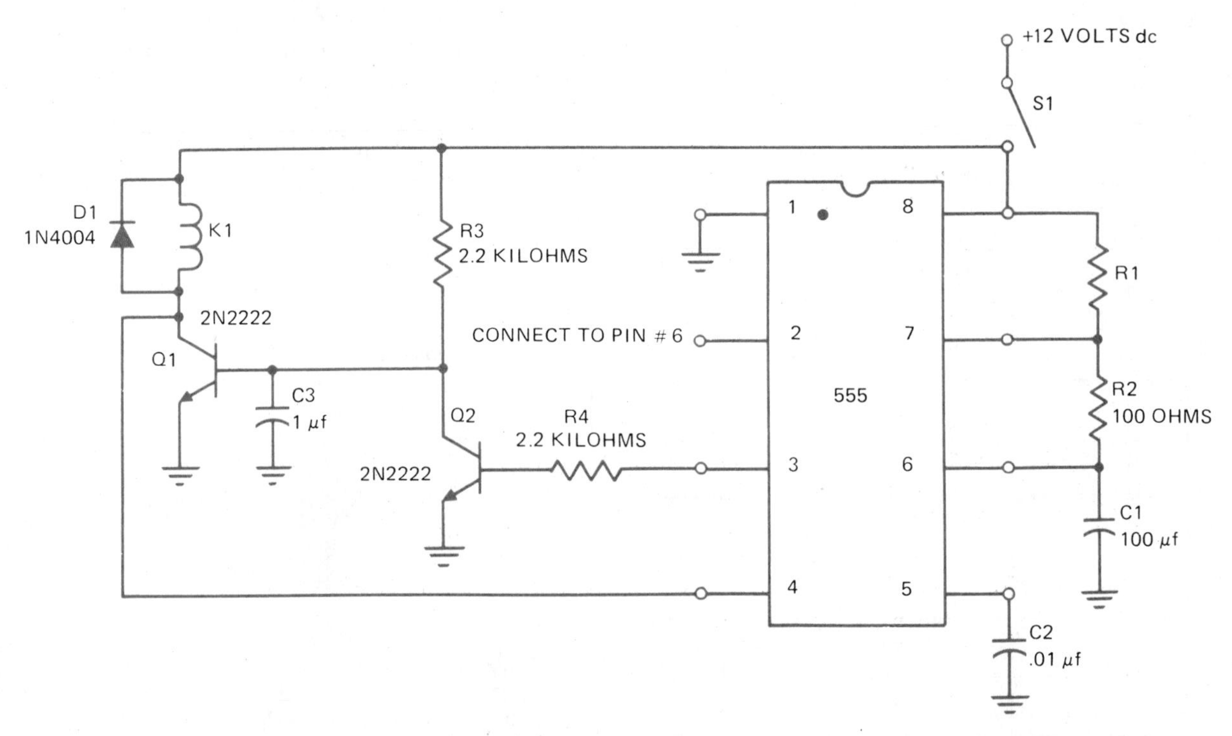

FIGURE 10-9 On-delay timer

resistor R4 limits the base current to transistor Q2.

Pin #4, the reset pin, is used as a latch in this circuit. When power is applied at Vcc, transistor Q1 is turned off. Since transistor Q1 is off, most of the applied voltage is dropped across the transistor, causing about 12 volts to appear at the collector of the transistor. Since pin #4 is connected to the collector of transistor Q1, 12 volts is applied to pin #4. For the timer to operate, pin #4 must be connected to a voltage that is greater than 2/3 of Vcc. When pin #4 is connected to a voltage that is less than 1/3 of Vcc, it turns on the discharge and keeps the timer from operating. When transistor Q1 turns on, the collector of the transistor drops to ground or 0 volts. Pin #4 is also connected to ground, which prevents the timer from further operation. Since the timer can no longer operate, the output remains turned off, which permits transistor Q1 to remain turned on.

Capacitor C1 and resistors R1 and R2 are used to set the amount of time delay. Resistor R2 should be kept at a value of about 100 ohms. The job of resistor R2 is to limit the current when capacitor C1 discharges. Resistor R2 has a relatively low value to enable capacitor C1 to discharge quickly. The time setting can be changed by changing the value of resistor R1.

To understand the operation of the circuit, assume that switch S1 is open and all capacitors are discharged. When switch S1 is closed, pin #2 which is connected to 0 volts, triggers the timer. When the timer is triggered, the output activates transistor Q2 which steals the base current from transistor Q1. Transistor Q1 remains off as long as transistor Q2 is on. When capacitor C1 has been charged to 2/3 of Vcc, the discharge turns on and the output of the timer turns off. When the output turns transistor Q2 off, transistor Q1 is supplied with base current through resistor R3 and turns on relay coil K1. When transistor Q1 is turned on, the voltage applied to the reset pin, #4, is changed from 12 volts to 0 volts. This causes the reset to lock the discharge on and the output off. Therefore, when transistor Q1 is turned on, switch S1 must be reopened to reset the circuit.

REVIEW QUESTIONS

1. How is pin #1 of an in-line, integrated circuit identified?
2. A 555 timer is connected to produce a pulse at the output once each second. The timer is connected to 12 volts dc. If the voltage is reduced to 8 volts dc, the 555 will continue to operate at the same pulse rate. Explain why the timer will operate at the same pulse rate when the voltage is reduced.
3. What is the range of voltage the 555 timer will operate on?
4. Explain the function of the control voltage, pin #5, when the timer is being used as an oscillator.
5. Explain what happens to the output and discharge pins of the 555 timer when the trigger, pin #2, is connected to a voltage that is less than 1/3 of Vcc.
6. Explain what happens to the output and discharge pins when the threshold, pin #6, is connected to a voltage that is greater than 2/3 of Vcc.
7. Refer to figure 10-6. The values of what components determine the length of time the output will be turned on?
8. The values of what components determine the amount of time the output will remain turned off?
9. Explain the operation of pin #4 on the 555 timer.
10. What is a stealer transistor?

UNIT 11

The Operational Amplifier

Objectives

After studying this unit, the student will be able to:

- **Discuss the operation of the operational amplifier (op amp)**
- **List the major types of connections for operational amplifiers**
- **Connect a level detector circuit using an op amp**
- **Connect an oscillator using an op amp**

The operational amplifier, like the 555 timer, has become a very common component in industrial electronic circuits. The operational amplifier, or op amp, is used in hundreds of applications. Different types of op amps are available for different types of circuits. Some op amps use bipolar transistors for input while others use field-effect transistors. The advantage of field-effect transistors is that they have an extremely high input impedance which can be several thousand megohms. As a result of this high input impedance, the amount of current needed to operate the amplifier is small. In fact, op amps that use field-effect transistors for the inputs are generally considered to require no input current.

The ideal amplifier would have an input impedance of infinity. With an input impedance of infinity, the amplifier would not drain power from the signal source; therefore, the strength of the signal source would not be affected by the amplifier. The ideal amplifier would also have zero output impedance. With zero output impedance, the amplifier could be connected to any load resistance without causing a voltage drop inside the amplifier. If it had no internal voltage drop, the amplifier would utilize 100% of its gain. Finally, the ideal amplifier would have unlimited gain. This would enable it to amplify any input signal as much as desired.

Although the ideal amplifier does not exist, the op amp is close. In this unit the operation of an old op amp, the 741, is described as typical of all operational amplifiers. Other op amps may have different characteristics of input and output impedance, but the basic theory of operation is the same for all of them.

The 741 op amp uses bipolar transistors for the inputs. The input impedance is about 2 megohms, the output impedance is about 75 ohms, and the open loop, or maximum gain, is about 200,000. The 741 is impractical for use with such a high gain so negative feedback, which will be discussed later, is used to reduce the gain. For example, assume that the amplifier has an output voltage of 15 volts. If the input signal voltage is greater than 1/200,000 of the output voltage, or 75 microvolts,

$$\left(\frac{15}{200,000} = .000075\right)$$

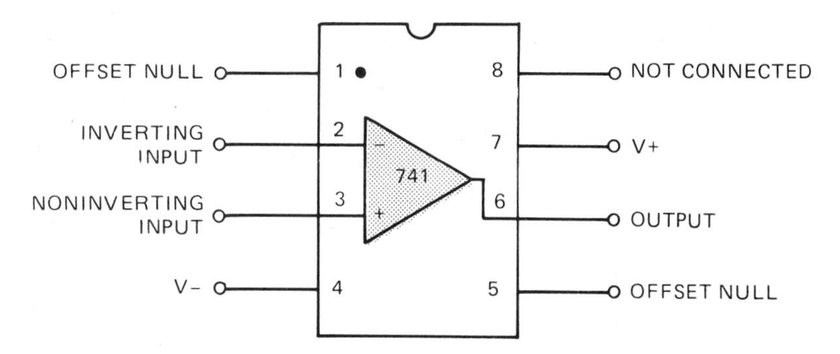

FIGURE 11-1 The 741 operational amplifier

the amplifier will be driven into saturation at which point it will not operate.

The 741 operational amplifier is usually housed in an eight-pin, in-line, integrated circuit package, figure 11-1. The op amp has two inputs called the inverting input and the noninverting input. These inputs are connected to a differential amplifier which amplifies the difference between the two voltages. If both of these inputs are connected to the same voltage, say by grounding both inputs, the output should be 0 volts. In actual practice, however, unbalanced conditions within the op amp may cause a voltage to be produced at the output. Since the op amp has a very high gain, a slight imbalance of a few microvolts at the input can produce several millivolts at the output. To counteract any imbalance, pins #1 and #5 are connected to the offset null which is used to produce 0 volts at the output. These pins are adjusted after the 741 is connected in a working circuit. To make the adjustments, a 10 kilohm potentiometer is connected across pins #1 and #5, and the wiper is connected to the negative voltage, figure 11-2.

Pin #2 is the inverting input. When a signal voltage is applied to this input, the output is inverted. For example, if a positive ac voltage is applied to the inverting input, the output will be a negative voltage, figure 11-3.

Pin #3 is the noninverting input. When a signal voltage is applied to the noninverting input, the output voltage is the same polarity. For example, if a positive ac voltage is applied to the noninverting input, the output voltage will be positive also, figure 11-4.

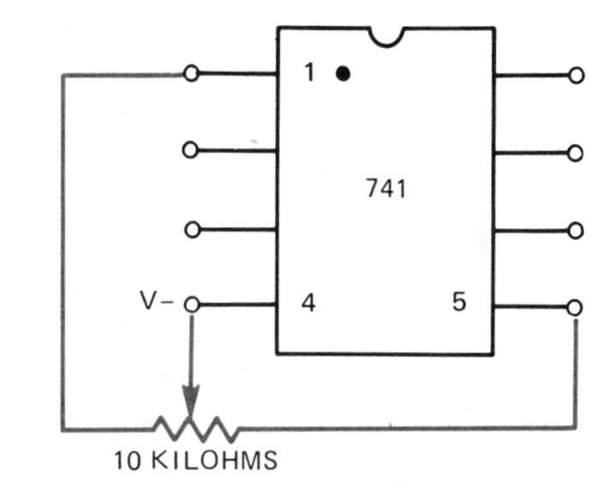

FIGURE 11-2 The offset null connection

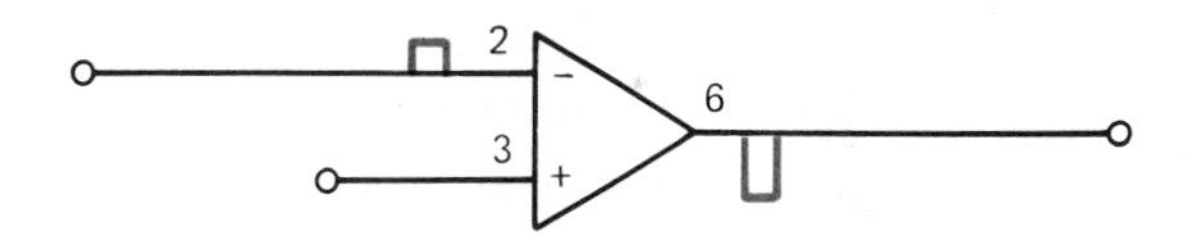

FIGURE 11-3 Inverted output

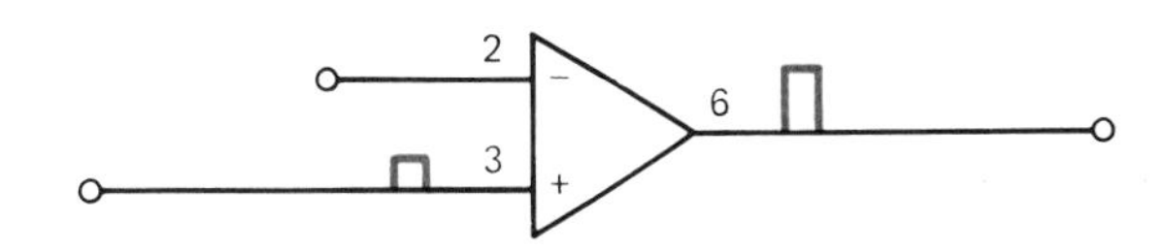

FIGURE 11-4 Noninverted output

Operational amplifiers are usually connected to above and below ground power supplies. Although there are some circuit connections that do not require an above and below ground power supply, these are the exception instead of the rule. Pins #4 and #7 are the voltage input pins. Pin #4 is connected to the negative, or below ground, voltage and pin #7 is connected to the positive, or above ground, voltage.

The 741 operates on voltages that range from about 4 volts to 16 volts. Generally, the operating voltage for the 741 is 12 to 15 volts plus and minus. The 741 has a maximum power output rating of about 500 milliwatts.

Pin #6 is the output and pin #8 is not connected.

As stated previously, the open loop gain of the 741 operational amplifier is about 200,000. Since this amount of gain is not practical for most applications, something must be done to reduce the gain to a reasonable level. One of the great advantages of the op amp is the ease with which the gain can be controlled, figure 11-5. The amount of gain is controlled by a negative feedback loop. This is accomplished by feeding a portion of the output voltage back to the inverting input. Since the output voltage is always opposite in polarity to the inverting input voltage, the amount of output voltage fed back to the input tends to reduce the input voltage. Negative feedback affects the operation of the amplifier in two ways: it reduces the gain; and it makes the amplifier more stable.

The gain of the amplifier is controlled by the ratio of resistor R2 to resistor R1. If a noninverting amplifier is used, the formula

$$\frac{R1 + R2}{R1}$$

is used to calculate the gain. If resistor R1 is 1 kilohm and resistor R2 is 10 kilohms, the gain of the amplifier is 11

$$\left(\frac{11{,}000}{1{,}000} = 11\right)$$

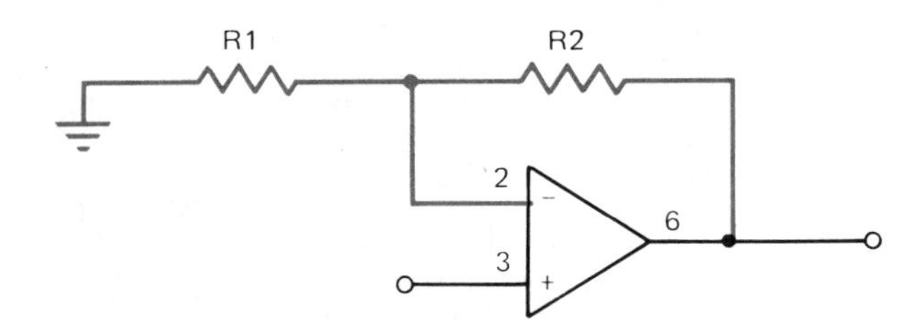

FIGURE 11-5 Negative feedback connection

If the op amp is connected as an inverting amplifier, the input signal will be out of phase with the feedback voltage of the output. This will cause a reduction in the input voltage applied to the amplifier and in the gain. The formula

$$\left(\frac{R2}{R1}\right)$$

is used to compute the gain of an inverting amplifier. If resistor R1 is 1 kilohm and resistor R2 is 10 kilohms, the gain of the inverting amplifier is 10

$$\left(\frac{10{,}000}{1{,}000} = 10\right)$$

As a general rule, the 741 operational amplifier is not operated above a gain of about 100 because it tends to become unstable at high gains. If more gain is desired, it is obtained by using more than one amplifier, figure 11-6. The output of one amplifier is fed into the input of another amplifier.

Another general rule for operating the 741 op amp is the total feedback resistance (R1 + R2) is kept at more than 1,000 ohms and less than 100,000 ohms. These rules apply to the 741 operational amplifier but may not apply to other operational amplifiers.

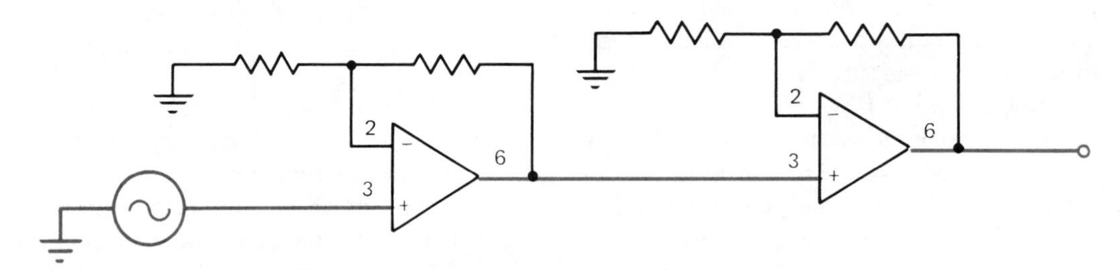

FIGURE 11-6 Two operational amplifiers are used to obtain a higher gain

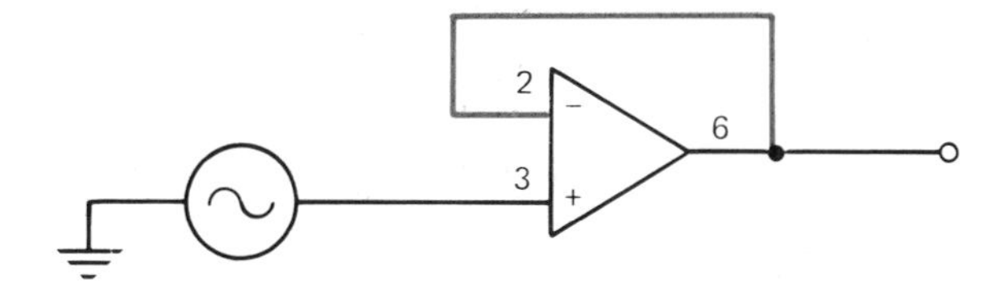

FIGURE 11-7 Voltage follower connection

BASIC CIRCUITS

Op amps are generally used in three basic circuits that are used to build other circuits. One of these basic circuits is the voltage follower. In this circuit, the output of the op amp is connected directly back to the inverting input, figure 11-7. Since there is a direct connection between the output of the amplifier and the inverting input, the gain of this circuit is 1. For example, if a signal voltage of .5 volts is connected to the noninverting input, the output voltage will be .5 volts also. You may wonder why anyone would want an amplifier that doesn't amplify. Actually, this circuit does amplify something. It amplifies the input impedance by the amount of the open loop gain. If the 741 has an open loop gain of 200,000 and an input impedance of 2 megohms, this circuit will give the amplifier an input impedance of 200 k × 2 meg or 400,000 megohms. This circuit connection is generally used for impedance matching purposes.

The second basic circuit is the noninverting amplifier, figure 11-8. In this circuit, the output voltage has the same polarity as the input voltage. If the input voltage is positive, the output voltage will be positive also. The formula

$$\frac{R1 + R2}{R1}$$

is used to calculate the amount of gain in the negative feedback loop.

The third basic circuit is the inverting amplifier, figure 11-9. In this circuit the output voltage is opposite in polarity to the input voltage. If the input signal is positive, the output voltage will be negative at the same instant in time. The formula

$$\frac{R2}{R1}$$

is used to calculate the amount of gain in this circuit.

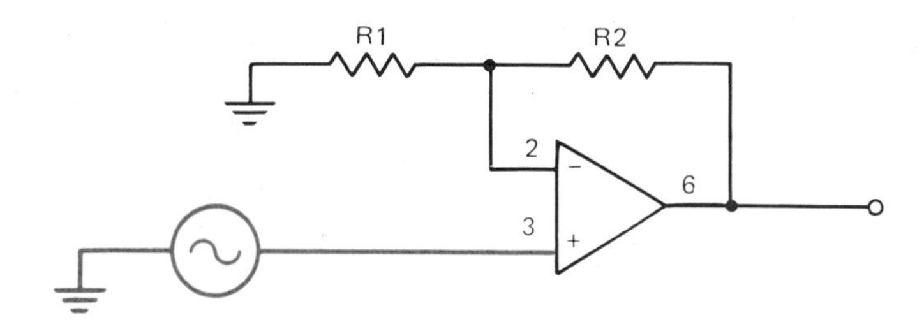

FIGURE 11-8 Noninverting amplifier connection

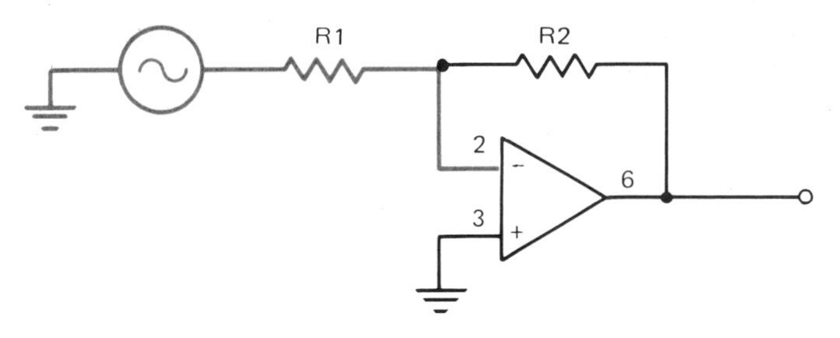

FIGURE 11-9 Inverting amplifier connection

CIRCUIT APPLICATIONS

The Level Detector

The operational amplifier is often used as a level detector or comparator. In this type of circuit, the 741 op amp is used as an inverted amplifier to detect when one voltage becomes greater than another, figure 11-10. This circuit does not use above and below ground power supplies. Instead, it is connected to a power supply that has a single positive and negative output.

During normal operation, the noninverting input of the amplifier is connected to a zener diode which produces a constant positive voltage at the noninverting input of the amplifier. This constant positive voltage is used as a reference. As long as the noninverting input is more positive than the inverting input, the output of the amplifier is high.

A light-emitting diode (LED), D1, is used to detect a change in the polarity of the output. As long as the output of the op amp is high, the LED is turned off. When the output of the amplifier is high, the LED has equal voltage applied to its anode and cathode. Since both the anode and cathode are connected to +12 volts, there is no potential difference and, therefore, no current flow through the LED.

If the voltage at the inverting input becomes more positive than the reference voltage applied to pin #3, the output voltage will fall to about +2.5 volts. The output voltage of the op amp will not fall to 0 or ground in this circuit because the op amp is not connected to a voltage that is below ground. To enable the output voltage to fall to 0 volts, pin #4 must be connected to a voltage below ground. When the output drops, a potential of about 9.5 volts (12 − 2.5 = 9.5) is produced across R1 and D1. The lowering of potential causes the LED to turn on, which indicates that the op amp's output has changed from high to low.

In this type of circuit, the op amp appears to be a digital device in that the output seems to have only two states, high and low. But, the op amp is not a digital device. This circuit only makes it appear to be digital. In figure 11-10, there is no negative feedback loop connected between the output and the inverting input. Therefore, the amplifier uses its open loop gain, which is about 200,000 for the 741, to amplify the voltage difference between the inverting input and the noninverting input. If the voltage applied to the inverting input becomes 1 millivolt more positive than the reference voltage applied to the noninverting input, the amplifier will try to produce an output that is 200 volts more negative than its high state voltage (.001 × 200,000 = 200). The output voltage of the amplifier cannot be driven 200 volts more negative, though, because only 12 volts are applied to the circuit. Therefore, the output voltage reaches the lowest voltage it can and goes into saturation. This causes the op amp to act like a digital device.

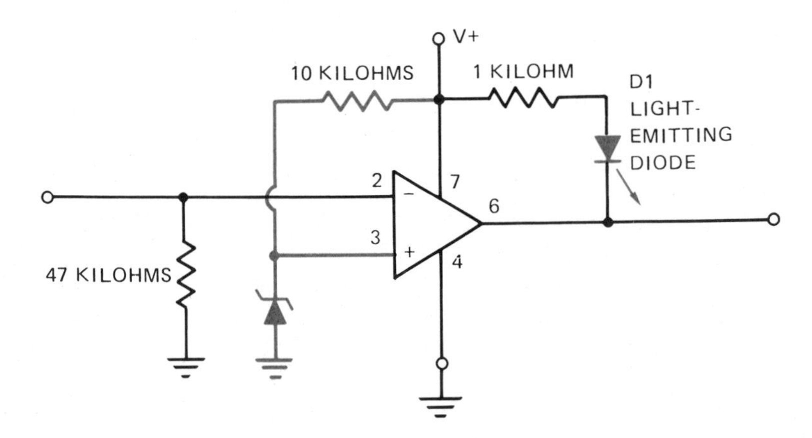

FIGURE 11-10 Inverting level detector

If the zener diode is replaced with a voltage divider as shown in figure 11-11, the reference voltage can be set to any value by adjusting the variable resistor. For example, if the voltage at the noninverting input is set for 3 volts, the output of the op amp will go low when the voltage applied to the inverting input becomes greater than +3 volts. If the voltage at the noninverting input is set for 8 volts, the output voltage will go low when the voltage applied to the inverting input becomes greater than +8 volts. In this circuit the output of the op amp can be manipulated through the adjustment of the noninverting input.

In the two circuits just described, the op amp's output shifted from a high level to a low level. There may be occasions, however, when the output must be changed from a low level to a high level. This can be accomplished by connecting the inverting input to the reference voltage, and the noninverting input to the voltage being sensed, figure 11-12. In this circuit, the zener diode is used to supply a positive reference voltage to the inverting input. As long as the voltage at the inverting input is more positive than the voltage at the noninverting input, the output voltage of the op amp will be low. If the voltage applied to the noninverting input becomes more positive than the reference voltage, the output of the op amp will become high.

Depending on the application, this circuit could cause a small problem. As stated previously, since this circuit does not use an above and below ground power supply, the low output voltage of the op amp is about +2.5 volts. This positive output voltage could cause any other devices connected to the op amp's output to be on when they

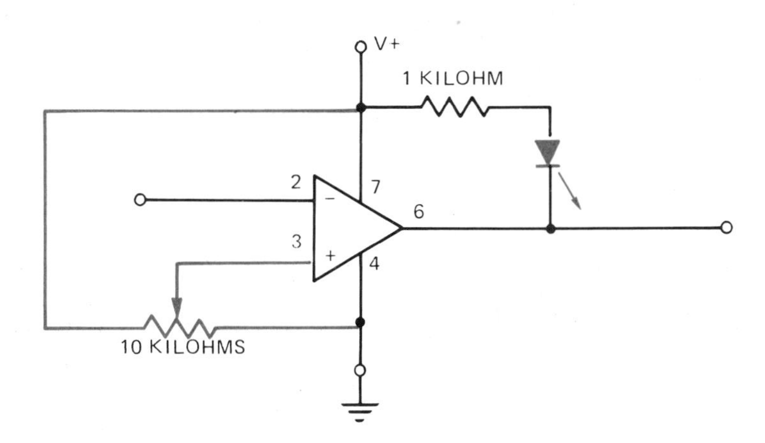

FIGURE 11-11 Adjustable inverting level detector

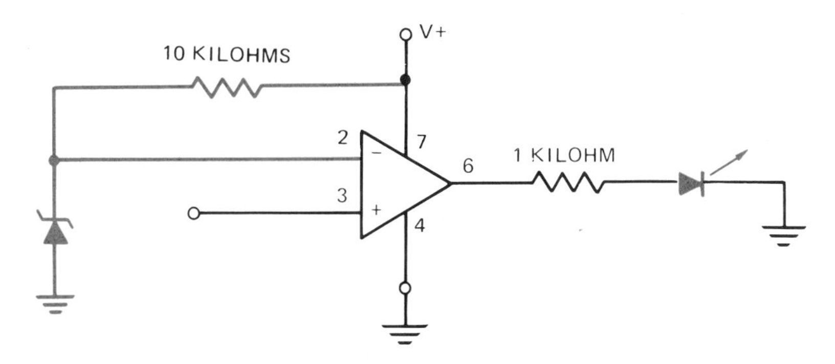

FIGURE 11-12 Noninverting level detector

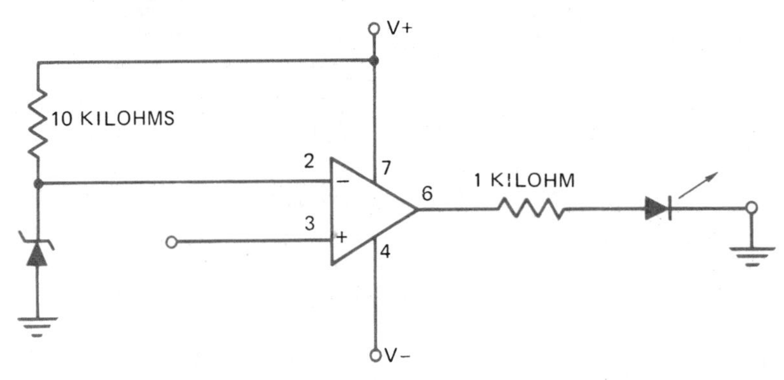

FIGURE 11-13 Below ground power connection permits the output voltage to become negative

should be off. For instance, if the LED shown in figure 11-12 is used, it will glow dimly even when the output is in the low state.

One way to correct this problem is to connect the op amp to an above and below ground power supply as shown in figure 11-13. In this circuit, the output voltage of the op amp is negative or below ground as long as the voltage applied to the inverting input is more positive than the voltage applied to the noninverting input. When the output voltage of the op amp is negative with respect to ground, the LED is reverse biased and cannot operate. If the voltage applied to the noninverting input becomes more positive than the voltage applied to the inverting input, the output of the op amp will become positive and the LED will turn on.

Another method of correcting the output voltage problem is shown in figure 11-14. In this circuit, the op amp is connected again to a power supply that has a single positive and negative output. A zener diode, D2, is connected in series with the output of the op amp and the LED. The voltage value of diode D2 is greater than the output voltage of the op amp in its low state, but less than the output voltage of the op amp in its high state. For instance, assume that the value of zener diode D2 is 5.1 volts. If the output voltage of the op amp in its low state is 2.5 volts, diode D2 will not conduct. If the output voltage becomes +12 volts when the op amp switches to its high state, diode D2 will turn on and conduct current to the LED. The zener diode, D2, keeps the LED completely off until the op amp switches to its high

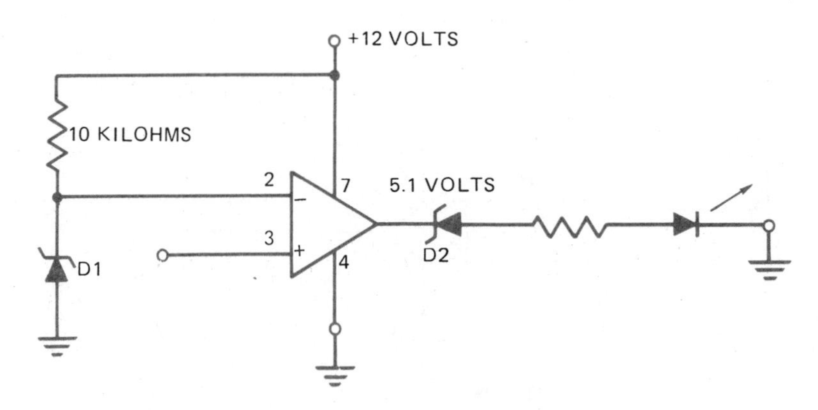

FIGURE 11-14 A zener diode is used to keep the output turned off

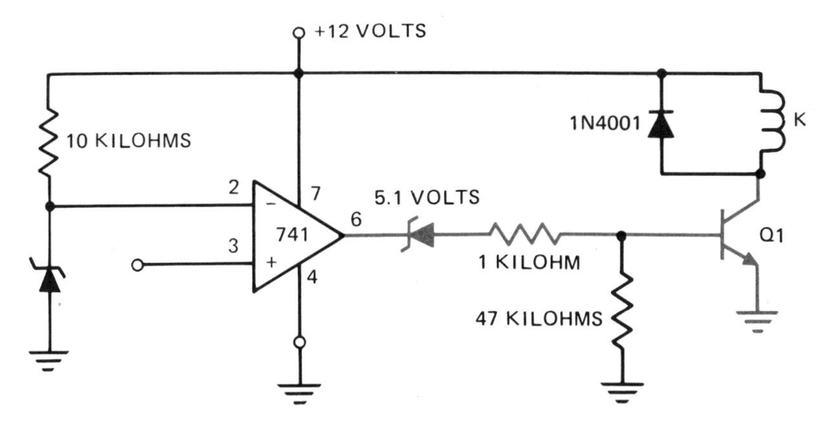

FIGURE 11-15 The operational amplifier supplies the base current for a switching transistor

state providing enough voltage to overcome the reverse voltage drop of the zener diode.

In the preceding circuits, an LED was used to indicate the output state of the amplifier. Keep in mind that the LED is used only as a detector, while the output of the op amp can be used to control almost anything. For example, the output of the op amp can be connected to the base of a transistor as shown in figure 11-15. The transistor can then control the coil of a relay which could, in turn, control almost anything.

The Oscillator

The operational amplifier can be used as an oscillator. The simple circuit shown in figure 11-16 produces a square wave output. However, this circuit is impractical because it depends on a slight imbalance in the op amp, or random circuit noise to start the oscillator. A voltage difference of a few millivolts between the two inputs is all that is needed to raise or lower the output of the amplifier. For example, if the inverting input becomes slightly more positive than the noninverting input, the output will go low or become negative. When the output is negative, capacitor CT charges through resistor RT to the negative value of the output voltage. When the voltage applied to the inverting input becomes slightly more negative than the voltage applied to the noninverting input, the output changes to a high, or positive, value of voltage. When the output is positive, capacitor CT charges through resistor RT toward the positive output voltage.

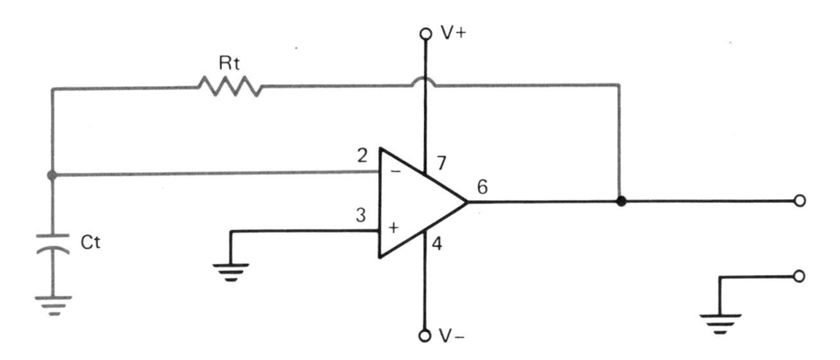

FIGURE 11-16 A simple square wave oscillator

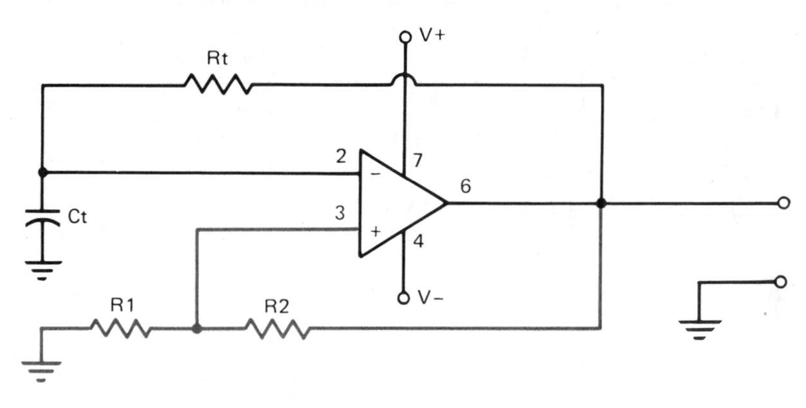

FIGURE 11-17 A square wave oscillator using a hysteresis loop

This circuit would work well if there were no imbalance in the op amp, and if the op amp were shielded from all electrical noise. In practical application, however, there is generally enough imbalance in the amplifier or enough electrical noise to send the op amp into saturation, which stops the operation of the circuit.

The problem with this circuit is that a millivolt difference between the two inputs is enough to drive the amplifier's output from one state to the other. This problem can be corrected by the addition of a hysteresis loop connected to the noninverting input as shown in figure 11-17. Resistors R1 and R2 form a voltage divider for the noninverting input. These resistors generally have equal value. To understand the circuit operation, assume that the inverting input is slightly more positive than the noninverting input. This causes the output voltage to be negative. Also assume that the output voltage is negative 12 volts as compared to ground. If resistors R1 and R2 have equal value, the noninverting input is driven to −6 volts by the voltage divider. Capacitor CT begins to charge through resistor RT to the value of the output voltage. When capacitor CT has been charged to a value slightly more negative than the −6 volts applied to the noninverting input, the op amp's output rises to +12 volts above ground. When the output of the op amp changes from −12 volts to +12 volts, the voltage applied to the noninverting input changes from −6 volts to +6 volts. Capacitor CT now begins to charge through resistor RT to the positive voltage of the output. When the voltage applied to the inverting input becomes more positive than the voltage applied to the noninverting input, the output changes to −12 volts. The voltage applied to the noninverting input is driven from +6 volts to −6 volts, and capacitor CT again begins to charge toward the negative output voltage of the op amp.

The addition of the hysteresis loop has greatly changed the operation of the circuit. The voltage differential between the two inputs is now volts instead of millivolts. The output frequency of the oscillator is determined by the values of CT and RT. The period of one cycle can be computed by using the formula $T = 2RC$.

The Pulse Generator

The operational amplifier can be used as a pulse generator. The difference between an oscillator and a pulse generator is the period of time the output is on compared to the period of time it is low or off. For instance, an oscillator is generally considered to produce a waveform that has positive and negative pulses of equal voltage and

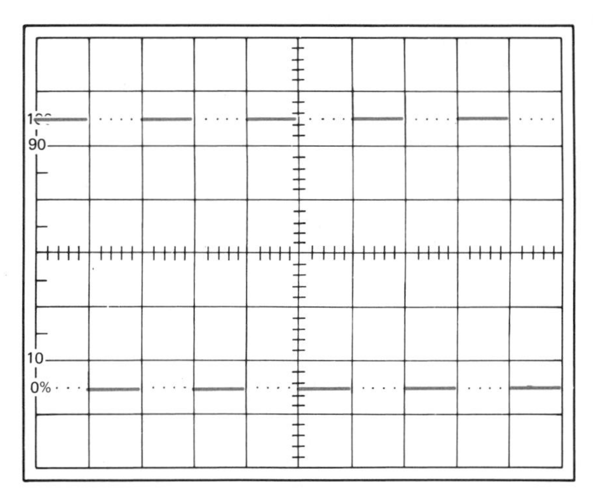

FIGURE 11-18 Output of an oscillator (Reproduced by permission of Tektronix, Inc., copyright © 1983)

time, figure 11-18. The positive value of voltage is the same as the negative value, and the positive and negative cycles are turned on for the same amount of time. This waveform is produced when an oscilloscope is connected to the output of a square wave oscillator.

If the oscilloscope is connected to a pulse generator, however, a waveform similar to the one shown in figure 11-19 will be produced. The positive value of voltage is the same as the negative value just as it was in figure 11-18, but the positive pulse is of a much shorter duration than the negative pulse.

The 741 operational amplifier can easily be changed from a square wave oscillator to a pulse generator, figure 11-20. The pulse generator cir-

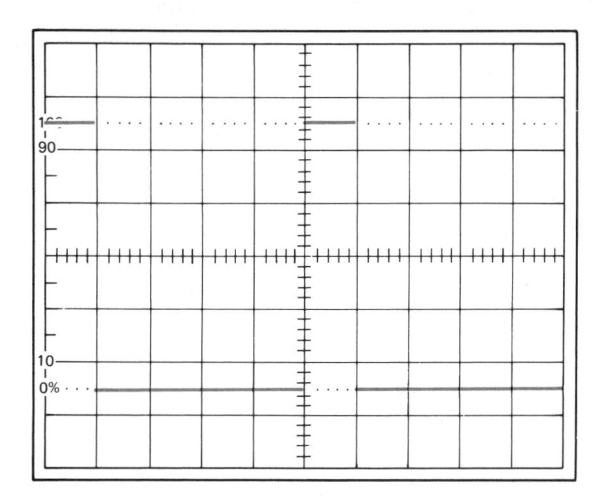

FIGURE 11-19 Output of a pulse generator (Reproduced by permission of Tektronix, Inc., copyright © 1983)

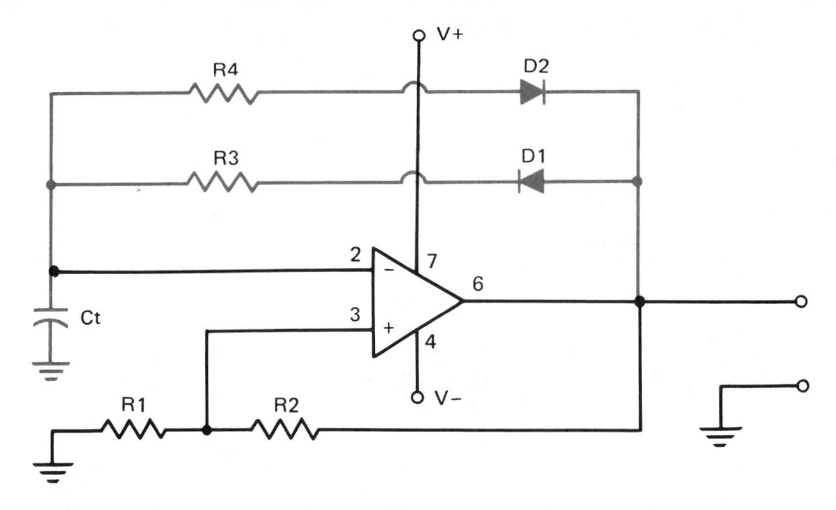

FIGURE 11-20 Pulse generator circuit

cuit is the same basic circuit as the square wave oscillator with the addition of resistors R3 and R4, and diodes D1 and D2. This circuit permits capacitor CT to charge at a different rate when the output is high, or positive, than when the output is low, or negative. For instance, assume that the voltage of the op amp's output is −12 volts. When the output voltage is negative, diode D1 is reverse biased and no current can flow through resistor R3. Therefore, capacitor CT must charge through resistor R4 and diode D2 which is forward biased. When the voltage applied to the inverting input becomes more negative than the voltage applied to the noninverting input, the output voltage of the op amp rises to +12 volts. When the output voltage is +12 volts, diode D2 is reverse biased and diode D1 is forward biased. Therefore, capacitor CT begins charging toward the +12 volts through resistor R3 and diode D1. The amount of time the output of the op amp is low is determined by the value of CT and R4, and the amount of time the output remains high is determined by the value of CT and R3. The ratio of the amount of time the output voltage is high to the amount of time it is low can be determined by the ratio of resistor R3 to resistor R4.

REVIEW QUESTIONS

1. When the voltage connected to the inverting input is more positive than the voltage connected to the noninverting input, will the output be positive or negative?
2. What is the input impedance of a 741 operational amplifier?
3. What is the average open loop gain of the 741 operational amplifier?
4. What is the average output impedance of the 741 operational amplifier?
5. Operational amplifiers are commonly used in what three connections?
6. When the operational amplifier is connected as a voltage follower, it has a gain of 1 (one). If the input voltage is not amplified, what is?
7. Name two effects of negative feedback.

8. Refer to figure 11-8. If resistor R1 is 200 ohms and resistor R2 is 10 kilohms, what is the gain of the amplifier?
9. Refer to figure 11-9. If resistor R1 is 470 ohms and resistor R2 is 47 kilohms, what is the gain of the amplifier?
10. What is the purpose of the hysteresis loop when the op amp is used as an oscillator?

SECTION 2

Motor Starters/Pilot Devices

UNIT 12

Fractional and Integral Horsepower Manual Motor Starters

Objectives *After studying this unit, the student will be able to:*

- Match simple schematic diagrams with the appropriate manual motor starters
- Connect manual fractional horsepower motor starters for automatic and manual operation
- Connect integral horsepower manual starters
- Explain the principles of operation of manual motor starters
- List common applications of manual starters
- Read and draw simple schematic diagrams
- Briefly explain how motors are protected electrically

FRACTIONAL HORSEPOWER MANUAL MOTOR STARTERS

One of the simplest types of motor starters is an on-off, hand-operated, snap action switch. A toggle lever is mounted on the front of the starter, figure 12-1. The motor is connected directly across the line voltage when the handle is turned to the START position. This situation usually is not objectionable with motors rated at one horsepower (hp) or less. Since a motor may draw up to a 600-percent current surge on starting, larger motors should not be connected directly across the line on start-up. Such a connection would result in large line surges that may disrupt power services or cause voltage fluctuations which impede the nor-

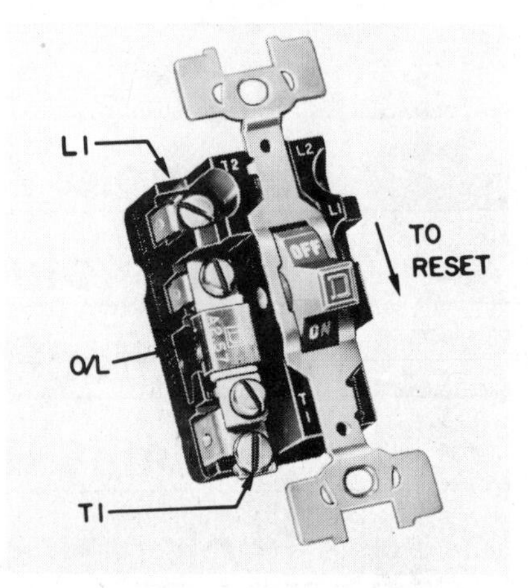

FIGURE 12-1 Open-type starter with overload heater (Courtesy Square D Co.)

mal operation of other equipment. Proper motor starters for larger motors are discussed later.

Fractional horsepower (FHP = 1hp or less) manual motor starters are used whenever it is desired to provide overload protection for a motor as well as "off" and "on" control of small alternating-current single-phase or direct-current motors. Electrical codes require that fractional horsepower motors be provided with overload protection whenever they are started automatically or by remote control. Basically, a manual starter is an *on-off* switch with motor overload protection.

Since manual starters are hand-operated mechanical devices (requiring no electrical coil), the contacts remain closed and the lever stays in the ON position in the event of a power failure. As a result, the motor automatically restarts when the power returns. Therefore, low-voltage protection (see Glossary) is not possible with manually operated starters. This automatic restart action is an advantage when the starter is used with motors that run continuously, such as those used on unattended pumps, blowers, fans, and refrigeration processes. This saves the maintenance electrician from running around the plant to restart all these motors after the power returns. On the other hand, the automatic restart feature is a disadvantage on lathes and machines that may be a danger to products, machinery, or people. It is definitely a safety factor to be observed.

The compact construction of the manual starter means that it requires little mounting space and can be installed on the driven machinery and in various other places where the available space is limited. The unenclosed, or open starter, can be mounted in a standard switch or conduit box installed in a wall and can be covered with a standard flush, single-gang switchplate. The ON and OFF positions are clearly marked on the operating lever, which is very similar to a standard lighting toggle switch lever, figure 12-1.

Application

Fractional horsepower manual starters have thermal overload protection, figures 12-1 and 12-2. When an overload occurs, the starter handle automatically moves to the center position to signify

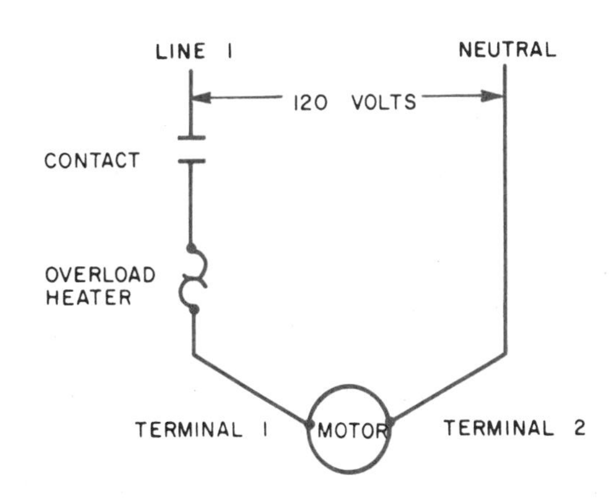

FIGURE 12-2 Diagram of single-pole manual starter shown in figure 12-1

that the contacts have opened and the motor is no longer operating. The starter contacts cannot be reclosed until the overload relay is reset manually. The relay is reset by moving the handle to the full OFF position after allowing about two minutes for the heater to cool. Should the circuit trip open again, the fault should be located and corrected.

Fractional horsepower manual starters are provided in several different types of enclosures as well as the open type to be installed in a switchbox, flush in the wall, or on the surface. Enclosures are obtained to shield the live starter circuit components from accidental contact, for mounting in machine cavities, to protect the starter from dust and moisture, figure 12-3, or to prevent the possibility of an explosion when the starter is used in hazardous locations. These different types of enclosures will be discussed in more detail in Unit 13.

AUTOMATIC AND REMOTE OPERATION

Common applications of manual starters provide control of small machine tools, fans, pumps, oil burners, blowers, and unit heaters. Almost any small motor should be controlled with a starter of this type. However, the contact capacity of the starter must be sufficient to make and break the full motor current. Automatic control devices such as pressure switches, float switches, or thermostats

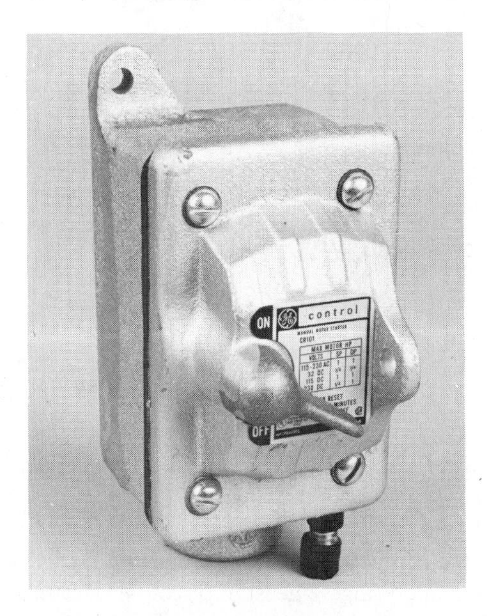

FIGURE 12-3 Water-tight and dust-tight manual starter enclosure (Courtesy General Electric Co.)

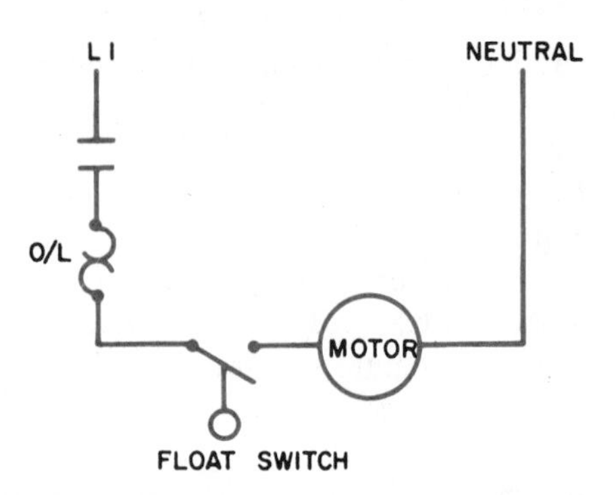

FIGURE 12-4 Automatic control with FHP-single-pole, single-phase manual starter for fractional hp motor without selector switch

rated to carry motor current may be used with fractional horsepower manual starters.

The schematic diagram shown in figure 12-4 illustrates a fractional horsepower motor controlled automatically by a float switch that is remotely connected in the small motor circuit as long as the manual starter contact is closed. When the float is up, the pump motor starts.

In figures 12-5 and 12-6, the selector switch must be turned to the automatic position if the float switch is to take over an automatic operation, such as sump pumping. A liquid-filled sump raises the float, closes the normally open electrical contact, and starts the motor. When the motor pumps the sump or tank empty, the float lowers and breaks electrical contact with the motor thus stopping the motor. This cycle of events will repeat when the sump fills again, automatically, without a human operator.

Note that a double-pole starter is used in figure 12-5. This type of starter is required when both lines to the motor must be broken such as for 230 volts, single phase. The double-pole

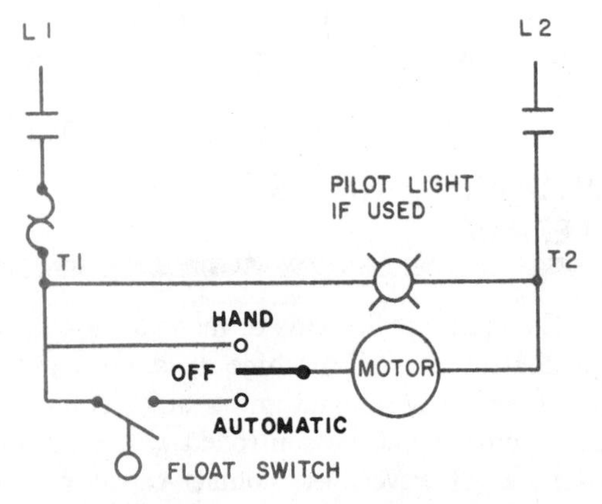

FIGURE 12-5 Manual FHP-double-pole, single-phase starter with automatic control for fractional hp motor using selector switch (See figure 12-6)

FIGURE 12-6 Fractional hp manual starter with selector switch, mounted in general-purpose enclosure (Courtesy Square D Co.)

starter is also recommended for heavy-duty applications because of its higher interrupting capacity and longer contact life, when using a two-pole motor starter on 115 volts. Normally, the single-pole motor starter is used for 115 volts.

MANUAL PUSH-BUTTON LINE VOLTAGE STARTERS

Manual push-button line voltage starters are integral horsepower motor starters (not fractional). Generally, manual push-button starters may be used to control single-phase motors rated up to 5 hp, polyphase motors rated up to 10 hp and direct-current motors rated up to 2 hp. They are available in two-pole for single-phase and three-pole for polyphase motors. A typical manual three-phase push-button starter and diagram are shown in figures 12-7 and 12-8.

When an overload relay trips, the starter mechanism unlatches, opening the contacts to stop the motor. The contacts cannot be reclosed until the starter mechanism has been reset by pressing the STOP button, after allowing time for the thermal unit to cool.

These starters are designed for infrequent starting of small ac motors. This manual starter provides overload protection also, but cannot be used where low or undervoltage protection is required or for remote or automatic operation.

FIGURE 12-7 Three-phase line voltage manual starter (Courtesy Square D Co.)

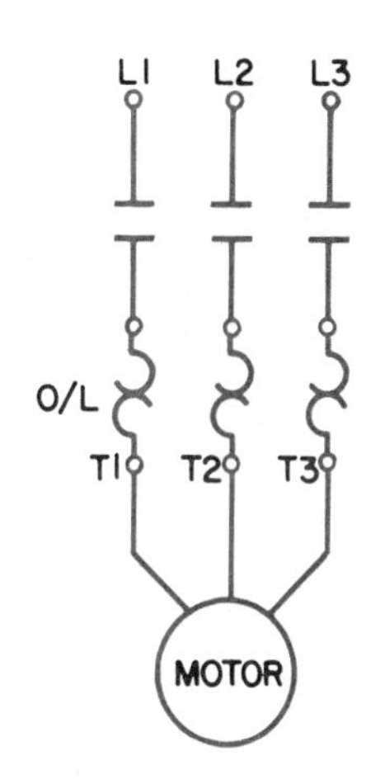

FIGURE 12-8 Wiring diagram for the three-pole line voltage, push-button manual starter seen in figure 12-7

Manual Starter with Low Voltage Protection

Integral horsepower manual starters with Low Voltage Protection (LVP) prevent automatic start-up of motors after a power loss. This is accomplished with a continuous-duty electrical solenoid which is energized whenever the line-side voltage is present, figure 12-9. If the line voltage is lost or disconnected, the solenoid de-energizes, opening the starter contacts. The contacts will not automatically re-close when the line voltage is restored. To close the contacts to restart the motor

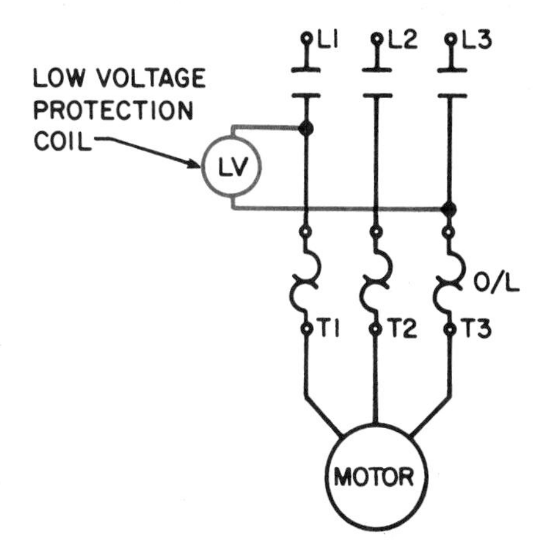

FIGURE 12-9A Wiring diagram for integral hp manual starter with low-voltage protection

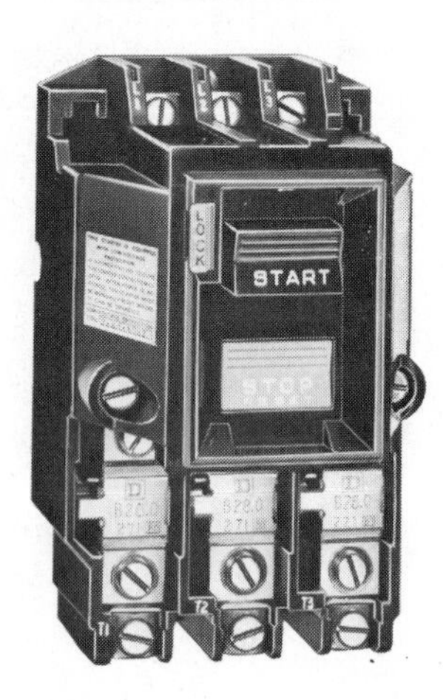

FIGURE 12-9B Manual starter with low-voltage protection

again, the device must be manually reset. This manual starter will not function unless the line terminals are energized. This starter should not be confused with magnetic starters described in the next unit. This is *not* a magnetic starter, but a lower cost starter.

Applications

Typical applications include conveyor lines, grinders, metal working machinery, mechanical power presses, mixers, woodworking machinery and wherever job specifications and standards require low-voltage protection, or wherever machine operator safety could be in jeopardy.

Therefore, this manually operated, push-button starter with low-voltage protection is a method of protecting an operator from injury using automatic restart of a machine upon resumption of voltage, after a power failure. This is normally accomplished with a magnetic starter with electrical (three-wire) control.

THERMAL OVERLOAD PROTECTION

Thermal overload units are widely used on both the fractional and integral horsepower manual starters for protection of motors from sustained electrical overcurrents that could result from overloading of the driven machine or from excessively low line voltage.

Heater elements which are closely calibrated to the full load current of the motor are used on the solder-pot and the thermo-bimetallic types of overload relays. On the solder-pot overload relays, the heating of the element causes alloy elements to melt when there is a motor overload due to excess current in the circuit. When the alloy melts, a spring-loaded ratchet is rotated and trips open a contact which then opens the supply circuit to the motor starter coil. This stops the motor. On the bimetallic overload relays, the heating from the thermal element causes the bimetallic switch to open, opening the supply circuit and thereby stopping the motor.

Normal motor starting currents and momentary overloads will not cause thermal relays to trip because of their inverse-time characteristics (see Glossary). However, continuous overcurrent through the heater unit raises the temperature of the alloy elements. When the melting point is reached, the ratchet is released and the switch mechanism is tripped to open the line or lines to the motor. The switch mechanism is trip-free, which means that it is impossible to hold the contacts closed against an overload.

Only one overload relay is required in either the single-pole or double-pole motor starter, since the starter is intended for use on dc or single-phase ac service. When the line current is excessively high, these relays offer protection against continued operation. Relays with meltable alloy elements are nontemperable and give reliable overload protection. Repeated tripping does not cause deterioration nor does it affect the accuracy of the trip point.

Many types of overload relay heater units are available so that the proper one can be selected on the basis of the actual full-load current rating of the motor. The applicable relay heater units for a particular overload relay are interchangeable and are accessible from the front of the starter. Since the motor current is connected in series with the heater element, the motor will not operate unless the relay unit has the heater installed. Overload units may be changed without disconnecting the wires from the starter or removing the starter from the enclosure. However, the disconnect switch and starter should be turned off first for safety reasons. Additional instruction on motor overload protection is given in Unit 13.

REVIEW QUESTIONS

1. If the contacts on a manual starter cannot be closed immediately after a motor overload has tripped them open, what is the probable reason?
2. If the handle of an installed motor starter is in the center position, what condition does this indicate?
3. How may a manual starter be installed or used for an automatic operation?
4. What does trip-free mean?
5. If overload heating elements are not installed in the starter, what is the result?

Select the *best* answer for each of the following.

6. A fractional horsepower manual motor starter
   a. starts and stops motors under one horsepower
   b. has a toggle switch type of handle and size
   c. has no low-voltage protection
   d. does all of these
7. An automatic operation is used with
   a. integral hp manual starters
   b. push-button manual starters
   c. FHP manual starters with a pressure switch
   d. fractional horsepower manual starters
8. Normally, low voltage (or under voltage) protection is achieved with
   a. a push-button manual starter
   b a toggle switch handle starter
   c. both a and b
   d. a push-button manual starter with energized coil
9. Low voltage protection is used mainly for
   a. fans
   b. air compressors
   c. pumps
   d. protection of the operator
10. Thermal overload protection
    a. aids motor warm-up
    b. prevents motors from freezing
    c. protects motors and conductors from mechanical damage
    d. provides electrical protection of motors

UNIT 13

Magnetic Line Voltage Starters

Objectives *After studying this unit, the student will be able to:*

- **Identify common magnetic motor starters and overload relays**
- **Describe the construction and operating principles of magnetic switches**
- **Describe the operating principle of a solenoid**
- **Troubleshoot magnetic switches**
- **Select starter protective enclosures for particular applications**

Magnetic control means the use of electromagnetic energy to close switches. Line voltage (across the line) magnetic starters are electromechanical devices that provide a safe, convenient, and economic means of full voltage starting and stopping motors. In addition, these devices can be controlled remotely. They are used when a full-voltage starting torque (see Glossary) may be applied safely to the driven machinery and when the current inrush resulting from across-the-line starting is not objectionable to the power system. Control for these starters is usually provided by pilot devices such as push buttons, float switches, timing relays, and more as discussed in Section 3. Automatic control is obtained from the use of some of these pilot devices.

MAGNETIC VS MANUAL STARTERS

Using manual control, the starter must be mounted so that it is easily within reach of the machine operator. With magnetic control, push-button stations are mounted nearby, but automatic control pilot devices can be mounted almost anywhere on the machine. The push buttons and automatic pilot devices can be connected by control wiring into the coil circuit of a remotely mounted starter, possibly closer to the motor to shorten the power circuit.

Operation

In the construction of a magnetic controller, the armature is mechanically connected to a set of contacts so that, when the armature moves to its closed position, the contacts also close. There are different variations and positions, but the operating principle is the same.

The simple up-and-down motion of a solenoid-operated, three-pole magnetic switch is shown in figure 13-1. Not shown are the motor overload relays and the maintaining and auxiliary electrical contacts. Double break contacts are used on this type of starter to cut the voltage in half on each contact, thus providing high arc rupturing capacity and longer contact life.

STARTER ELECTROMAGNETS

The operating principle that makes a magnetic starter different from a manual starter is the use of an electromagnet. Electrical control equipment makes extensive use of a device called a solenoid. This electromechanical device is used to operate motor starters, contactors, relays, and valves. By placing a coil of many turns of wire around a soft iron core, the magnetic flux set up by the energized coil tends to be concentrated; therefore, the magnetic field effect is strengthened. Since the iron core is the path of least resistance to the magnetic lines of force, magnetic attraction concentrates according to the shape of the magnet core.

There are several different variations in design of the basic solenoid magnetic core and coil. Figure 13-2 shows a few examples. As shown in the solenoid design of figure 13-2C, linkage to the movable contacts assembly is obtained through a hole in the movable plunger. The plunger is shown in the open de-energized position.

The center leg of each of the E-Shaped magnet cores in figures 13-2B and C is ground shorter than the outside legs to prevent the magnetic switch from accidentally staying closed (due to residual magnetism) when power is disconnected.

Figure 13-3 shows a manufactured magnet structure and how the starter contacts are mounted on the armature.

When a magnetic motor starter coil is energized and the armature has sealed in, it is held tightly against the magnet assembly. A small air gap is always deliberately placed in the center leg, iron circuit. When the coil is de-energized, a small amount of magnetism remains. If it were not for this gap in the iron circuit, the residual magnetism might be enough to hold the moveable armature in the sealed-in position. This knowledge can be important to the electrician when troubleshooting a motor that will not stop.

The OFF or OPEN position is obtained by de-energizing the coil and allowing the force of gravity or spring tension to release the plunger from the magnet body, thereby opening the electrical contacts. The actual contact surfaces of the plunger and core body are machine ground to insure a high degree of flatness on the contact surfaces so that operation on alternating current is quieter. Improper alignment of the contacting surfaces and foreign matter between the surfaces may cause a noisy hum on alternating-current magnets.

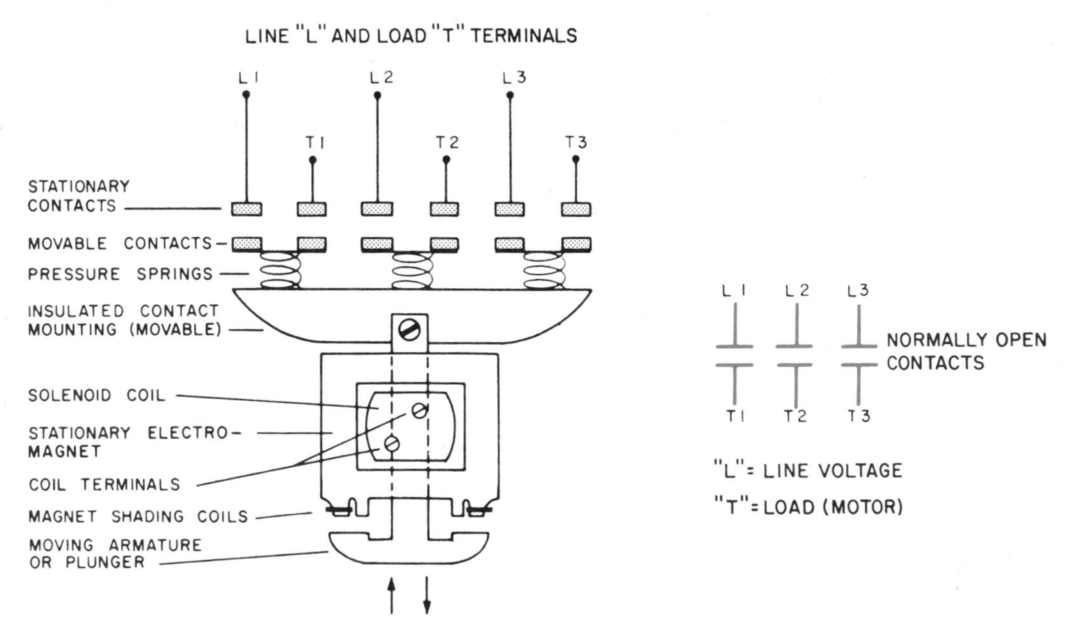

FIGURE 13-1 Three-pole, solenoid-operated magnetic switch (contactor) and electrical wiring symbols

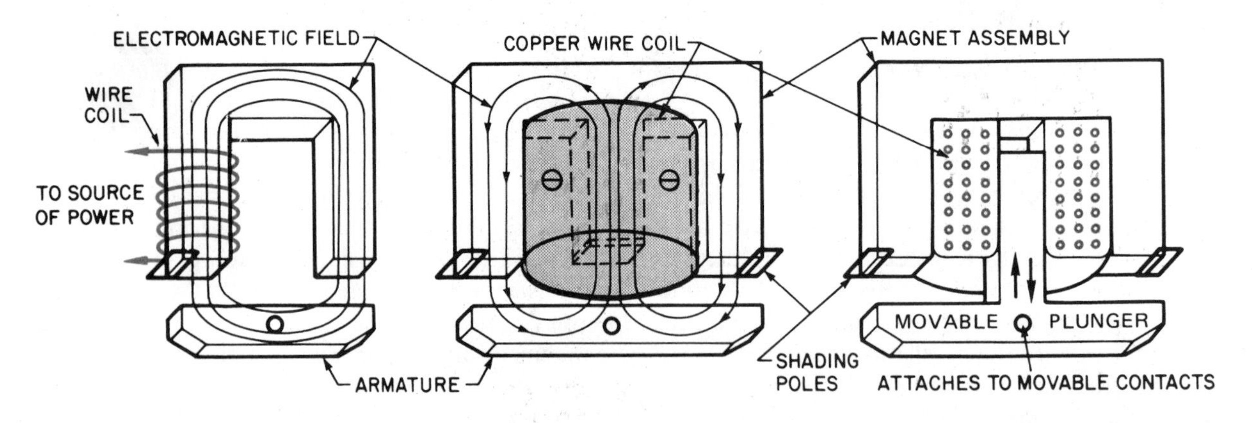

FIGURE 13-2 Some variations of basic magnet core and coil configurations of electromagnets

Another source of noise is loose laminations. The magnet body and plunger (armature) are made up of thin sheets of iron laminated and riveted together to reduce *eddy currents* and hysteresis, iron losses showing up as heat (see figure 13-4). Eddy currents are shorted currents induced in the metal by the transformer action of an ac coil. Although these currents are small, they heat up the metal, create an iron loss, and contribute to inefficiency. At one time, laminations in magnets were insulated from each other by a thin, nonmagnetic coating; however, it was found that the normal oxidation of the metallic laminations reduces the effects of eddy currents to a satisfactory degree, thus eliminating the need for a coating.

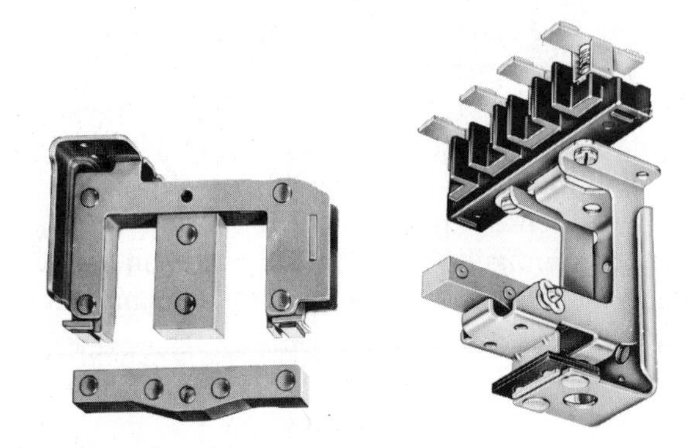

FIGURE 13-3 Magnet structure (left) and movable contacts and armature guide assembly (right) of a four-pole magnetic switch (Courtesy Square D Co.)

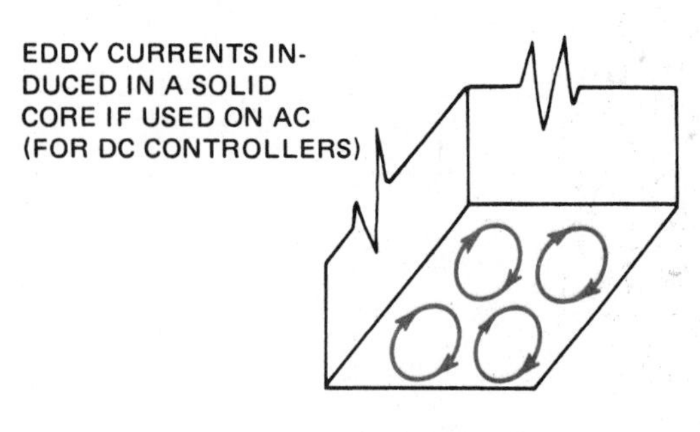

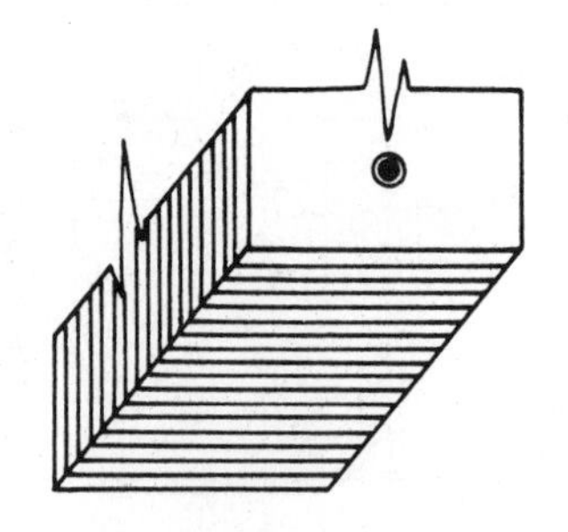

FIGURE 13-4 Types of magnet cores

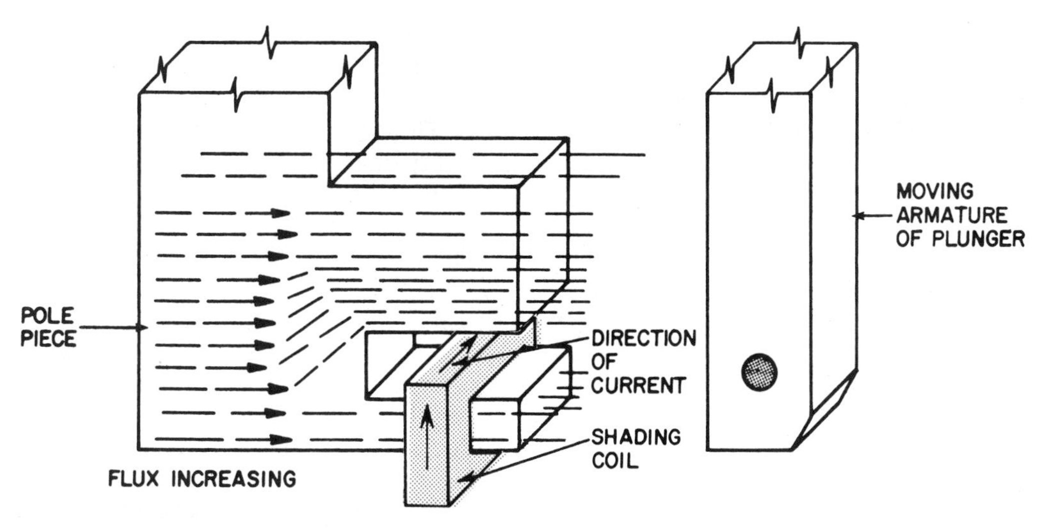

FIGURE 13-5 Pole face section with shading coil; current is in the clockwise direction for increasing flux

SHADED POLE PRINCIPLE

The shaded pole principle is used to provide a time delay in the decay of flux in dc coils, and to prevent chatter and wear in the moving parts of ac magnets. Figure 13-5 shows a copper band or short-circuited coil (shading coil) of low resistance connected around a portion of a magnet pole piece. When the flux is increasing in the pole piece from left to right, the induced current in the shading coil is in a clockwise direction.

The magnetic flux produced by the shading coil opposes the direction of the flux of the main field. Therefore, with the shading coil in place, the flux density in the shaded portion of the magnet will be considerably less, and the flux density in the unshaded portion of the magnet will be more than if the shading coil were not in place.

Figure 13-6 shows the magnet pole with the flux direction still from left to right, but now the flux is decreasing in value. The current in the coil is in a counterclockwise direction. As a result, the

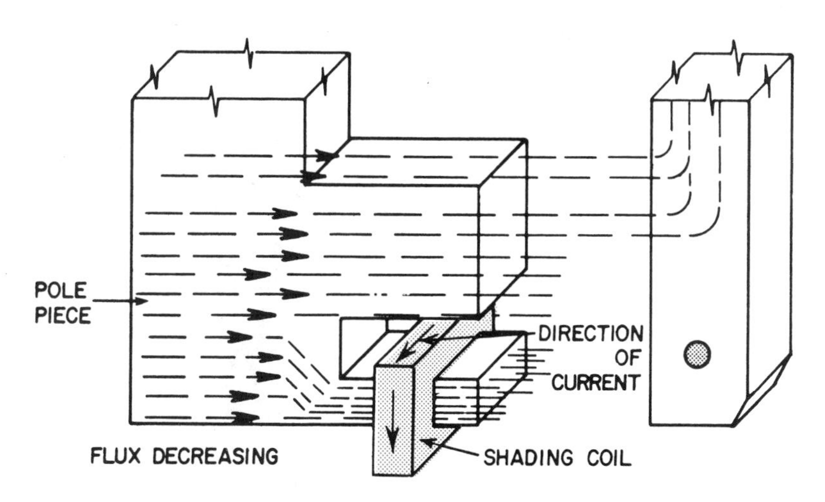

FIGURE 13-6 Pole face section; current is in the counterclockwise direction for decreasing flux

magnetic flux produced by the coil is in the same direction as the main field flux. With the shading coil in place, the flux density in the shaded portion of the magnet will be larger and that in the unshaded portion will be less than if the shading coil were not used.

Thus, when the electric circuit of a coil is opened, the current decreases rapidly to zero, but the flux decreases much more slowly because of the action of the shading coil. This produces a more stable magnetic pull on the armature as the ac waveform alternates from maximum to minimum values and helps prevent chatter and ac hum.

Use of the Shading Pole to Prevent Wear and Noise

The attraction of an electromagnet operating on alternating current is pulsating and equals zero twice during each cycle. The pull of the magnet on its armature also drops to zero twice during each cycle. As a result, the sealing surfaces of the magnet tend to separate each time the flux is zero and then contact again as the flux builds up in the opposite direction. This continual making and breaking of contact will result in a noisy starter and wear on the moving parts of the magnet. The noise and wear can be eliminated in ac magnets by the use of shaded poles. As shown previously, by shading a pole tip, the flux in the shaded portion lags behind the flux in the unshaded portion. The diagram shows the flux variations with time in both the shaded and unshaded portions of the magnet.

The two flux waves are made as near 90 degrees apart as possible. Pull produced by each flux is also shown. If flux waves are exactly 90° apart, the pulls will be 180° apart, and the resultant pull will be constant. However, with fluxes *nearly* 90° apart, the resulting pull varies only a small amount from its average value and never goes through zero. Voltage induced in the shading coil causes flux to exist in the electromagnet, even when the main coil current instantaneously passes through a zero point. As a result, contact between the sealing surfaces of the magnet is not broken and chattering and wear are prevented.

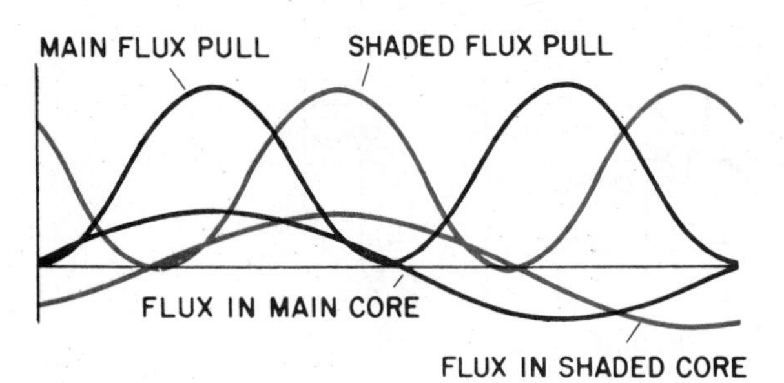

MAGNET COIL

The magnet coil has many turns of insulated copper wire that is tightly wound on a spool. Most coils are protected by a tough epoxy molding which makes them very resistant to mechanical damage, figure 13-7.

Above Normal Voltage Effects

The manufacturer makes available coils of practically any desired control voltage. Some starters are designed with dual-voltage coils.

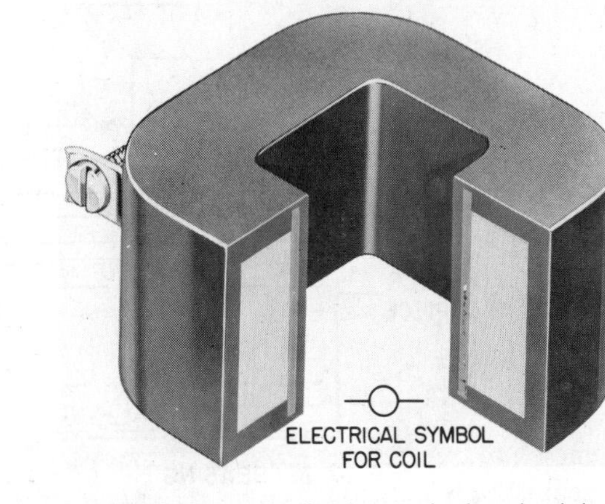

FIGURE 13-7 Magnet coil cut away to show insulated copper wire wound on a spool and protected by a molding

NEMA standards require that the magnetic switch operates properly at varying control voltages from a high of 110% to a low of 85% of the rated coil voltage. This range of required operation is then designed by the manufacturer. It insures that the coil will withstand elevated temperatures at voltages up to 10% over rated voltage and that the armature will pick up and seal in, even though the voltage may drop to 15% under the rating. Normally, power company service voltages are very reliable. Plant voltages may vary due to other loaded, operating machines, and other reasons affecting the electrical distribution system. If the voltage applied to the coil is too high, the coil will draw too much current. Excessive heat will be produced and may cause the coil insulation to break down and burn out. The magnetic pull will be too high and will cause the armature to slam in with too much force. The magnet pole faces will wear faster, leading to a shortened life for the controller. In addition, reduced contact life may result from excessive contact bounce.

Below Normal Voltage Effects

Undervoltage produces low coil currents thereby reducing the magnetic pull. On common starters the magnet may pick up (start to move) but not seal. The armature must sit against the pole faces of the magnet to operate satisfactorily. Without this condition, the coil current will not fall to the sealed value because the magnetic circuit is open, decreasing impedance (ac resistance). As the coil is not designed to continuously carry a current greater than its sealed current, it will quickly get very hot and burn out. The armature will also chatter. In addition to the noise, there is excessive wear on the magnet pole faces. If the armature does not seal, the contacts may touch but not close with enough pressure, creating another problem. Excessive heat, with arcing and possible welding of the contacts, will occur as the controller attempts to carry a motor-starting current with insufficient contact pressure.

POWER (OR MOTOR) CIRCUIT OF THE MAGNETIC STARTER

The number of poles refers to the number of power contacts, determined by the electrical supply service. For example, in a three-phase, three-wire system, a three-pole starter is required. The power circuit of a starter includes the main stationary and movable contacts and the thermal unit or heater unit of the overload relay assembly.This can be seen in figure 13-8 (and in figure 13-1, less the thermal overload relay assembly).

MOTOR OVERHEAT

An electric motor does not know enough to quit when the load gets too much for it. It keeps going until it burns out. If a motor is subjected, over a period of time, to internal or external heat levels that are high enough to destroy the insulation on the motor windings, it will fail—burn out.

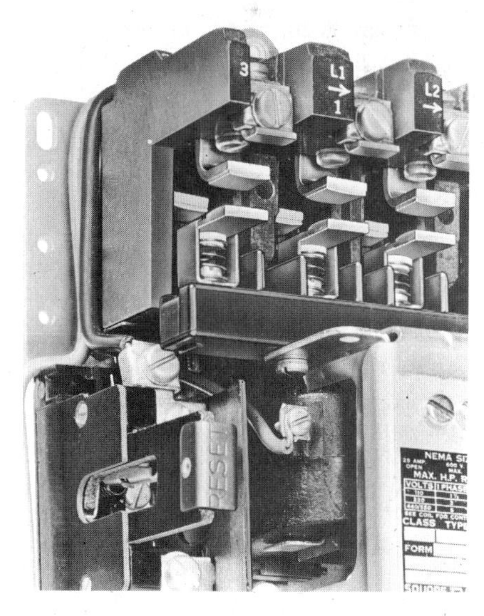

FIGURE 13-8 Ac magnetic starter with contact arcing chamber removed. Note overload relay (lower left) and coil and magnet (lower center). (Courtesy Square D Co.)

(EFFECTS OF HIGH AMBIENT TEMPERATURE)

A solution to this problem *might* be to install a larger motor whose capacity is in excess of the normal horsepower required. This isn't too practical since there are other reasons for a motor to overheat besides excess loads. A motor will run cooler in winter snow country than in summer hot tropical weather. A high, surrounding air temperature (*ambient temperature*) has the same effect as higher-than-normal current flow through a motor—it tends to deteriorate the insulation on the motor windings.

High ambient temperature is also created by *poor ventilation* of the motor. Motors must get rid of their heat so any obstructions to this must be avoided. High inrush currents of *excessive starting* create heat within the motor. The same is true with *starting heavy loads*. There are several other related causes that generate heat within a motor such as *voltage unbalance*, *low voltage*, and single phasing. In addition, when the rotating member of the motor will not turn (a condition called *locked rotor*), heat is generated. It must be impossible to design a motor that will adjust itself for all the various changes in total heat that can occur. Some device is needed to protect a motor against expected overheating.

Motor Overload Protection

The ideal overload protection for a motor is an element with current sensing properties very similar to the heating curve of the motor. This would act to open the motor circuit when full load is exceeded. The operation of the protective device is ideal if the motor is allowed to carry small, short and harmless overloads, but is quickly disconnected from the line when an overload has persisted too long. Dual element, or time-delay, fuses may provide motor overload protection, but they have the disadvantage of being nonrenewable and must be replaced.

An overload relay is added to the magnetic switch that was shown in figure 13-1. Now it is called a motor starter. The overload relay assembly is the heart of motor protection. A typical solid-state overload relay is shown in figure 13-9. The motor can do no more work than the overload relay permits. Like the dual element fuse, the overload relay has characteristics permitting it to hold in during the motor accelerating period when the inrush current is drawn. Nevertheless, it still provides protection on small overloads above full-load current when the motor is running. Unlike the fuse, the overload relay can be reset. It can

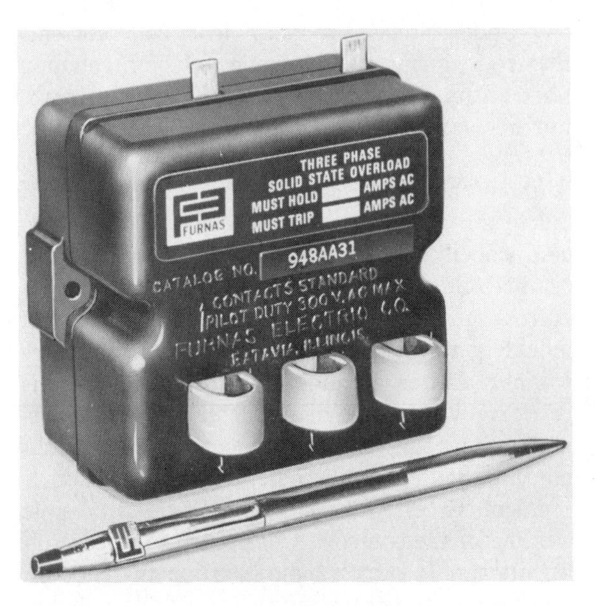

FIGURE 13-9 Three-phase, solid-state overload relay (Courtesy Furnas Electric Company)

withstand repeated trip and reset cycles without need of replacement. It is emphasized that the overload relay does *not* provide short circuit protection. This is the function of overcurrent protective equipment like fuses and circuit breakers, generally located in the disconnecting switch enclosure.

Current drawn by a motor is a convenient and accurate measure of the motor load and motor heating. Therefore, the device used for overload protection, the overload relay, is usually connected with the motor current. It is provided as part of the starter or controller. As the relay carries the motor current, it is affected by that current. If a dangerous over-current condition occurs, it operates or trips the relay to open the control circuit of the magnetic starter and disconnect the motor from the line; this helps insure the maximum operating life of the motor. In a manual starter, an overload trips a mechanical latch causing the starter contacts to open and disconnect the motor from the line.

To provide *overload* or *running protection* to keep a motor from overheating, overload relays are used on starters to limit the amount of current drawn to a predetermined value. The NEC and local electrical codes determine the size of protective overload relays and heating elements which are properly sized to the motor.

The controller normally is installed in the same room or area as the motor. This makes it subject to the same ambient temperature as the motor. The tripping characteristic of the proper thermal overload relay will then be affected by room temperature exactly as the motor is affected. This is done by selecting a thermal relay element (from a chart provided by the manufacturer) that trips at the danger temperature for the motor windings. When excessive current is drawn, the relay de-energizes the starter and stops the motor.

Overload relays can be classified as being either *thermal* or *magnetic*. Magnetic overload relays react only to current excesses, and are not affected by temperature. As the name implies, *thermal overload relays* depend on the rising ambient temperature and temperatures caused by the overload current to trip the overload mechanism.

Thermal overload relays can be further subdivided into two types—melting alloy and bimetallic.

Melting Alloy Thermal Units

The melting alloy assembly consisting of a heater element and solder pot is shown in figure 13-10. The solder pot holds the ratchet wheel in one position. Excessive motor current passes through the heater element and melts an alloy solder pot. Since the ratchet wheel is then free to turn in the molten pool, it trips a set of normally closed contacts which is in the starter control circuit; this stops the motor, figure 13-11. A cooling-off period is required to allow the solder pot to become solid again before the overload relay can be reset and motor service restored.

Melting alloy thermal units are interchangeable. They have a one-piece construction which insures a constant relationship between the heater element and the solder pot. As a result, this unit can be factory calibrated, to make it virtually tamper-proof in the field. These important features are not possible with any other type of overload relay construction. To obtain appropriate tripping current for motors of different sizes, a wide selection of interchangeable thermal units (heaters) is available. They give exact overload protection to motors of different full-load current ratings. Thermal units are rated in amperes and are selected on the basis of motor full-load current. For most accurate overload heater selection, the manufacturer publishes a number of rating tables keyed to the controller in which the overload

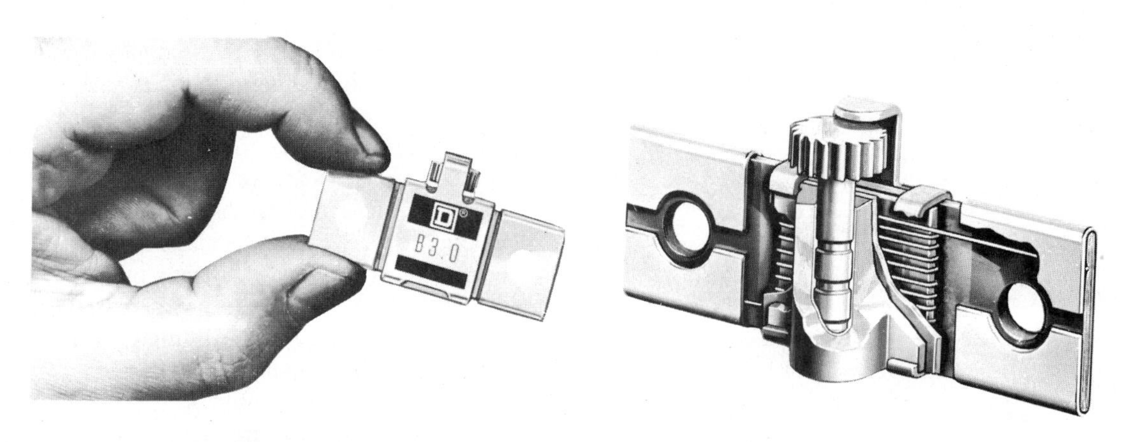

FIGURE 13-10 Melting alloy type overload heater. Cutaway view (right) shows construction of heater. (Courtesy Square D Co.)

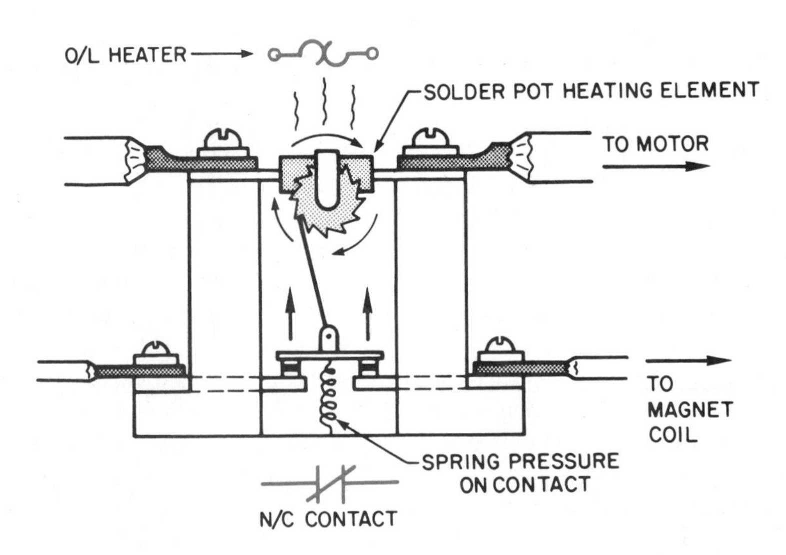

FIGURE 13-11 Melting alloy thermal overload relay. Spring pushes contact open as heat melts alloy allowing ratchet wheel to turn freely. Note electrical symbols for overload heater and normally closed control contact.

relay is used. The units are easily mounted into the overload relay assembly and held in place with two screws. Being in series with the motor circuit, the motor will not operate without these heating elements installed in the starter.

Bimetallic Overload Relays

Bimetallic overload relays are designed specifically for two general types of application: the automatic reset and bimetallic relay. The automatic reset feature means that the devices can be mounted in locations that are not easily accessible for manual reset operation and may be set in the automatic position by the electrician.

In the automatic reset position, the relay contacts, after tripping, will automatically reclose after the relay has cooled down. This is an advantage when the reset button is hard to reach. Automatic reset overload relays are not normally recommended when used with automatic (two-wire) pilot control devices. With this control arrangement, when the overload relay contacts reclose after an overload trip, the motor will restart. Unless the cause of the overload has been removed, the overload relay will trip again. This event will repeat. Soon the motor will burn out because of the accumulated heat from the repeated high inrush and the overload current. (An overload-indicating light or alarm can be installed to call attention before this happens.) **Caution:** The more important point to consider is the possible danger to personnel. This unexpected restarting of the machine may find the operator or electrician in a hazardous situation, as attempts are made to find out why this machine has stopped. *The NEC prohibits this later installation.*

Most bimetallic relays can be adjusted to trip within a range of 85 to 115 percent of the nominal trip rating of the heater unit. This feature is useful when the recommended heater size may result in unnecessary tripping, while the next larger size will not give adequate protection. Ambient temperatures affect thermally-operated overload relays.

This ambient-compensated bimetallic overload relay is recommended for installations when the motor is located in a different ambient temperature from the motor starter. If the controller is located in a changing temperature, the overload relay can be adjusted to compensate for these temperature changes. This thermal overload relay is always affected by the surrounding temperature. If a standard thermal overload relay were used, it would not trip consistently at the same level of motor current whenever the controller temperature changed.

The tripping of the control circuit in the bimetallic relay results from the difference of expansion of two dissimilar metals fused together. Movement occurs if one of the metals expands more than the other when subjected to heat. A U-shaped bimetallic strip is used to calibrate this type of relay, figure 13-12. The U-shaped strip and a heater element inserted in the center of the U compensate for possible uneven heating due to variations in the mounting location of the heater element. Since a motor starter is installed in series with the load, the starter must have the heating element (bimetallic and solder pot) installed in the overload relay before a motor will start.

Magnetic Overload Relays

The magnetic overload relay coil is connected in series directly with the motor or is indirectly connected by current transformers (as in circuits with large motors). As a result, the coil of the magnetic relay must be wound with wire large enough in size to pass the motor current. These

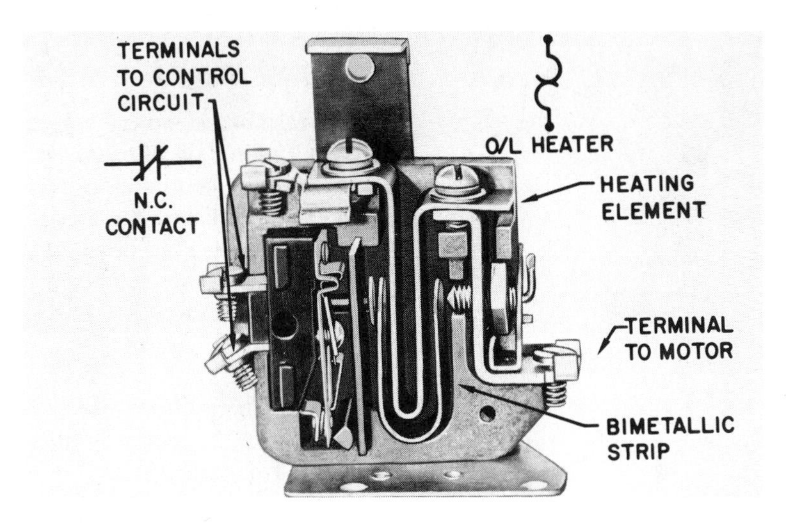

FIGURE 13-12 Cutaway view showing construction of bimetallic overload relay and electrical symbols representing elements of the relay (Courtesy Square D Co.)

overload relays operate by current intensity and not heat.

Magnetic overload relays are used when an electrical contact must be opened or closed as the actuating current rises to a certain value. In some cases, the relay may also be used so that it is actuated when the current falls to a certain value. Magnetic overload relays are used to protect large motor windings against continued overcurrent. Typical applications are: to stop a material conveyor when conveyors ahead become overloaded, and to limit torque reflected by the motor current.

Time Limit Overload Relays

Time delay overload relays, figure 13-13, make use of the oil dashpot principle. Motor current passing through the coil of the relay exerts a magnetic pull on a plunger. The magnetic flux set up inside the coil tends to raise the plunger which is attached to a piston immersed in oil. As the current increases in the relay coil, so does the magnetic flux. The force of gravity is overcome and the plunger and piston move upward. During this upward movement, oil is forced through bypass holes in the piston. As a result, the operation of the contacts is delayed. A valve disc is turned to open or close bypass holes of various sizes in the

FIGURE 13-13 Time limit overload magnetic relay (Courtesy Square D Co.)

piston. This action changes the rate of oil flow and so adjusts the time delay factor. The rate of upward travel—of the core and piston—depends directly upon the degree of overload. The greater the current load, the faster the upward movement. As the rate of upward movement increases, the relay tripping time decreases.

This inverse time characteristic prevents the relay from tripping on the normal starting current or on harmless momentary overloads. In these cases, the line current drops to its normal value before the operating coil is able to lift the core and piston far enough to operate the overload control contacts. However, if the overcurrent continues for a prolonged period, the core is pulled far enough to operate the contacts. As the line current increases, the relay tripping time decreases. Tripping current adjustment is achieved by adjusting the plunger core with respect to the overload relay coil. Quick tripping is obtained through the use of a light trade dashpot oil and by adjustment of the oil bypass holes.

A valve in the piston allows almost instantaneous resetting of the circuit to restart the motor. The current must then be reduced to a very low value before the relay will reset. This action is accomplished automatically when the tripping of the relay disconnects the motor from the line. Magnetic overload relays are available with either automatic reset contacts or hand reset contacts.

Instantaneous Trip Current Relays

Instantaneous trip current relays are used to take a motor off the line as soon as a predetermined load condition is reached. For example, when a blockage of material on a woodworking machine causes a sudden high current, an instantaneous trip relay can cut off the motor quickly. After the cause of the blockage is removed, the motor can be restarted immediately because the relay resets itself as soon as the overload is removed. This type of relay is also used on conveyors to stop the motor before mechanical breakage results from a blockage.

The instantaneous trip current relay does not have the inverse time characteristic. Thus, it must not be used in ordinary applications requiring an overload relay. The instantaneous trip current relay should be considered as a special-purpose relay.

The operating mechanism of the trip relay in figure 13-14 consists of a solenoid coil through which the motor current flows. There is a movable iron core within the coil. Mounted on top of the solenoid frame is a snap-action precision switch that has connections for either a normally open or normally closed contact. The motor current exerts a magnetic pull upward on the iron core. Normally, however, the pull is not sufficient to lift the core. If an overcurrent condition causes the core to be lifted, the snap-action precision switch is operated to trip the control contact of the relay.

The tripping value of the relay can be set over a wide range of current ratings by moving the plunger core up and down on the threaded stem. As a result, the position of the core in the solenoid is changed. By lowering the core, the magnetic flux is weakened and a higher current is required to lift the core and trip the relay.

Number of Overload Relays Needed to Protect a Motor

The National Electrical Code requires three overload relays for three-phase starters on new installations. This helps maintain a balanced supply voltage for polyphase load installations.

A single-phase load on a three-phase circuit can produce serious unbalanced motor currents. A large three-phase motor on the same feeder with a small three-phase motor may not be protected if a single-phase condition occurs, figure 13-15.

A defective line fuse, an open "leg" through a circuit breaker, a loose or broken wire anywhere in the conduit system or in a motor lead can result in single-phase operation. This will show up as a sluggish, hot-running motor. The motor will not start at all but will produce a distinct magnetic hum when it is energized. The three-phase motor may continue to operate (at reduced torque) when single phasing occurs. But once stopped, it will not restart. This is also a sign of a single-phase condition in a three-phase motor.

Unbalanced single-phase loads on three-phase panel boards must be avoided. Problems may occur on distribution systems where one or more large motors may feed back power to smaller motors under open-phase conditions.

FIGURE 13-14 Instantaneous trip current relay (Courtesy Allen-Bradley Co.)

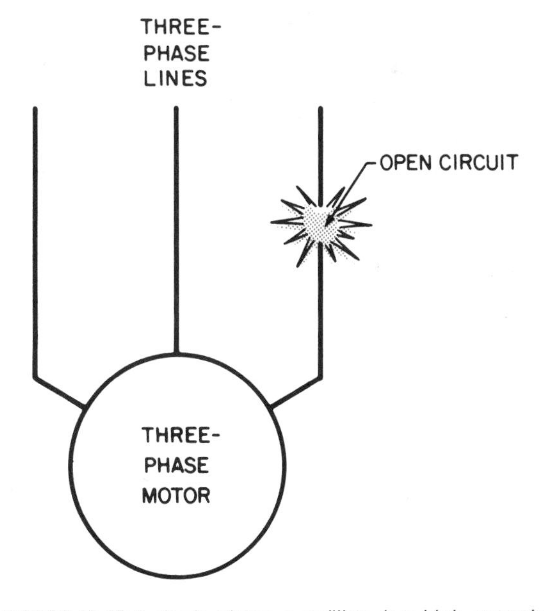

FIGURE 13-15 A single-phase condition: two high currents, one zero current

(A) 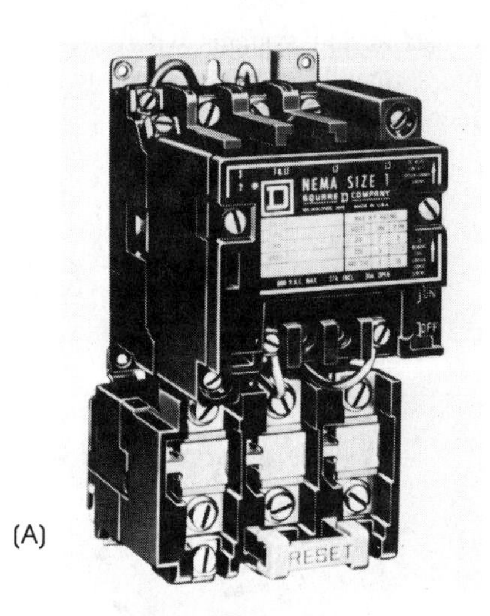

(B) 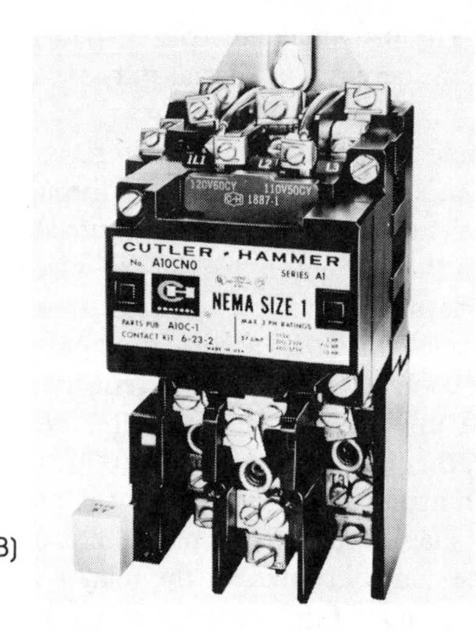

FIGURE 13-16 Three-phase ac magnetic motor starters (A) Courtesy Square D Co., (B) Courtesy EATON Corp., Cutler-Hammer Products

THE AC MAGNETIC STARTER

An ac three-phase magnetic motor starter is shown in figure 13-16(A). It is also called a full-voltage or across-the-line starter.

The overload reset button can be seen on the bottom center of the photo. It is usual practice to build motor controllers with manual reset overload relays. This encourages the machine operator to remove the cause of the overload. It also enforces at least a little cooling off period after tripping.

Three overload heating elements for three-phase operation are installed in the relays above the reset button. The contacts are under the insulating arcing block cover, easily accessible for inspection with the removal of two screws. The starter must be mounted in an enclosure for installation. Another type of three-phase ac magnetic motor starter is shown in figure 13-16(B).

Starter Sizes

Magnetic starters are available in many sizes as shown in Table 13-1. Each size has been assigned a horsepower rating that applies when the motor used with the starter is operated for normal starting duty. All starter ratings comply with the National Electrical Manufacturers Association Standards. The capacity of a starter is determined by the size of its contacts and the wire cross-sectional area. The size of the power contacts is reduced when the voltage is double because the current is halved for the same power ($P = I \times E$). Power circuit contacts handle the motor load.

Three-pole starters are used with motors operating on three-phase, three-wire ac systems. Two-pole starters are used for single-phase motors.

The number of *poles* refers to the power contacts, or the motor load contacts, and does not include control contacts for control circuit wiring.

AC COMBINATION STARTERS

The circuit breakers and fuses of the motor feeders and branch circuits are normally selected for overcurrent, short-circuit, or ground-fault protection.

With minor exceptions, the National Electrical Code and some local codes also require that every motor has a disconnect means. This means may be an attachment cord cap and receptacle, a nonfusible isolation disconnect safety switch, a fusible disconnect motor switch or a combination starter. A combination starter (figure 13-17) con-

TABLE 13-1 Motor Starter Sizes and Ratings

NEMA SIZE	LOAD VOLTS	MAXIMUM HORSEPOWER RATING—NONPLUGGING AND NONJOGGING DUTY	
		Single Phase	Poly-Phase
00	115	1/2	. . .
	200	. . .	1 1/2
	230	1	1 1/2
	380	. . .	1 1/2
	460	. . .	2
	575	. . .	2
0	115	1	. . .
	200	. . .	3
	230	2	3
	380	. . .	5
	460	. . .	5
	575	. . .	5
1	115	2	. . .
	200	. . .	7 1/2
	230	3	7 1/2
	380	. . .	10
	460	. . .	10
	575	. . .	10
* 1P	115	3	. . .
	230	5	. . .
2	115	3	. . .
	200	. . .	10
	230	7 1/2	15
	380	. . .	25
	460	. . .	25
	575	. . .	25
3	115	7 1/2	. . .
	200	. . .	25
	230	15	30
	380	. . .	50
	460	. . .	50
	575	. . .	50
4	200	. . .	40
	230	. . .	50
	380	. . .	75
	460	. . .	100
	575	. . .	100
5	200	. . .	75
	230	. . .	100
	380	. . .	150
	460	. . .	200
	575	. . .	200
6	200	. . .	150
	230	. . .	200
	380	. . .	300
	460	. . .	400
	575	. . .	400
7	230	. . .	300
	460	. . .	600
	575	. . .	600
8	230	. . .	450
	460	. . .	900
	575	. . .	900

Tables are taken from NEMA Standards.
(*1 3/4, 10 hp is available)

sists of an across-the-line starter and a disconnect means wired together in a common enclosure. Combination starters may have a blade-type disconnect switch, either fusible or nonfusible, or a thermal-magnetic trip circuit breaker. The starter may be controlled remotely with push buttons or selector switches, or these devices may be installed in the cover of the starter enclosure. The combination starter takes little mounting space and makes compact electrical installation possible.

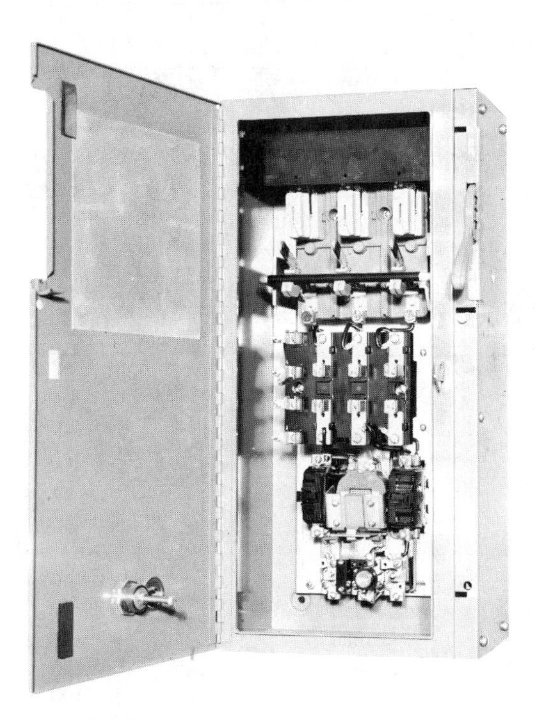

FIGURE 13-17 Combination starter with fuses removed. Disconnect switch must be open before door can be open.

A combination starter provides safety for the operator because the cover of the enclosure is interlocked with the external, operating handle of the disconnecting means. The door cannot be opened while the disconnecting means is closed. When the disconnecting means is open, all parts of the starter are accessible; however, the hazard is reduced since the readily accessible parts of the starter are not connected to the power line. This safety feature is not available on separately enclosed starters. In addition, the starter enclosure is provided with a means for padlocking the disconnect in the OFF position. Controller enclosures are available for every purpose and application.

Protective Enclosures

The selection and installation of the correct enclosure can contribute to useful, safe service and freedom from trouble in operating electromagnetic control equipment.

An enclosure is the surrounding controller case, cabinet or box. Generally, this electrical equipment is enclosed for one or more of the following reasons:

(A) To shield and protect workers and other personnel from accidental contact with electrically live parts, thereby preventing electrocution.

(B) To prevent other conducting equipment from coming into contact with live electrical parts, thereby preventing unnecessary electrical outages and indirectly protecting personnel from electrical contact.

(C) To protect the electrical controller from harmful atmospheric or environmental conditions, such as the presence of dust or moisture, to prevent corrosion and interference of operation.

(D) To contain the electrical arc of switching within the enclosure, to prevent explosions and fires which may occur with flammable gases or vapors within the area.

You may readily understand why some form of enclosure is necessary and required. The most frequent requirement is usually met by a general-purpose, sheet steel cabinet. The conduit is installed with lock-nuts and bushings. The presence of dust, moisture, or explosive gases often makes it necessary to use a special enclosure to protect the controller from corrosion or the surrounding equipment from possible explosions. Conduit access is through threaded openings, hubs, or flanges. In selecting and installing control apparatus, it is necessary to carefully consider the conditions under which the apparatus must operate. There are many applications where a general-purpose sheet steel enclosure does not give sufficient protection.

Water-tight and dust-tight enclosures are used for the protection of control apparatus. Dirt, oil, or excessive moisture are destructive to insulation and frequently form current-carrying paths that lead to short circuits or grounded circuits.

Special enclosures for hazardous locations are used for the protection of life and property. Explosive vapors or dusts exist in some departments of many industrial plants as well as in grain elevators, refineries, and chemical plants. The National Electrical Code and local codes describe hazardous locations. The Underwriters' Laboratories have defined the requirements for protective enclosures according to the hazardous conditions. The National Electrical Manufacturers Association (NEMA) has standardized enclosures from these requirements. Some examples are as follows.

General-purpose enclosures (NEMA 1) These enclosures are constructed of sheet steel, and are designed primarily to prevent accidental contact with live parts. Covers have latches with provisions for padlocking, figure 13-18. Enclosures are intended for use indoors, in areas where unusual service conditions do not exist. They do provide protection from light splash, dust and falling debris such as dirt.

Watertight enclosures (NEMA 4) These enclosures are made of cast construction or of sheet metal of suitable rigidity and are designed to pass a hose test with no leakage of water. Watertight enclosures are suitable for outdoor applications, on ship docks, in dairies, breweries, and other locations where the apparatus is subjected to dripping or splashing liquids, figure 13-19. Enclosures that meet requirements for more than one NEMA type may be designated by a combination of type numbers, for example, Type 3-4, dust-tight and water-tight.

FIGURE 13-18 General-purpose enclosure (NEMA 1) (Courtesy Square D Co.)

Dust-tight enclosures (NEMA 12) These enclosures are constructed of sheet steel and are provided with cover gaskets to exclude dust, lint, dirt, fibers and flyings. Dust-tight enclosures are suitable for use in steel and knitting mills, coke plants, and similar locations where nonhazardous dusts are present. Mounting is by means of outside flanges or mounting feet.

FIGURE 13-19 Watertight enclosure (NEMA 4) (Courtesy Square D Co.)

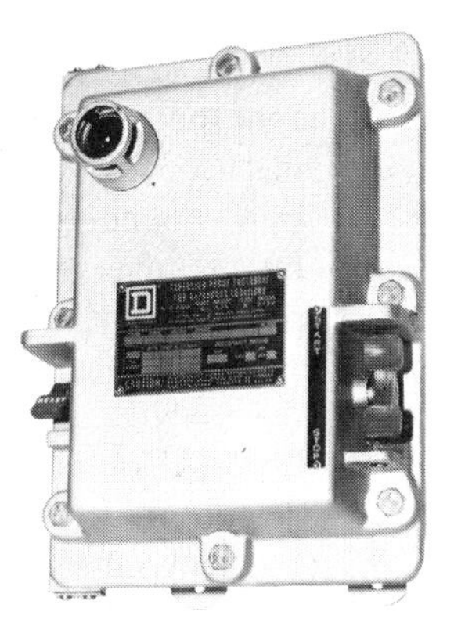

FIGURE 13-20 Hazardous location enclosure (NEMA 7) (Courtesy Square D Co.)

Hazardous locations (NEMA 7) Class 1 enclosures are designed for use in hazardous locations where atmospheres containing gasoline, petroleum, naphtha, alcohol, acetone, or lacquer solvent vapors are present or may be encountered. Enclosures are heavy, grey iron castings, machined to provide a metal-to-metal seal, figure 13-20.

NOTE: Applicable and enforced National, State, or local electrical codes and ordinances should be consulted to determine the safe way to make any installation.

REVIEW QUESTIONS

1. What is a magnetic line voltage motor starter?
2. How many poles are required on motor starters for the following motors: (a) 240-volt, single-phase induction motor, (b) 440-volt, three-phase induction motor?
3. If a motor starter is installed according to directions but will not start, what is a common cause for the failure to start?
4. Using the time limit overload or the dashpot overload relay, how are the following achieved: time delay characteristics; tripping current adjustments?
5. What is meant by chattering of an ac magnet?
6. What is the phase relationship between the flux in the main pole of a magnet and the flux in the shaded portion of the pole?
7. In what devices is the principle of the shaded pole used?
8. What does the electrician look for to remedy the following conditions: loud or noisy hum; chatter?
9. What type of protective enclosure is used most commonly?
10. Why is a disconnect fuse switch or circuit breaker installed with a motor starter?
11. What safety feature does the type of assembly given in question 10 provide that individual starter assemblies do not?
12. List the probable causes if the armature does not release after the magnetic starter is de-energized.
13. How is the size of the overload heaters selected for a particular installation?
14. What type of motor starter enclosure is recommended for an installation requiring safe operation around an outside flammable paint filling pump?

Select the *best* answer for each of the following.

15. The magnetic starter is held closed
    a. mechanically
    b. by 15% undervoltage
    c. by 15% overvoltage
    d. magnetically
16. When a motor starter coil is de-energized
    a. the contacts stay closed
    b. it is held closed mechanically
    c. gravity and spring tension open contacts
    d. it must cool for a restart
17. An ac magnet may hum excessively due to
    a. improper alignment
    b. foreign matter between contact surfaces

c. loose laminations
d. all of these

18. Ac magnets are made of *laminated* iron
a. for better induction
b. to reduce heating effect
c. for ac and dc use
d. to prevent chattering

19. The purpose of overload protection on a motor is to protect the
a. motor from sustained overcurrents
b. wire from high currents
c. motor from sustained overvoltage
d. motor from short circuits

20. The number of magnetic starter poles refers to
a. the number of power load contacts
b. the number of control contacts
c. the number of north and south poles
d. all of these

21. Motors may burn out because of
a. excessive heat from within and without
b. overloads
c. high ambient temperatures
d. poor ventilation

22. The purpose of a shading coil on an ac electromagnetic pole tip is to
a. prevent overheating of the coil
b. limit the tripping current
c. limit the closing current
d. prevent chattering

23. The current drawn by a motor is
a. low on starting
b. an accurate measurement of motor load
c. an inaccurate measurement of motor load
d. none of these

24. Thermal overload relays depend on
a. rising ambient and temperatures due to current overload
b. heavy mechanical loads
c. heavy electrical loads
d. rising starting currents

25. The thermal relay heating element is selected
a. 15% under voltage
b. 10% over voltage
c. from a table by the manufacturer
d. by ambient temperature

26. When the reset button does not re-establish the control circuit after an overload, the probable cause is
a. the overload heater is too small
b. the overload trip has not cooled sufficiently
c. the auxiliary contacts are defective
d. the overload heater is burned out

27. If an operator pushes a start button on a three-phase induction motor and the motor starts to hum, but not run, the probable trouble is
    a. one fuse is blown and the motor is single phasing
    b. the overload trip needs resetting
    c. the auxiliary contact is shorted
    d. one phase is grounded
28. A combination starter provides
    a. disconnecting means
    b. overload protection
    c. short circuit protection
    d. all of these

UNIT 14

Push Buttons and Control Stations

Objectives *After studying this unit, the student will be able to:*

- Describe the difference between closed and open push buttons
- Draw the wiring diagram symbols for push buttons and pilot lights
- Draw simple circuits using normally open and closed push buttons
- Connect combination push buttons in simple circuits
- Explain single and double break contacts
- List type of control operators on control stations
- Draw simple circuits using a selector switch
- Wire simple circuits with a selector switch

PUSH BUTTONS

A push-button station is a device that provides control of a motor through a motor starter by pressing a button which opens or closes contacts. It is possible to control a motor from as many places as there are stations—through the same magnetic controller. This can be done by using more than one push-button station. A single circuit push-button station is shown in figure 14-1.

Two sets of momentary contacts are usually provided with push buttons so that when the button is pressed, one set of contacts is opened and the other set is closed. The use of this combination push button is simply illustrated in the wiring diagram of figure 14-2. When the push button is in its normal position as shown in figure 14-2(A), current flows from L1 through the normally closed contacts, through the red pilot light (on top) to L2 to form a complete circuit; this lights

FIGURE 14-1 Push-button control station with general-purpose enclosure, usually made of molded plastic or sheet metal. Corresponding wire diagram symbols are shown for single-circuit push buttons. (Courtesy Square D Co.)

the lamp. Since the lower push-button circuit is opened by the normally open contact, the green pilot lamp does not glow. Note that this situation is reversed when the button is pushed, see figure 14-2(B). Current now flows from L1 through the pushed closed contact and lights the green pilot lamp when completing the circuit to L2. The red lamp is now out of the circuit and does not glow. However, because the push buttons are momentary contact (spring loaded), we return to the position shown in figure 14-2(A) when pressure on the button is released. Thus, by connecting to the proper set of contacts, either a normally open or a normally closed situation is obtained. *Normally open* and *normally closed* mean that the contacts are in a rest position, held there by spring tension, and are not subject to either mechanical or electrical external forces. (See the Glossary for more detailed definitions.)

Push-button stations are made for two types of service: *standard duty* stations for normal applications safely passing coil currents of motor starters up to size 4, and *heavy duty* stations, when the push buttons are to be used frequently and subjected to hard or rough usage. Heavy duty push-button stations have high contact ratings.

The push-button station enclosure containing the contacts is usually made of molded plastic or sheet metal. Some double-break contacts are made of copper. However, in most push buttons, silver-to-silver contact surfaces are provided for better electrical conductivity and longer life.

Figure 14-3 shows a combination, normally open and normally closed, push-button unit with a mushroom head for fast and easy access such as may be required in a safety circuit. This is also called a *palm-operated push button,* especially when it is used as a safety device for both operator and machine, such as a punch-press or emergency stop in a control station. The push-button terminals in figure 14-2 represent one-half of the terminals shown in figures 14-3 and 14-4, for a double-pole, double-throw push button.

Since control push buttons are subject to high momentary voltages caused by the inductive effect of the coils to which they are connected, good clearance between the contacts and insulation to ground and operator is provided.

The push-button station may be mounted adjacent to the controller or at a distance from it. The amount of current broken by a push button is usually small. As a result, operation of the con-

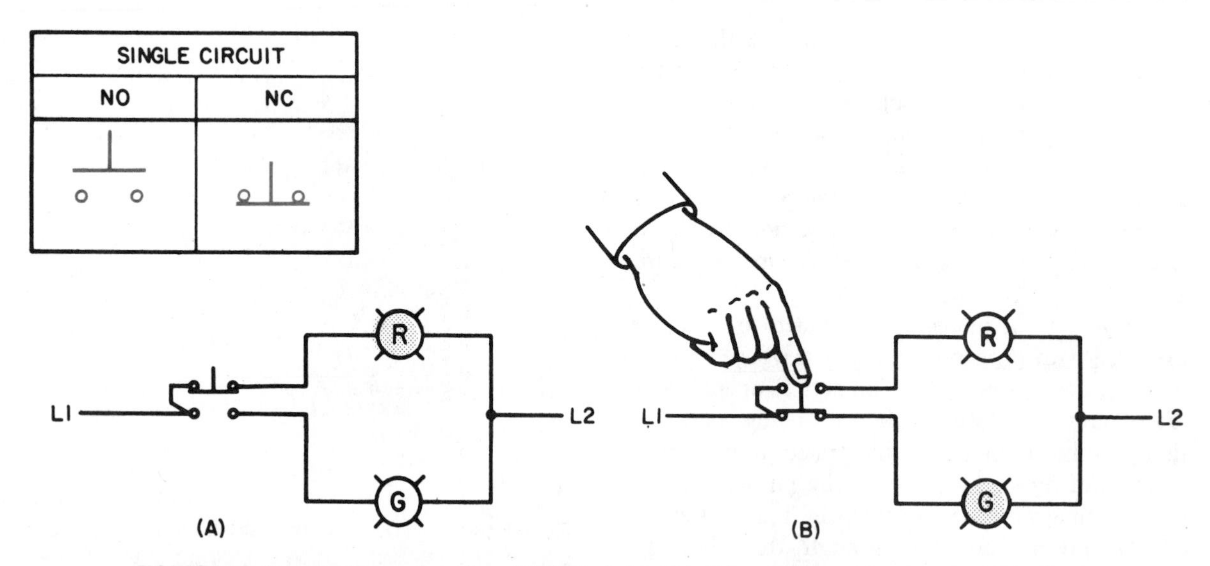

FIGURE 14-2 Combination push button: (A) Red pilot light is lit through normally closed push-button contact. (B) Green pilot light is lit when the momentary contact button is pushed.

troller is hardly affected by the length of the wires leading from the controller to a remote push-button station.

Push buttons can be used to control any or all of the many operating conditions of a motor, such as *start, stop, forward, reverse, fast,* and *slow.* Push buttons also may be used as remote stop buttons with manual controllers equipped with potential trip or low-voltage protection.

SELECTOR SWITCHES

Selector switches, as can be seen in figure 14-5, are usually "maintained" contact positions, with three and sometimes two selector positions. Selector switch positions are made by turning the operator knob—not pushing it. Figure 14-5(A) is a single-break contact arrangement, and figure 14-5(B) is a double-break contact, disconnecting the control circuit at two points. These switches may also have a spring return to give momentary contact operation.

Figure 14-6A shows a single break contact selector switch connected to two lights. The red light may be selected to glow by *turning* the switch to the red position. Current now flows from line 1 through the selected red position of the switch, through the red lamp to line 2, completing the circuit. Note this switch has an *off* position in the center. In figure 14-6(B), there are two position

FIGURE 14-3 Open the push-button unit with mushroom head (Courtesy Allen-Bradley Co.)

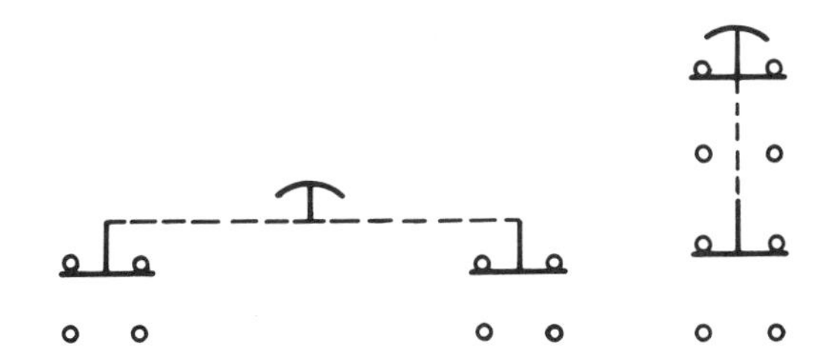

FIGURE 14-4 Terminal configurations used by different manufacturers of figure 14-3 push button. Both configurations represent the same push button.

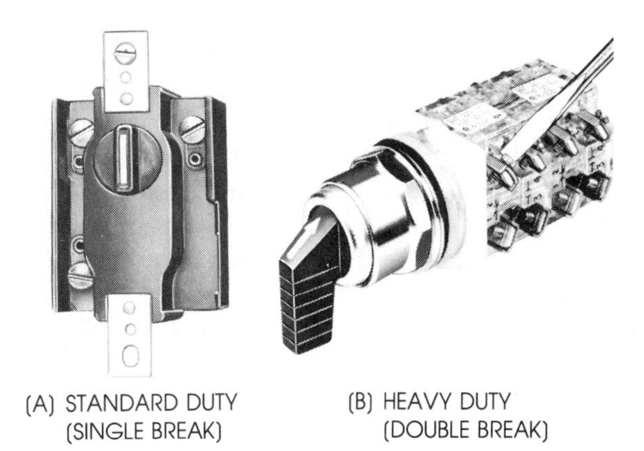

FIGURE 14-5 Open selector switches (A) Courtesy EATON Corp., Cutler-Hammer Products, (B) Courtesy Square D Co.

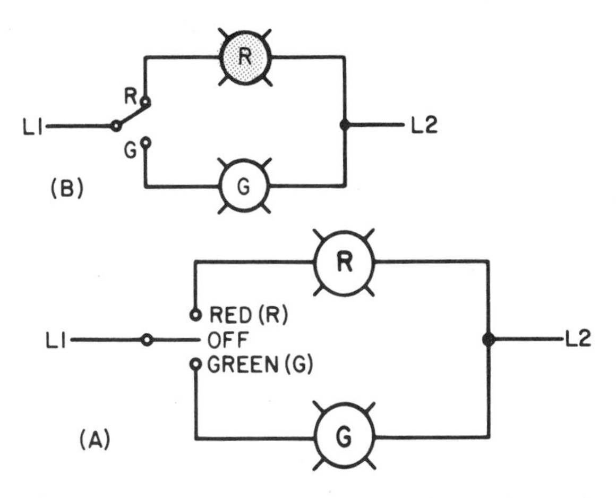

FIGURE 14-6 Elementary diagram using (A) A single-break, three-position selector switch and (B) A two-position, single-break switch

selector switches available, with no *off* position. Here, whichever light is selected would burn continuously.

Figure 14-7(A) is an elementary diagram of the heavy duty, double-break selector switch similarly shown in figure 14-5. Note that the red light will glow (see A) with the two-position switch in this position, the green indicating lamp is de-energized, and there is no "off" position. Figure 14-7(B) illustrates a three-position selector switch containing an "off" position. Both lamps may be turned off using this switch, but not in (A).

CONTROL STATIONS

A control station may contain push buttons alone, figure 14-8A, or a combination of push buttons, selector switches and pilot lights (indicating lights), figure 14-8B. Indicating lights may be mounted in the enclosure. These lights are usually red or green and are used for communication and safety purposes. Other common colors available are amber, blue, white, and clear. They indicate when the line is energized, the motor is running, or any other condition is designated.

Control stations may also include switches that are key, coin, or hand operated wobble sticks. A wobble stick is a stem-operated push button, for operation from any direction. There are ball lever push button operators for a gloved hand or for frequent operations.

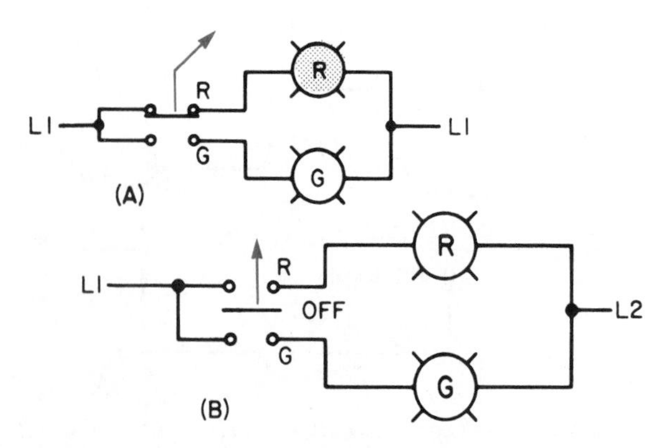

FIGURE 14-7 Heavy-duty, double-break selector switch for (A) Two position and (B) Three position switches

Name plates are installed to designate each control operation. These can be seen in figures 14-1 and 14-8. These control operators are commonly used in control circuits of magnetic devices in factory production machinery.

Combination indicating light, nameplate, push-button units are available. These illuminated push buttons and indicating lights are designed to save space in a wide variety of applications such as con-

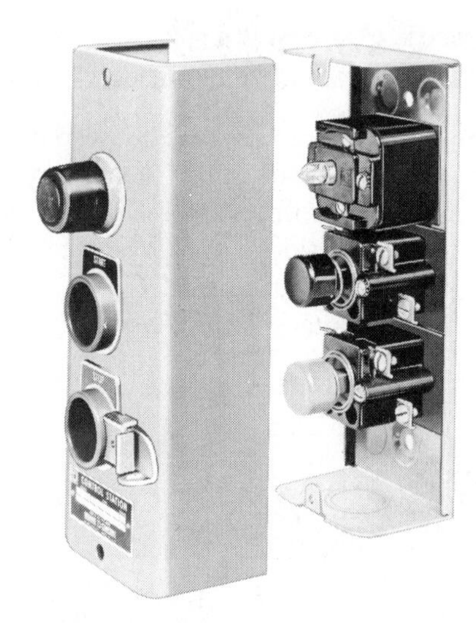

FIGURE 14-8A Push-button control station with pilot light (Courtesy Square D Co.)

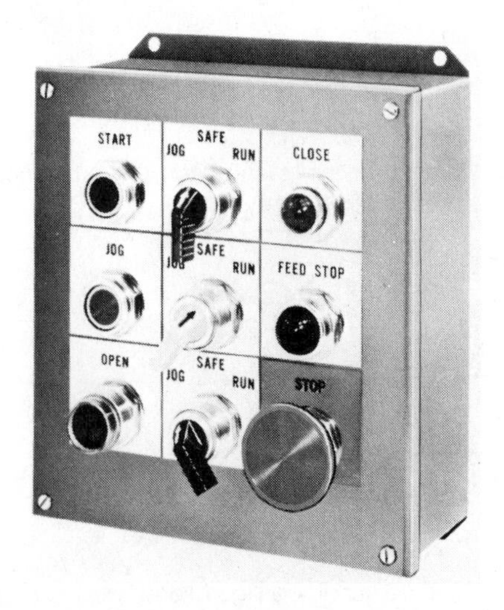

FIGURE 14-8B Control station with push buttons, selector switches, and indicating lights (Courtesy Square D Co.)

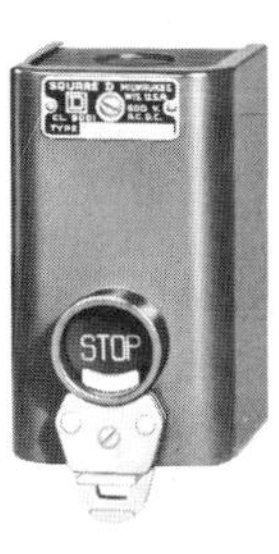

FIGURE 14-9 Mechanical lockout-on-stop push button (Courtesy Square D Co.)

trol and instrument panels, laboratory instruments, and computers. Miniature buttons are also used for this purpose.

Standard control station enclosures are available for normal general-purpose conditions, while special enclosures are used in situations requiring watertight, dust-tight, oiltight, explosion-proof, or submersible protection. Provisions are often made for *padlocking* stop buttons in the open position (for safety purposes), figure 14-9. Relays, contactors and starters cannot be energized while an electrician is working on them with the stop button in this position.

The Push-Pull Operator

Another type of push-button control is the push-pull operator. This control contains two sets of contacts in one unit. One set of contacts is operated by pulling outward on the button and the other set is operated by pushing the button. The head of the push-pull operator is made in such a manner as to permit a machine operator to pull outward on the button by placing two fingers behind it. Figure 14-10 shows a push-pull operator control. The head of the control can be pushed to operate the other set of contacts.

There are two types of push-pull operators. The type used is determined by the requirements of the circuit. One type of control contains two normally open momentary contacts. Figure 14-11 shows a schematic drawing of this control. In its normal position, neither movable contact A or B connects with either of the stationary contacts. In figure 14-12, the control has been pulled outward. This causes movable contact A to connect with two of the stationary contacts. Movable contact B does not connect with its stationary contacts.

If the push-pull operator is pressed, movable contact B connects with a set of stationary contacts as shown in figure 14-13. Movable contact A does not connect with its stationary contacts.

FIGURE 14-10 Push-pull operator with light unit (Courtesy EATON Corp., Cutler-Hammer Products)

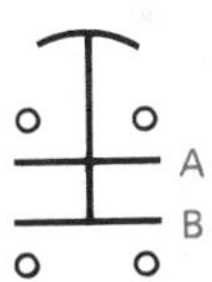

FIGURE 14-11 Push-pull operator with two normally open contacts

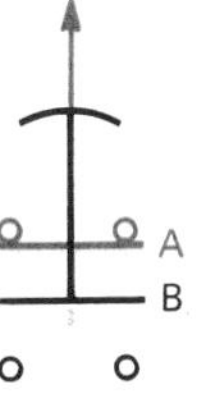

FIGURE 14-12 Movable contact A connects with the stationary contact

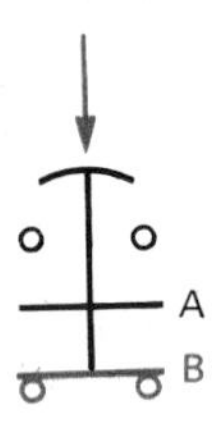

FIGURE 14-13 Movable contact B connects with the stationary contact

FIGURE 14-15 Both movable contacts connect with their stationary contacts

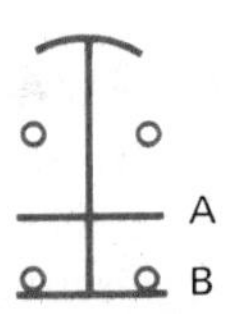

FIGURE 14-14 Movable contact A is normally open and movable contact B is normally closed

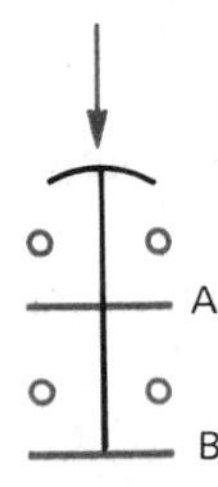

FIGURE 14-16 Both movable contacts break connection with their stationary contacts

This type of push-pull operator control can be used with run-jog controls. An example of this type of control is shown in Unit 41.

The second type of push-pull operator has one normally open contact and one normally closed contact. Figure 14-14 shows a schematic drawing of this type of control. When this type of push-pull operator is in its normal position, movable contact A is open and movable contact B is closed.

When the button is pulled outward, contact A connects with its stationary contacts, and contact B maintains connection with its stationary contacts. Figure 14-15 illustrates this condition. When the button is released, the movable contacts return to their normal position.

When the push button is pressed, contact B breaks the connection with its stationary contact. Figure 14-16 shows this condition.

This type of push-pull operator control can be used as a start-stop motor control station. The advantage of this type of control is that it requires the space of only one push-button control element instead of two. Also, since the control must be pulled outward to start the motor and pressed inward to stop the motor, the possibility of accidently pushing the wrong button is eliminated.

A start-stop control circuit using the push-pull operator is shown in figure 14-17. When the push-pull operator is pulled outward, contact A completes a circuit to motor starter coil M. Normally open contact M, connected parallel to contact A, closes to maintain the circuit to coil M when the button is released. When the push-pull operator is pressed, movable contact B breaks the circuit to motor starter coil M and the circuit de-energizes.

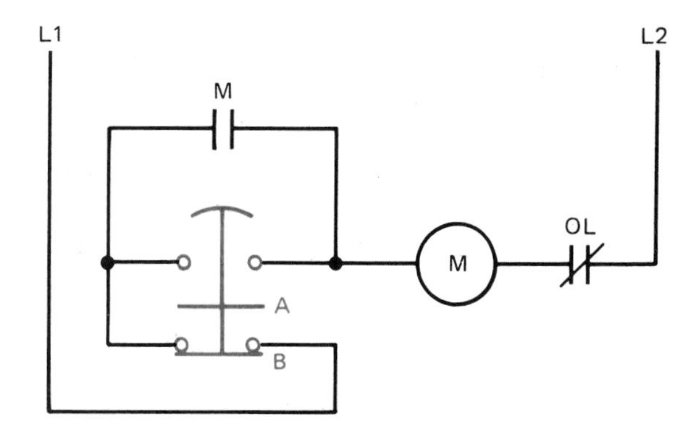

FIGURE 14-17 Push-pull operator used as a start-stop motor control

REVIEW QUESTIONS

1. What is meant by normally open contacts and normally closed contacts?
2. Why is it that normally open and normally closed contacts cannot be closed simultaneously?
3. How are colored pilot lights indicated in wiring diagrams?

Select the *best* answer for each of the following.

4. A single-break control is a
   a. heavy duty selector switch
   b. single-circuit push button
   c. standard duty selector switch
   d. double-circuit push button
5. Control stations may contain
   a. push buttons
   b. selector switches
   c. indicating lights
   d. all of these
6. A wobble stick is
   a. operated from any direction
   b. knob controlled
   c. palm controlled
   d. glove operated
7. Common selector switches are
   a. one position
   b. two positions
   c. three positions
   d. two and three positions

8. Most push buttons are
   a. momentary contact
   b. single contact
   c. double contact
   d. a combination
9. Control station enclosures are
   a. general purpose
   b. explosion proof
   c. watertight
   d. all of these
10. The diagram that illustrates a single circuit, normally closed push button is

a. 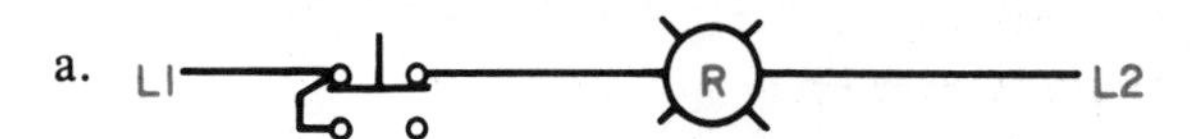

b. 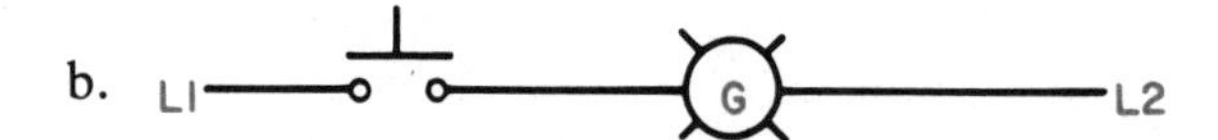

c. 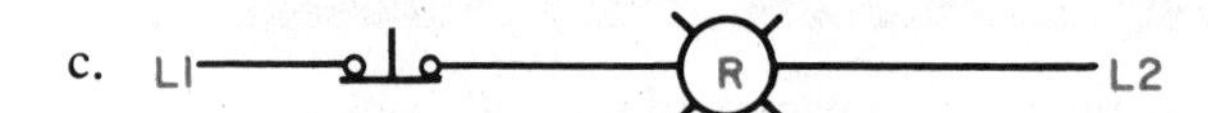

d. 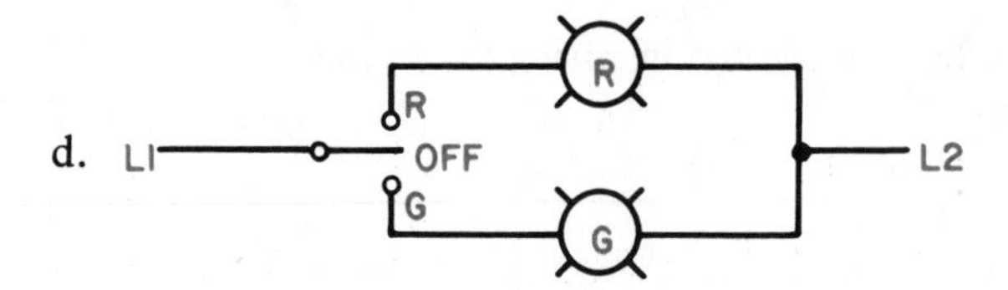

11. One type of push-pull operator contains
    a. Two normally closed momentary contacts
    b. Two normally closed maintained contacts
    c. Two normally open momentary contacts
    d. Two normally open maintained contacts
12. Another type of push-pull operator contains
    a. Two momentary normally closed contacts
    b. One maintained normally open contact and one momentary normally closed contact
    c. Two maintained normally open contacts
    d. One maintained normally closed contact and one momentary normally open contact

UNIT 15

Relays and Contactors

Objectives *After studying this unit, the student will be able to:*

- **Tell how magnetic relays differ from contactors**
- **List the principal uses of each magnetic relay and contactor**
- **Describe the operation of magnetic blowout coils and how they provide arc suppression**
- **Identify single and double throw contacts; single and double break contacts**
- **Draw the wiring diagram symbols for contactors and relays**
- **Describe the operation and use of mechanically held relays**
- **Draw elementary diagrams of control and load for mechanically held relays and contactors**
- **Connect wiring for mechanically held relays and contactors**

CONTROL RELAYS

Control magnetic relays are used as auxiliary devices to switch control circuits and large motor starter and contactor coils, and to control small loads such as small motors, solenoids, electric heaters, pilot lights, audible signal devices and other relays, figure 15-1.

A magnetically held relay is operated by an electromagnet which opens or closes electrical contacts when the electromagnet is energized. The position of the contacts changes by spring and gravity action when the electromagnet is de-energized.

Relays are generally used to enlarge or amplify the contact capability, or multiply the switching functions of a pilot device by adding more contacts to the circuit, figures 15-2A and B.

Most relays are used in control circuits; therefore, their lower ratings (0-15 amperes maximum to 600 volts) show the reduced current levels at which they operate.

Magnetic relays do not provide motor overload protection. This type of relay ordinarily is used in a two-wire control system (any electrical contact-making device with two wires). Whenever it is desired to use momentary contact pilot devices, such as push buttons, any available normally open contact can be wired as a holding circuit in a three-wire system (see the Glossary). The contact arrangement and a description of the magnetic structure of relays was presented in Unit 13, as motor starters. Starters, contactors, and relays are similar in construction and operation but are not identical.

Control relays are available in single-or double-throw arrangements with various combinations of normally open (NO) and normally closed (NC) contact circuits. While there are some single-break contacts used in industrial relays, most of the re-

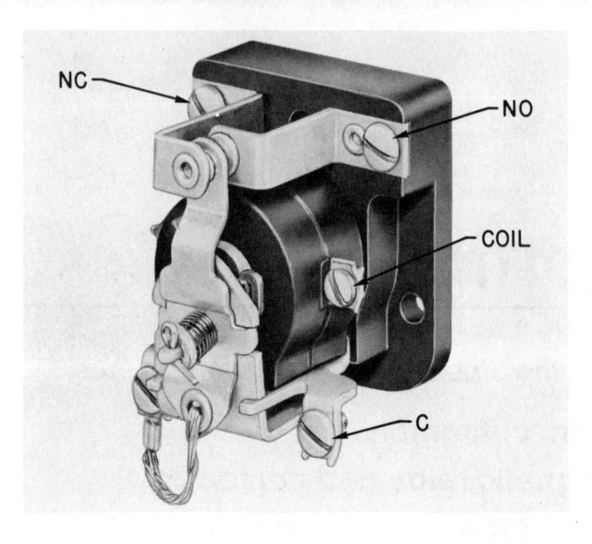

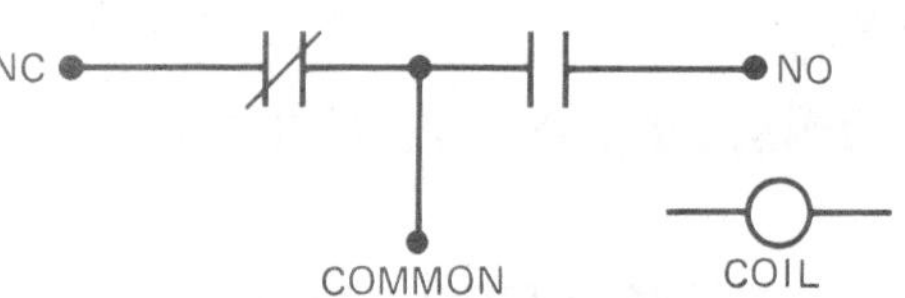

FIGURE 15-1 Single-pole, double-throw, single-break ac control relay with wiring symbols (Courtesy Square D Co.)

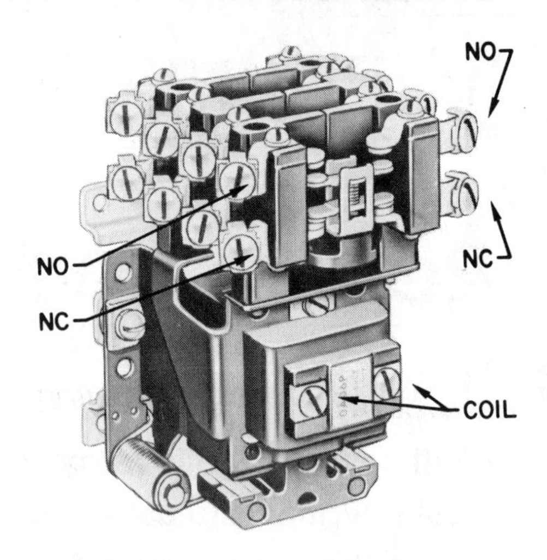

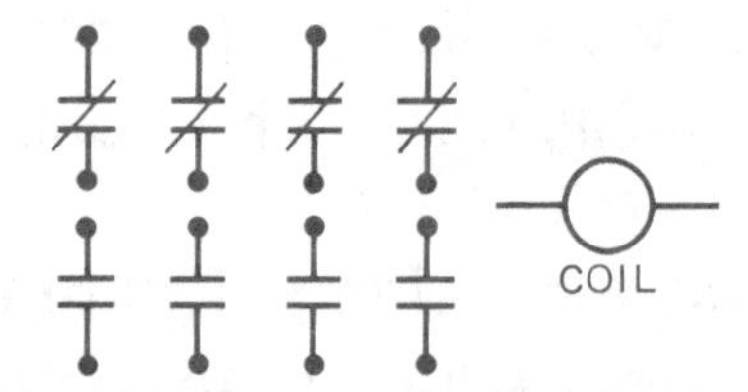

FIGURE 15-2A Eight-pole control relay with four normally open and four normally closed contacts. To change from the open to the closed condition, wire connections to the respective poles are changed. (Courtesy Allen-Bradley Co.)

lays used in machine tool control have double-break contacts. The comparison can be seen in figures 15-1 and 15-2A. Looking at the relay contacts in figure 15-2A, note the upper contact being open at two points, making it a double break. The lower contact is normally closed with a circuit from the left-hand terminal screw, through the double-break contact to the right-hand terminal screw. One set of normally closed and one of normally open contacts represents this description with the wiring symbols. This is also a single-throw contact because it has no common connection between the normally open and normally closed contacts, such as can be seen in figure 15-1. The common terminal between the normally closed and normally open contacts makes this a "single-pole, double-throw" relay.

It may be of particular interest to an electrician to know about changing contacts that are normally open to normally closed, or the other way around, NC to NO. Most machine tool relays have some means to make this change. It ranges from a simple flip-over contact to removing the contacts and relocating with spring location changes.

Also, by overlapping contacts in this case, one contact can be arranged to operate at a different time relative to another contact on the same relay. For example, the normally open contact closes before the normally closed contact opens.

Relays differ in voltage ratings, number of contacts, contact rearrangement, physical size, and in attachments to provide accessory functions such as mechanical latching and timing. (These are discussed later in this unit.)

In using a relay for a particular application, one of the first steps should be to determine the control (coil) voltage at which the relay will operate. The necessary contact rating must be made, as well as the number of contacts and other characteristics needed. Because of the variety of styles of relays available, it is possible to select the correct relay for almost any application.

Relays are used more often to open and close control circuits than to operate power circuits.

FIGURE 15-2B Multipole relays (Courtesy EATON Corp., Cutler-Hammer Products)

Typical applications include the control of motor starter and contactor coils, the switching of solenoids, and the control of other relays. A relay is a small but vital switching component of many complex control systems. Low-voltage relay systems are used extensively in switching residential and commercial lighting circuits and individual lighting fixtures.

While control relays from various manufacturers differ in appearance and construction, they are interchangeable in control wiring systems if their specifications are matched to the requirements of the system. Different types of relays are shown in figure 15-2C.

CONTACTORS

Magnetic contactors are electromagnetically operated switches that provide a safe and convenient means for connecting and interrupting branch circuits. The principal difference between a contactor and a motor starter is that the contactor does not contain overload relays. Contactors are used in combination with pilot control devices to switch lighting and heating loads and to control ac motors in those cases where overload protection is provided separately. The larger contactor sizes are used to provide remote control of relatively high-current circuits where it is too expensive to run the power leads to the remote controlling location, figure 15-3. This flexibility is one of the main advantages of electromagnetic control over manual control. Pilot devices such as push buttons, float switches, pressure switches, limit switches, and thermostats are provided to operate the contactors.

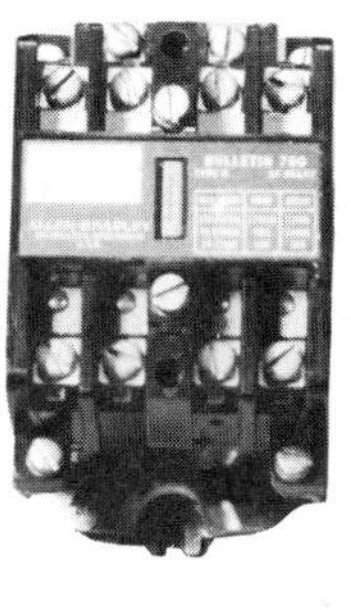

FIGURE 15-2C Different types of control relays found throughout industry

Magnetic Blowout

The contactors shown in figures 15-4 and 15-5 operate on alternating current. Heavy-duty contact arc-chutes are provided on most of these larger contactors, figure 15-4. The chutes contain heavy copper coils called blowout coils, mounted above the contacts in series with the load to provide better arc suppression. These magnetic blowout coils help to extinguish an electric arc at contacts opening under alternating-current and direct-current loads. These arcs may be similar in intensity to the electric arc welding process. An arc-quenching device is used to assure longer contact life. Since the hot arc is transferred from the contact tips very rapidly, the contacts remain cool and, thereby, last longer.

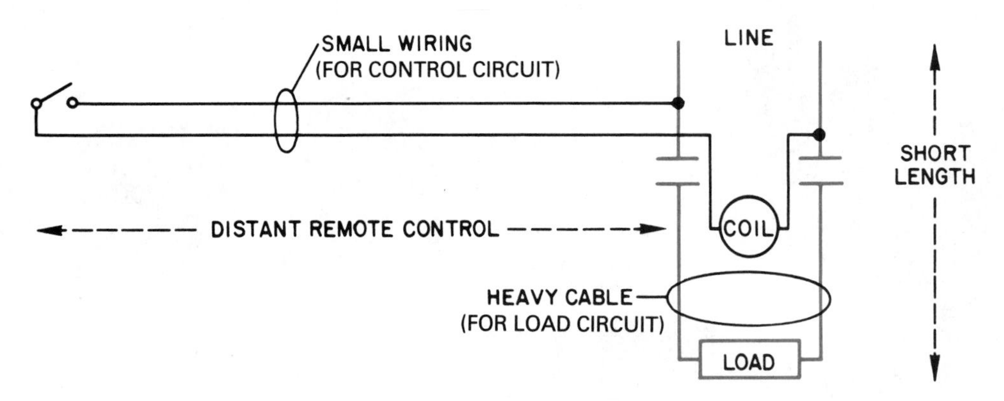

FIGURE 15-3 An advantage of a remote control load

FIGURE 15-4 Open magnetic contactor "clapper" action, Size 6 (Courtesy Square D Co.)

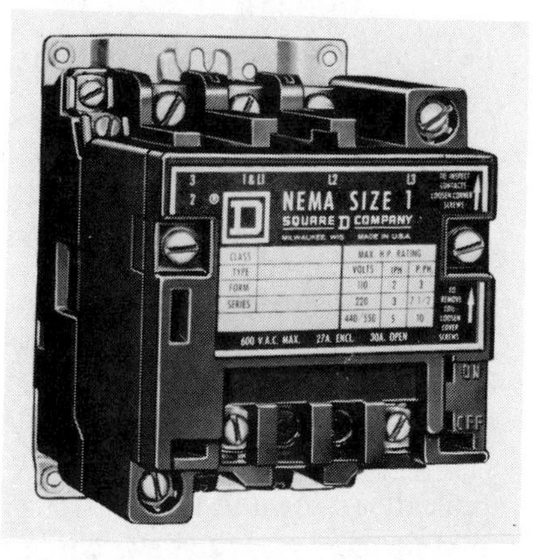

FIGURE 15-5 Contactor, Size 1. Contacts are accessible by removing the two front screws. (Courtesy Square D Co.)

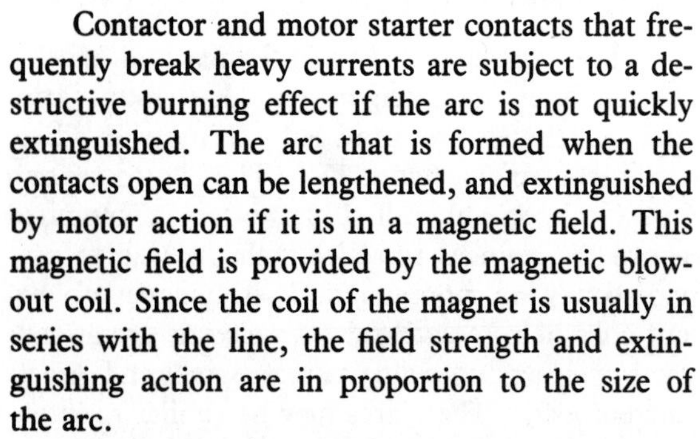

Contactor and motor starter contacts that frequently break heavy currents are subject to a destructive burning effect if the arc is not quickly extinguished. The arc that is formed when the contacts open can be lengthened, and extinguished by motor action if it is in a magnetic field. This magnetic field is provided by the magnetic blowout coil. Since the coil of the magnet is usually in series with the line, the field strength and extinguishing action are in proportion to the size of the arc.

Figure 15-6 is a sketch of a blowout magnet with a straight conductor (ab) located in the field and in series with the magnet. This figure can represent either dc polarity or instantaneous alternating current. With alternating current, the blowout coil magnetic field and the conductor (arc) magnetic field will reverse simultaneously. According to Fleming's left-hand rule, motor action will tend to force the conductor in an upward direction. The application of the right-hand rule for a single conductor shows that the magnetic field around the conductor aids the main field on the bottom and opposes it on the top, thus producing an upward force on the conductor.

Figure 15-7 shows a section of figure 15-6 with the wire (ab) replaced by a set of contacts. The contacts have started to open and there is an arc between them.

Figure 15-8 shows what happens because of the magnetic action. Part A shows the beginning deflection of the arc because of the effect of the

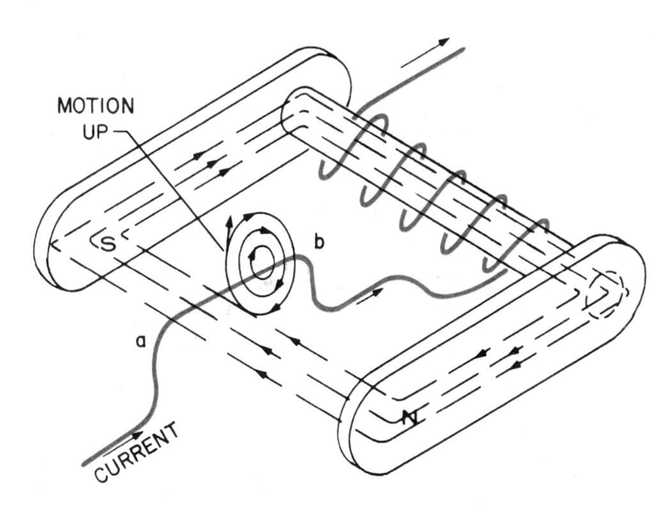

FIGURE 15-6 Illustration of the magnetic blowout principle. Straight conductor simulates arc.

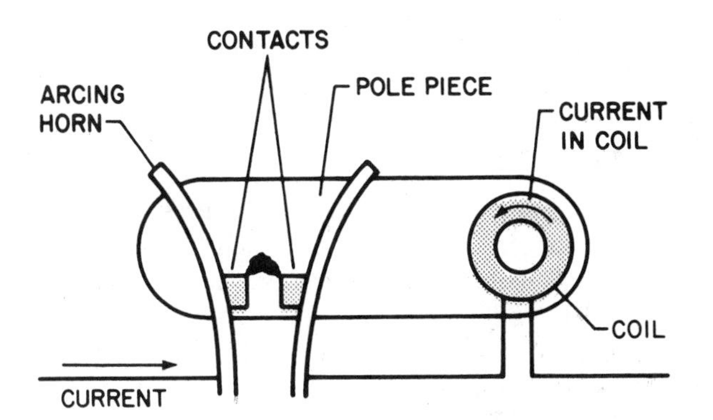

FIGURE 15-7 Section of blowout magnet with straight conductor replaced by a set of contacts. An arc is conducting between the contacts.

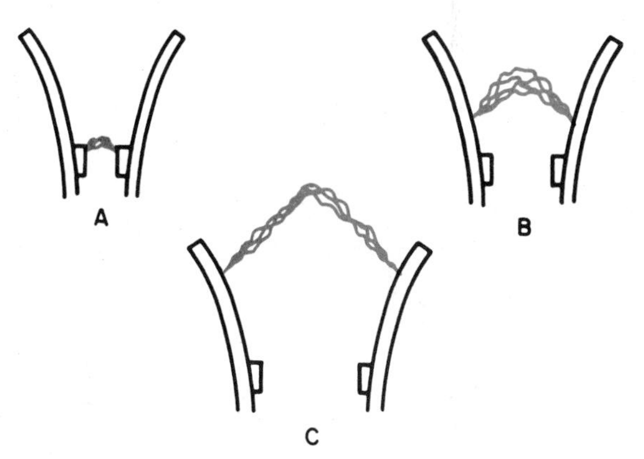

FIGURE 15-8 Arc deflection between contacts

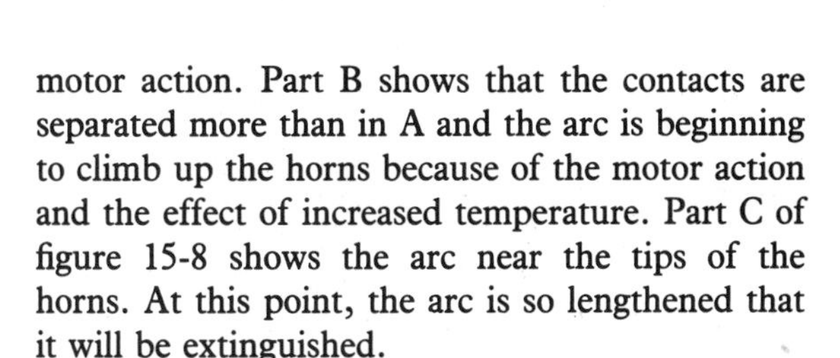

motor action. Part B shows that the contacts are separated more than in A and the arc is beginning to climb up the horns because of the motor action and the effect of increased temperature. Part C of figure 15-8 shows the arc near the tips of the horns. At this point, the arc is so lengthened that it will be extinguished.

The function of the blowout magnet is to move the arc upward at the same time that the contacts are opening. As a result, the arc is lengthened at a faster rate than would normally occur because of the opening of the contacts alone. It is evident that the shorter the time the arc is allowed to exist, the less damage it will do to the contacts. Most arc quenching action is based upon this principle.

AC MECHANICALLY HELD CONTACTORS AND RELAYS

A mechanically held relay, or contactor, is operated by electromagnets but the electromagnets are automatically disconnected by contacts within the relay. Accordingly, these relays are mechanically held in position and no current flows through the operating coils of these electromagnets after switching. It is apparent, therefore, that near continuous operation of multiple units of substantial size will lower the electrical energy requirements. Also, the magnetically held relay, in comparison, will change contact position upon loss of voltage to the electromagnet, whereas the mechanically held relay will respond only to the action of the control device.

Sequence of Operation

Referring to figure 15-9A, when the "on" push button is pressed momentarily, current flows from L1 through the *on* push-button contact energizing the M coil through the now closed clearing contact, to L2. The relay now closes and latches mechanically. At the same time it closes M contacts (in figure 15-9B), lighting a bank of lamps when the circuit breaker is closed. To unlatch the relay, thereby turning the lamps off, the *off* button is pressed momentarily, unlatching the relay and opening contacts M, turning off the lamps. Most operating coils are not designed for continuous duty. Therefore, they are disconnected automatically by contacts to prevent an accidental coil burnout. These coil clearing contacts change position alternately with a change in contactor latching position.

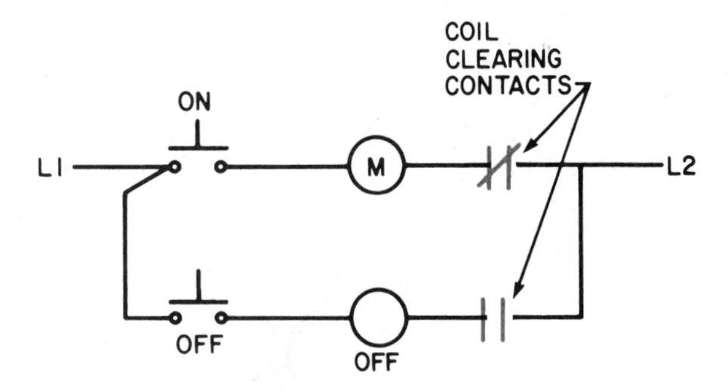

FIGURE 15-9A Mechanically held relay control circuit

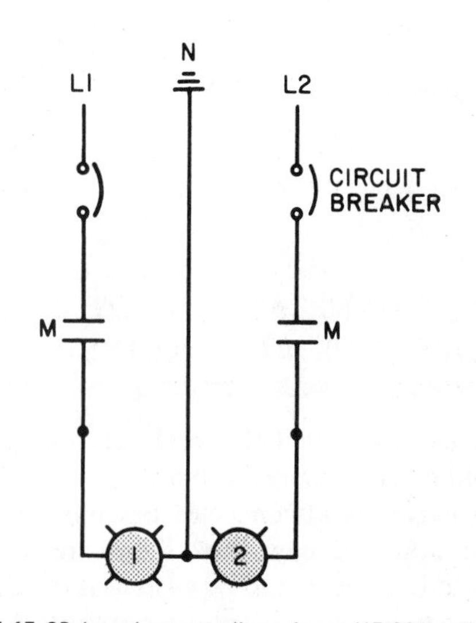

FIGURE 15-9B Load connections for a 115/230-volt, three-wire load

Figure 15-10 shows a three-phase power load application using one main contactor to disconnect the distribution panel. Selective single, or three-phase, branch circuits may be switched independently by other mechanically held contactors or relays.

These mechanically held contactors and relays are electromechanical devices, figure 15-11. They provide a safe and convenient means of switching circuits where *quiet operation, energy efficiency,* and *continuity of circuit connection* are requirements of the installation. For example, circuit continuity during power failures is often important in automatic processing equipment, where a sequence of operations must continue from the point of interruption after power is

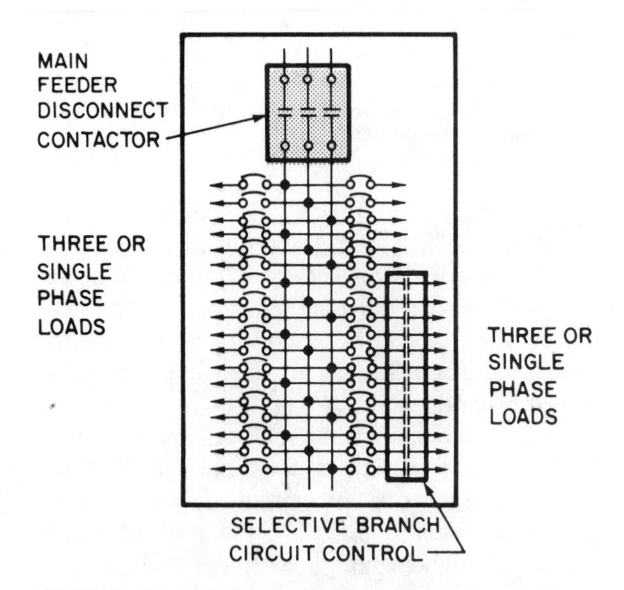

FIGURE 15-10 Mechanically held contactor loads for three-phase power

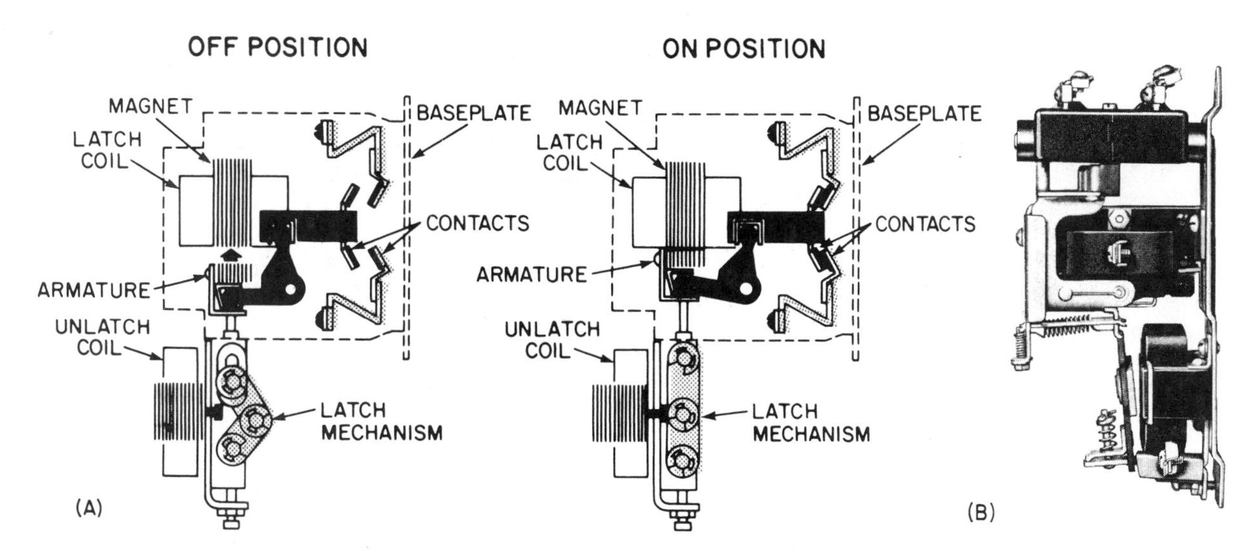

FIGURE 15-11 Two types of latched-in or mechanically-held relays in service. The upper coil is energized momentarily to close contacts, and the lower coil is energized momentarily to open the contact circuit. The momentary energizing of the coil is an energy-saving feature. (Courtesy Square D Co.)

resumed—rather than return to the beginning of the sequence. Quiet operation of contactors and relays is required in many control systems used in hospitals, schools, and office buildings. Mechanically held contactors and relays are generally used in locations where the slight hum, characteristic of alternating-current magnetic devices, is objectionable.

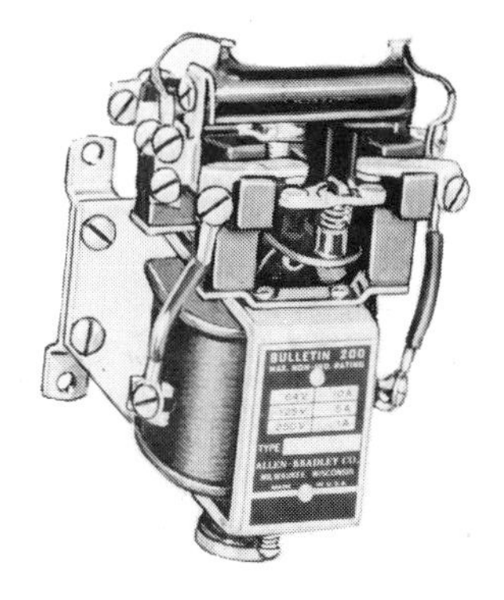

FIGURE 15-12 Thermostat relay (Courtesy Allen-Bradley Co.)

In addition, mechanically held relays are often used in machine tool control circuits. These relays can be latched and unlatched through the operation of limit switches, timing relays, starter interlocks, timeclocks, photoelectric cells, other control relays, or push buttons. Generally, mechanically held relays are available in 10- and 15-ampere sizes; mechanically held contactors are also available in sizes ranging from 30 amperes to 1200 amperes.

THERMOSTAT RELAY

Thermostat-type relays, figure 15-12, are used with three-wire, gauge-type thermostat controls or other pilot controls having a slowly moving element which makes a contact for both the closed and open positions of the relay. The contacts of a thermostat control device usually cannot handle the current to a starter coil; therefore, a

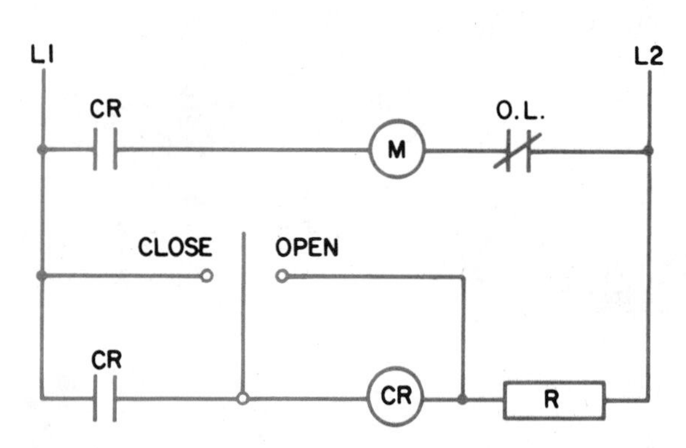

FIGURE 15-13 Starter coil (M) is controlled by thermostat relay

thermostat relay must be used between the thermostat control and the starter, figure 15-13.

When the moving element of the thermostat control touches the closed contact, the relay closes and is held in this position by a maintaining contact. When the moving element touches the open contact, the current flow bypasses the operating coil through a small resistor and causes the relay to open. The resistor is usually built into the relay and serves to prevent a short circuit.

The thermostat contacts must not overlap or be adjusted too closely to one another as this may result in the resistance unit being burned out. It is also advisable to compare the inrush current of the relay with the current rating of the thermostat.

REVIEW QUESTIONS

1. What are control relays?
2. What are typical uses for control relays?
3. Why are contactors described as both control pilot devices and large magnetic switches controlled by pilot devices?
4. What is the principal difference between a contactor and a motor starter?
5. What causes the arc to move upward in a blowout magnet?
6. Why is it desirable to extinguish the arc as quickly as possible?
7. What will happen if the terminals of the blowout coil are reversed?
8. Why will the blowout coil also operate on alternating current?
9. What are the advantages of a mechanically held relay?
10. Why are mechanically held relays energy-saving devices?
11. Why does coil (CR) in figure 15-13 drop out when the thermostat touches the open position?
12. Match each item in the left-hand column with the appropriate item in the right-hand column.

a. Double-break contact	1. C
b. Relay	2.
c. Magnetic blowout coil	3.
d. Single pole, double throw	4.

e. Normally closed contact

f. Latched-in relay control

g. Single-break contact

h. Relay coil

i. Normally open contact

j. Contactor

5. Arc suppression

6.

7. 0-15 amperes

8. Three-phase power

9.

10. 1200 amperes

11.

12.

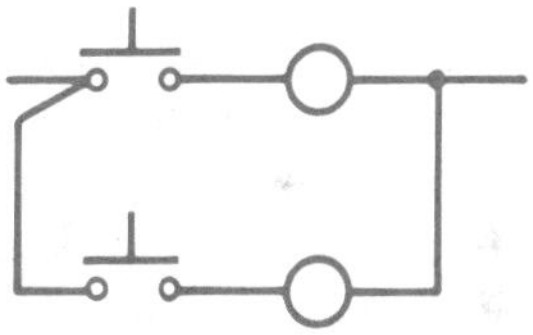

UNIT 16

The Solid-State Relay

Objectives *After studying this unit, the student will be able to:*

- **Discuss the differences between dc solid-state relays and ac solid-state relays**
- **Discuss opto-isolation**
- **Discuss magnetic isolation**
- **Connect a solid-state relay in a circuit**

The solid-state relay is a device that has become increasingly popular for switching applications. The solid-state relay has no moving parts, it is resistant to shock and vibration, and it is sealed against dirt and moisture. The greatest advantage of the solid-state relay, however, is the fact that the control input voltage is isolated from the line device the relay is intended to control, figure 16-1.

Solid-state relays can be used to control dc and ac loads. If the relay is designed to control a dc load, a power transistor is used to connect the load to the line as shown in figure 16-2. The relay shown in figure 16-2 has a *light-emitting diode* (LED) connected to the input or control voltage. When the input voltage turns the LED on, a photodetector connected to the base of the transistor turns the transistor on and connects the load to the line. This optical coupling is a very common method used with solid-state relays. The relays that use this method of coupling are said to be *opto-isolated,* which means the load side of the relay is optically isolated from the control side of the relay. Since a light beam is used as the control medium, no voltage spikes or electrical noise produced on the load side of the relay can be transmitted to the control side of the relay.

Solid-state relays intended for use as ac controllers have a triac, rather than a power transistor, connected to the load circuit, figure 16-3. As in figure 16-2, an LED is used as the control device in this example. When the photodetector "sees" the LED, it triggers the gate of the triac and connects the load to the line.

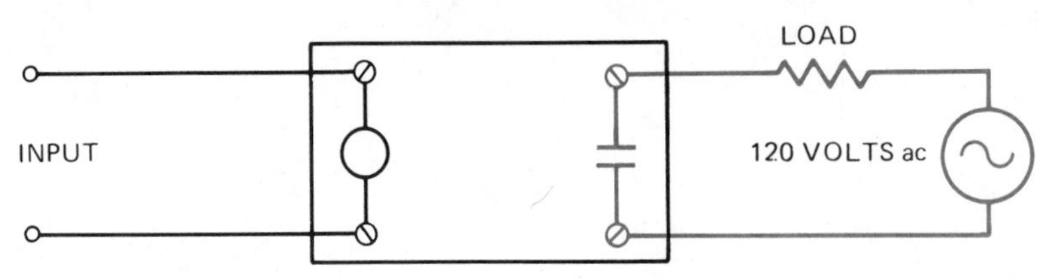

FIGURE 16-1 Solid-state relay

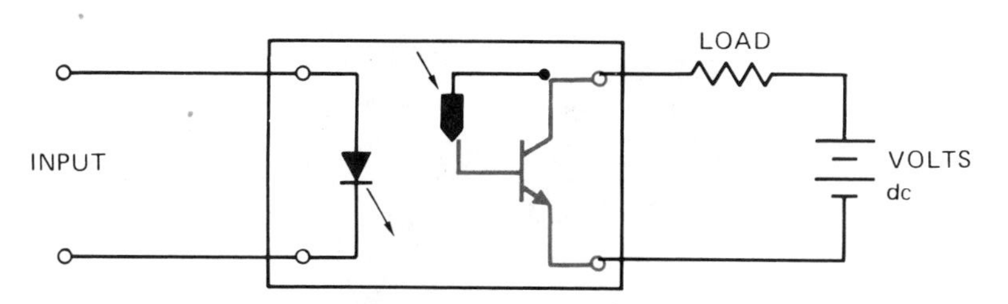

FIGURE 16-2 Power transistor used to control a dc load

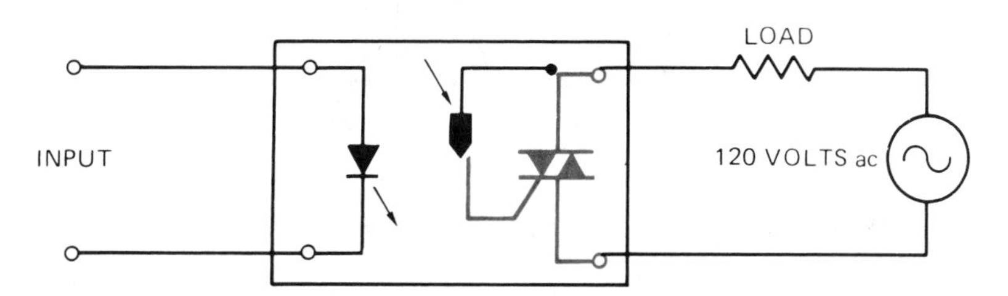

FIGURE 16-3 Triac used to control an ac load

Although opto-isolation is probably the most commonly used method for the control of a solid-state relay, it is not the only method used. Some relays use a small reed relay to control the output, figure 16-4. A small set of reed contacts are connected to the gate of the triac. The control circuit is connected to the coil of the reed relay. When the control voltage causes a current to flow through the coil, a magnetic field is produced around the coil of the relay. This magnetic field closes the reed contacts, causing the triac to turn on. In this type of solid-state relay a magnetic field, rather than a light beam, is used to isolate the control circuit from the load circuit.

The control voltage for most solid-state relays ranges from about 3 to 32 volts and can be dc or ac. If a triac is used as the control device, load voltage ratings of 120 to 240 volts ac are common and current ratings can range from 5 to 25 amps. Many solid-state relays have a feature known as *zero switching*. Zero switching means that if the relay is told to turn off when the ac voltage is in the middle of a cycle, it will continue to conduct until the ac voltage drops to a zero level, and will then turn off. For example, assume that the ac voltage is at its positive peak value when the gate tells the triac to turn off. The triac will continue to conduct until the ac voltage drops to a zero level before it actually turns off. Zero switching can be a great advantage when used with some inductive loads such as transformers. The core material of a transformer can be left saturated on one end of the flux swing if power is removed from the primary winding when the ac voltage is at its positive or

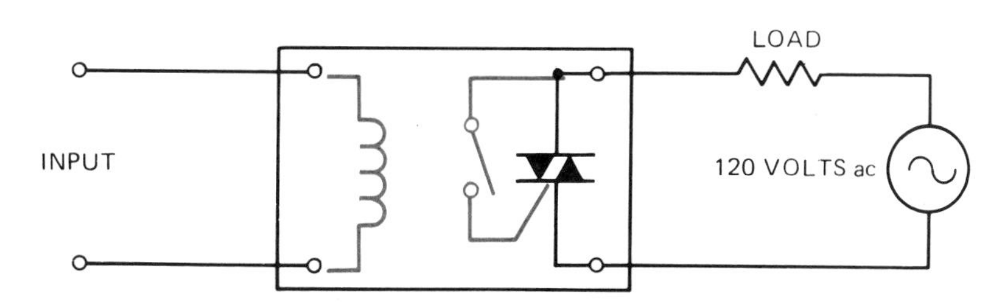

FIGURE 16-4 Reed relay controls the output of a solid-state relay

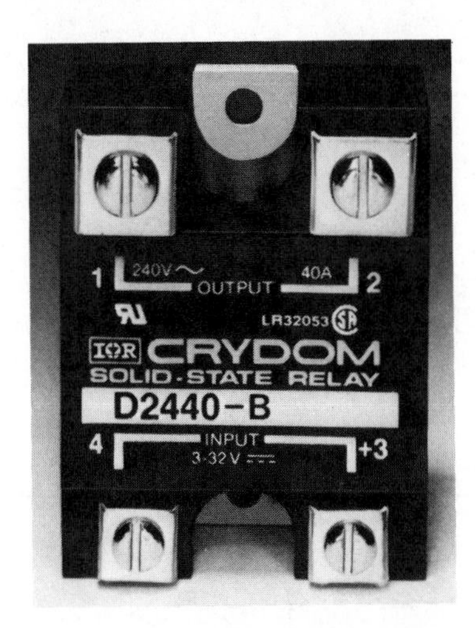

FIGURE 16-5 Solid-state relay (Courtesy International Rectifier)

negative peak. This can cause inrush currents of up to 600% of the normal operating current when power is restored to the primary winding.

Solid-state relays are available in different case styles and power ratings. Figure 16-5 shows a typical solid-state relay. Some solid-state relays are designed to be used as time delay relays. One of the most common uses for the solid-state relay is the I/O (eye-oh) track of a programmable controller, which will be covered in a later unit.

REVIEW QUESTIONS

1. What electronic component is used to control the output of a solid-state relay used to control a dc voltage?
2. What electronic component is used to control the output of a solid-state relay used to control an ac voltage?
3. Explain opto-isolation.
4. Explain magnetic isolation.
5. What is meant by zero switching?

UNIT 17

Timing Relays

Objectives *After studying this unit, the student will be able to:*

- **Identify the primary types of timing relays**
- **Explain the basic steps in the operation of the common timing relays**
- **List the factors that affect the selection of a timing relay for a particular use**
- **List applications of several types of timing relays**
- **Draw simple circuit diagrams using timing relays**
- **Identify *on-* and *off-*delay timing wiring symbols**

Time delay relays can be divided into two general classifications: the on-delay relay, and the off-delay relay. The on-delay relay is often referred to as DOE which stands for "Delay On Energize." The off-delay relay is often referred to as DODE which stands for "Delay On De-Energize."

Timer relays are similar to other control relays in that they use a coil to control the operation of some number of contacts. The difference between a control relay and a timer relay is that the contacts of the timer relay delay changing their position when the coil is energized or de-energized. When power is connected to the coil of an on-delay timer, the contacts delay changing position for some period of time. For this example assume that the timer has been set for a delay of 10 seconds. Also assume that the contact is normally open. When voltage is connected to the coil of the on-delay timer, the contacts will remain in the open position for 10 seconds and then close. When voltage is removed and the coil is de-energized, the contact will immediately change back to its normally open position. The contact symbols for an on-delay relay are shown in figure 17-1.

The operation of the off-delay timer is the opposite of the operation of the on-delay timer. For this example, again assume that the timer has been set for a delay of 10 seconds, and also assume that the contact is normally open. When voltage is applied to the coil of the off-delay timer, the contact will change immediately from open to closed. When the coil is de-energized, however, the contact will remain in the closed position for 10 seconds before it reopens. The contact symbols for an off-delay relay are shown in figure 17-2. Time-delay relays can have normally open, normally closed, or a combination of normally open and normally closed contacts.

Although the contact symbols shown in figures 17-1 and 17-2 are standard NEMA symbols

FIGURE 17-1 On-delay normally open and normally closed contacts

FIGURE 17-2 Off-delay normally open and normally closed contacts

FIGURE 17-3 Time closing contact

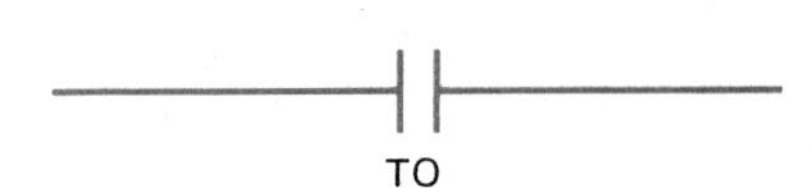

FIGURE 17-4 Time opening contact

for on-delay and off-delay contacts, some control schematics may use a different method of indicating timed contacts. The abbreviations TO and TC are used with some control schematics to indicate a time-operated contact. *TO stands for time opening, and TC stands for time closing.* If these abbreviations are used with standard contact symbols, their meaning can be confusing. Figure 17-3 shows a standard normally open contact symbol with the abbreviation TC written beneath it. This contact must be connected to an on-delay relay if it is to be time delayed when closing. Figure 17-4 shows the same contact with the abbreviation TO beneath it. If this contact is to be time delayed when opening, it must be operated by an off-delay timer. These abbreviations can also be used with standard NEMA symbols as shown in figure 17-5.

FIGURE 17-5 Contact A is an on-delay contact with the abbreviation NOTC (normally open time closing). Contact B is an off-delay contact with the abbreviation NOTO (normally open time opening).

DASHPOT TIMERS

Although timers are divided into two basic classifications, the on-delay and the off-delay, several methods are used to obtain these time delays. One of these is the dashpot timer shown in figure 17-6. This timer operates by forcing a fluid to flow through orifices in a piston. The operation of a dashpot timer is the same as the operation of the dashpot overload discussed in Unit 13. The only real difference between the dashpot timer and the dashpot overload relay is the type of coil used. The dashpot timer uses a voltage operated coil and the dashpot overload relay uses a current operated coil.

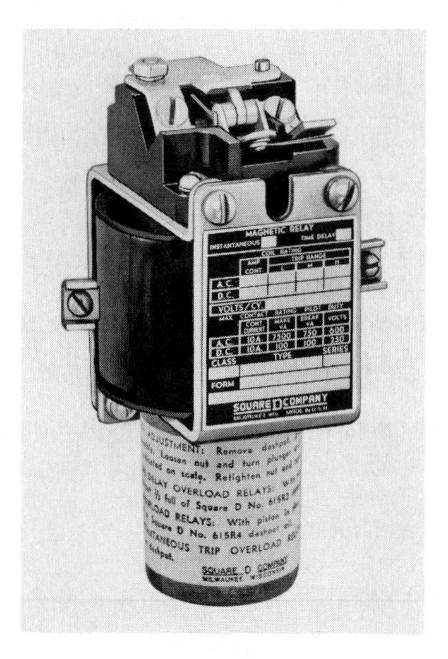

FIGURE 17-6 Fluid dashpot timing relay (Courtesy Square D Co.)

PNEUMATIC TIMERS

Pneumatic, or air timers, operate by restricting the flow of air through an orifice to a rubber

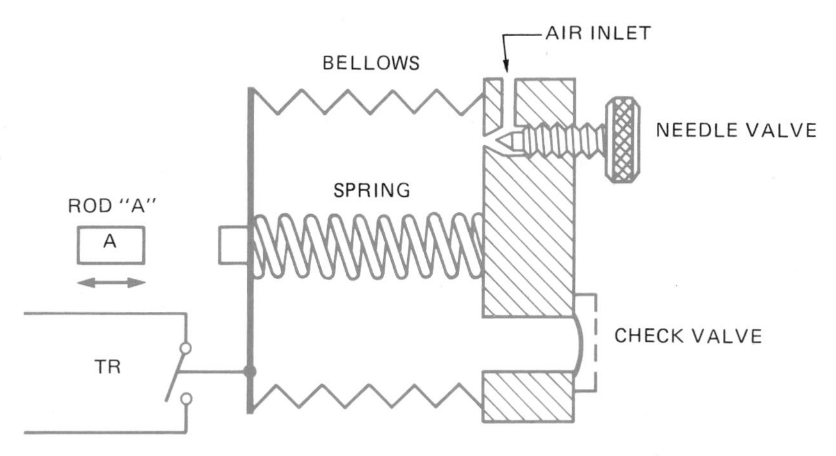

FIGURE 17-7 Bellows-operated pneumatic timer

bellows or diaphragm. Figure 17-7 illustrates the principle of operation of a simple bellows timer. If rod "A" pushes against the end of the bellows, air is forced out of the bellows through the check valve as the bellows contracts. When the bellows is moved back, contact TR changes from an open to a closed contact. When rod "A" is pulled away from the bellows, the spring tries to return the bellows to its original position. Before the bellows can be returned to its original position, however, air must enter the bellows through the air inlet port. The rate at which the air is permitted to enter the bellows is controlled by the needle valve. When the bellows returns to its original position, contact TR returns to its normally open position.

Pneumatic timers are popular throughout industry because they have the following characteristics:

A. They are unaffected by variations in ambient temperature or atmospheric pressure.

B. They are adjustable over a wide range of time periods.

C. They have good repeat accuracy.

D. They are available with a variety of contact and timing arrangements.

Some pneumatic timers are designed to permit the timer to be changed from on-delay to off-delay, and the contact arrangement to be changed to normally opened or normally closed, figure 17-8. This type of flexibility is another reason for the popularity of pneumatic timers.

Many timers are made with contacts that operate with the coil as well as time delayed contacts. When these contacts are used, they are generally referred to as *instantaneous contacts* and indicated on a schematic diagram by the abbreviation, inst., printed below the contact, figure 17-9. These instantaneous contacts change their positions immediately when the coil is energized and change back to their normal positions immediately when the coil is de-energized.

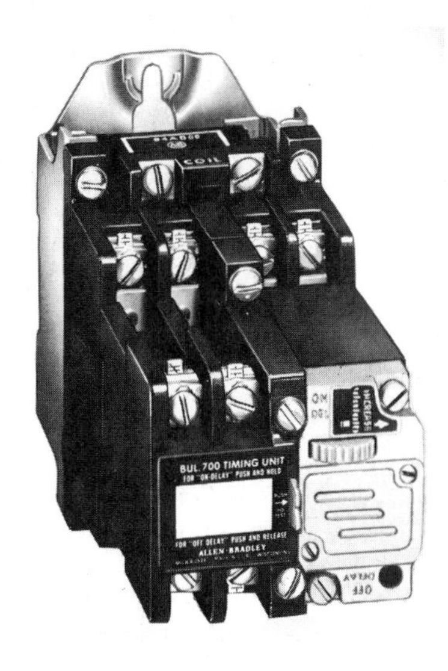

FIGURE 17-8 Pneumatic timer (Courtesy Allen-Bradley Co.)

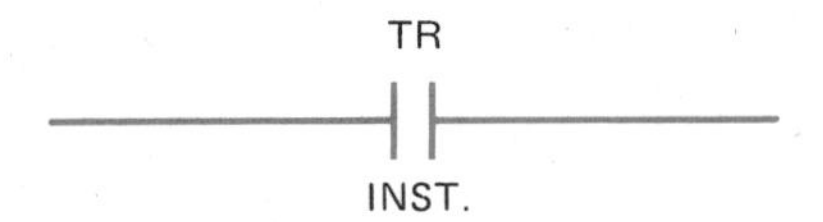

FIGURE 17-9 Normally open instantaneous contact of a timer relay

FIGURE 17-10 Clock driven timer (Courtesy Eagle Signal Controls)

CLOCK TIMERS

Another timer frequently used is the clock timer, figure 17-10. Clock timers use a small ac synchronous motor similar to the motor found in a wall clock to provide the time measurement for the timer. The length of time of one clock timer may vary greatly from the length of time of another. For example, one timer may have a full range of 0 to 5 seconds and another timer may have a full range of 0 to 5 hours. The same type of timer motor could be used with both timers. The gear ratio connected to the motor would determine the full range of time for the timer. Some advantages of clock timers are:

A. They have extremely high repeat accuracy.

B. Readjustment of the time setting is simple and can be done quickly. Clock timers are generally used when the machine operator must make adjustments of the time length.

Min. Time Delay: 0.05 second

Max. Time Delay: 3 minutes

Minimum Reset Time: .075 second

Accuracy: ±10 percent of setting

Contact Ratings:

Ac

6.0 A, 115 V
3.0 A, 230 V
1.5 A, 460 V
1.2 A, 550 V

Dc

1.0 A, 115 V
0.25 A, 230 V

Operating Coils: Coils can be supplied for voltages and frequencies up to 600 volts, 60 hertz ac and 250 volts dc

Types of Contacts: One normally open and one normally closed. Cadmium silver alloy contacts.

FIGURE 17-11 Typical specifications

MOTOR-DRIVEN TIMERS

When a process has a definite on and off operation, or a sequence of successive operations, a motor-driven timer is generally used, figures 17-11 and 17-12. A typical application of a motor-driven timer is to control laundry washers where the loaded motor is run for a given period in one direction, reversed, and then run in the opposite direction.

Generally, this type of timer consists of a small, synchronous motor driving a cam-dial assembly on a common shaft. A motor-driven timer successively closes and opens switch contacts which are wired in circuits to energize control relays or contactors to achieve desired operations.

CAPACITOR TIME LIMIT RELAY

Assume that a capacitor is charged by connecting it momentarily across a dc line and then the capacitor direct current is discharged through a relay coil. The current induced in the coil will decay slowly, depending on the relative values of

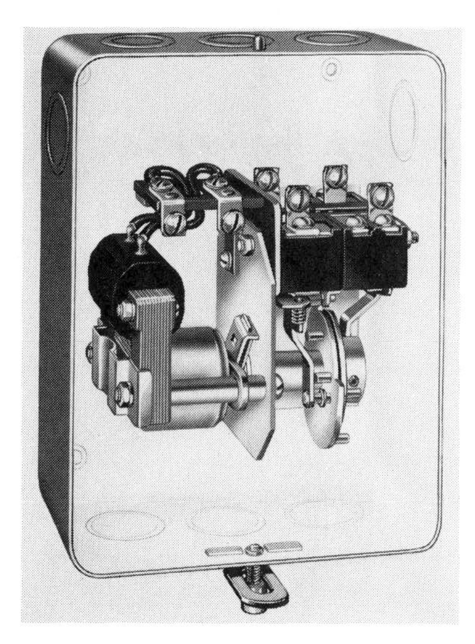

FIGURE 17-12 Motor-driven process timer in a general-purpose enclosure (Courtesy Allen-Bradley Co.)

capacitance, inductance, and resistance in the discharge circuit.

If a relay coil and a capacitor are connected parallel to a dc line, figure 17-13, the capacitor is charged to the value of the line voltage and a current appears in the coil. If the coil and capacitor combination is now removed from the line, the current in the coil will start to decrease along the curve shown in figure 17-13.

If the relay is adjusted so that the armature is released at current i_1, a time delay of t_1 is obtained. The time delay can be increased to a value of t_2 by adjusting the relay so that the armature will not be released until the current is reduced to a value of i_2. Figure 17-14 shows a relay used for this type of time control.

A potentiometer is used as an adjustable resistor to vary the time. This resistance-capacitance (RC) theory is used in industrial electronic and solid-state controls also. This timer is highly accurate and is used in motor acceleration control and in many industrial processes.

ELECTRONIC TIMERS

Electronic timers use solid-state components to provide the time delay desired. Some of these timers use an RC time constant to obtain the time base and others use quartz clocks as the time base, figure 17-15. RC time constants are inexpensive and have good repeat times. The quartz timers, however, are extremely accurate and can often be set for .1 second times. These timers are generally housed in a plastic case and are designed to be plugged into some type of socket. An electronic timer that is designed to be plugged into a standard eight-pin tube socket is shown in figure 17-16. The length of the time delay can be set by adjusting the control knob shown on top of the timer.

The schematic for a simple on-delay timer is shown in figure 17-17. The timer operates as follows: When switch S1 is closed, current flows through resistor RT and begins charging capacitor

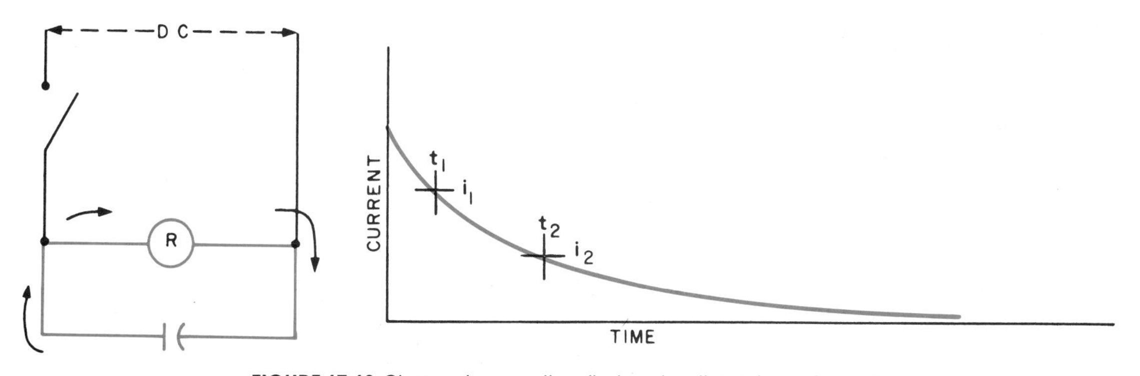

FIGURE 17-13 Charged capacitor discharging through a relay coil. The graph at the right illustrates the current decrease in the coil.

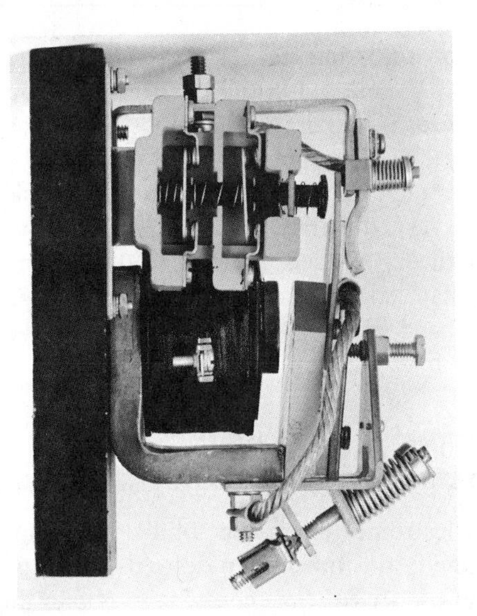

FIGURE 17-14 Time-delay contactor (Courtesy General Electric Co.)

FIGURE 17-15 Digital clock timer (Courtesy Eagle Signal Controls)

C1. When capacitor C1 has been charged to the trigger value of the unijunction transistor, the UJT turns on and discharges capacitor C1 through resistor R2 to ground. The sudden discharge of capacitor C1 causes a spike voltage to appear

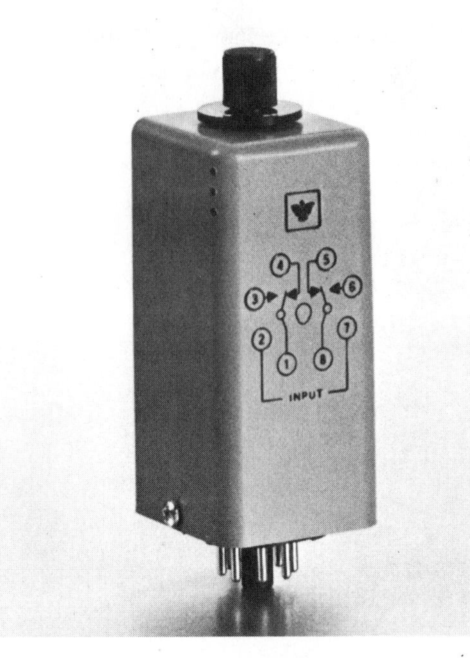

FIGURE 17-16 Electronic timer (Courtesy Eagle Signal Controls)

across resistor R2. This voltage spike travels through capacitor C2 and fires the gate of the SCR. When the SCR turns on, current is provided to the coil of relay K1.

Resistor R1 limits the current flow through the UJT. Resistor R3 is used to keep the SCR turned off until the UJT provides the pulse to fire the gate. Diode D1 is used to protect the circuit from the spike voltage produced by the collapsing magnetic field around coil K1 when the current is turned off.

By adjusting resistor RT, capacitor C1 can be charged at different rates. In this manner, the relay can be adjusted for time. Once the SCR has turned on, it will remain on until switch S1 is opened.

Programmable controllers, which will be discussed in Unit 64, contain "internal" electronic timers. Most programmable controllers (PCs) use a quartz-operated clock as the time base. When the controller is programmed, the timers can be set in time increments of .1 second. This, of course, provides very accurate time delays for the controller.

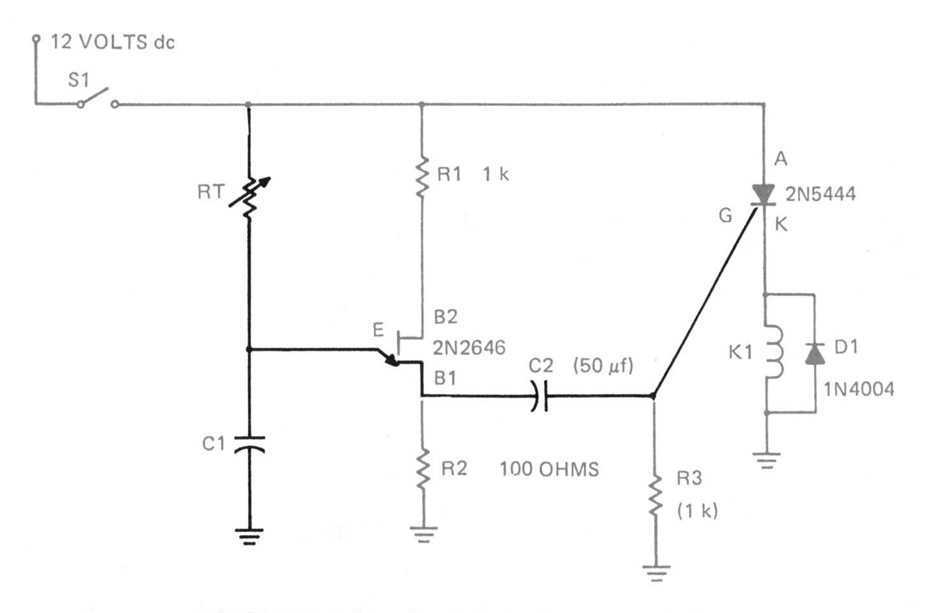

FIGURE 17-17 Schematic of electronic on-delay timer

REVIEW QUESTIONS

1. What are the two basic classifications of timers?
2. Explain the operation of an on-delay relay.
3. Explain the operation of an off-delay relay.
4. What are instantaneous contacts?
5. Dashpot timers differ from dashpot overloads in that the timer has a ________ operated coil and the overload has a ________________ operated coil.
6. How are pneumatic timers adjusted?
7. Name two methods used by electronic timers to obtain their time base.

UNIT 18

Pressure Switches and Regulators

Objectives ***After studying this unit, the student will be able to:***

- **Describe how pressure switches, vacuum switches, and pressure regulators may control motors**
- **List the adjustments that can be made to pressure switches**
- **Identify wiring symbols used for pressure switches**

Any industrial application that has a pressure sending requirement can use a pressure switch, figure 18-1. A large variety of pressure switches are available to cover the wide range of control requirements for pneumatic or hydraulic machines such as welding equipment, machine tools, high pressure lubricating systems, and motor-driven pumps and air compressors.

The pressure ranges over which pressure switches can maintain control also vary widely. For example, a diaphragm-actuated switch can be used when a sensitive response is required to small pressure changes at low-pressure ranges. A metal bellows-actuated control is used for pressures up to 2000 pounds per square inch. Piston-operated hydraulic switches are suitable for pressures up to 15,000 psi. In all of these pressure controlled devices, a set of contacts is operated.

The most commonly used pressure switches are single-pole switches. Two-pole switches are also used for some applications. Field adjustments of the range and the differential pressure (or the difference between the cut-in and cut-out pressures) can be made for most pressure switches. The spring pressure determines the pressures at which the switch closes and opens its contacts.

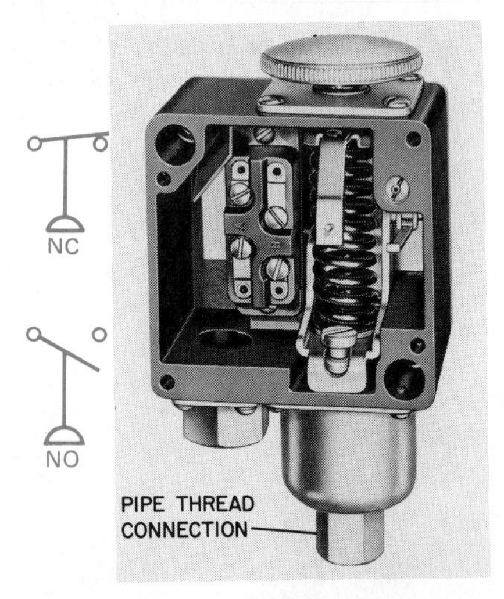

FIGURE 18-1 Industrial pressure switch with cover removed. Note operating knob. Also note wiring diagram symbols for normally closed and normally open contacts. (Courtesy Square D Co.)

Pressure regulators provide accurate control of pressure or vacuum conditions for systems. When they are used as pilot control devices with magnetic starters, pressure regulators are able to control the operation of liquid pump or air compressor motors in a manner similar to that of pressure switches. Reverse action regulators can be used on pressure system interlocks to prevent the start of an operation until the pressure in the system has reached the desired level.

Pressure regulators consist of a Bourdon-type pressure gauge and a control relay. Delicate contacts on the gauge energize the relay and cause it to open or close. The relay contacts are used to control a large motor starter in order to avoid damage or burning of the gauge contacts. Standard regulators will open a circuit at high pressure and close it at low pressure. Special reverse operation regulators will close the circuit at high pressure and open the circuit at low pressure.

PRESSURE SENSORS

Pressure sensors are designed to produce an output voltage or current that is dependent on the amount of pressure being sensed. Piezoresistive sensors are very popular because of their small size, reliability, and accuracy, figure 18-2. These sensors are available in ranges from 0–1 psi (pound per square inch) and 0–30 psi. The sensing element is a silicon diaphragm integrated with an integrated circuit chip. The chip contains four implanted piezoresistors connected to form a bridge circuit, figure 18-3. When pressure is applied to the diaphragm, the resistance of piezoresistors change proportionally to the applied pressure, which changes the balance of the bridge. The voltage across V0 changes in proportion to the applied pressure (V0 = V4 − V2 [when referenced to V3]) Typical millivolt outputs and pressures are shown below:

1 psi = 44 mV

5 psi = 115 mV

15 psi = 225 mV

30 psi = 315 mV

FIGURE 18-2 Piezoresistive pressure sensor (Courtesy Micro Switch, a Honeywell Division)

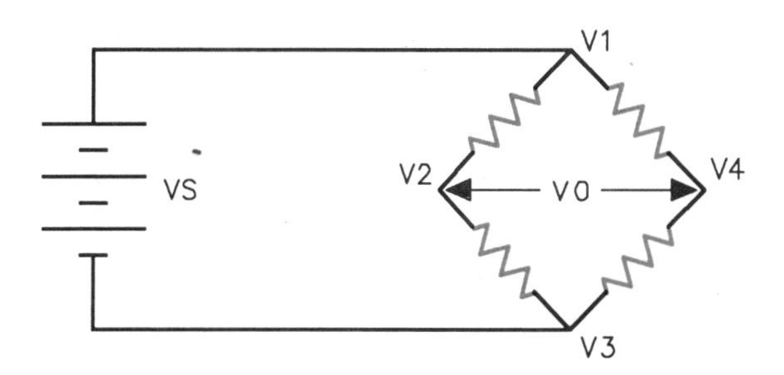

FIGURE 18-3 Piezoresistor bridge

Another type of piezoresistive sensor is shown in figure 18-4. This particular sensor can be used to sense absolute, gage, or differential pressure. Units are available which can be used to sense vacuum. Sensors of this type can be obtained to sense pressure ranges of 0–1, 0–2, 0–5, 0–15, 0–30, and 0–(− 15[vacuum]). The sensor contains an internal operational amplifier and can provide an output voltage proportional to the pressure. Typical supply voltage for this unit is 8 VDC. The *regulated* voltage output for this unit is 1–6 volts. Assume for example that the sensor is intended to sense a pressure range of 0–15 psi. At 0 psi the sensor would produce an output voltage of 1 volt. At 15 psi the sensor would produce an output voltage of 6 volts.

Sensors can also be obtained that have a ratiometric output. The term ratiometric means that the output voltage will be proportional to the supply voltage. Assume that the supply voltage increases by 50% to 12 VDC. The output voltage would increase by 50% also. The sensor would now produce a voltage of 1.5 volts at 0 psi and 9 volts at 15 psi.

Other sensors can be obtained that produce a current output of 4 to 20 mA, instead of a regulated voltage output, figure 18-5. One type of pressure to current sensor, which can be used to sense pressures as high as 250 psi, is shown in figure 18-6. This sensor can also be used as a set point detector to provide a normally open or normally closed output. Sensors that produce a proportional output current instead of voltage have fewer problems with induced noise from surrounding magnetic fields, and with voltage drops due to long wire runs.

FIGURE 18-5 Pressure to current sensor for low pressures (Courtesy Micro Switch, a Honeywell Division)

FIGURE 18-4 Differential pressure sensor (Courtesy Micro Switch, a Honeywell Division)

FIGURE 18-6 Pressure to current sensor for high pressures (Courtesy Micro Switch, a Honeywell Division)

REVIEW QUESTIONS

1. Describe how pressure switches are connected to start and stop (a) small motors and (b) large motors.
2. Draw the schematic symbol for a pressure switch. Draw both the normally open and the normally closed contact.

UNIT 19

Float Switches

Objectives

After studying this unit, the student will be able to:

- **Describe the operation of float switches**
- **List the sequence of operation for sump pumping or tank filling**
- **Draw wiring symbols for float switches**

A float switch is used when a pump motor must be started and stopped according to changes in the water (or other liquid) level in a tank or sump. Float switches are designed to provide automatic control of ac and dc pump motor magnetic starters and automatic direct control of light motor loads.

The operation of a float switch is controlled by the upward or downward movement of a float placed in a water tank. The float movement causes a rod-operated, figure 19-1, or chain and counterweight, figure 19-2, assembly to open or close electrical contacts. The float switch contacts may be either normally open or normally closed and may not be submerged. Float switches may be connected to a pump motor for tank or sump pumping operations or tank filling, depending on the contact arrangement.

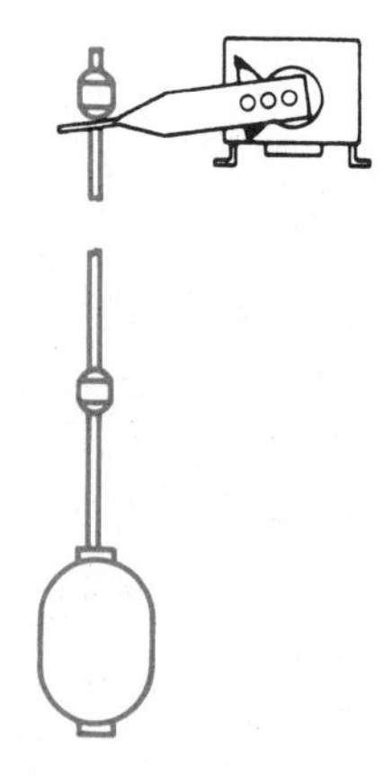

FIGURE 19-1 Rod-operated float switch

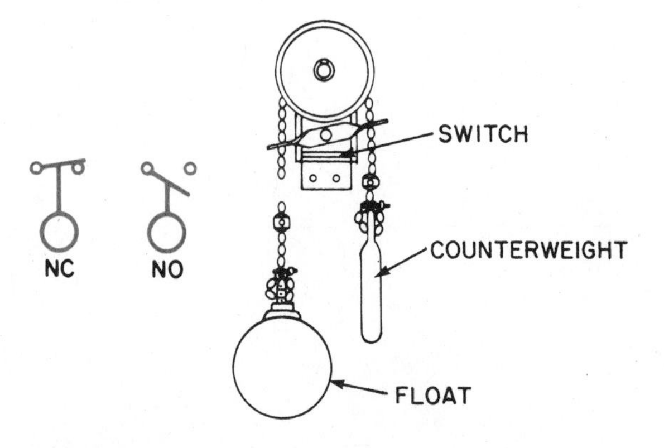

FIGURE 19-2 Chain-operated float switch with normally closed (NC) and normally open (NO) wiring symbols

REVIEW QUESTIONS

1. Describe the sequence of operations required to (a) pump sumps and (b) fill tanks.
2. Draw the normally open and normally closed contact symbols for a float switch.

Flow Switches and Sensors

Objectives *After studying this unit, the student will be able to:*

- **Describe the purpose and functions of flow switches**
- **Connect a flow switch to other electrical devices**
- **Draw and read wiring diagrams of systems using flow switches**

A flow switch is a device that can be inserted in a pipe so that when liquid or air flows against a part of the device called a paddle, a switch is activated (figure 20-1). This switch either closes or opens a set of electrical contacts. The contacts may be connected to energize motor starter coils, relays, or indicating lights. In general, a flow switch contains both normally open and normally closed electrical contacts, figure 20-2.

Figure 20-3 shows a flow switch installed in a pipe line tee. Half couplings are welded into larger pipes for flow switch installations.

Typical applications of flow switches are shown in figures 20-4 through 20-7. These applications are commonly found in the chemical and petroleum industries. Vaporproof electrical connections must be used with vaporproof switches. The insulation of the wire leading to the switches must be adequate to withstand the high tempera-

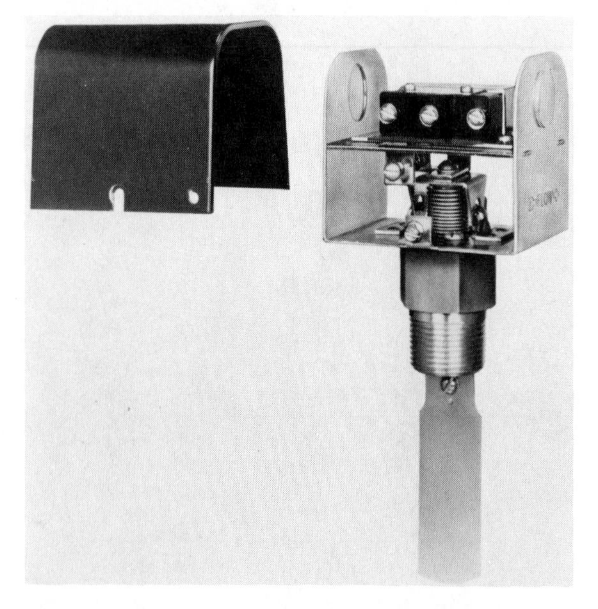

FIGURE 20-1 Flow switch (Courtesy McDonnell & Miller ITT, Fluid Handling Division)

TOP VIEW OF SWITCH

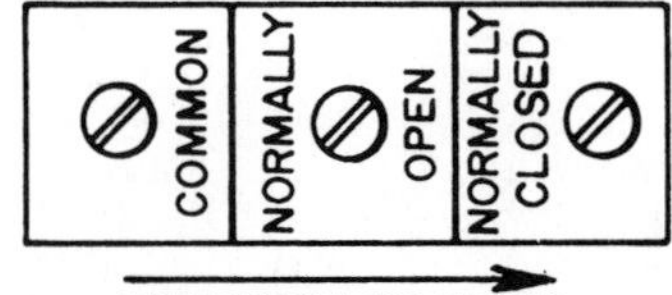

DIRECTION OF FLOW

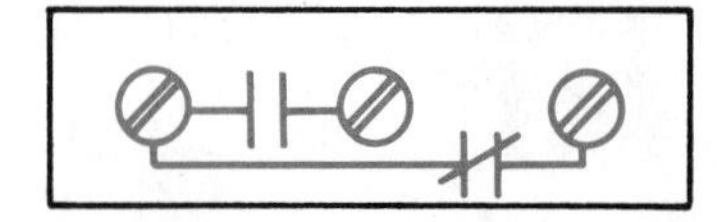

NO FLOW CONTACT POSITION

FIGURE 20-2 Electrical terminals and contact arrangement of a flow switch

ture of the liquid inside the pipe. (Consult the National Electric Code for insulation temperature ratings.)

Airflow or sail switches are also used in ducts in air conditioning systems. Another use of these switches is to prevent duct heaters from energizing when there is no air movement in the duct. While the construction of airflow switches is different from that of liquid flow switches, the electrical connections are similar.

The schematic symbols for a flow switch are shown in figure 20-8.

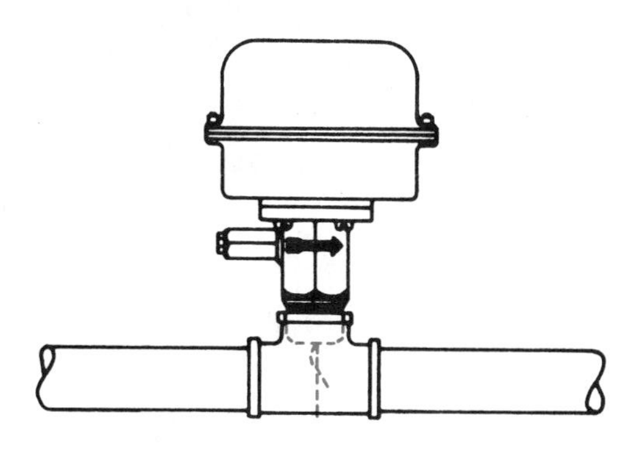
FIGURE 20-3 Flow switch installed

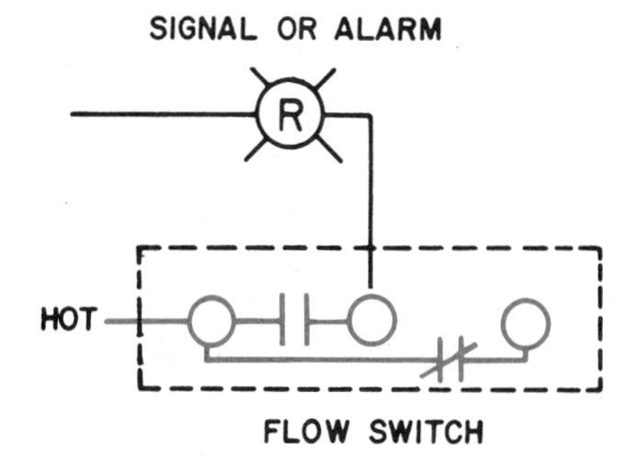

FIGURE 20-4 Flow switch used to sound alarm or light signal when flow occurs.

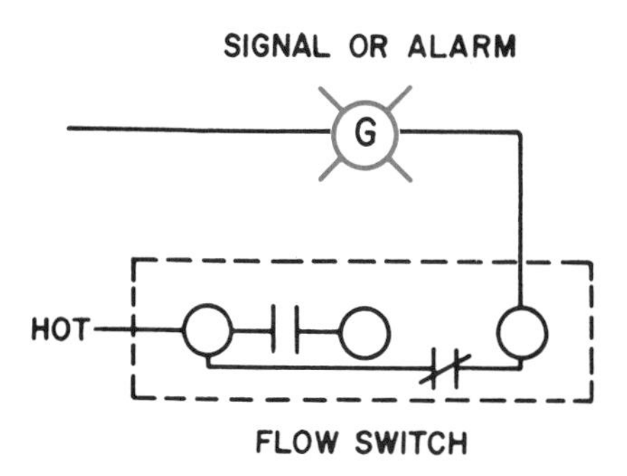

FIGURE 20-5 Flow switch used to sound alarm or light signal when there is no flow

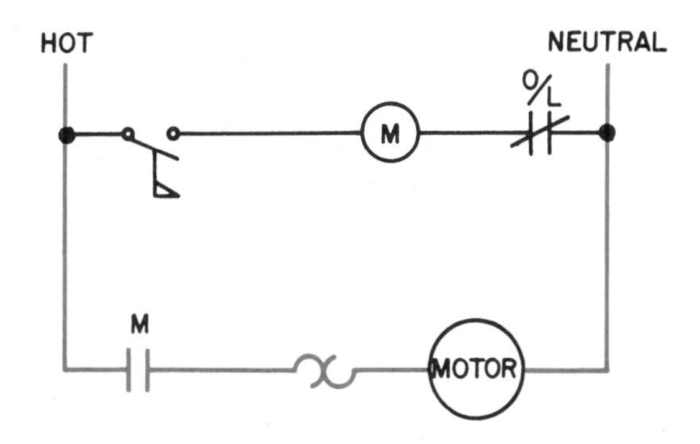

FIGURE 20-6 Flow switch used with single-phase circuit; starts motor when flow occurs, stops motor when there is no flow.

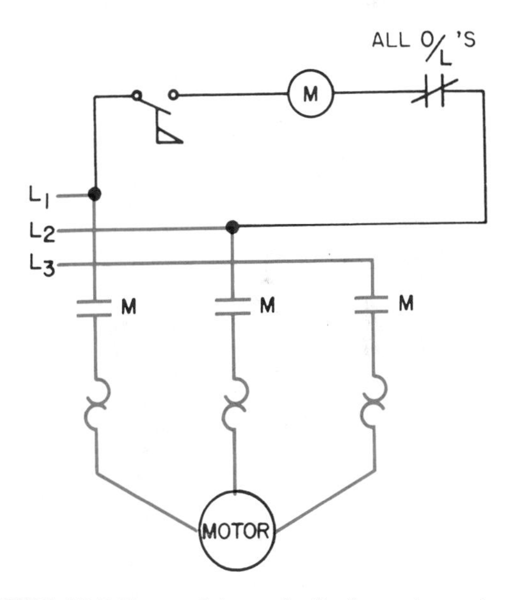

FIGURE 20-7 Flow switch used with three-phase circuit; starts motor when flow occurs, stops motor when there is no flow.

FIGURE 20-8 Schematic symbols for flow switches

FLOW SENSORS

Flow switches are used to detect liquid flowing through a pipe or air flowing through a duct. Flow switches, however, can not detect the amount of liquid or air flow. To detect the amount of liquid or air flow, a *transducer* must be used. A transducer is a device that converts one form of energy into another. In this case, the kinetic energy of a moving liquid or gas is converted into electrical energy. Many flow sensors are designed to produce an output current of 4 to 20 mA. This current can be used as the input signal to a programmable controller or as the input to a meter designed to measure the flow rate of the liquid or gas being metered, figure 20-9.

Liquid Flow Sensors

There are several methods that can be used to measure the flow rate of a liquid in a pipe. One method uses a *turbine* type sensor, figure 20-10.

FIGURE 20-9 Meter used to measure the flow rate of a liquid (Courtesy Sparling Instruments Co. Inc.)

FIGURE 20-10 Turbine used to measure liquid flow (Courtesy Sparling Instruments Co. Inc.)

The turbine sensor consists of a turbine blade which must be inserted inside the pipe containing the liquid. The moving liquid causes the turbine blade to turn. The speed at which the blade turns is proportional to the amount of flow in the pipe. The sensor's electrical output is determined by the speed of the turbine blade. One disadvantage of the turbine type sensor is that the turbine blade offers some resistance to the flow of the liquid.

Electromagnetic Flow Sensors

Another type of flow sensor is the *electromagnetic* flow sensor. These sensors operate on the principle of Faraday's Law concerning conductors moving through a magnetic field. This law states that when a conductor moves through a magnetic field, a voltage will be induced into the conductor. The amount of induced voltage is proportional to the strength of the magnetic field and the speed of the moving conductor. In the case of the electromagnetic flow sensor, the moving liquid is the conductor. As a general rule, liquids should have a minimum conductivity of about 20 micromhos per centimeter.

Flow rate is measured by small electrodes mounted inside the pipe of the sensor. The electrodes measure the amount of voltage induced in the liquid as it flows through the magnetic field produced by the sensor. Since the strength of the magnetic field is known, the induced voltage will be proportional to the flow rate of the liquid. Several different designs of electromagnetic flow sensors are shown in figures 20-11A, 20-11B, and 20-11C.

FIGURE 20-11A Electromagnetic flow sensor and meter (Courtesy Sparling Instruments Co. Inc.)

FIGURE 20-11B Electromagnetic flow sensors with transducer for producing a 4 to 20 milliampere current (Courtesy Sparling Instruments Co. Inc.)

FIGURE 20-11C Electromagnetic flow sensor with ceramic liner. Sensor is shown with meter. (Courtesy Sparling Instruments Co. Inc.)

Airflow Sensors

Large volumes of air flow can be sensed by prop-driven devices similar to the liquid flow sensor shown in figure 20-10. Solid state devices similar to the one shown in figure 20-12 are com-

FIGURE 20-12 Solid state airflow sensor (Courtesy Micro Switch, a Honeywell Division)

monly used to sense smaller amounts of air or gas flow. This device operates on the principle that air or gas flowing across a surface causes heat transfer. The sensor contains a thin film thermally isolated bridge with a heater and temperature sensors. The output voltage is dependent on the temperature of the sensor surface. Increased air flow through the inlet and outlet ports will cause a greater amount of heat transfer, reducing the surface temperature of the sensor.

REVIEW QUESTIONS

1. What are typical uses of flow switches?
2. Draw a line diagram to show that a green light will glow when liquid flow occurs.
3. Draw a one-line diagram showing a bell that will ring in the absence of flow. Include a switch to turn off the bell manually.
4. What is a transducer?
5. What is the most common output current for flow sensors?
6. What is Faraday's Law concerning conductors moving through a magnetic field?
7. What type of flow sensors used Faraday's Law as their principle of operation?
8. What is the operating principle of the solid state airflow sensor described in this text?

UNIT 21

Limit Switches

Objectives ***After studying this unit, the student will be able to:***

- **Explain the use of limit switches in the automatic operation of machines and machine tools**
- **Wire a simple two-wire circuit using a limit switch**
- **Read and draw normally open (NO) and normally closed (NC) wiring symbols**

The automatic operation of machinery requires the use of switches that can be activated by the motion of the machinery. The repeat accuracy of the switches must be reliable and the response virtually instantaneous.

The size, operating force, stroke, and manner of mounting are all critical factors in the installation of limit switches due to mechanical limitations in the machinery. The electrical ratings of the switches must be carefully matched to the loads to be controlled.

In general, the operation of a limit switch begins when the moving machine or moving part of a machine strikes an operating lever which actuates the switch, figure 21-1. The limit switch, in turn, affects the electrical circuit controlling the machine and its movement.

Limit switches are used as pilot devices in the control circuits of magnetic starters to start, stop, speed up, slow down or reverse electric motors. Limit switches may be used either as control devices for regular operation or as emergency switches to prevent the improper functioning of machinery. They may be momentary contact (spring return) or maintained contact types.

Limit switch contacts are often drawn differently than the symbols shown in figure 21-1 on control schematics. The contact symbols shown in figure 21-1 are the standard NEMA symbols for normally open and normally closed limit switch contacts. The contact symbol shown in figure 21-2A shows a limit switch that is "normally open held closed." This means the contact is wired as a normally open contact, but when the circuit is in

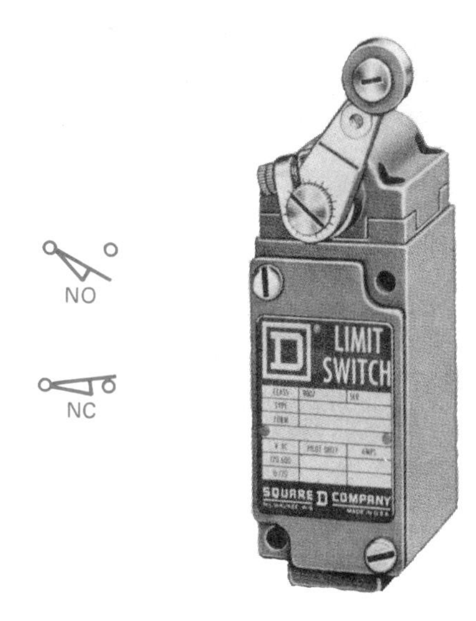

FIGURE 21-1 Limit switch shown with wiring symbols (Courtesy Square D Co.)

NORMALLY OPEN
HELD CLOSED

A

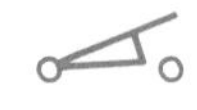

NORMALLY CLOSED
HELD OPEN

B

FIGURE 21-2 Other limit switch symbols

its normal off state, some part of the machine holds the contact closed. This symbol can be recognized as normally open because the movable contact arm is shown below the stationary contact point.

Figure 21-2B shows a limit switch wired "normally closed held open." The contact symbol is normally closed because the movable contact arm is drawn above the stationary contact point, but some part of the machine is holding the contact open.

Other contacts such as pressure switches, float switches, and flow switches can also be connected in this manner. The electrician is more likely to encounter limit switches used in this manner, however, because of the number of limit switches used in industry and the manner in which they are used.

FIGURE 21-3 Micro limit switch (Courtesy Micro Switch, a Honeywell Division)

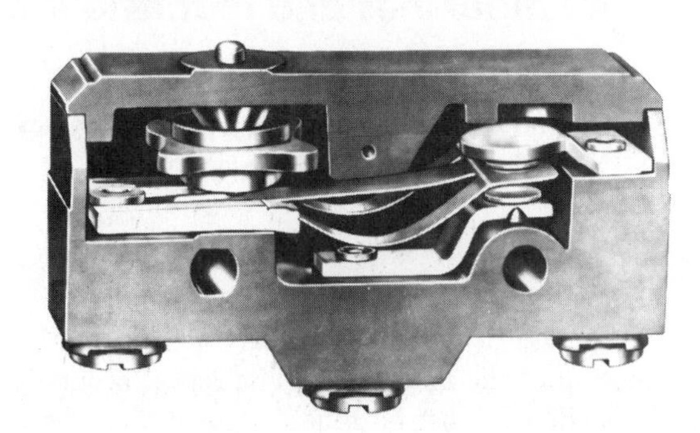

FIGURE 21-4 Spring loaded contacts of a basic micro switch (Courtesy Micro Switch, a Honeywell Division)

MICRO LIMIT SWITCHES

Another type of limit switch often used in different types of control circuits is the micro limit switch or *micro switch.* Micro switches are much smaller in size than the limit switch shown in figure 21-1, which permits them to be used in small spaces that would never be accessible to the larger device. Another characteristic of the micro switch is that the actuating plunger requires only a small amount of travel to cause the contacts to change position. The micro switch shown in figure 21-3 has an activating plunger located at the top of the switch. This switch requires that the plunger be depressed approximately 0.015 inch or 0.38 mm. Switching the contact position with this small amount of movement is accomplished by spring loading the contacts as shown in figure 21-4. A small amount of movement against the spring will cause the movable contact to snap from one position to another.

Electrical ratings for the contacts of the basic micro switch are generally in the range of 250 volts ac and 10 to 15 amps depending on the type of switch. The basic micro switch can be obtained with a variety of different activating arms as shown in figure 21-5.

SUBMINIATURE MICRO SWITCHES

The *Subminiature micro switch* employs a similar spring contact arrangement as the basic micro switch, figure 21-6. The subminiature switches are approximately ½ to ¼ the size of the basic switch, depending on the model. Due to their reduced size, the contact rating of subminiature switches

FIGURE 21-5 Micro switches can be obtained with different types of activating arms (Courtesy Micro Switch, a Honeywell Division)

FIGURE 21-6 Subminiature micro switches employ a similar set of spring loaded contacts (Courtesy Micro Switch, a Honeywell Division)

range from about 1 ampere to about 7 amperes depending on the switch type. Two different types of subminiature micro switches are shown in figure 21-7.

FIGURE 21-7 Subminiature micro switches (Courtesy Micro Switch, a Honeywell Division)

REVIEW QUESTIONS

1. Draw a simple circuit showing how a red pilot light is energized when a limit switch is operated by a moving object.
2. Draw the schematic symbol for a limit switch that is normally open, but held closed.

UNIT 22

Phase Failure Relays

Objectives ***After studying this unit, the student will be able to:***

- **Explain the purpose of phase failure relays**
- **List the hazards of phase failure and phase reversal**

If two phases of the supply to a three-phase induction motor are interchanged, the motor will reverse its direction of rotation. This action is called phase reversal. In the operation of elevators and in many industrial applications, phase reversal may result in serious damage to the equipment and injury to people using the equipment. In other situations, if a fuse blows or a wire to a motor breaks while the motor is running, the motor will continue to operate on single phase but will experience serious overheating. To protect motors against these conditions, phase failure and reversal relays are used.

One type of phase failure relay, figure 22-1, uses coils connected to two lines of the three-phase supply. The currents in these coils set up a rotating magnetic field that tends to turn a copper disc clockwise. This clockwise torque actually is the result of two torques. One polyphase torque tends to turn the disc clockwise, and one single-phase torque tends to turn the disc counterclockwise.

The disc is kept from turning in the clockwise direction by a projection resting against a stop. However, if the disc begins to rotate in the counterclockwise direction, the projecting arm will move a toggle mechanism to open the line contactors and remove the motor from the line. In other words, if one line is opened, the polyphase torque disappears and the remaining single-phase torque rotates the disc counterclockwise. As a result, the motor is removed from the line. In case of phase reversal, the polyphase torque helps the single-phase torque turn the disc counterclockwise, and again, the motor is disconnected from the line.

Other designs of phase failure and phase reversal relays are available to protect motors, machines, and personnel from the hazards of open

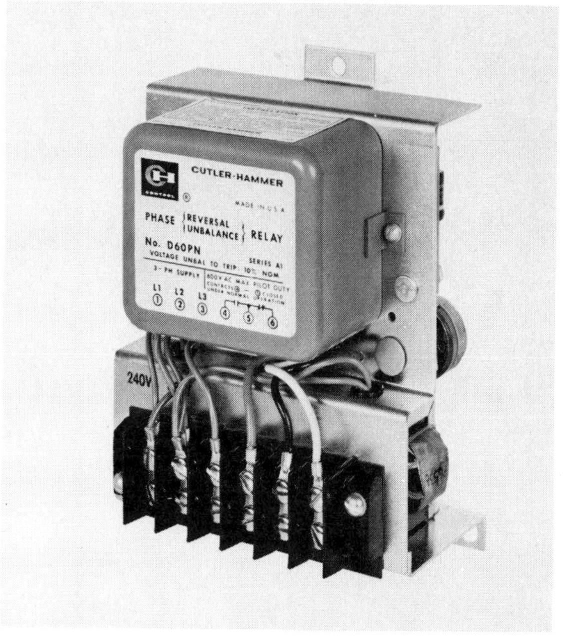

FIGURE 22-1 Phase monitoring relay (Courtesy EATON Corp., Cutler-Hammer Products)

phase or reverse phase conditions. For example, one type of relay consists of a static, current-sensitive network connected in series with the line and a switching relay connected in the coil circuit of the starter. The sensing network continuously monitors the motor line currents. If one phase opens, the sensing network immediately detects it and causes the relay to open the starter coil circuit to disconnect the motor from the line. A built-in delay of five cycles prevents nuisance dropouts caused by fluctuating line voltages.

A solid-state phase monitoring relay is shown in figure 22-2. This relay provides protection in the event of a voltage unbalance or a phase reversal. The unit automatically resets after the correct voltage conditions return. Indicating lights show when the relay is activated.

FIGURE 22-2 Solid-state phase monitor relay (Courtesy EATON Corp., Cutler-Hammer Products)

REVIEW QUESTIONS

1. What is the purpose of phase failure relays?
2. What are the hazards of phase failure and phase reversal?

UNIT 23

Solenoid Valves

Objectives *After studying this unit, the student will be able to:*

- **Describe the purpose and operation of two-way solenoid valves**
- **Describe the purpose and operation of four-way solenoid valves**
- **Connect and troubleshoot solenoid valves**
- **Read and draw wiring symbols for solenoid valves**

Valves are mechanical devices designed to control the flow of fluids such as oil, water, air, and other gases. Many valves are manually operated, but electrically operated valves are most often used in industry because they can be placed close to the devices they operate, thus minimizing the amount of piping required. Remote control is accomplished by running a single pair of control wires between the valve and a control device such as a manually operated switch or an automatic device.

A solenoid valve is a combination of two basic units: an assembly of the solenoid (the electromagnet) and plunger (the core), and a valve containing an opening in which a disc or plug is positioned to regulate the flow. The valve is opened or closed by the movement of the magnetic plunger. When the coil is energized, the plunger (core) is drawn into the solenoid (electromagnet). The valve operates when current is applied to the solenoid. The valve returns automatically to its original position when the current ceases.

Most control pilot devices operate a single-pole switch, contact, or solenoid coil. The wiring diagrams of these devices are not difficult to understand and the actual devices can be connected easily into systems. It is recommended that the electrician know the *purpose* of and understand the *action* of the *total* industrial system for which various electrical control elements are to be used. In this way, the electrician will find it easier to design or assist in designing the electrical control system. It will also be easier for the electrician to install and maintain the control system.

TWO-WAY SOLENOID VALVES

Two-way (in and out) solenoid valves, figure 23-1, are magnetically operated valves which are used to control the flow of Freon, methyl chloride, sulphur dioxide, and other liquids in refrigeration and air conditioning systems. These valves can also be used to control water, oil, and air flow.

Standard applications of solenoid valves generally require that the valve be mounted directly in line in the piping with the inlet and outlet connections directly opposite each other. Simplified valve mounting is possible with the use of a bottom outlet which eliminates elbows and bends. In the bottom outlet arrangement, the normal side outlet is closed with a standard pipe plug.

The valve body is usually a special brass forging which is carefully checked and tested to insure that there will be no seepage due to porosities. The armature, or plunger, is made from a high

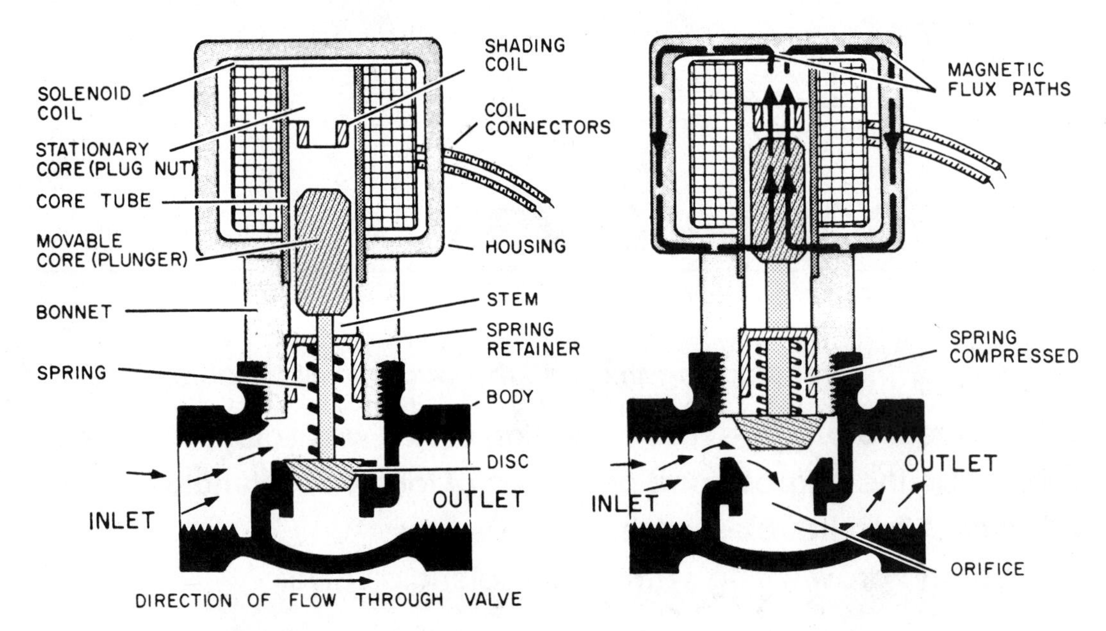

FIGURE 23-1 Two-way solenoid valve (Courtesy Automatic Switch Co.)

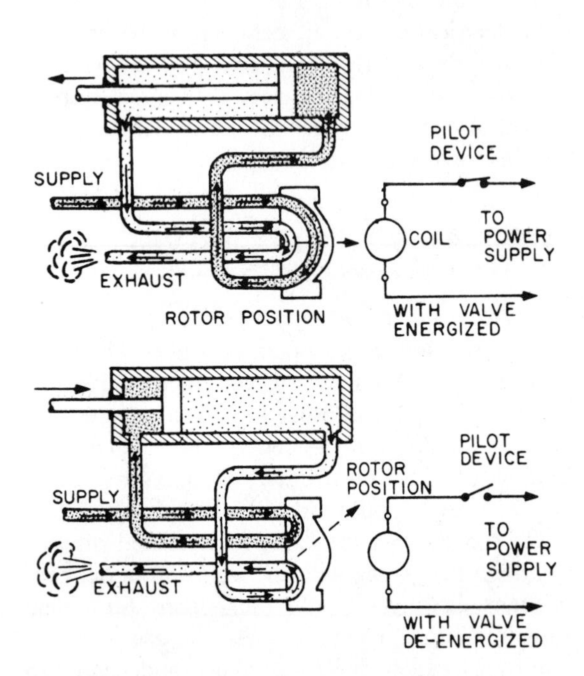

FIGURE 23-2 Control of double-acting cylinder by a four-way, electrically operated valve shown with elementary diagrams

grade stainless steel. The effects of residual magnetism are eliminated by the use of a kickoff pin and spring which prevent the armature from sticking. A shading coil insures that the armature will make a complete seal with the flat surface above it to eliminate noise and vibration.

It is possible to obtain dc coils with a special winding that will prevent the damage that normally results from an instantaneous voltage surge when the circuit is broken. Surge capacitors are not required with this type of coil.

To insure that the valve will always seat properly it is recommended that strainers be used to prevent grit or dirt from lodging in the orifice or valve seat. Dirt in these locations will cause leakage. The inlet and outlet connections of the valve must not be reversed. The tightness of the valve depends to a degree on pressure acting downward on the sealing disc. This pressure is possible only when the inlet is connected to the proper point as indicated on the valve.

FOUR-WAY SOLENOID VALVES

Electrically operated, four-port, four-way air valves are used to control a double-acting cylinder, figure 23-2.

When the coil is de-energized, one side of the piston is at atmospheric pressure, and the other side is acted upon by the line pressure. When the valve magnet coil is energized, the valve exhausts the high pressure side of the piston to atmospheric pressure. As a result, the piston and its associated load reciprocate in response to the valve movement.

Four-way valves are used extensively in industry to control the operation of the pneumatic cylinders used on spot welders, press clutches, machine and assembly jig clamps, tools, and lifts.

REVIEW QUESTIONS

1. Why is it important to understand the purpose and action of the total operational system when working with controls?
2. If an electrically controlled, two-way solenoid valve is leaking, what is the probable cause?
3. What is the difference between a two-way solenoid valve and a flow switch?

UNIT 24

Temperature Sensing Devices

Objectives ***After studying this unit, the student will be able to:***

- **Describe different methods for sensing temperature**
- **Discuss different devices intended to be operated by a change of temperature**
- **List several applications for temperature sensing devices**
- **Read and draw the NEMA symbols for temperature switches**

There are many times when the ability to sense temperature is of great importance. The industrial electrician will encounter some devices designed to change a set of contacts with a change of temperature and other devices used to sense the amount of temperature. The method used depends a great deal on the applications of the circuit and the amount of temperature that must be sensed.

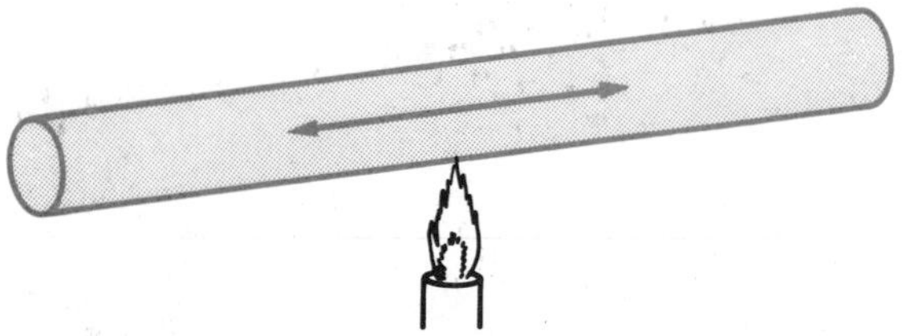

FIGURE 24-1 Metal expands when heated (From Herman and Sparkman, *Electricity and Controls for Heating, Ventilating, and Air Conditioning*, 2e, copyright 1991 by Delmar Publishers Inc.)

EXPANSION OF METAL

A very common and reliable method for sensing temperature is by the expansion of metal. It has long been known that metal expands when heated. The amount of expansion is proportional to two factors:

1. The type of metal used
2. The amount of temperature

Consider the metal bar shown in figure 24-1. When the bar is heated, its length expands. When the metal is permitted to cool, it will contract. Although the amount of movement due to contractions and expansion is small, a simple mechanical principle can be used to increase the amount of movement, figure 24-2.

The metal bar is mechanically held at one end. This permits the amount of expansion to be in one direction only. When the metal is heated and the bar expands, it pushes against the mechanical arm. A small movement of the bar causes a great amount of movement in the mechanical arm. This increased movement in the arm can be used to indicate the temperature of the bar by attaching a pointer and scale, or to operate a switch as shown. It should be understood that illustrations are used to convey a principle. In actual practice, the switch shown in figure 24-2 would be spring loaded to provide a "snap" action for the contacts. Electrical contacts must never be permitted to open or close slowly. This produces poor contact pressure and will cause the contacts to burn or will cause erratic operation of the equipment they are intended to control.

Hot-wire Starting Relay

A very common device that uses the principle of expanding metal to operate a set of contacts is the *hot-wire starting relay* found in the refrigeration industry. The hot-wire relay is so named because it uses a length of resistive wire connected in series with the motor to sense motor current. A diagram of this type relay is shown in figure 24-3.

When the thermostat contact closes, current can flow from line L1 to terminal L of the relay. Current then flows through the resistive wire, the movable arm, and the normally closed contacts to the run and start windings. When current flows through the resistive wire, its temperature increases. This increase of temperature causes the wire to expand in length. When the length increases, the movable arm is forced downward. This downward pressure produces tension on the springs of both contacts. The relay is so designed that the start contact will snap open first, disconnecting the motor start winding from the circuit. If the motor current is not excessive, the wire will never become hot enough to cause the overload contact to open. If the motor current should become too great, however, the temperature of the resistive wire will become high enough to cause the wire to expand to the point that it will cause the overload contact to snap open and disconnect the motor run winding from the circuit.

The Mercury Thermometer

Another very useful device that works on the principle of contraction and expansion of metal is the *mercury thermometer*. Mercury is a metal that remains in a liquid state at room temperature. If the Mercury is confined in a glass tube as shown in figure 24-4, it will rise up the tube as it expands due to an increase in temperature. If the tube is calibrated correctly, it provides an accurate measurement for temperature.

The Bimetal Strip

The *bimetal strip* is another device that operates by the expansion of metal. It is probably the most common heat sensing device used in the production of room thermostats and thermometers. The bimetal strip is made by bonding two dissimilar types of metal together, figure 24-5. Since these two metals are not alike, they have different expansion rates. This causes the strip to bend or warp when heated, figure 24-6. A bimetal strip is

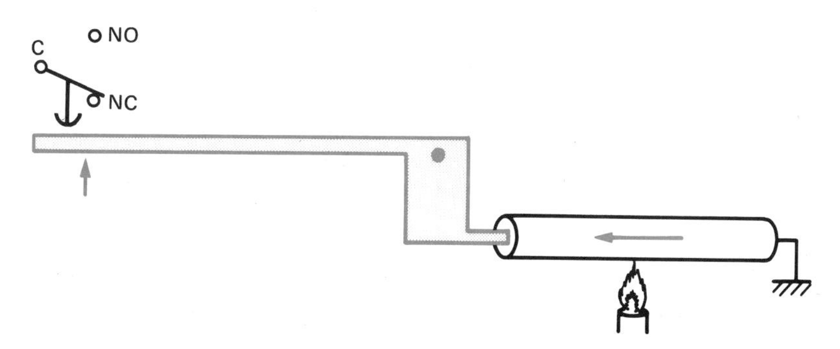

FIGURE 24-2 Expanding metal operates a set of contacts (From Herman and Sparkman, *Electricity and Controls for Heating, Ventilating, and Air Conditioning*, 2e, copyright 1991 by Delmar Publishers Inc.)

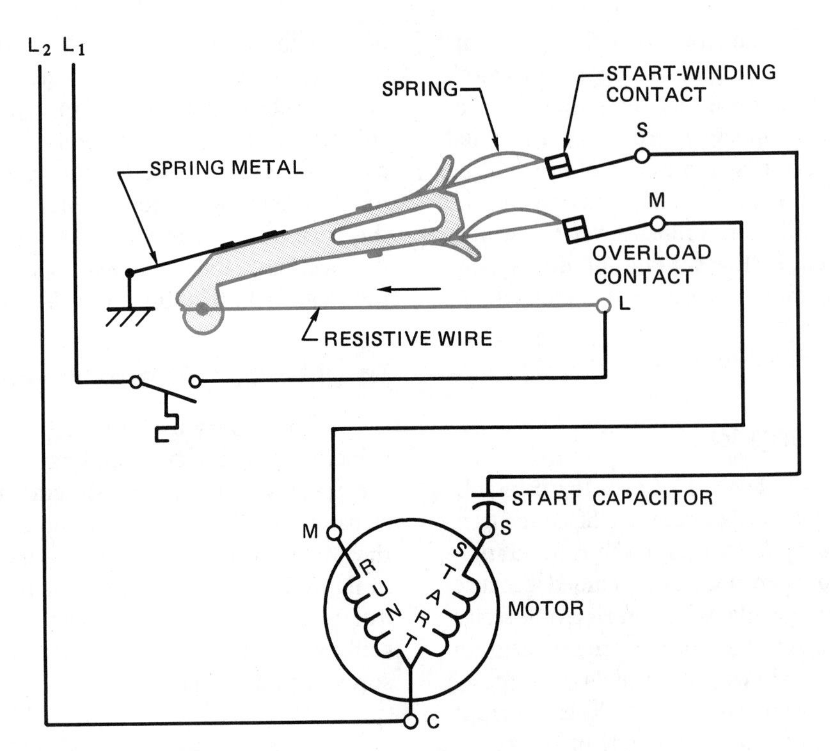

FIGURE 24-3 Hot-wire relay connection (From Herman and Sparkman, *Electricity and Controls for Heating, Ventilating, and Air Conditioning,* 2e, copyright 1991 by Delmar Publishers Inc.)

FIGURE 24-5 A bimetal strip (From Herman and Sparkman, *Electricity and Controls for Heating, Ventilating, and Air Conditioning,* 2e, copyright 1991 by Delmar Publishers Inc.)

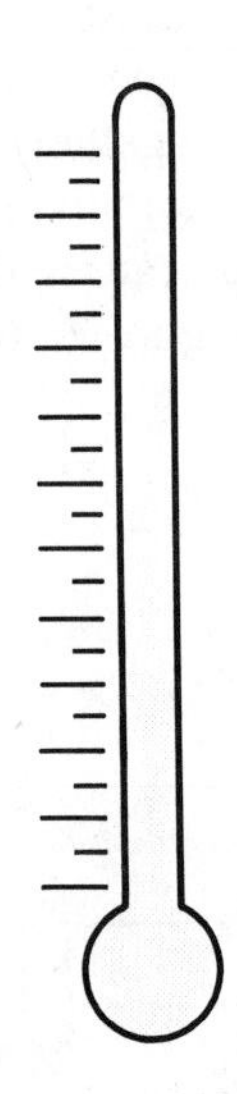

FIGURE 24-4 A mercury thermometer operates by the expansion of metal (From Herman and Sparkman, *Electricity and Controls for Heating, Ventilating, and Air Conditioning,* 2e, copyright 1991 by Delmar Publishers Inc.)

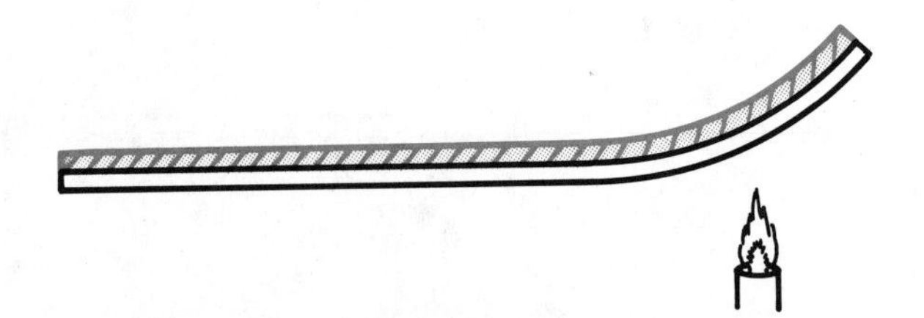

FIGURE 24-6 A bimetal strip warps with a change of temperature (From Herman and Sparkman, *Electricity and Controls for Heating, Ventilating, and Air Conditioning,* 2e, copyright 1991 by Delmar Publishers Inc.)

often formed into a spiral shape as shown in figure 24-7. The spiral permits a longer bimetal strip to be used in a small space. A long bimetal strip is desirable because it exhibits a greater amount of movement with a change of temperature.

If one end of the strip is mechanically held and a pointer is attached to the center of the spiral, a change in temperature will cause the pointer to rotate. If a calibrated scale is placed behind the pointer, it becomes a thermometer. If the center of the spiral is held in position and a contact is attached to the end of the bimetal strip, it becomes a thermostat. A small permanent magnet is used to provide a snap action for the contacts, figure 24-8. When the moving contact reaches a point that is close to the stationary contact, the magnet attracts the metal strip and causes a sudden closing of the contacts. When the bimetal strip cools, it pulls away from the magnet. When the force of the bimetal strip becomes strong enough, it overcomes the force of the magnet and the contacts snap open.

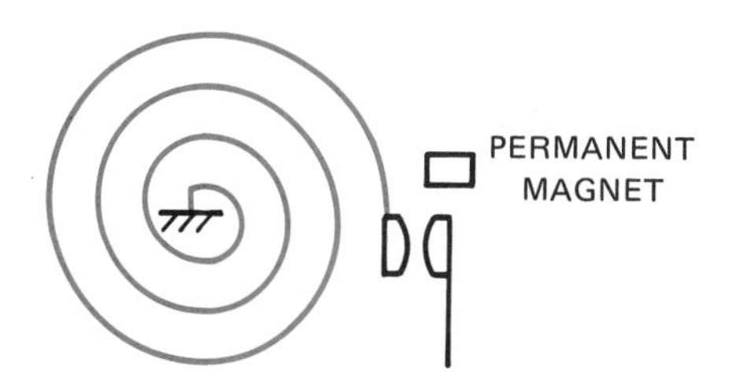

FIGURE 24-8 A bimetal strip used to operate a set of contacts (From Herman and Sparkman, *Electricity and Controls for Heating, Ventilating, and Air Conditioning,* 2e, copyright 1991 by Delmar Publishers Inc.)

Thermocouples

In 1822, a German scientist named Seebeck discovered that when two dissimilar metals are joined at one end, and that junction is heated, a voltage is produced, figure 24-9. This is known as the *Seebeck effect.* The device produced by the joining of two dissimilar metals for the purpose of producing electricity with heat is called a *thermocouple.* The amount of voltage produced by a thermocouple is determined by:

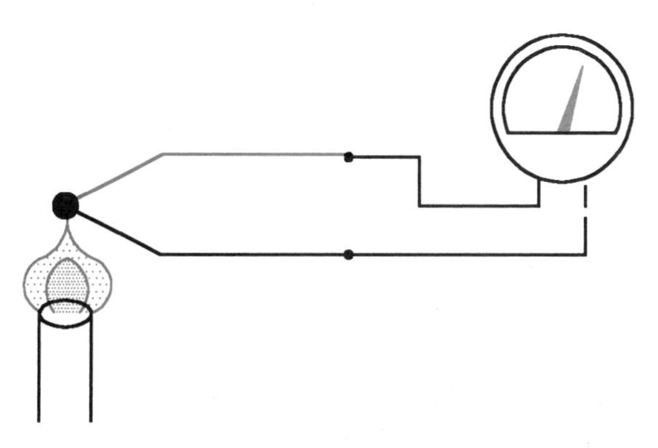

FIGURE 24-9 Thermocouple

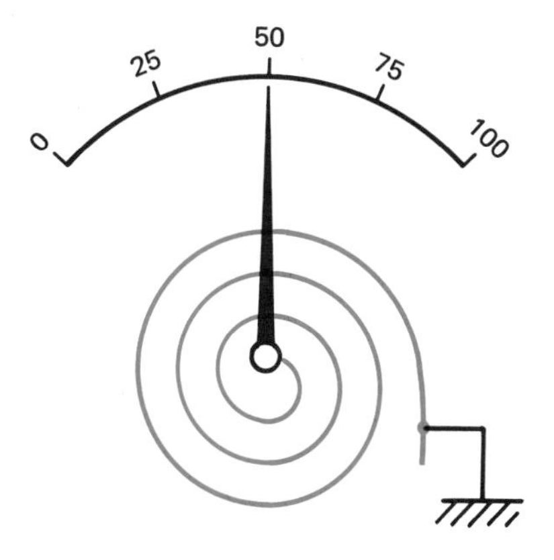

FIGURE 24-7 A bimetal strip used as a thermometer (From Herman and Sparkman, *Electricity and Controls for Heating, Ventilating, and Air Conditioning,* 2e, copyright 1991 by Delmar Publishers Inc.)

1. The type materials used to produce the thermocouple.
2. The temperature difference of the two junctions.

The chart in figure 24-10 shows common types of thermocouples. The different metals used in the construction of thermocouples is shown as well as their normal temperature ranges.

The amount of voltage produced by a thermocouple is small, generally in the order of millivolts (1 millivolt = 0.001 volt). The polarity of the voltage of some thermocouples is determined by the temperature. For example, a type "J" thermocouple produces zero volt at about 32°F. At temperatures above 32°F, the iron wire is positive

TYPE	MATERIAL		DEGREES F	DEGREES C
J	IRON	CONSTANTAN	−328 to +32 +32 to +1432	−200 to 0 0 to 778
K	CHROMEL	ALUMEL	−328 to +32 +32 to 2472	−200 to 0 0 to 1356
T	COPPER	CONSTANTAN	−328 to +32 +32 to 752	−200 to 0 0 to 400
E	CHROMEL	CONSTANTAN	−328 to +32 +32 to 1832	−200 to 0 0 to 1000
R	PLATINUM 13% RHODIUM	PLATINUM	+32 to + 3232	0 to 1778
S	PLATINUM 10% RHODIUM	PLATINUM	+32 to +3232	0 to 1778
B	PLATINUM 30% RHODIUM	PLATINUM 6% RHODIUM	+992 to +3352	533 to 1800

FIGURE 24-10 Thermocouple chart

and the constantan wire is negative. At temperatures below 32°F, the iron wire becomes negative and the constantan wire becomes positive. At a temperature of +300°F, a type "J" thermocouple will produce a voltage of about +7.9 millivolts. At a temperature of −300°F, it will produce a voltage of about −7.9 millivolts.

Since thermocouples produce such low voltages, they are often connected in series as shown in figure 24-11. This connection is referred to as a *thermopile*. Thermocouples and thermopiles are generally used for making temperature measurements and are sometimes used to detect the presence of a pilot light in appliances which operate with natural gas. The thermocouple is heated by the pilot light. The current produced by the thermocouple is used to produce a magnetic field that holds a gas valve open and permits gas to flow to the main burner. If the pilot light should go out, the thermocouple ceases to produce current and the valve closes, figure 24-12.

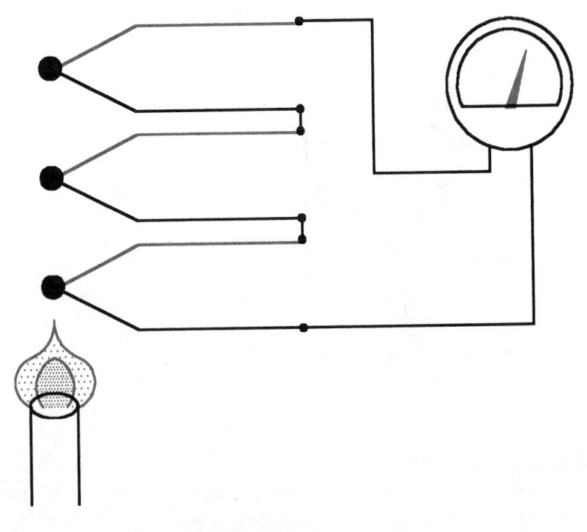

FIGURE 24-11 Thermopile

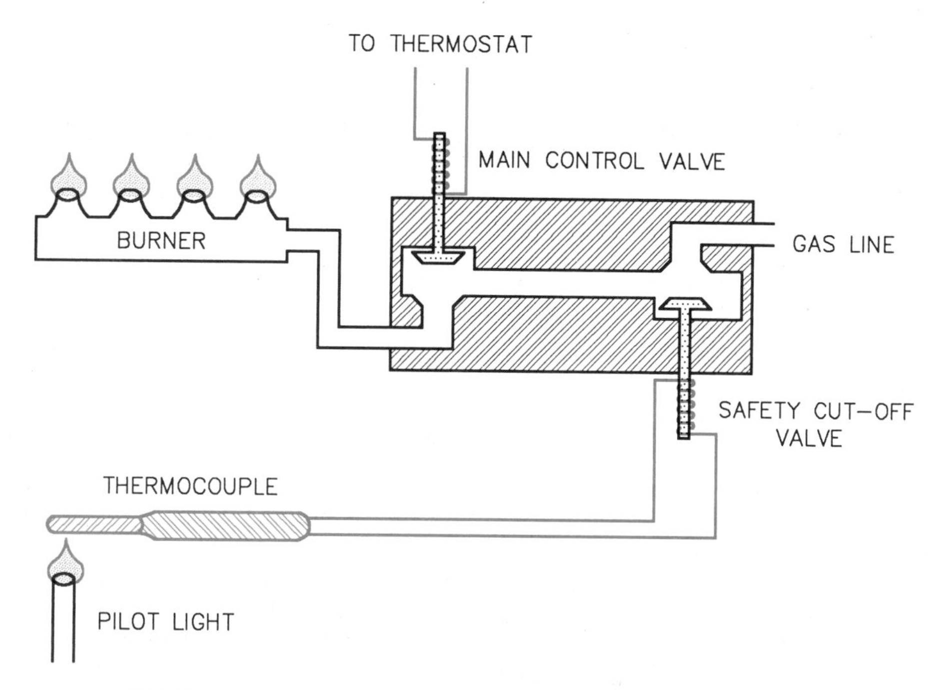

FIGURE 24-12 A thermocouple provides power to the safety cut-off valve

RESISTANCE TEMPERATURE DETECTORS

The *resistance temperature detector* (RTD) is made of platinum wire. The resistance of platinum changes greatly with temperature. When platinum is heated, its resistance increases at a very predictable rate; this makes the RTD an ideal device for measuring temperature very accurately. RTDs are used to measure temperatures that range from −328 to +1166 degrees Fahrenheit (−200° to +630 C°). RTDs are made in different styles to perform different functions. Figure 24-13 illustrates a typical RTD used as a probe. A very small coil of platinum wire is encased inside a copper tip. Copper is used to provide good thermal contact. This permits the probe to be very fast-acting. The chart in figure 24-14 shows resistance versus temperature for a typical RTD probe. The temperature is given in degrees Celsius and the resistance is given in ohms. RTDs in several different case styles are shown in figure 24-15.

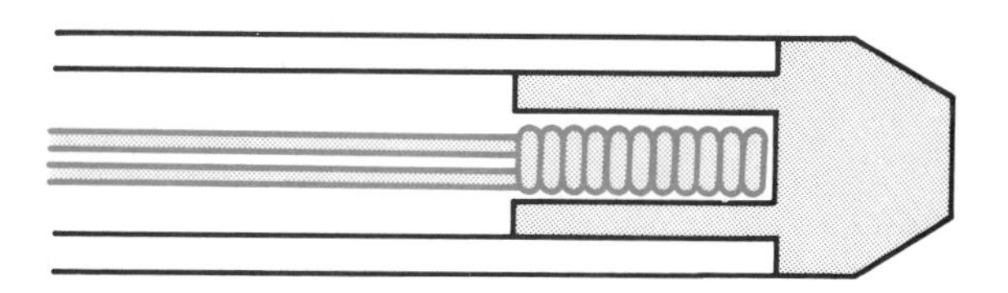

FIGURE 24-13 Resistance temperature detector (From Herman and Sparkman, *Electricity and Controls for Heating, Ventilating, and Air Conditioning,* 2e, copyright 1991 by Delmar Publishers Inc.)

Thermistors

The term *thermistor* is derived from the words "thermal resistor." Thermistors are actually thermally sensitive semi-conductor devices. There are two basic types of thermistors: one type has a negative temperature coefficient (NTC) and the other has a positive temperature coefficient (PTC). A thermistor that has a negative temperature coefficient will decrease its resistance as the temperature increases. A thermistor that has a positive temperature coefficient will increase its resistance as the temperature increases. The NTC thermistor is the most widely used.

DEGREES C	RESISTANCE
0	100
50	119.39
100	138.5
150	157.32
200	175.84
250	194.08
300	212.03
350	229.69
400	247.06
450	264.16
500	280.93
550	297.44
600	313.65

FIGURE 24-14 Temperature and resistance for a typical RTD (From Herman and Sparkman, *Electricity and Controls for Heating, Ventilating, and Air Conditioning*, 2e, copyright 1991 by Delmar Publishers Inc.)

Thermistors are highly nonlinear devices. For this reason they are difficult to use for measuring temperature. Devices that measure temperature with a thermistor must be calibrated for the particular type of thermistor being used. If the thermistor is ever replaced, it has to be an exact replacement or the circuit will no longer operate correctly. Because of their nonlinear characteristics, thermistors are often used as *set point detectors* as opposed to actual temperature measurement. A set point detector is a device that activates some process or circuit when the temperature reaches a certain level. For example, assume a thermistor has been placed inside the stator winding of a motor. If the motor should become overheated, the windings could become severely damaged or destroyed. The thermistor can be used to detect the temperature of the windings. When the temperature reaches a certain point, the resistance value of the thermistor changes enough to cause the starter coil to drop out and disconnect the motor from the line. Thermistors can be operated in temperatures that range from about −100° to +300 F°.

One common use for thermistors is in the solid-state starting relays used with small refrigeration compressors, figure 24-16. Starting relays are used with hermetically sealed motors to disconnect the start windings from the circuit when the motor reaches about 0.75% of its full speed. Thermistors can be used for this application because they exhibit an extremely rapid change of resistance with a change of temperature. A schematic diagram showing the connection for a solid state relay is shown in figure 24-17.

When power is first applied to the circuit, the thermistor is cool and has a relatively low resistance. This permits current to flow through both the start and run windings of the motor. The temperature of the thermistor increases because of the current flowing through it. The increase of tem-

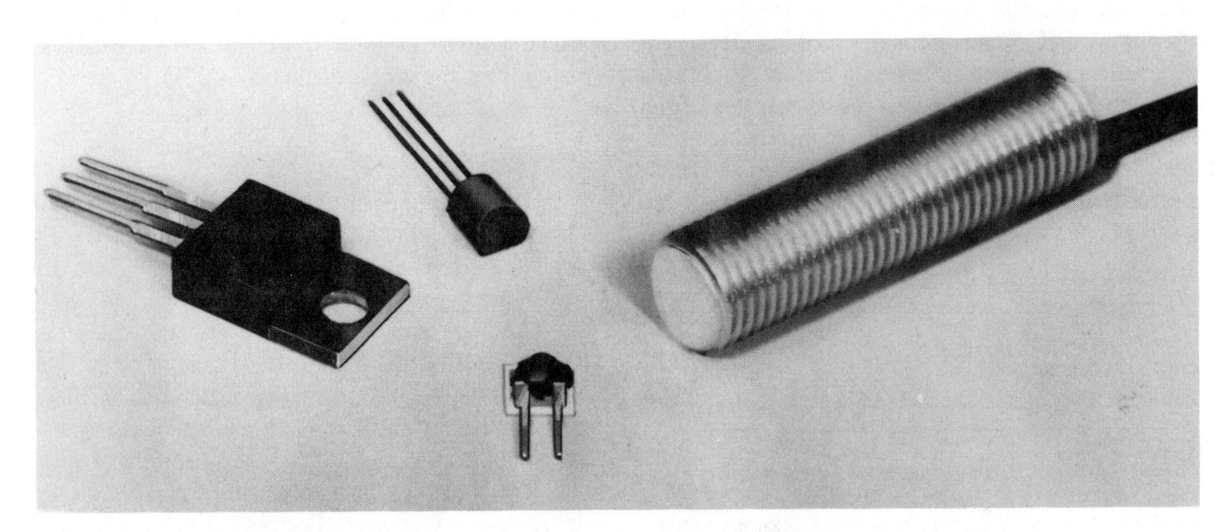

FIGURE 24-15 RTDs in different case styles (Courtesy Micro Switch, a Honeywell Division)

FIGURE 24-16 Solid-state starting relay (From Herman and Sparkman, *Electricity and Controls for Heating, Ventilating, and Air Conditioning*, 2e, copyright 1991 by Delmar Publishers Inc.)

perature causes the resistance to change from a very low value of 3 or 4 ohms to several thousand ohms. This increase of resistance is very sudden and has the effect of opening a set of contacts connected in series with the start winding. Although the start winding is never completely disconnected from the power line, the amount of current flow through it is very small, typically 0.03 to 0.05 amps, and does not affect the operation of the motor. This small amount of *leakage current* maintains the temperature of the thermistor and prevents it from returning to a low resistance. After power has been disconnected from the motor, a cool-down period of about 2 minutes should be allowed before restarting the motor. This cool-down period is needed for the thermistor to return to a low value of resistance.

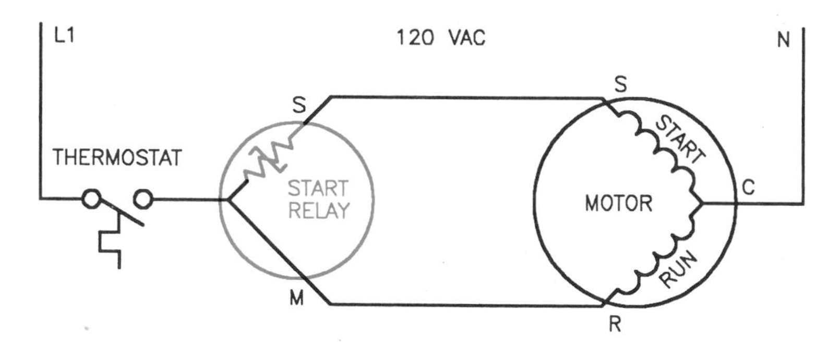

FIGURE 24-17 Connection of solid-state starting relay. (From Herman and Sparkman, *Electricity and Controls for Heating, Ventilating, and Air Conditioning*, 2e, copyright 1991 by Delmar Publishers Inc.)

The PN Junction

Another device that has the ability to measure temperature is the PN junction or diode. The diode is becoming a very popular device for measuring temperature because it is accurate and linear.

When a silicon diode is used as a temperature sensor, a constant current is passed through the diode. Figure 24-18 illustrates this type of circuit. In this circuit, resistor R1 limits the current flow through the transistor and sensor diode. The value of R1 also determines the amount of current that flows through the diode. Diode D1 is a 5.1 volt zener used to produce a constant voltage drop between the base and emitter of the PNP transistor. Resistor R2 limits the amount of current flow through the zener diode and the base of the transistor. D1 is a common silicon diode. It is being used as the temperature sensor for the circuit. If a digital voltmeter is connected across the diode, a voltage drop between 0.8 and 0 volts can be seen. The amount of voltage drop is determined by the temperature of the diode.

Another circuit that can be used as a constant current generator is shown in figure 24-19. In this circuit, a field effect transistor (FET) is used to produce a current generator. Resistor R1 determines the amount of current that will flow through the diode. Diode D1 is the temperature sensor.

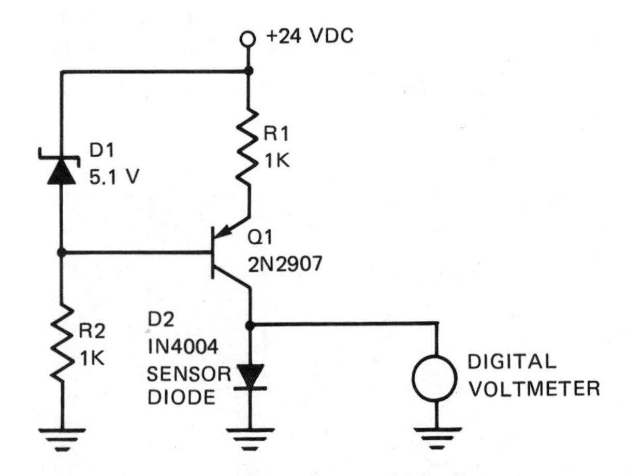

FIGURE 24-18 Constant current generator (From Herman and Sparkman, *Electricity and Controls for Heating, Ventilating, and Air Conditioning*, 2e, copyright 1991 by Delmar Publishers Inc.)

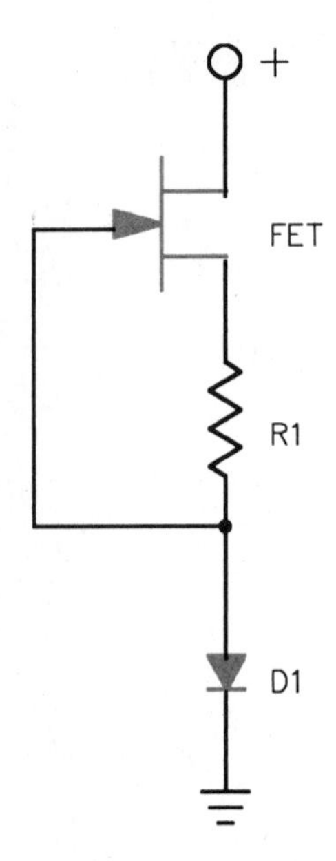

FIGURE 24-19 Field effect transistor used to produce a constant current generator

If the diode is subjected to a lower temperature, say by touching it with a piece of ice, the voltage drop across the diode will increase. If the diode temperature is increased, the voltage drop will decrease because the diode has a negative temperature coefficient. As its temperature increases, its voltage drop becomes less.

In figure 24-20, two diodes connected in a series are used to construct an electronic thermostat. Two diodes are used to increase the amount of voltage drop as the temperature changes. A field effect transistor and resistor are used to provide a constant current to the two diodes used as the heat sensor. An operational amplifier is used to turn a solid state relay on or off as the temperature changes. In the example shown, the circuit will operate as a heating thermostat. The output of the amplifier will turn on when the temperature decreases sufficiently. The circuit can be converted to a cooling thermostat by reversing the connections of the inverting and noninverting inputs of the amplifier.

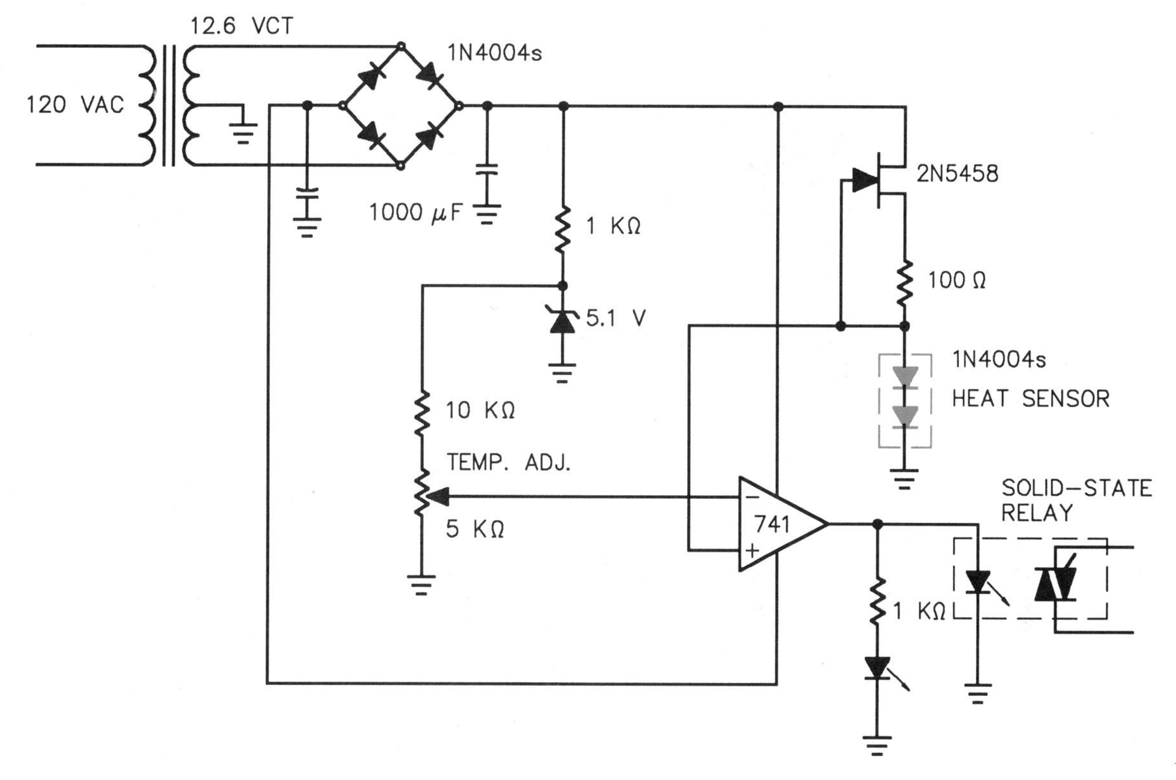

FIGURE 24-20 Solid-state thermostat using diodes as heat sensors

EXPANSION DUE TO PRESSURE

Another common method of sensing a change of temperature is by the increase of pressure of some chemicals. Refrigerants confined in a sealed container, for example, will increase the pressure in the container with an increase of temperature. If a simple bellows is connected to a line containing refrigerant, figure 24-21, the bellows will expand as the pressure inside the sealed system increases. When the surrounding air temperature decreases, the pressure inside the system decreases and the bellows contracts. When the air temperature increases, the pressure increases and the bellows expands. If the bellows controls a set of contacts, it becomes a bellows type thermostat. A bellows thermostat and the standard NEMA symbols used to represent a temperature operated switch are shown in figure 24-22.

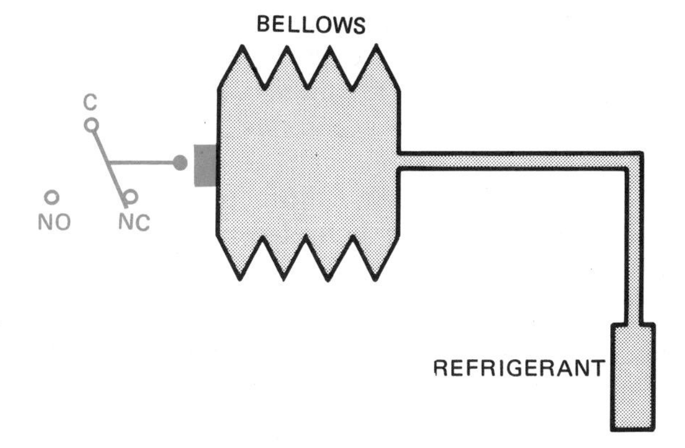

FIGURE 24-21 Bellows contracts and expands with a change of refrigerant pressure (From Herman and Sparkman, *Electricity and Controls for Heating, Ventilating, and Air Conditioning,* 2e, copyright 1991 by Delmar Publishers Inc.)

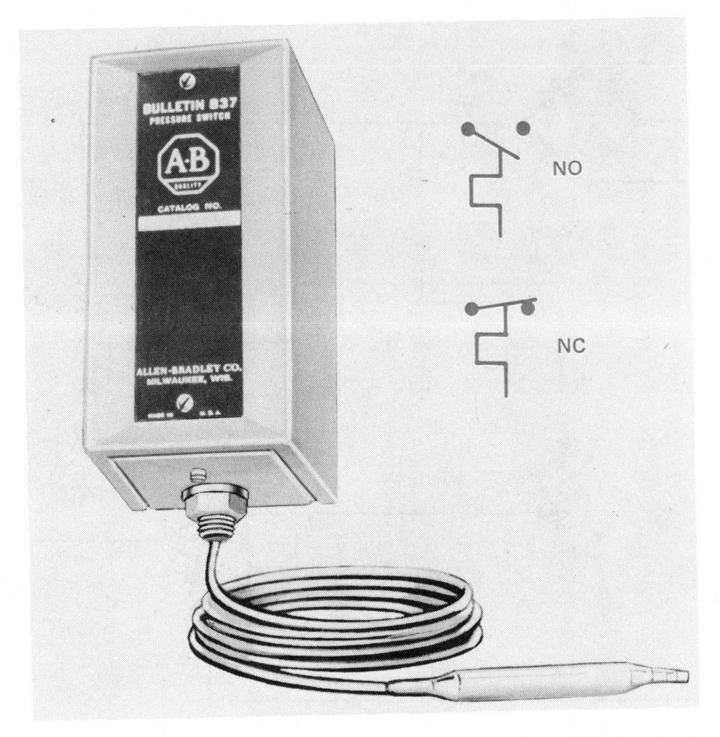

FIGURE 24-22 Industrial temperature switch with expansion bulb (Courtesy Allen-Bradley Co.)

REVIEW QUESTIONS

1. Should a metal bar be heated or cooled to make it expand?
2. What type of metal remains in a liquid state at room temperature?
3. How is a bimetal strip made?
4. Why are bimetal strips often formed into a spiral shape?
5. Why should electrical contacts never be permitted to open or close slowly?
6. What two factors determine the amount of voltage produced by a thermocouple?
7. What is a thermopile?
8. What do the letters RTD stand for?
9. What type of wire is used to make an RTD?
10. What material is a thermistor made of?
11. Why is it difficult to measure temperature with a thermistor?
12. If the temperature of a NTC thermistor increases, will its resistance increase or decrease?
13. How can a silicon diode be made to measure temperature?
14. Assume that a silicon diode is being used as a temperature detector. If its temperature increases, will its voltage drop increase or decrease?
15. What type of chemical is used to cause a pressure change in a bellows type thermostat?

UNIT 25

Hall Effect Sensors

Objectives ***After studying this unit, the student will be able to:***

- **Describe the Hall effect**
- **Discuss the principles of operation of a Hall generator**
- **Discuss applications in which Hall generators can be used**

PRINCIPLES OF OPERATION

The Hall effect is a simple principle that is widely used in industry today. The Hall effect was discovered by Edward H. Hall at Johns Hopkins University in 1879. Mr. Hall originally used a piece of pure gold to produce the Hall effect, but today a piece of semiconductor material is used because semiconductor material works better and is less expensive to use. The device is often referred to as the Hall generator.

Figure 25-1 illustrates how the Hall effect is produced. A constant current power supply is connected to opposite sides of a piece of semiconductor material. A sensitive voltmeter is connected to the other two sides. If the current flows straight through the semiconductor material, no voltage is produced across the voltmeter connection.

Figure 25-2 shows the effect of bringing a magnetic field near the semiconductor material. The magnetic field causes the current flow path to

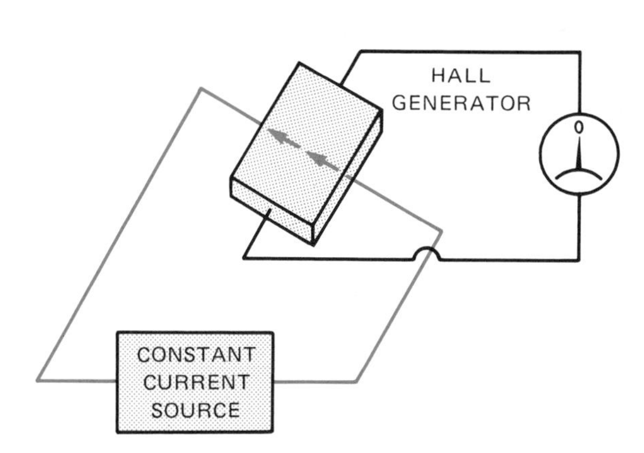

FIGURE 25-1 Constant current flows through a piece of semiconductor material

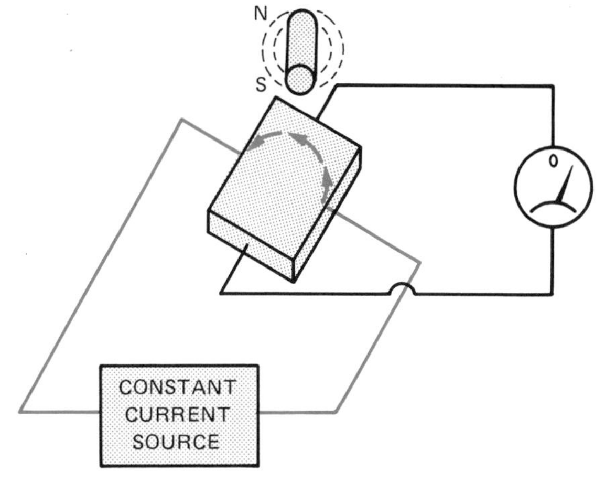

FIGURE 25-2 A magnetic field deflects the path of current flow through the semiconductor

be deflected to one side of the material. This causes a potential or voltage to be produced across the opposite sides of the semiconductor material.

If the polarity of the magnetic field is reversed, the current path is deflected in the opposite direction as shown in figure 25-3. This causes the polarity of the voltage produced by the Hall generator to change. Two factors determine the polarity of the voltage produced by the Hall generator:

1. the direction of current flow through the semiconductor material; and
2. the polarity of the magnetic field used to deflect the current.

The amount of voltage produced by the Hall generator is determined by

1. the amount of current flowing through the semiconductor material; and
2. the strength of the magnetic field used to deflect the current path.

The Hall generator has many advantages over other types of sensors. Since it is a solid-state device, it has no moving parts or contacts to wear out. It is not affected by dirt, oil, or vibration. The Hall generator is an integrated circuit which is mounted in many different types and styles of cases.

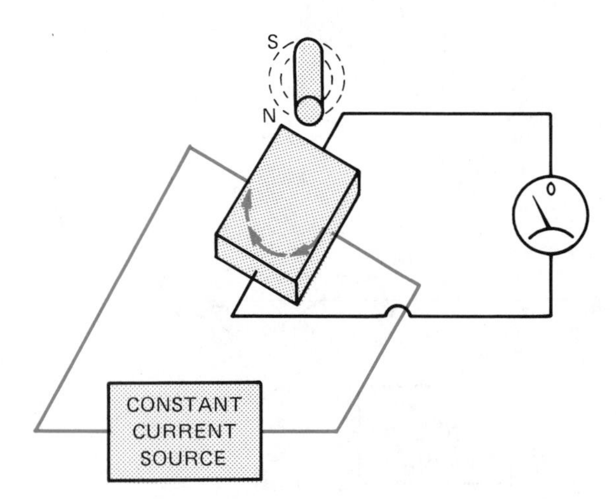

FIGURE 25-3 The current path is deflected in the opposite direction

HALL GENERATOR APPLICATIONS

Motor Speed Sensor

The Hall generator can be used to measure the speed of a rotating device. If a disk with magnetic poles around its circumference is attached to a rotating shaft, and a Hall sensor is mounted near the disk, a voltage will be produced when the shaft turns. Since the disk has alternate magnetic polarities around its circumference, the sensor will produce an ac voltage. Figure 25-4 shows a Hall generator used in this manner. Figure 25-5 shows the ac waveform produced by the rotating disk. The frequency of the ac voltage is proportional to the number of magnetic poles on the disk and the speed of rotation.

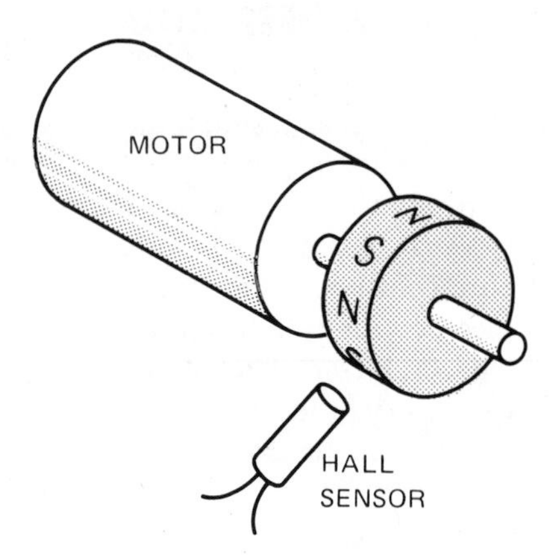

FIGURE 25-4 Speed is measured by spinning a magnetic disk

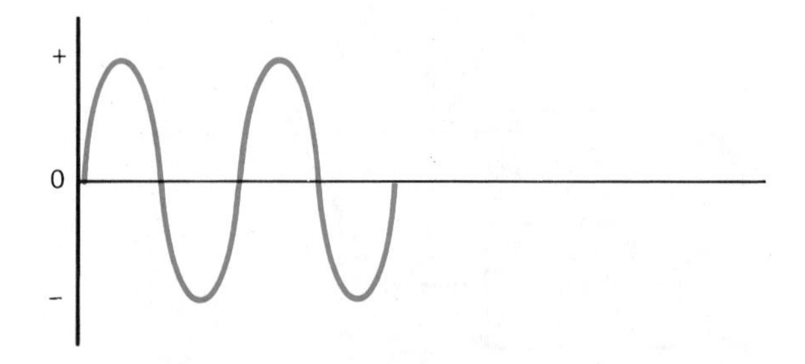

FIGURE 25-5 An ac voltage is produced by the rotating magnetic disk

Another method for sensing speed is to use a *reluctor*. A reluctor is a ferrous metal disk used to shunt a magnetic field away from some other object. This type of sensor uses a notched metal disk attached to a rotating shaft. The disk separates a Hall sensor and a permanent magnet, figure 25-6. When the notch is between the sensor and the magnet, a voltage is produced by the Hall generator. When the solid metal part of the disk is between the sensor and magnet, the magnetic field is shunted away from the sensor. This causes a significant drop in the voltage produced by the Hall generator.

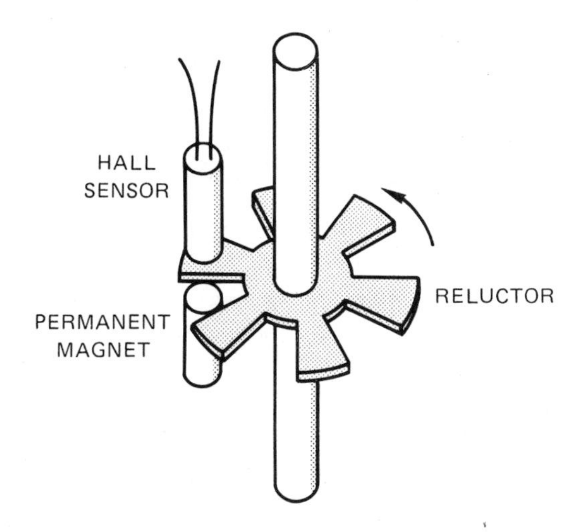

FIGURE 25-6 Reluctor shunts magnetic field away from sensor

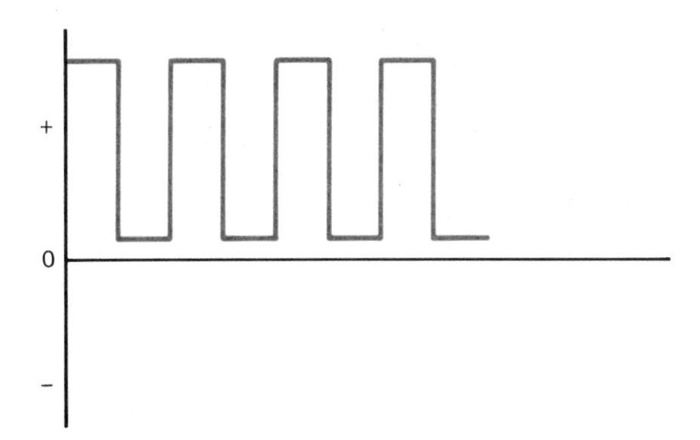

FIGURE 25-7 Square wave pulses produced by the Hall generator

Since the polarity of the magnetic field does not change, the voltage produced by the Hall generator is pulsating direct current instead of alternating current. Figure 25-7 shows the dc pulses produced by the generator. The number of pulses produced per second is proportional to the number of notches on the reluctor and the speed of the rotating shaft.

Position Sensor

The Hall generator can be used in a manner similar to a limit switch. If the sensor is mounted beside a piece of moving equipment, and a permanent magnet is attached to the moving equipment, a voltage will be produced when the magnet moves near the sensor, figure 25-8. The advantages of the Hall sensor are that it has no lever arm or contacts to wear like a common limit switch, and it can operate through millions of operations of the machine.

Several different types of Hall effect position sensors are shown in figure 25-9. Notice that these sensors vary in size and style to fit almost any application. Position sensors operate as a digital device in that they sense the presence or absence of magnetic field. They do not have the ability to sense the intensity of the field.

Hall Effect Limit Switches

Another Hall effect device used in a very similar application is the Hall effect limit switch, figure 25-10. This limit switch uses a Hall gener-

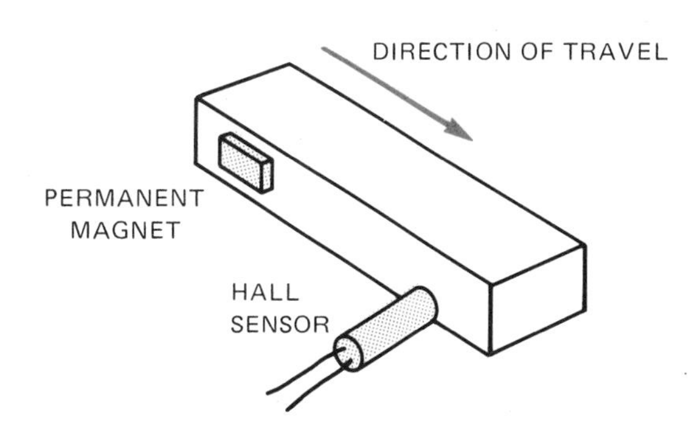

FIGURE 25-8 Hall generator used to sense position of moving device

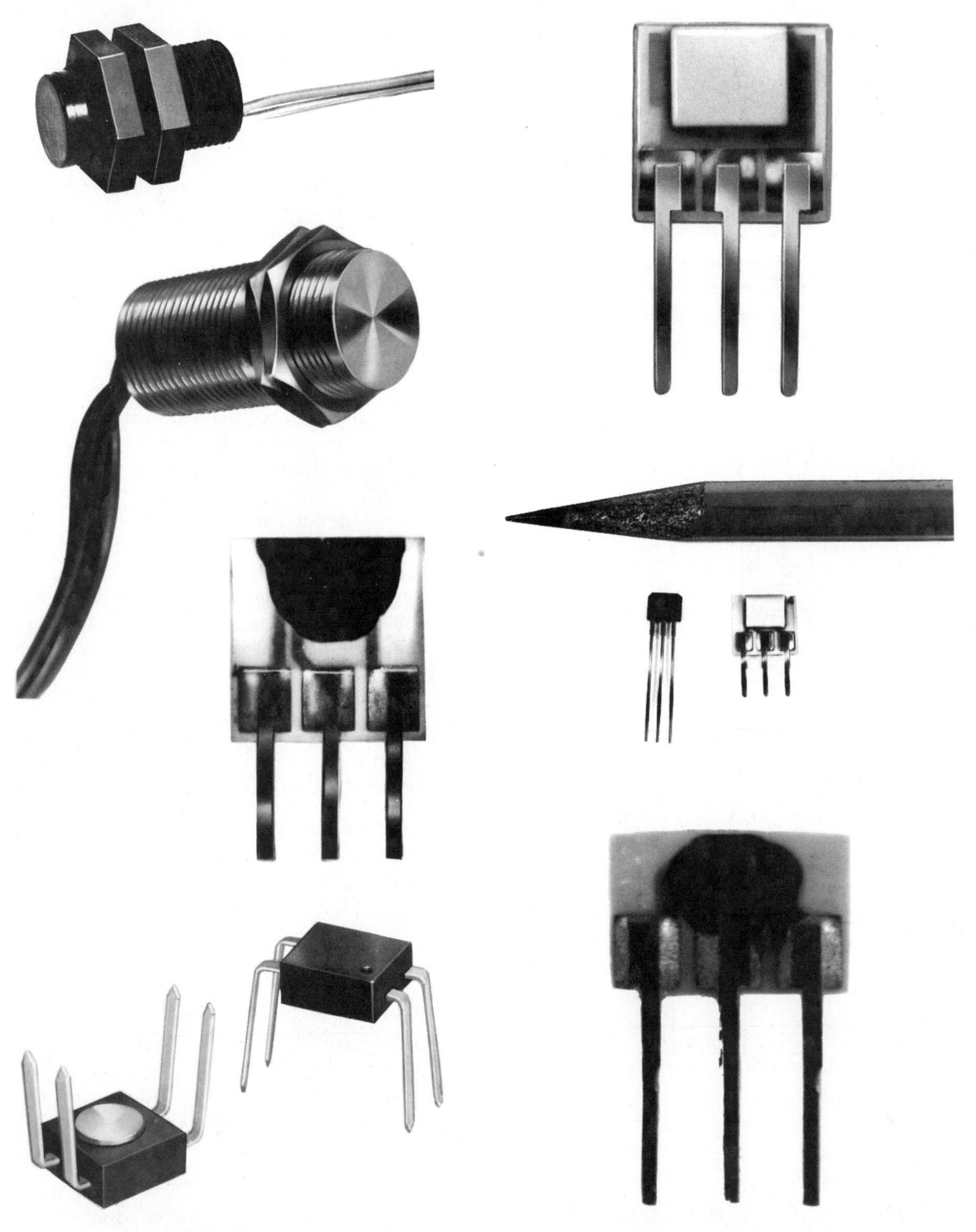

FIGURE 25-9 Hall effect position sensors (Courtesy Micro Switch, a Honeywell Division)

FIGURE 25-10 Hall effect limit switch (Courtesy Micro Switch, a Honeywell Division)

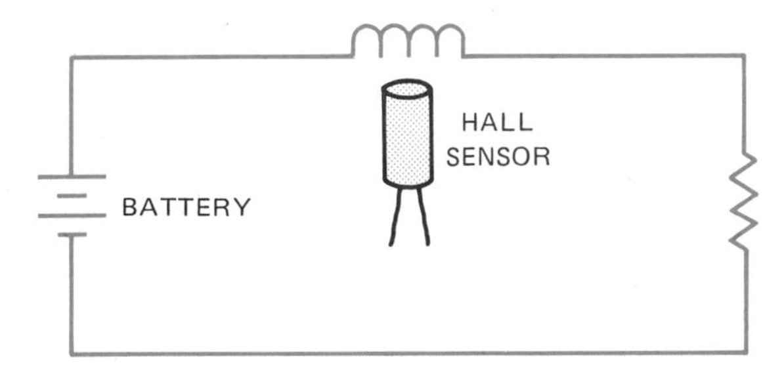

FIGURE 25-11 Hall sensor detects when dc current flows through the circuit

ator instead of a set of contacts. A magnetic plunger is mechanically activated by the small button. Different types of levers can be fitted to the switch, which permits it to be used for many applications. These switches are generally intended to be operated by a 5 volt dc supply for TTL logic applications, or by a 6 to 24 volt dc supply for interface with other types of electronic controls or to provide input for programmable controllers.

Current Sensor

Since the current source for the Hall generator is provided by a separate power supply, the magnetic field does not have to be moving or changing to produce an output voltage. If a Hall sensor is mounted near a coil of wire, a voltage will be produced by the generator when current flows through the wire. Figure 25-11 shows a Hall sensor used to detect when a dc current flows through a circuit. Hall effect sensors are shown in figure 25-12.

The Hall generator is being used more and more in industrial applications. Since the signal rise and fall time of the Hall generator is generally less than 10 microseconds, it can operate at pulse rates as high as 100,000 pulses per second. This makes it especially useful in industry.

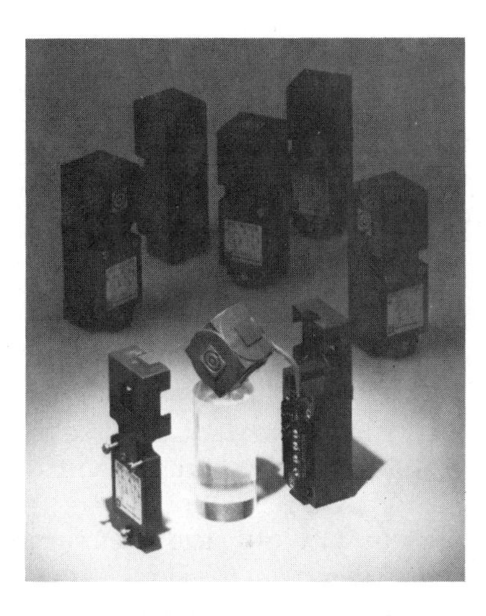

FIGURE 25-12 Hall effect sensors (Courtesy Telemecanique Inc.)

Linear Transducers

Linear transducers are designed to produce an output voltage which is proportional to the strength of a magnetic field. Input voltage is typically 8 to 16 volts, but the amount of output voltage is determined by the type of transducer used. Hall effect linear transducers can be obtained that have two types of outputs. One type has a *regulated* output and produces voltages of 1.5 to 4.5 volts. The other type has a *ratiometric* output and produces an output voltage which is 25% to 75% of the input voltage. A Hall effect linear transducer is shown in figure 25-13.

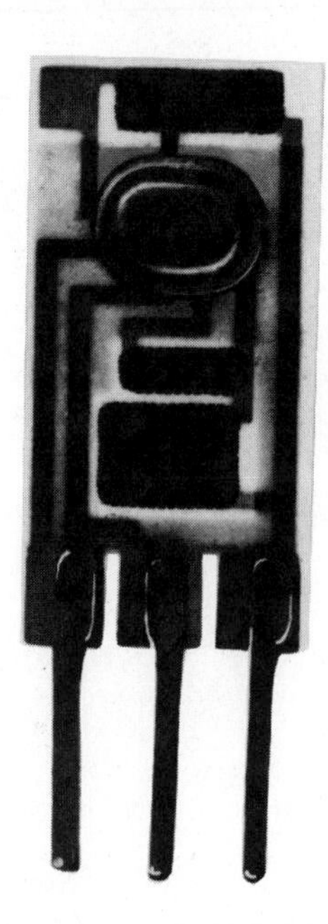

FIGURE 25-13 Hall effect linear transducer (Courtesy Micro Switch, a Honeywell Division)

REVIEW QUESTIONS

1. What material was used to make the first Hall generator?
2. What two factors determine the polarity of the output voltage produced by the Hall generator?
3. What two factors determine the amount of voltage produced by the Hall generator?
4. What is a reluctor?
5. Why does a magnetic field not have to be moving or changing to produce an output voltage in the Hall generator?

UNIT 26

Proximity Detectors

Objectives ***After studying this unit, the student will be able to:***

- **Describe the operation of proximity detectors**
- **Describe different types of proximity detectors**

APPLICATIONS

Proximity detectors are basically metal detectors. They are used to detect the presence or absence of metal without physically touching it. This prevents wear on the unit and gives the detector the ability to sense red hot metals. Most proximity detectors are designed to detect ferrous metals only, but there are some units that detect all metals.

CIRCUIT OPERATION

There are several methods used to make proximity detectors. One method is shown in figure 26-1. This is a very simple circuit intended to illustrate the principle of operation of a proximity detector. The sensor coil is connected through a series resistor to an oscillator. A voltage detector, in this illustration a voltmeter, is connected across the resistor. Since ac voltage is applied to this circuit, the amount of current flow is determined by the resistance of the resistor and the inductive reactance of the coil. The voltage drop across the resistor is proportional to its resistance and the amount of current flow.

If ferrous metal is placed near the sensor coil, its inductance increases in value. This causes an increase in inductive reactance, and a decrease in the amount of current flow through the circuit. When the current flow through the resistor is decreased, the voltage drop across the resistor decreases also, figure 26-2. The drop in voltage can be used to turn relays or other devices on or off.

This method of detecting metal does not work well for all conditions. Another method which is more sensitive to small amounts of metal

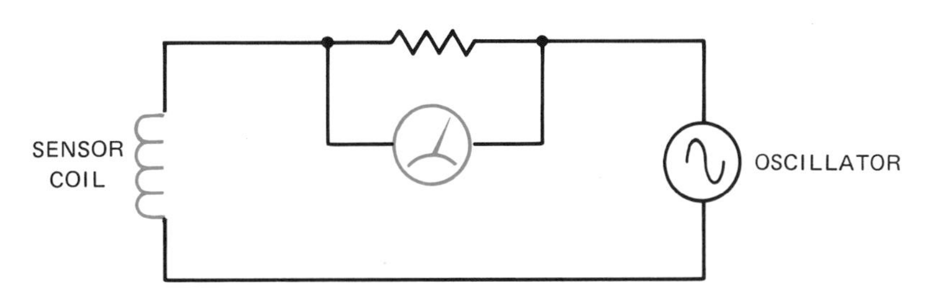

FIGURE 26-1 Simple proximity detector

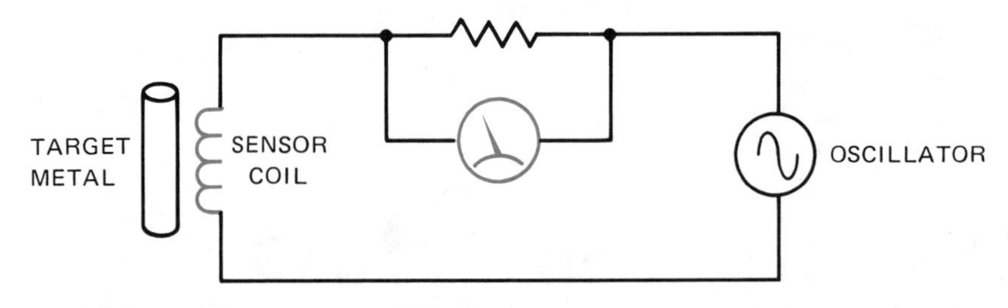

FIGURE 26-2 The presence of metal causes a decrease of voltage drop across the resistor

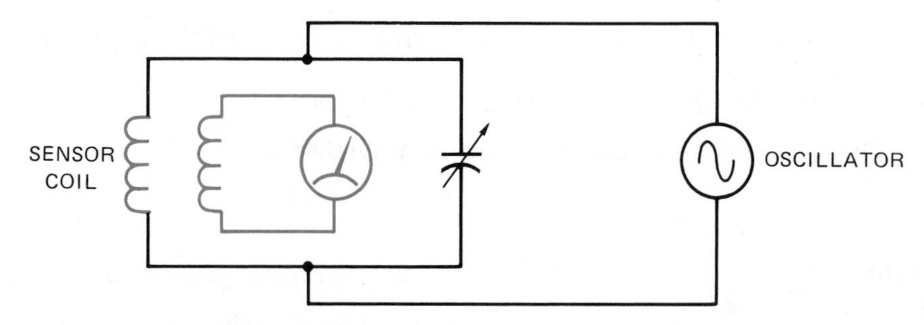

FIGURE 26-3 Tuned tank circuit used to detect metal

is shown in figure 26-3. This detector uses a tank circuit tuned to the frequency of the oscillator. The sensor head contains two coils instead of one. This type of sensor is a small transformer. When the tank circuit is tuned to the frequency of the oscillator, current flow around the tank loop is high. This causes a high voltage to be induced into the secondary coil of the sensor head.

When ferrous metal is placed near the sensor as shown in figure 26-4, the inductance of the coil increases. When the inductance of the coil changes, the tank circuit no longer resonates to the frequency of the oscillator. This causes the current flow around the loop to decrease significantly. The decrease of current flow through the sensor coil causes the secondary voltage to drop also.

Notice that both types of circuits depend on a ferrous metal to change the inductance of a coil. If a detector is to be used to detect nonferrous metals, some means other than changing the inductance of the coil must be used. An all-metal detector uses a tank circuit as shown in figure 26-5. All-metal detectors operate at radio frequencies,

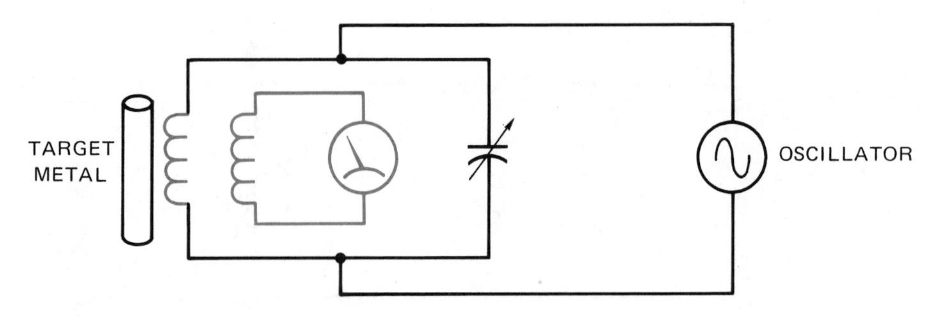

FIGURE 26-4 The presence of metal detunes the tank circuit

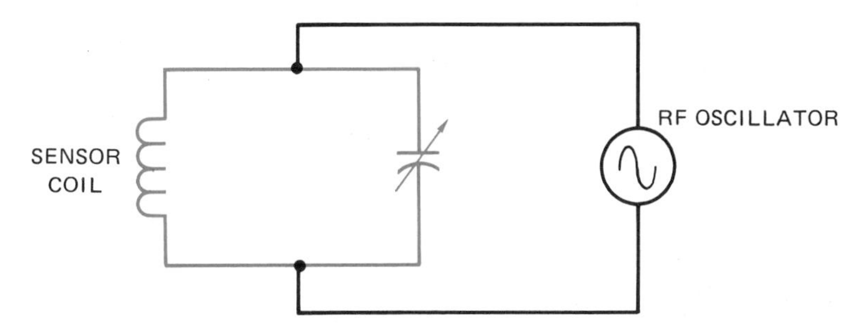

FIGURE 26-5 Balance of the tank circuit permits the oscillator to operate

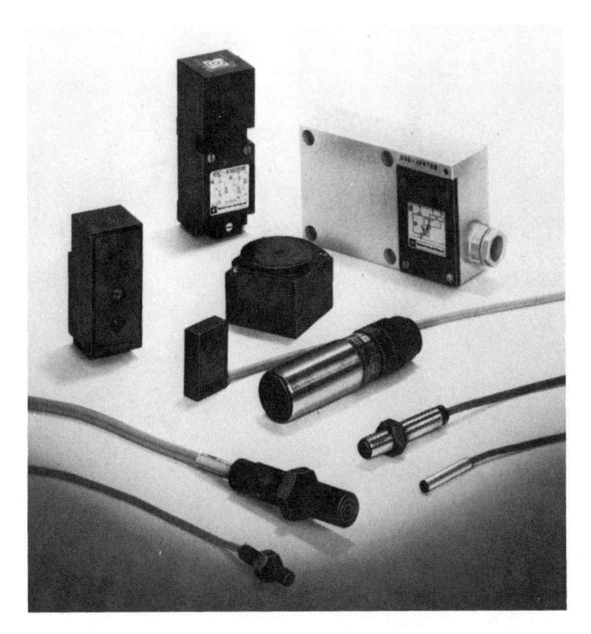

FIGURE 26-6 Proximity detectors (Courtesy Telemecanique Inc.)

and the balance of the tank circuit is used to keep the oscillator running. If the tank circuit becomes unbalanced, the oscillator stops operating. When a nonferrous metal, such as aluminum, copper, or brass, is placed near the sensor coil, eddy currents are induced into the surface of the metal. The induction of eddy currents into the metal causes the tank circuit to become unbalanced and the oscillator to stop operating. When the oscillator stops operating, some other part of the circuit signals an output to turn on or off.

Proximity detectors used to sense all types of metals will sense ferrous metals better than nonferrous. A ferrous metal can be sensed at about three times the distance of a nonferrous metal.

MOUNTING

Some proximity detectors are made as a single unit. Other detectors use a control unit which can be installed in a relay cabinet and a sensor which is mounted at a remote location. Figure 26-6 shows different types of proximity detectors. Regardless of the type of detector used, care and forethought should be used when mounting the sensor. The sensor must be near enough to the target metal to provide a strong positive signal, but it should not be so near that there is a possibility of the sensor being hit by the metal object. One advantage of the proximity detector is that no physical contact is necessary between the detector and the metal object for the detector to sense the object.

Sensors should be mounted as far away from other metals as possible. This is especially true for sensors used with units designed to detect all types of metals. In some cases it may be necessary to mount the sensor unit on a nonmetal surface such as wood or plastic. If proximity detectors are to be used in areas that contain metal shavings or metal dust, an effort should be made to place the sensor in a position that will prevent the shavings or dust from collecting around it. In some installations it may be necessary to periodically clean the metal shavings or dust away from the sensor.

CAPACITIVE PROXIMITY DETECTORS

Although proximity detectors are generally considered to be metal detectors, there are other types that sense the presence of objects that do not contain metal of any kind. One type of these detectors operate on a change of capacitance. When an object is brought into the proximity of one of these detectors, a change of capacitance causes the detector to activate. Several different types of capacitive proximity detectors is shown in figure 26-7.

Since capacitive proximity detectors do not depend on metal to operate, they will sense virtually any material such as wood, glass, concrete, plastic, and sheet rock. They can even be used to sense liquid levels through a sight glass. One disadvantage of capacitive proximity detectors is that they have a very limited range. Most cannot sense objects over approximately one inch or 25 millimeters away. Many capacitive proximity detectors are being used to replace mechanical limit switches since they do not have to make contact with an object to sense its position. Most can be operated with a wide range of voltages such as 2 – 250 VAC, or 20 – 320 VDC.

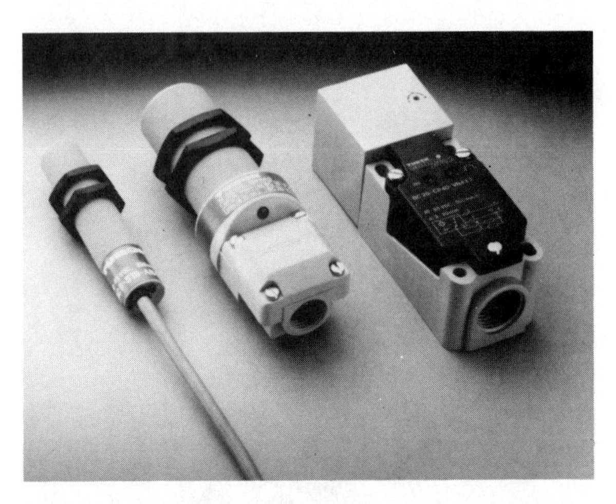

FIGURE 26-7 Capacitive proximity detectors (Courtesy Turck Inc.)

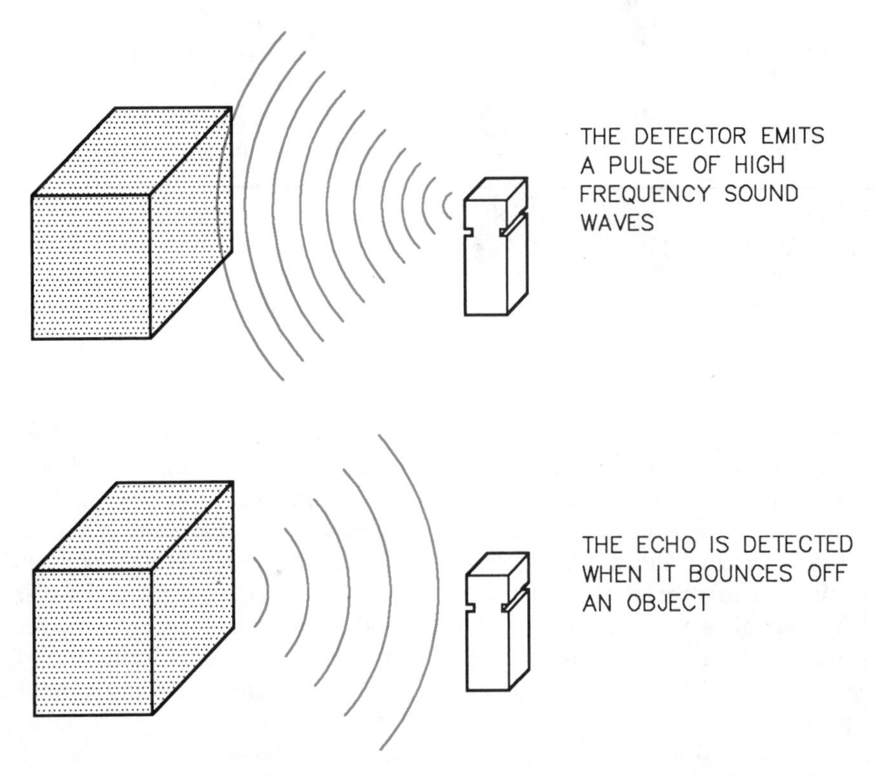

FIGURE 26-8 Ultrasonic proximity detectors operate by emitting high frequency sound waves

ULTRASONIC PROXIMITY DETECTORS

Another type of proximity detector that does not depend on the presence of metal for operation is the *ultrasonic detector*. Ultrasonic detectors operate by emitting a pulse of high frequency sound and then detecting the echo when it bounces off an object, figure 26-8. These detectors can be used to determine the distance to the object by measuring the time interval between the emission of the pulse and the return of the echo. Many ultrasonic sensors have an analog output of voltage or current, the value of which is determined by the distance to the object. This feature permits them to be used in applications where it is necessary to sense the position of an object, figure 26-9. An ultrasonic proximity detector is shown in figure 26-10.

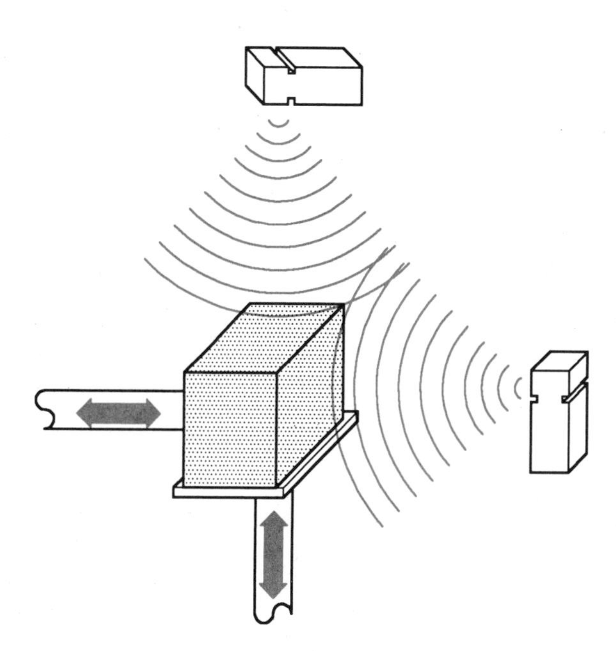

FIGURE 26-9 Ultrasonic proximity detectors used as position sensors

FIGURE 26-10 Ultrasonic proximity detector (Courtesy of Turck Inc.)

REVIEW QUESTIONS

1. Proximity detectors are basically ________________ ________________.
2. What is the basic principle of operation used with detectors designed to detect only ferrous metals?
3. What is the basic principle of operation used with detectors designed to detect all types of metals?
4. What type of electric circuit is used to increase the sensitivity of the proximity detector?
5. What type of proximity detector uses an oscillator that operates at radio frequencies?

6. Name two types of proximity detectors that can be used to detect objects not made of metal.
7. What is the maximum range at which most capacitive proximity detectors can be used to sense an object?
8. How is it possible for an ultrasonic proximity detector to measure the distance to an object?

UNIT 27

Photodetectors

Objectives *After studying this unit, the student will be able to:*

- **List different devices used as light sensors**
- **Discuss the advantages of photo-operated controls**
- **Describe different methods of installing photodetectors**

APPLICATIONS

Photodetectors are widely used in today's industry. They can be used to sense the presence or absence of almost any object. Photodetectors do not have to make physical contact with the object they are sensing, so there is no mechanical arm to wear out. Many photodetectors can operate at speeds that cannot be tolerated by mechanical contact switches. They are used in almost every type of industry, and their uses are increasing steadily.

TYPES OF DETECTORS

Photo-operated devices fall into one of three categories: photovoltaic, photoemissive, and photoconductive.

Photovoltaic

Photovoltaic devices are more often called solar cells. They are made of silicon and have the ability to produce a voltage in the presence of light. The amount of voltage produced by a cell is determined by the material it is made of. When silicon is used, the solar cell produces .5 volts in the presence of direct sunlight. If there is a complete circuit connected to the cell, current will flow through the circuit. The amount of current produced by a solar cell is determined by the surface area of the cell. For instance, assume a solar cell has a surface area of 1 square inch, and another cell has a surface area of 4 square inches. If both cells are made of silicon, both will produce .5 volts when in direct sunlight. The larger cell, however, will produce four times as much current as the small one.

Figure 27-1 shows the schematic symbol for a photovoltaic cell. Notice that the symbol is the same as the symbol used to represent a single cell battery except for the arrow pointing toward it. The battery symbol means the device has the ability to produce a voltage, and the arrow means that it must receive light to do so.

Photovoltaic cells have the advantage of being able to operate electrical equipment without external power. Since silicon solar cells produce only .5 volt, it is often necessary to connect several of

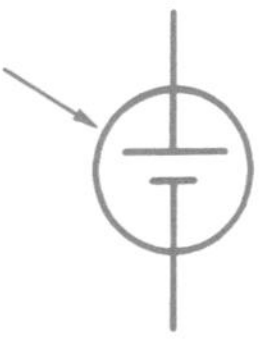

FIGURE 27-1 Schematic symbol for a photovoltaic cell

them together to obtain enough voltage and current to operate the desired device. For example, assume that solar cells are to be used to operate a dc relay coil that requires 3 volts at 250 milliamps. Now assume that the solar cells to be used have the ability to produce .5 volt at 150 milliamps. If six solar cells are connected in series, they will produce 3 volts at 150 milliamps, figure 27-2. The voltage produced by the connection is sufficient to operate the relay, but the current capacity is not. Therefore, six more solar cells must be connected in series. This connection is then connected parallel to the first connection producing a circuit that has a voltage rating of 3 volts and a current rating of 300 milliamps, which is sufficient to operate the relay coil.

Photoemissive Devices

Photoemissive devices emit electrons when in the presence of light. They include such devices as the phototransistor, the photodiode, and the photo-SCR. The schematic symbols for these devices are shown in figure 27-3. The emission of electrons is used to turn these solid-state components on. The circuit in figure 27-4 shows a phototransistor used to turn on a relay coil. When the phototransistor is in darkness, no electrons are emitted by the base junction, and the transistor is turned off. When the phototransistor is in the presence of light, it turns on and permits current to flow through the relay coil. The diode connected parallel to the relay coil is known as a kickback or freewheeling diode. Its function is to pre-

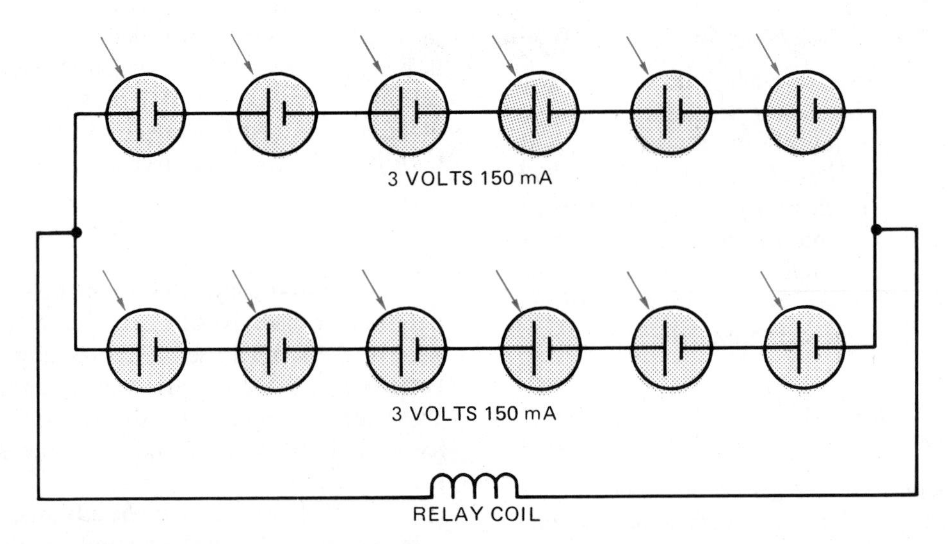

FIGURE 27-2 Series-parallel connection of solar cells produces 3 volts at 300 milliamps

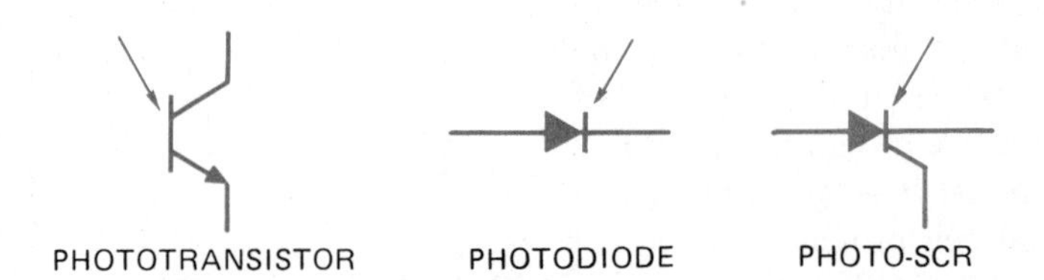

FIGURE 27-3 Schematic symbols for the phototransistor, the photodiode, and the photo-SCR

vent induced voltage spikes from occurring when the current suddenly stops flowing through the coil and the magnetic field collapses.

In the circuit shown in figure 27-4, the relay coil will turn on when the phototransistor is in the presence of light, and turn off when the phototransistor is in darkness. Some circuits may require the reverse operation. This can be accomplished by adding a resistor and a junction transistor to the circuit, figure 27-5. In this circuit a common junction transistor is used to control the current flow through the relay coil. Resistor R1 limits the current flow through the base of the junction transistor. When the phototransistor is in darkness, it has a very high resistance. This permits current to flow to the base of the junction transistor and turn it on. When the phototransistor is in the presence of light, it turns on and connects the base of the junction transistor to the negative side of the battery. This causes the junction transistor to turn off. The phototransistor in the circuit is used as a *stealer* transistor. A stealer transistor steals the base current away from some other transistor to keep it turned off.

Some circuits may require the phototransistor to have a higher gain than it has under normal conditions. This can be accomplished by using the phototransistor as the driver for a Darlington amplifier circuit, figure 27-6. A Darlington amplifier circuit generally has a gain of over 10,000.

Photodiodes and photo-SCRS are used in circuits similar to those shown for the phototransistor. The photodiode will permit current to flow through it in the presence of light. The photo-SCR has the same operating characteristics as a common junction SCR. The only difference is that light is used to trigger the gate when using a photo-SCR.

Regardless of the type of photoemissive device used, or the type circuit it is used in, the greatest advantage of the photoemissive device is speed. A photoemissive device can turn on or off in a few microseconds. Photovoltaic or photoconductive devices generally require several milliseconds to turn on or off. This makes the use of photoemissive devices imperative in high speed switching circuits.

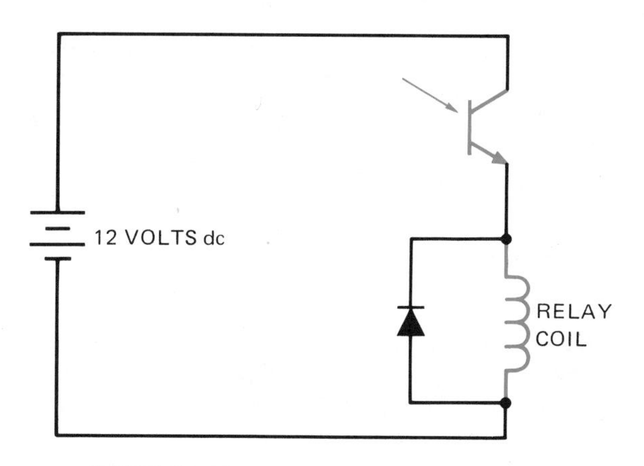

FIGURE 27-4 Phototransistor controls relay coil

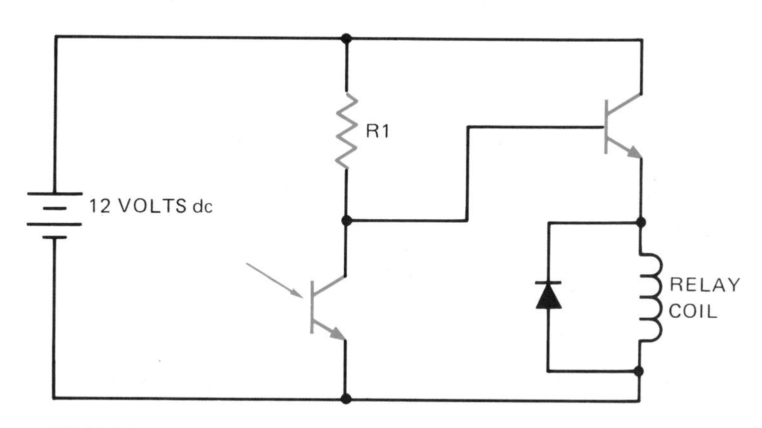

FIGURE 27-5 The relay turns on when the phototransistor is in darkness

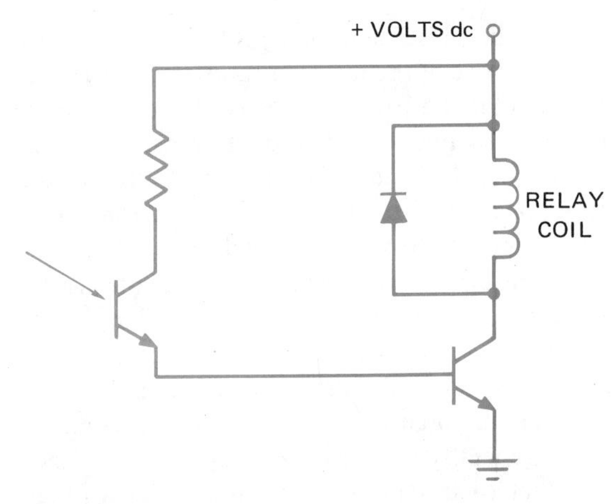

FIGURE 27-6 The phototransistor is used as the driver for a Darlington Amplifier

Photoconductive Devices

Photoconductive devices exhibit a change of resistance due to the presence or absence of light. The most common photoconductive device is the cadmium sulfide cell or cad cell. The cad cell has a resistance of about 50 ohms in direct sunlight and several hundred thousand ohms in darkness. It is generally used as a light sensitive switch. The schematic symbol for a cad cell is shown in figure 27-7. Figure 27-8 shows a typical cad cell.

Figure 27-9 shows a basic circuit of a cad cell being used to control a relay. When the cad cell is in darkness, its resistance is high. This prevents the amount of current needed to turn the relay on from flowing through the circuit. When the cad cell is in the presence of light, its resistance is low. The amount of current needed to operate the relay can now flow through the circuit.

Although this circuit will work if the cad cell is large enough to handle the current, it has a couple of problems.

1. There is no way to adjust the sensitivity of the circuit. Photo-operated switches are generally located in many different areas of a plant. The surrounding light intensity can vary from one area to another. It is, therefore, necessary to be able to adjust the sensor for the amount of light needed to operate it.
2. The sense of operation of the circuit cannot be changed. The circuit shown in figure 27-9 permits the relay to turn on when the cad cell is in the presence of light. There may be conditions that would make it desirable to turn the relay on when the cad cell is in darkness.

Figure 27-10 shows a photodetector circuit that uses a cad cell as the sensor and an operational amplifier as the control circuit. The circuit operates as follows. Resistor R1 and the cad cell form a voltage divider circuit which is connected

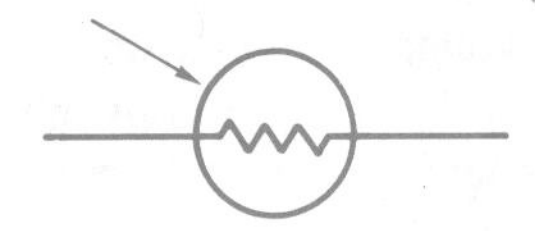

FIGURE 27-7 Schematic symbol for a cad cell

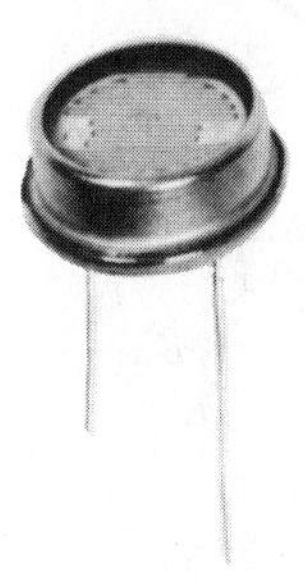

FIGURE 27-8 Cad cell (Courtesy EG&G Vactec, Inc.)

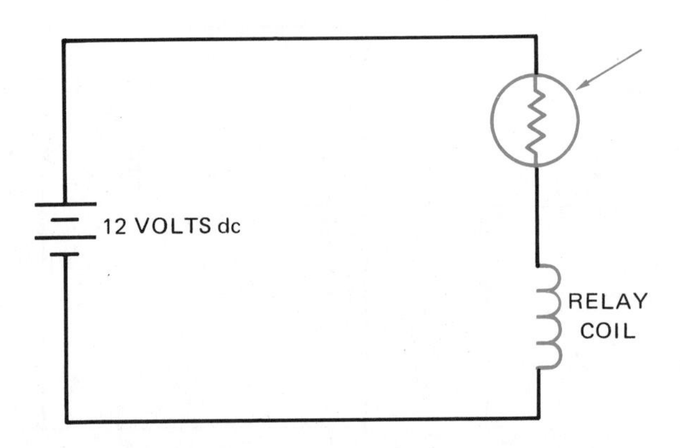

FIGURE 27-9 Cad cell controls relay coil

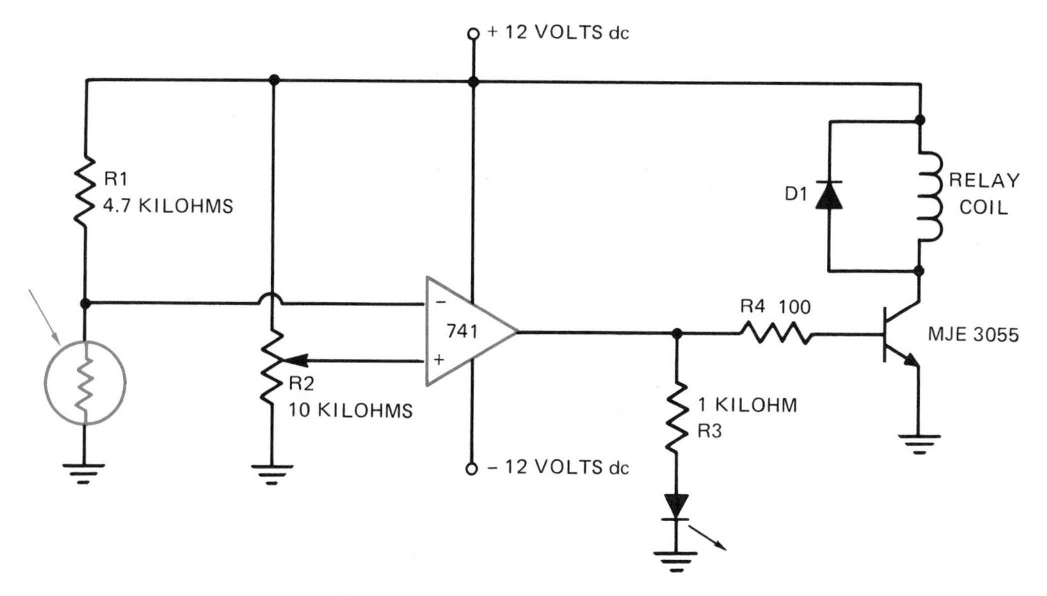

FIGURE 27-10 The relay coil is energized when the cad cell is in the presence of light

to the inverting input of the amplifier. Resistor R2 is used as a potentiometer to preset a positive voltage at the noninverting input. This control adjusts the sensitivity of the circuit. Resistor R3 limits the current to a light-emitting diode (LED). The LED is mounted on the outside of the case of the photodetector and is used to indicate when the relay coil is energized. Resistor R4 limits the base current to the junction transistor. The junction transistor is used to control the current needed to operate the relay coil. Many op amps do not have enough current rating to control this amount of current. Diode D1 is used as a kickback diode.

Assume that Resistor R2 has been adjusted to provide a potential of 6 volts at the noninverting input. When the cad cell is in the presence of light, it has a low resistance and a potential less than 6 volts is applied to the inverting input. Since the noninverting input has a higher positive voltage connected to it, the output is high also. When the output of the op amp is high, the LED and the transistor are turned on.

When the cad cell is in the presence of darkness, its resistance increases. When its resistance becomes greater than 4.7 kilohms, a voltage greater than 6 volts is applied to the inverting input. This causes the output of the op amp to change from a high state to a low state, and turn the LED and transistor off. Notice in this circuit that the relay is turned on when the cad cell is in the presence of light, and turned off when it is in darkness.

Figure 27-11 shows a connection that will reverse the operation of the circuit. The potentiometer has been reconnected to the inverting input, and the voltage divider circuit has been connected to the noninverting input. To understand the operation of this circuit, assume that a potential of 6 volts has been preset at the inverting input.

When the cad cell is in the presence of light, it has a low resistance and a voltage less than 6 volts is applied to the noninverting input. Since the inverting input has a greater positive voltage connected to it, the output is low and the LED and the transistor are turned off.

When the cad cell is in darkness, its resistance becomes greater than 4.7 kilohms and a voltage greater than 6 volts is applied to the noninverting input. This causes the output of the op amp to change to a high state which turns on the LED and transistor. Notice that this circuit turns

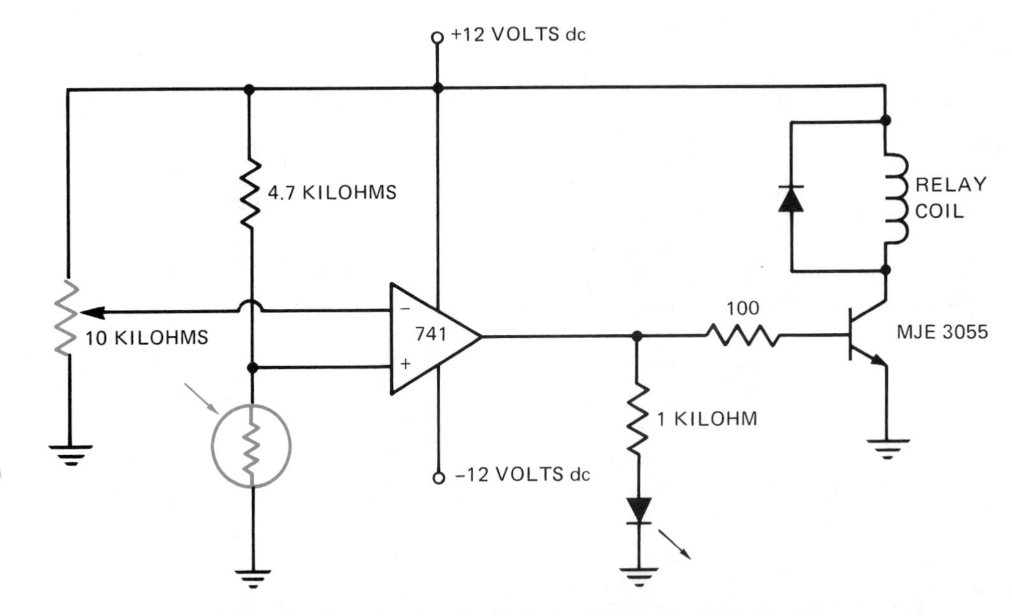

FIGURE 27-11 The relay is energized when the cad cell is in darkness

the relay on when the cad cell is in darkness and off when it is in the presence of light.

MOUNTING

Photodetectors designed for industrial use are made to be mounted and used in different ways. There are two basic types of photodetectors: one type has separate transmitter and receiver units; the other type has both units mounted in the same housing. The type used is generally determined by the job requirements. The transmitter section is the light source which is generally a long life incandescent bulb. There are photodetectors, however, that use an infrared transmitter. These cannot be seen by the human eye and are often used in burglar alarm systems. The receiver unit houses

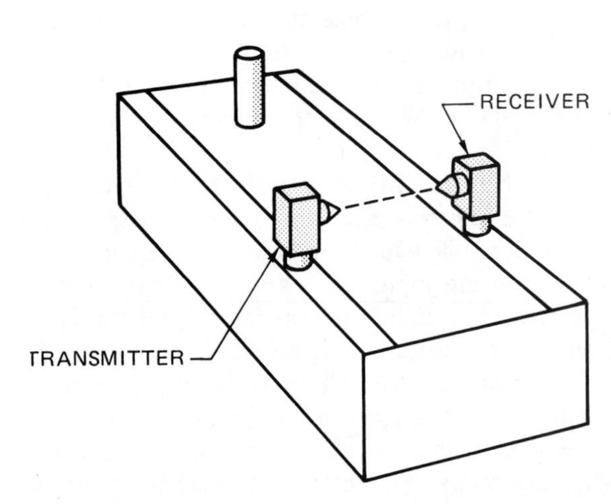

FIGURE 27-12 Photodetector senses presence of object on conveyor line

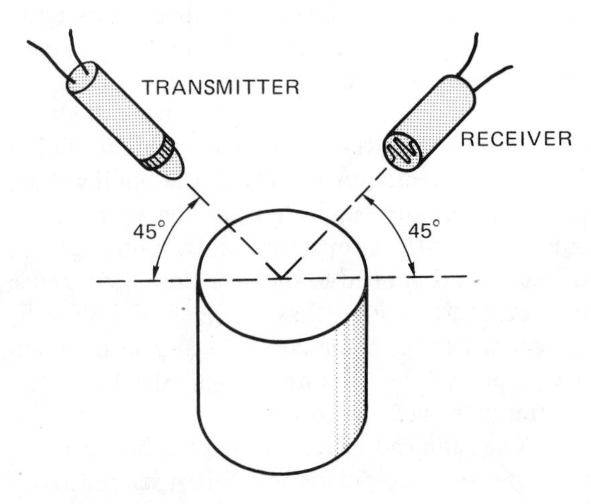

FIGURE 27-13 Object is sensed by reflecting light off a shiny surface

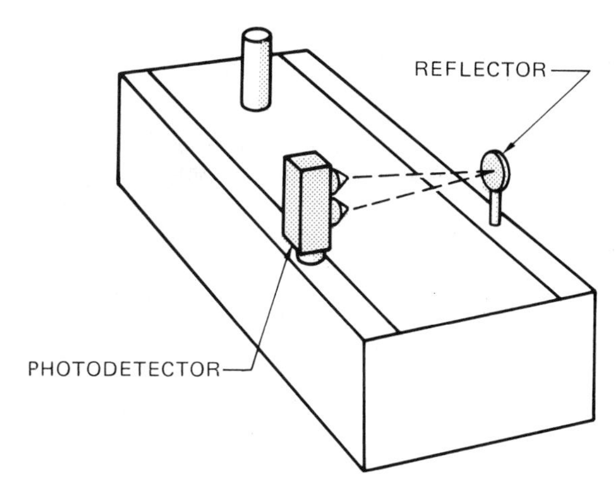

FIGURE 27-14 The object is sensed when it passes between the photodetector and the reflector

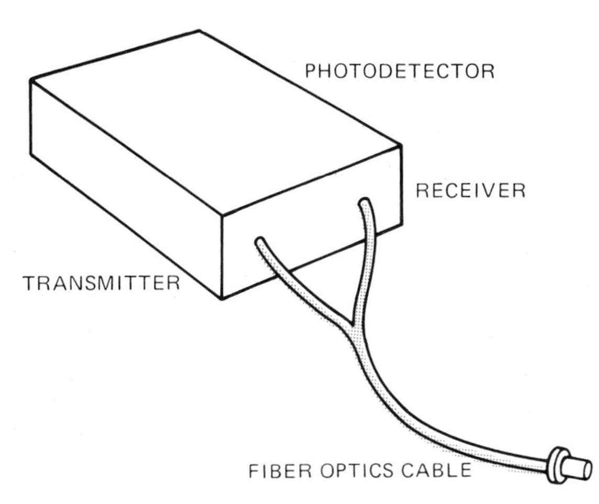

FIGURE 27-15 Optical cable is used to transmit and receive light

the photodetector and, generally, the circuitry required to operate the system.

Figure 27-12 shows a photodetector used to detect the presence of an object on the conveyor line. When the object passes between the transmitter and receiver units, the light beam is broken and the detector activates. Notice that no physical contact was necessary for the photodetector to sense the presence of the object.

Figure 27-13 illustrates another method of mounting the transmitter and receiver. In this example, an object is sensed by reflecting light off of a shiny surface. Notice that the transmitter and receiver must be mounted at the same angle with respect to the object to be sensed. This type of mounting will only work with objects that have the same height, such as cans on a conveyor line.

Photodetectors that have both the transmitter and the receiver units mounted in the same housing depend on a reflector for operation. Figure 27-14 shows this type of unit mounted on a conveyor line. The transmitter is aimed at the reflector. The

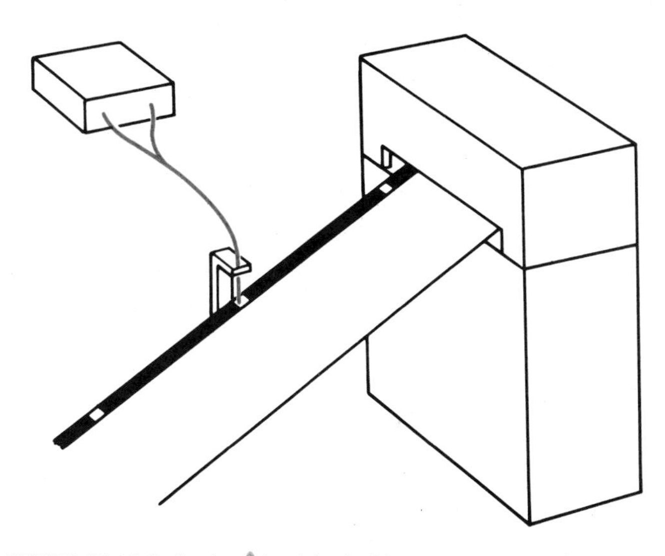
FIGURE 27-16 Optical cable detects shiny area on one side of label

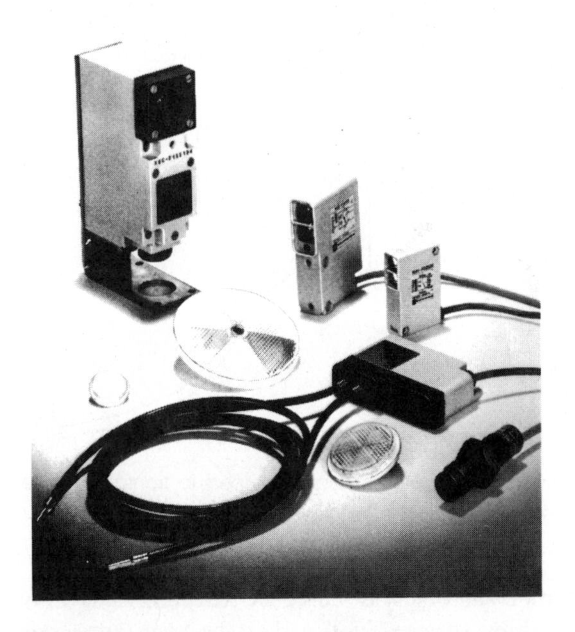

FIGURE 27-17 Different types of photodetectors are shown. Some operate with reflectors and others operate with optical fiber cable. (Courtesy Telemecanique Inc.)

light beam is reflected back to the receiver. When an object passes between the photodetector unit and the reflector, the light to the receiver is interrupted. This type of unit has the advantage of needing electrical connection at only one piece of equipment. This permits easy mounting of the photodetector unit, and mounting of the reflector in hard to reach positions that would make running control wiring difficult. Many of these units have a range of 20 feet and more.

Another type of unit that operates on the principle of reflected light uses an optical fiber cable. The fibers in the cable are divided in half. One half of the fibers is connected to the transmitter, and the other half is connected to the receiver, figure 27-15. This unit has the advantage of permitting the transmitter and the receiver to be mounted in a very small area. Figure 27-16 illustrates a common use for this type of unit. The unit is used to control a label cutting machine. The labels are printed on a large roll and must be cut for individual packages. The label roll contains a narrow strip on one side which is dark colored except for shiny sections spaced at regular intervals. The optical fiber cable is located above this narrow strip. When the dark surface of the strip is passing beneath the optical cable, no reflected light returns to the receiver unit. When the shiny section passes beneath the cable, light is reflected back to the receiver unit. The photodetector sends a signal to the control circuit and tells it to cut the label.

Photodetectors are very dependable and have an excellent maintenance and service record. They can be used to sense almost any object without making physical contact with it, and can operate millions of times without damage or wear. A variety of photodetectors is shown in figure 27-17.

REVIEW QUESTIONS

1. List the three major categories of photodetectors.
2. In which category does the solar cell belong?
3. In which category do phototransistors and photodiodes belong?
4. In which category does the cad cell belong?
5. The term cad cell is a common name for what device?
6. What is the function of the transmitter in a photodetector unit?
7. What is the advantage of a photodetector that uses a reflector to operate?
8. An object is to be detected by reflecting light off a shiny surface. If the transmitter is mounted at a 60 degree angle, at what angle must the receiver be mounted?
9. How much voltage is produced by a silicon solar cell?
10. What determines the amount of current a solar cell can produce?

UNIT 28

The Control Transformer

Objectives *After studying this unit, the student will be able to:*

- **Discuss the use of control transformers in a control circuit**
- **Connect a control transformer for operation on a 240- or 480-volt system**

Most industrial motors operate on voltages that range from 240 to 480 volts. Magnetic control systems, however, generally operate on 120 volts. A control transformer is used to step the 240 or 480 volts down to 120 volts to operate the control system. There is really nothing special about a control transformer except that most of them are made with two primary windings and one secondary winding. Each primary winding is rated at 240 volts and the secondary winding is rated at 120 volts. This means there is a turns ratio of 2:1 (2 to 1) between each primary winding and the secondary winding. For example, assume that each primary winding contains 200 turns of wire and the secondary winding contains 100 turns. There are two turns of wire in each primary winding for every one turn of wire in the secondary.

One of the primary windings of the control transformer is labeled H1 and H2. The other primary winding is labeled H3 and H4. The secondary winding is labeled X1 and X2. If the transformer is to be used to step 240 volts down to 120 volts, the two primary windings are connected parallel to each other as shown in figure 28-1. Notice that in figure 28-1 the H1 and H3 leads are connected together, and the H2 and H4 leads are connected together. Since the voltage applied to each primary winding is the same, the effect is the same as having only one primary winding with 200 turns of wire in it. This means that when the transformer is connected in this manner, the turns ratio is 2:1. When 240 volts are connected to the primary winding, the secondary voltage is 120 volts.

If the transformer is to be used to step 480 volts down to 120 volts, the primary windings are connected in series as shown in figure 28-2. With the windings connected in series, the primary

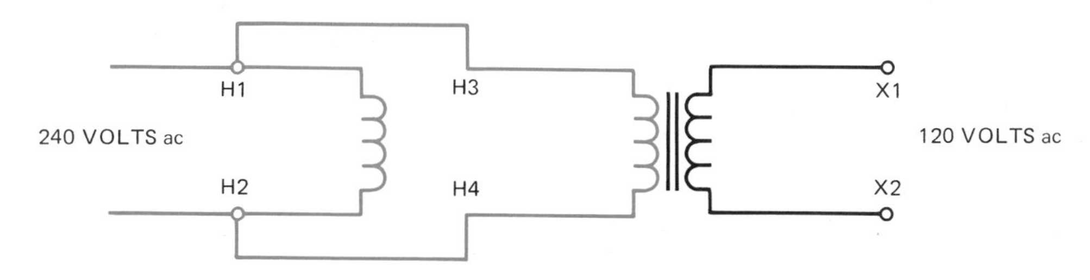

FIGURE 28-1 Primaries connected in parallel for 240-volt operation

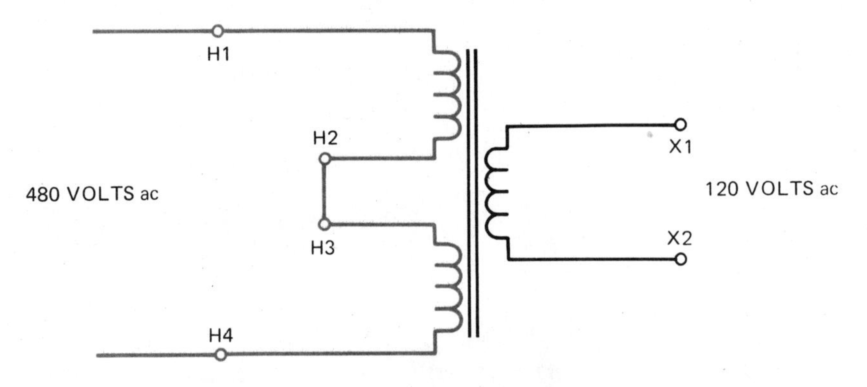

FIGURE 28-2 Primaries connected in series for 480-volt operation

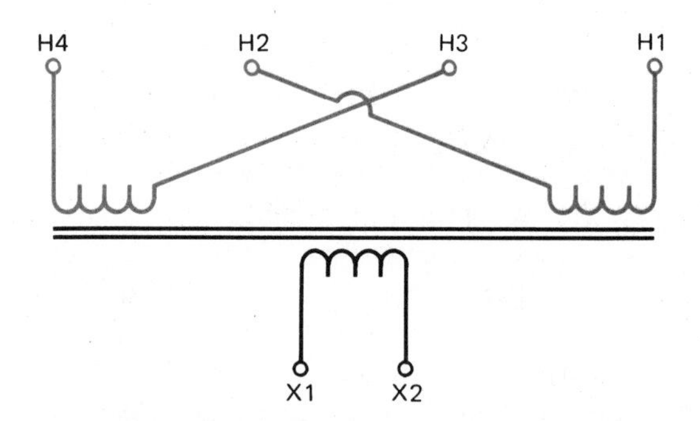

FIGURE 28-3 Primary leads are crossed

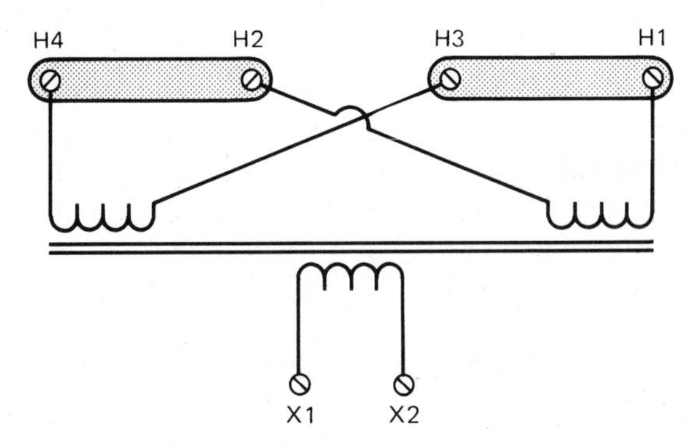

FIGURE 28-4 Metal links used to make a 240-volt connection

winding now has a total of 400 turns of wire, which makes a turns ratio of 4:1. When 480 volts is connected to the primary winding, the secondary winding has an output of 120 volts.

Control transformers generally have screw terminals connected to the primary and secondary leads. The H2 and H3 leads are crossed to make connection of the primary winding easier, figure 28-3. For example, if the transformer is to be connected for 240-volt operation, the two primary windings must be connected parallel to each other as shown in figure 28-1. This connection can be made on the transformer by using one metal link to connect leads H1 and H3, and another metal link to connect H2 and H4, figure 28-4.

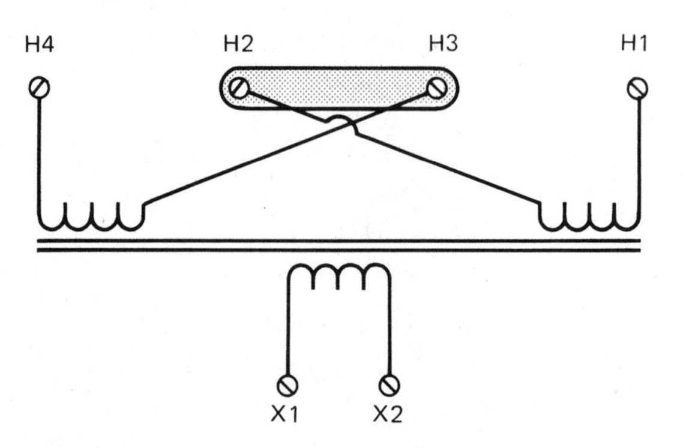

FIGURE 28-5 Metal link used to make a 480-volt connection

If the transformer is to be used for 480-volt operation, the primary windings must be connected in series as shown in figure 28-2. This connection can be made on the control transformer by using a metal link to connect H2 and H3 as shown in figure 28-5. A typical control transformer is shown in figure 28-6.

FIGURE 28-6 Control transformer with fuse protection added to the secondary winding (Courtesy Hevi-Duty Electric Co.)

REVIEW QUESTIONS

1. What is the operating voltage of most magnetic control systems?
2. How many primary windings do control transformers have?
3. How are the primary windings connected when the transformer is to be operated on a 240-volt system?
4. How are the primary windings connected when the transformer is to be operated on a 480-volt system?
5. Why are two of the primary leads crossed on a control transformer?

SECTION 3

Control Circuits

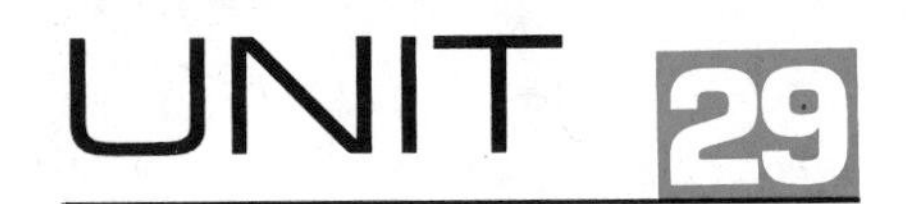

Basic Control Circuits

Objectives *After studying this unit, the student will be able to:*

- **Describe the operation of a two-wire circuit**
- **Describe the operation of a three-wire circuit**

TWO-WIRE CONTROLS

Magnetic control circuits are divided into two basic types: the two-wire control circuit, and the three-wire control circuit. Two-wire control circuits are operated by manual control devices such as the open starter in figure 12-1 or the manual push-button starter in figure 12-9B. This type of control circuit provides overload protection for the motor connected to it and provides under-voltage or no-voltage release. Figure 29-1 shows a typical two-wire control circuit. Notice that as long as the two-wire control device is closed, power can be supplied to the coil of the controller. If the motor is stopped by a power interruption, the two-wire control device will not open. Since the control device does not open, the motor will restart when power is restored to the system.

Two-wire control circuits are used in applications where this self restarting characteristic is desirable. This enables devices such as blower fans and pumps to restart after a power failure without an electrician having to walk around the plant and restart each device. Since two-wire control does permit automatic restarting of equipment, it could become a safety hazard to people working around the equipment. For this reason, two-wire controls should be used only when there is little or no danger of a person being injured if the equipment should suddenly restart after a power failure.

THREE-WIRE CONTROLS

Three-wire control circuits are generally operated by momentary contact pilot devices. The simplest example of a three-wire control circuit is probably the start-stop, push-button control shown in figure 29-2. A set of auxiliary contacts controlled by M coil is connected parallel to the start button. These contacts are generally referred to as *maintaining, sealing,* or *holding* contacts. The job of this contact is to hold the coil in the circuit after the start button has been used to energize the relay.

If the start button in figure 29-2 is pressed, a circuit is completed between points 2 and 3, which allows current to flow through the motor starter coil and the normally closed overload contact to line 2 (L2). When current flows through the M relay coil, the relay energizes and closes all M contacts. Since the contact that is connected parallel to the start button is now closed, the start button can be returned to its normally open position. Contact M maintains a current path around the start button to keep coil M energized.

Three-wire control is used to a much greater extent than two-wire control because of its flexibility. Pilot devices such as push buttons can be located at remote locations such as control panels, while motor starters and control relays are housed in separate cabinets. Three-wire circuits also per-

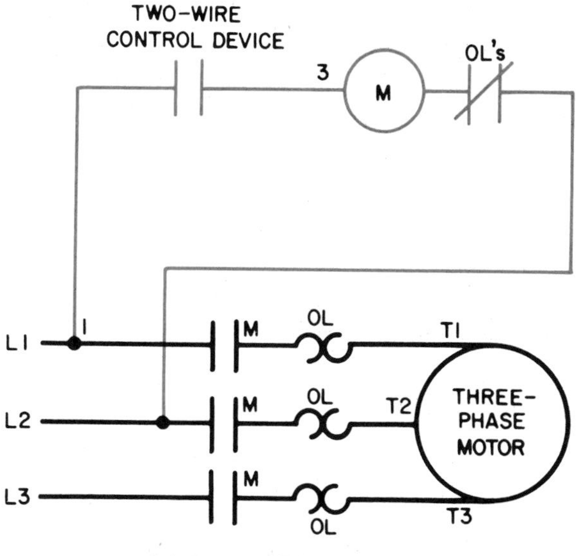

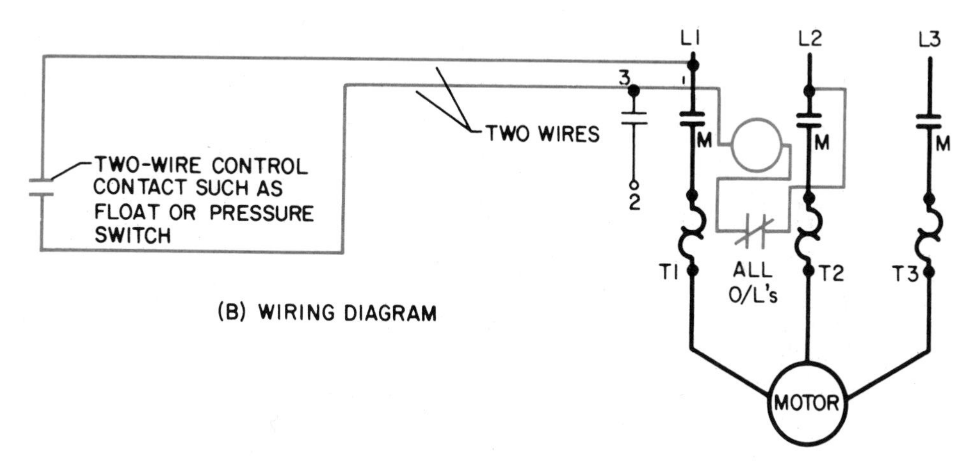

FIGURE 29-1 Basic two-wire control circuit

mit the use of different types of pilot devices such as float switches, pressure switches, and limit switches.

Since three-wire control circuits use holding contacts to maintain the circuit, if power is interrupted, the equipment will not restart when power is restored. If the power supplying the circuit shown in figure 29-2 is stopped, M contacts will return to their normally open position. When power is restored, the start button must again be pressed to re-energize the M coil.

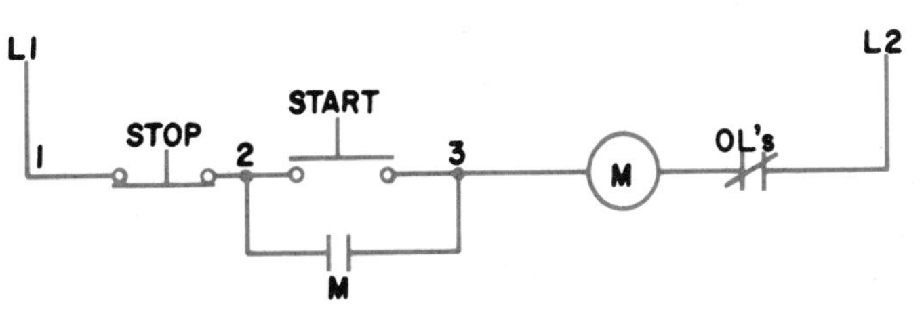

FIGURE 29-2 Basic three-wire control circuit

REVIEW QUESTIONS

1. What are some advantages of using two-wire control circuits?
2. What is a possible safety hazard of two-wire control circuits?
3. What is the advantage of a three-wire control system compared to a two-wire system?
4. What are holding contacts?

UNIT 30

Schematics and Wiring Diagrams

Objectives ***After studying this unit, the student will be able to:***

- **Interpret schematic diagrams**
- **Interpret wiring diagrams**
- **Connect control circuits using schematic and wiring diagrams**

Schematic and wiring diagrams are the written language of control circuits. If a maintenance electrician is going to install control equipment, or troubleshoot existing control circuits, he must be able to interpret schematic and wiring diagrams. Schematic diagrams are also known as line diagrams and ladder diagrams. *Schematic diagrams show components in their electrical sequence without regard to physical location.* Schematics are used more than any other type of diagram to connect or troubleshoot a control circuit.

Wiring diagrams show a picture of the control components with connecting wires. Wiring diagrams are sometimes used to install new control circuits, but they are seldom used for troubleshooting existing circuits. Figure 30-1A shows a schematic diagram of a start-stop, push-button circuit. Figure 30-1B shows a wiring diagram of the same circuit.

When reading schematic diagrams, the following rules should be remembered.

A. Read a schematic as you would a book—from top to bottom and from left to right.

B. Contact symbols are shown in their de-energized or off position.

C. When a relay is energized, all the contacts controlled by that relay change position. If a contact is shown normally open on the schematic, it will close when the coil controlling it is energized.

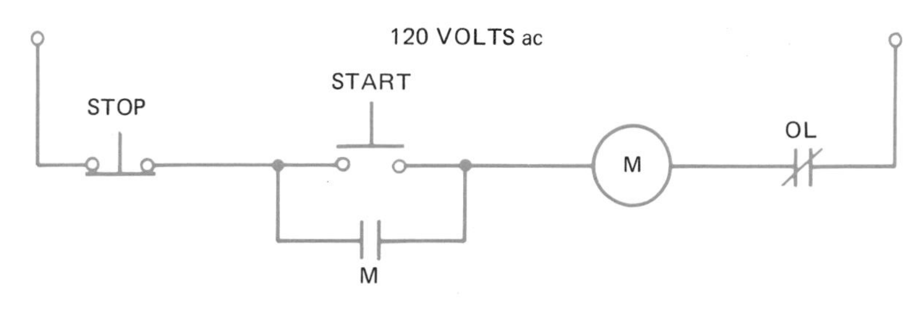

FIGURE 30-1A

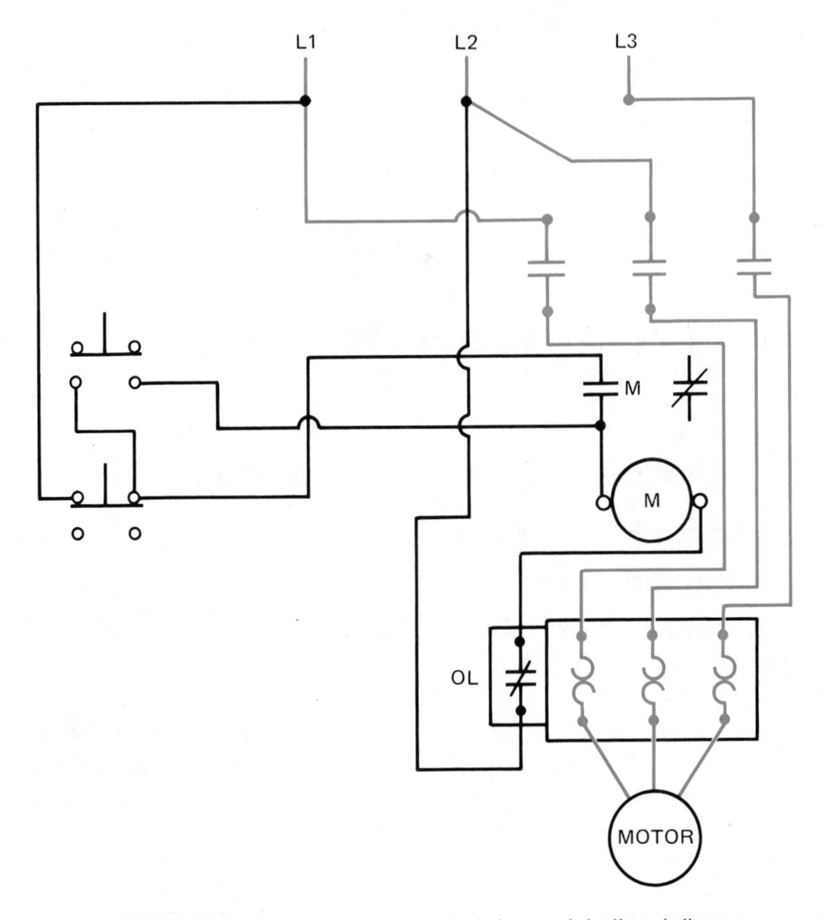

FIGURE 30-1B Wiring diagram of a start-stop push-button station

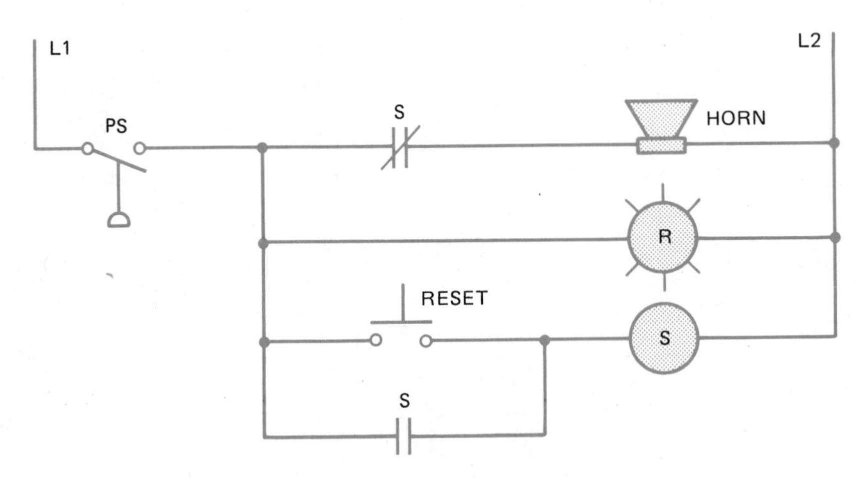

FIGURE 30-2A Alarm silencing circuit

The following circuits are used to illustrate how to interpret the logic of a control circuit using a schematic diagram.

Circuit #1

The circuit shown in figure 30-2A is an alarm silencing circuit. The purpose of the circuit is to sound a horn and turn on a red warning light when the pressure of a particular system becomes too great. After the alarm has sounded, the reset button can be used to turn the horn off, but the red warning light must remain on until the pressure in the system drops to a safe level. Notice that no current can flow in the system because of the open pressure switch, PS.

If the pressure rises high enough to cause pressure switch PS to close, current can flow through the normally closed S contact to the horn. Current can also flow through the red warning light. Current cannot, however, flow through the normally open reset button or the normally open S contact, figure 30-2B.

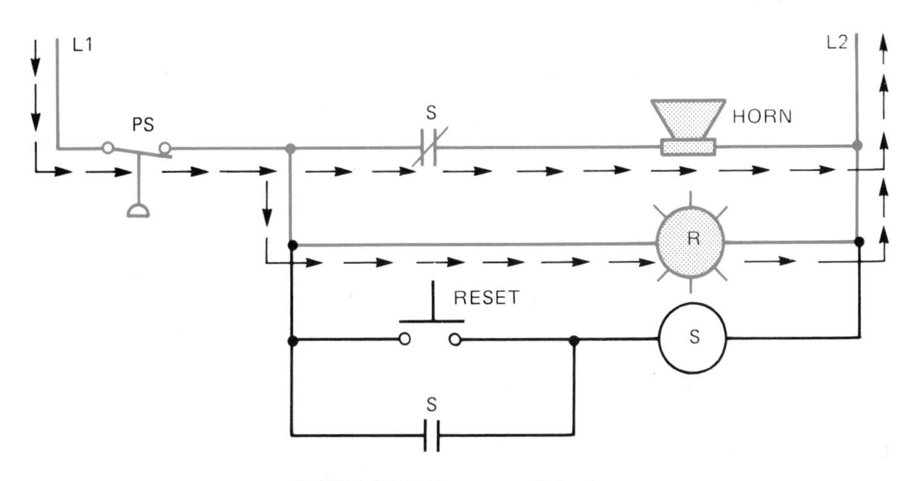

FIGURE 30-2B Pressure switch closes

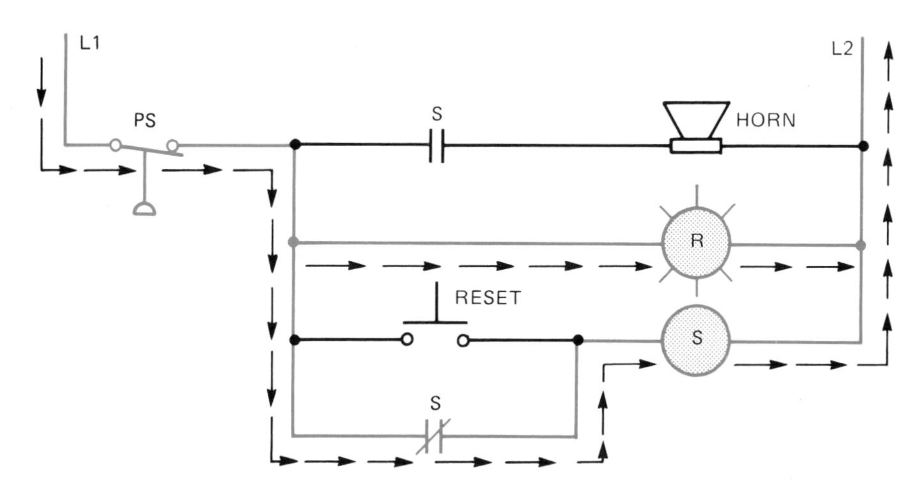

FIGURE 30-2C The alarm has been silenced but the warning light remains on

If the reset button is pushed, a circuit is completed through the S relay coil. When relay coil S energizes, the normally closed S contact opens and the normally open S contact closes. When the normally closed S contact opens, the circuit to the horn is broken. This causes the horn to turn off. The normally open S contact is used as a holding contact to maintain current to the coil of the relay when the reset button is released, figure 30-2C.

The red warning light will remain turned on until the pressure switch opens again. When the pressure switch opens, the circuit is broken and current flow through the system stops. This causes the red warning light to turn off, and it de-energizes the coil of relay S. When relay S de-energizes, both of the S contacts return to their original position. The circuit is now back to the same condition it was in in figure 30-2A.

REVIEW QUESTIONS

1. Define a schematic diagram.
2. Define a wiring diagram.
3. Referring to circuit 30-2A, explain the operation of the circuit if pressure switch PS was connected normally closed instead of normally open.

UNIT 31

Timed Starting for Three Motors (Circuit #2)

Objectives *After studying this unit, the student will be able to:*

- **Discuss the operation of circuit #2**
- **Troubleshoot circuit #2 using the schematic**

A machine contains three large motors. The current surge to start all three motors at the same time is too great for the system. Therefore, when the machine is to be started, there must be a delay of 10 seconds between the starting of each motor. The circuit shown in figure 31-1 is a start-stop, push-button control which controls three motor starters and two time-delay relays. The circuit is designed so that an overload on any motor will stop all motors.

When the start button is pressed, a circuit is completed through the start button, M1 motor starter coil, and TR1 relay coil. When coil M1 energizes, motor #1 starts and auxiliary contact M1, which is parallel to the start button, closes. This contact maintains the current flow through the circuit when the start button is released, figure 31-2.

After a 10-second interval, contact TR1 closes. When this contact closes, a circuit is completed through motor starter coil M2 and timer relay coil TR2. When coil M2 energizes, motor #2 starts, figure 31-3.

Ten seconds after coil TR2 energizes, contact TR2 closes. When this contact closes, a circuit is completed to motor starter coil M3, which causes motor #3 to start, figure 31-4.

If the stop button is pressed, the circuit to coils M1 and TR1 is broken. When motor starter M1 de-energizes, motor #1 stops and auxiliary contact M1 opens. TR1 is an on-delay relay; therefore, when coil TR1 is de-energized, contact TR1 opens immediately.

When contact TR1 opens, motor starter M2 de-energizes, which stops motor #2, and coil TR2 de-energizes. Since TR2 is an on-delay relay, contact TR2 opens immediately. This breaks the circuit to motor starter M3. When motor starter M3 de-energizes, motor #3 stops. Although it takes several seconds to explain what happens when the stop button is pressed, the action of the relays is almost instantaneous. If one of the overload contacts opens while the circuit is energized, the effect is the same as pressing the stop button. After the circuit stops, all contacts return to their normal positions and the circuit is the same as the original circuit shown in figure 31-1.

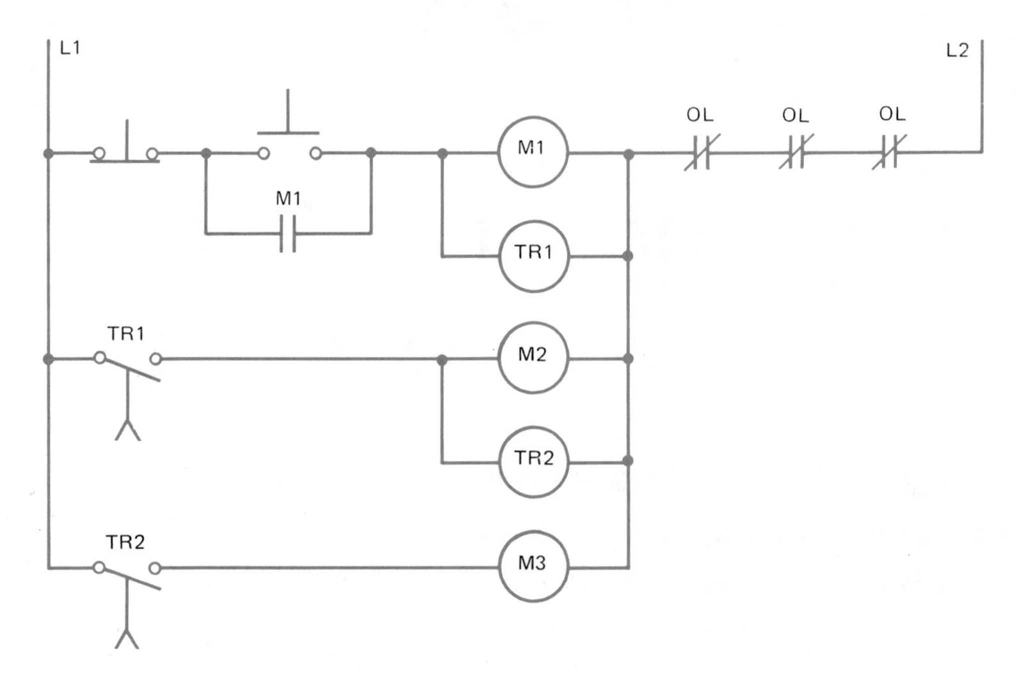

FIGURE 31-1 Time delay starting for three motors

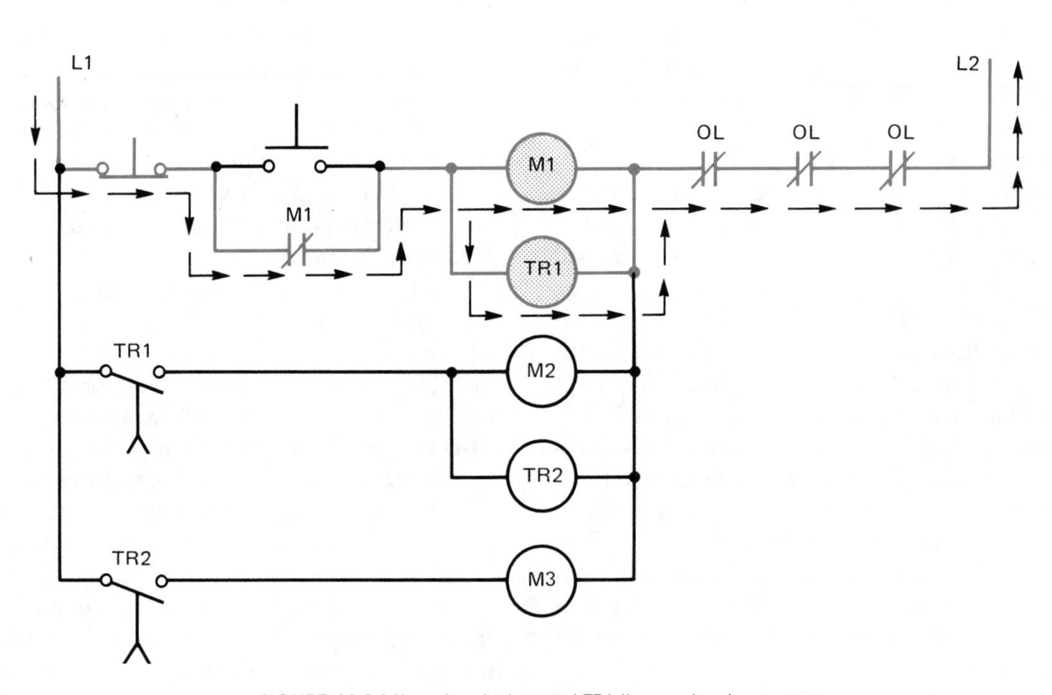

FIGURE 31-2 M1 motor starter and TR1 timer relay turn on

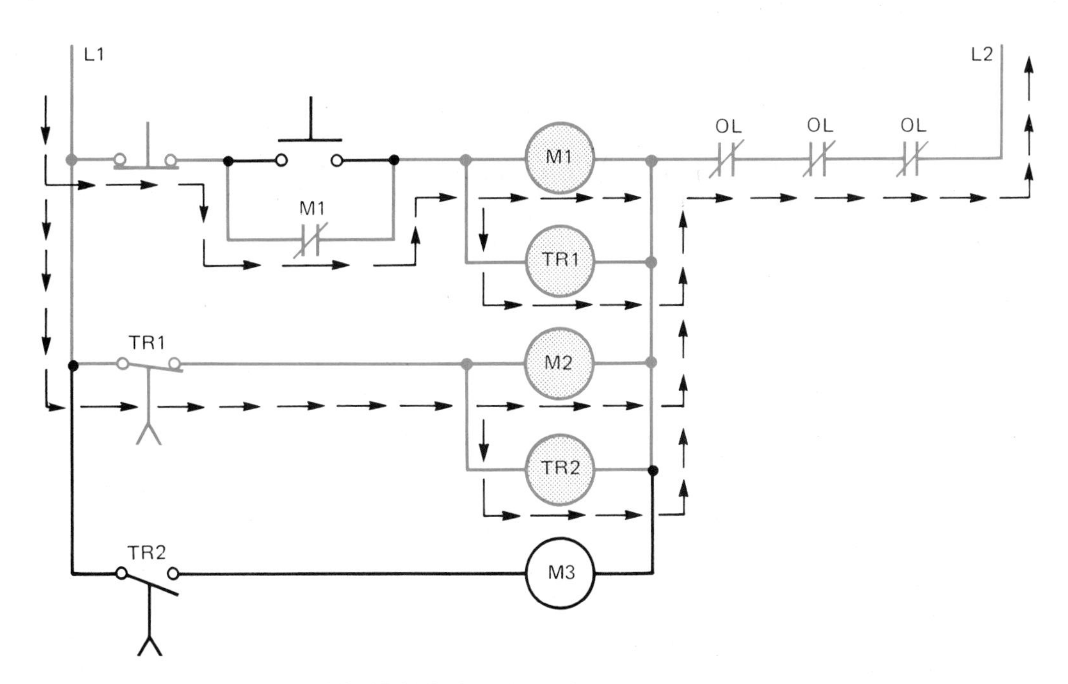

FIGURE 31-3 Motor 2 and TR2 have energized

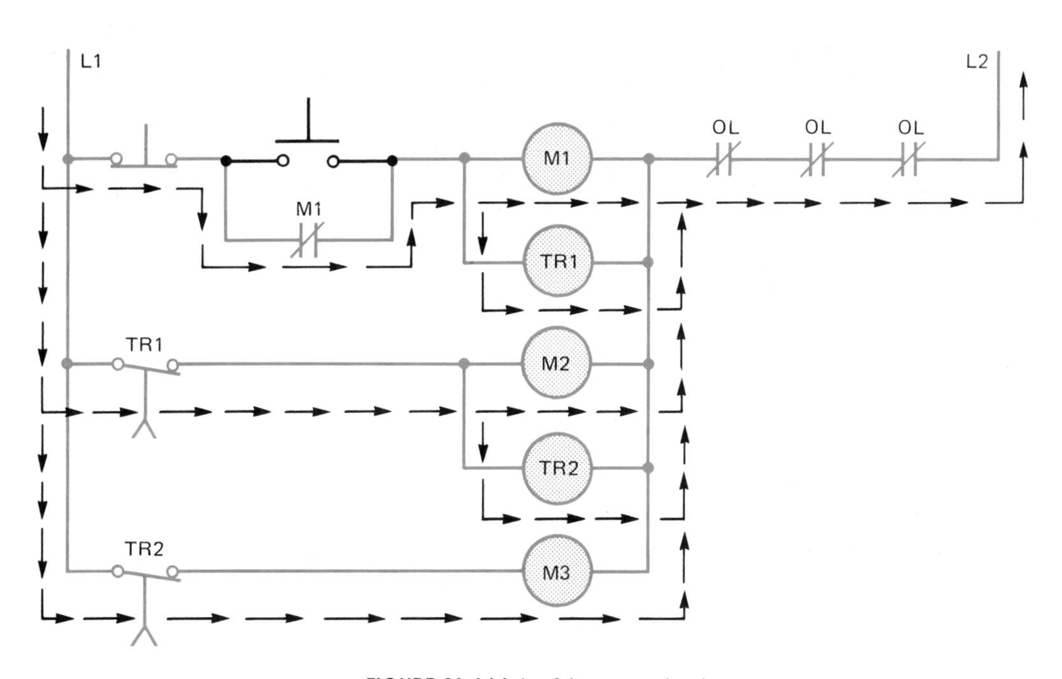

FIGURE 31-4 Motor 3 has energized

REVIEW QUESTIONS

(Refer to circuit 31-1.)

1. Explain the operation of circuit 31-1 if contact M1 did not close.
2. Explain the operation of circuit 31-1 if relay coil TR2 were burned out.

Float Switch Control of a Pump and Pilot Lights (Circuit #3)

Objectives ***After studying this unit, the student will be able to:***

- **Discuss the operation of circuit #3**
- **Troubleshoot circuit #3 using the schematic**

In this circuit, a float switch is used to operate a pump motor. The pump is used to fill a tank with water. When the tank is low on water, the float switch activates the pump motor and turns a red pilot light on. When the tank is filled with water, the float switch turns the pump motor and red pilot light off, and turns an amber pilot light on to indicate that the pump motor is not running. If the pump motor becomes overloaded, an overload relay stops the pump motor only.

The requirements for this circuit indicate that a float switch is to be used to control three different items: a red pilot light, a motor starter, and an amber pilot light. However, most pilot devices, such as float switches, pressure switches, and limit switches, seldom contain more than two contacts. When the circuit requires these pilot devices to use more contacts than they contain, it is common practice to let a set of contacts on the pilot device operate a control relay. The contacts of the control relay can be used as needed to fulfill the requirements of the circuit.

The float switch in figure 32-1 is used to operate a control relay labeled FSCR. The contacts of the control relay are used to control the motor starter and the two pilot lights.

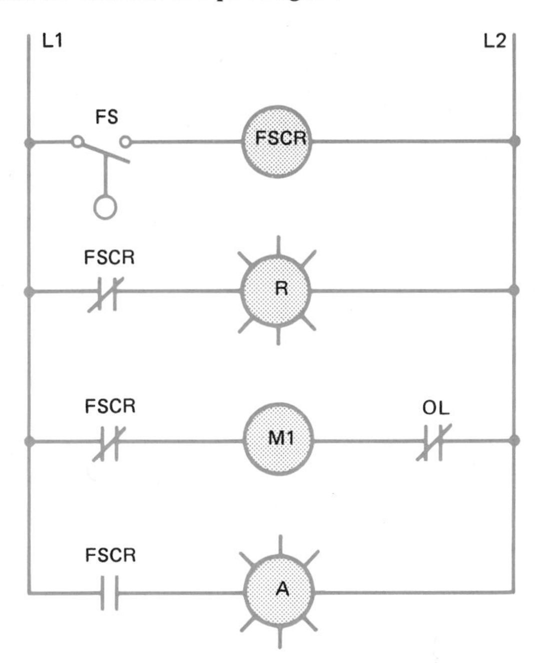

FIGURE 32-1 Float switch used to operate a control relay

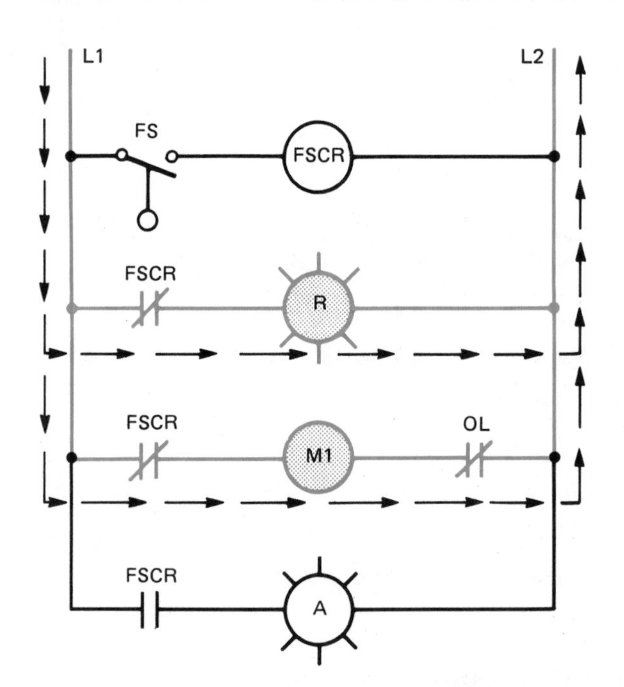

FIGURE 32-2 Warning light and pump motor have energized

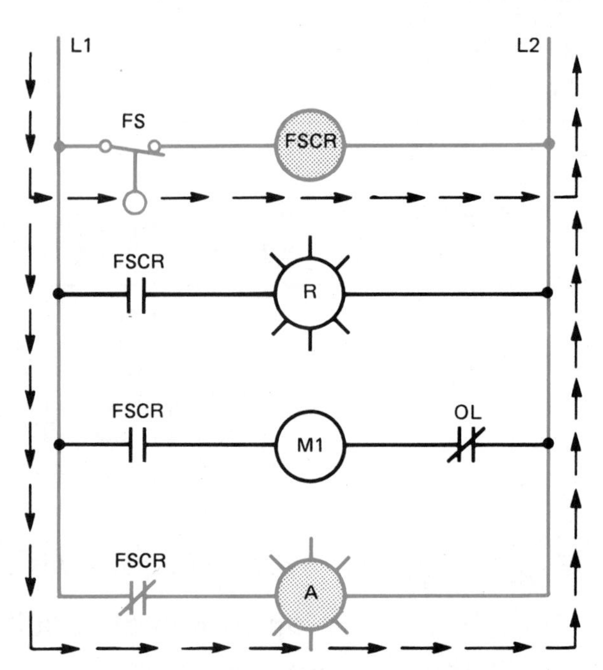

FIGURE 32-3 Float switch energizes FSCR relay

In the circuit shown in figure 32-2, current can flow through the normally closed FSCR contact to the red pilot light, and through a second normally closed FSCR contact to the coil of motor starter M1. When motor starter M1 energizes, the pump motor starts and begins to fill the tank with water. As water rises in the tank, the float of float switch FS rises also. When the tank is sufficiently filled, the float switch contact closes and energizes relay FSCR, figure 32-3.

When the coil of relay FSCR energizes, all FSCR contacts change. The normally closed contacts open and the normally open contact closes. When the normally closed contacts open, the circuits to the red pilot light and to coil M1 are broken. When motor starter M1 de-energizes, the pump motor stops. When the normally open FSCR contact closes, current flows to the amber pilot light. When the pump motor turns off, the water level begins to drop in the tank. When the water level drops low enough, the float switch opens and de-energizes relay coil FSCR. When relay FSCR de-energizes, all FSCR contacts return to their normal positions as shown in figure 32-1. If the pump motor is operating and the overload relay opens the overload contact, only the motor starter will be de-energized. The pilot lights will continue to operate.

REVIEW QUESTIONS

(Refer to circuit 32-1.)

1. Explain the operation of circuit 32-1 if float switch FS were connected normally closed instead of normally open.
2. Explain the operation of circuit 32-1 if relay coil M1 were burned out.

UNIT 33

Developing a Wiring Diagram (Circuit #1)

Objectives *After studying this unit, the student will be able to:*

- **Interpret a wiring diagram**
- **Develop a wiring diagram from a schematic diagram**
- **Connect a control circuit using a wiring or schematic diagram**

Wiring diagrams will now be developed for the three circuits just discussed. The method used for developing wiring diagrams is the same as the method used for installing new equipment. To illustrate this principle, the components of the system will be placed on paper and connections will be made to the various contacts and coils. Using a little imagination, it will be possible to visualize actual relays and contacts mounted in a panel, and wires connecting the various components.

Figure 33-1 shows the schematic for the alarm silencing circuit, and figure 33-2 shows the components of the system. The connection of the circuit is more easily understood with the aid of a simple numbering system. The rules for this system are as follows:

A. Each time a component is crossed the number must change.

B. Number all connected components with the same number.

C. Never use a number set more than once.

Figure 33-3 shows the schematic of the alarm silencing circuit with numbers placed beside each component. Notice that a 1 has been placed beside L1 and one side of the pressure switch. The pressure switch is a component. Therefore, the number must change when the pressure switch is crossed. The other side of the pressure switch is

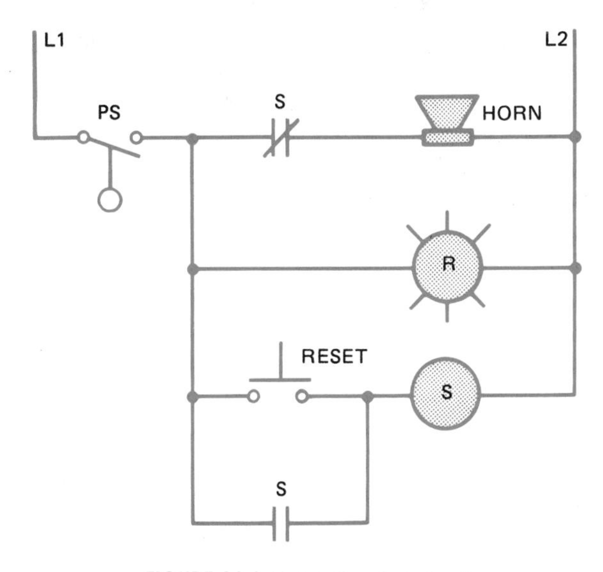

FIGURE 33-1 Alarm silencing circuit

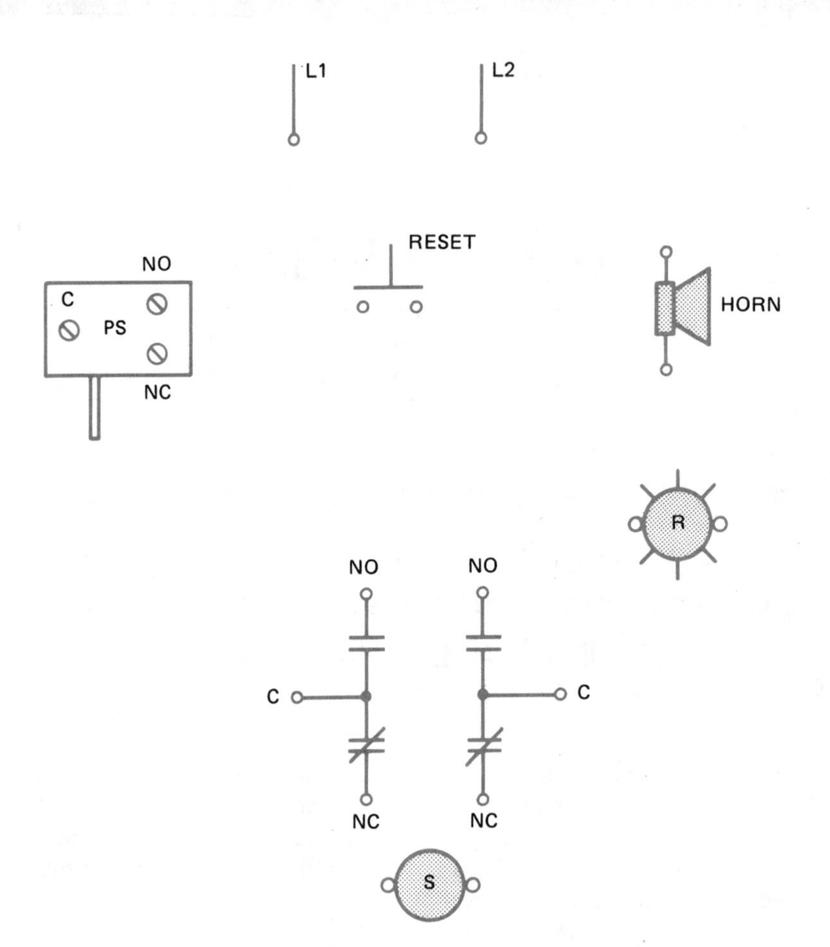

FIGURE 33-2 Circuit components

numbered with a 2. A 2 is also placed beside one side of the normally closed S contact, one side of the red warning light, one side of the normally open reset push button, and one side of the normally open S contact. All of these components are connected electrically; therefore, each has the same number.

When the normally closed S contact is crossed, the number is changed. The other side of the normally closed S contact is now a 3, and one side of the horn is a 3. The other side of the horn is connected to L2. The other side of the red warning light and one side of relay coil S is also connected to L2. All of these points are labeled with a 4. The other side of the normally open reset button, the other side of the normally open S contact, and the other side of relay coil S are numbered with a 5.

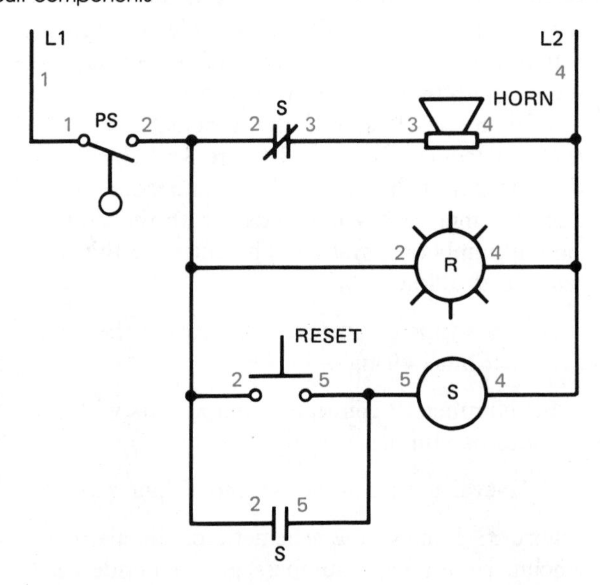

FIGURE 33-3 Numbers aid in circuit connection

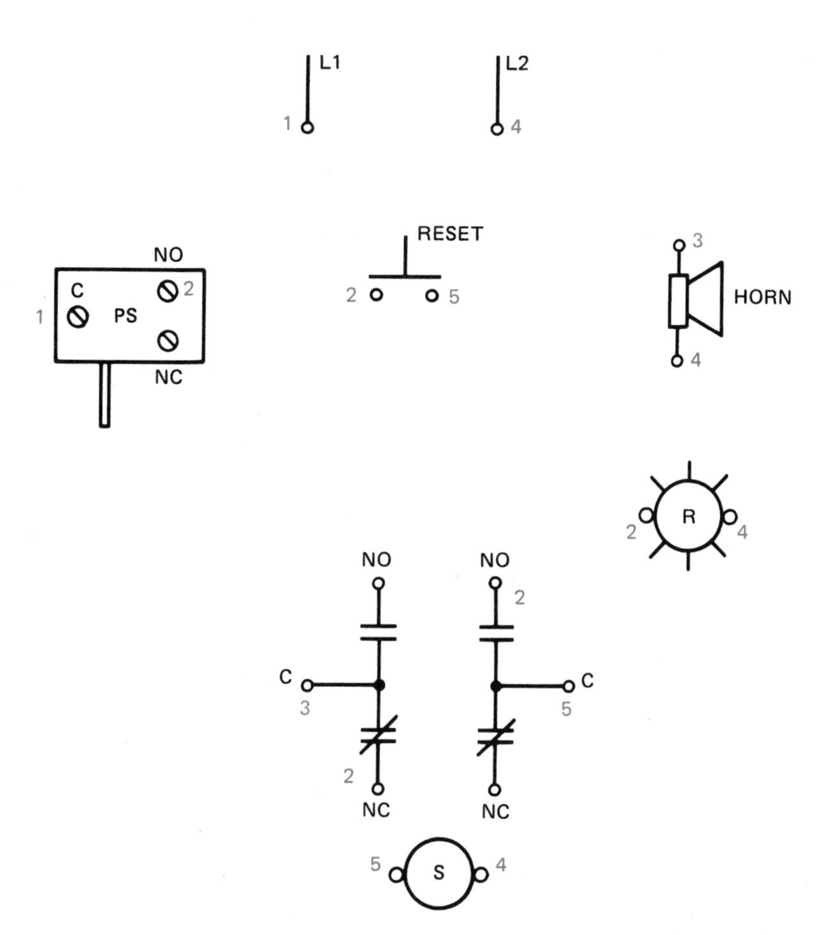

FIGURE 33-4 Circuit components have been numbered to match the schematic

The same numbers that are used to label the schematic in figure 33-3 are used to label the components shown in figure 33-4. L1 in the schematic is labeled with a 1; therefore, 1 is used to label L1 on the wiring diagram in figure 33-4. One side of the pressure switch in the schematic is labeled with a 1 and the other side is labeled with a 2. The pressure switch in the wiring diagram is shown with three terminals. One terminal is labeled C for common, one is labeled NO for normally open, and one is labeled NC for normally closed. This is a common contact arrangement used on many pilot devices and control relays (see figure 15-1). In the schematic the pressure switch is connected as a normally open device; therefore, terminals C and NO will be used. A 1 is placed by terminal C and a 2 is placed beside terminal NO. Notice that a 2 has also been placed beside one side of the normally open reset button, one side of the normally closed contact located on relay S, one side of the normally open contact located on relay S, and one side of the red warning light. A 3 is placed beside the common terminal of relay contact S which is used to produce a normally closed contact, and beside one of the terminal connections of the horn. A 4 is placed beside L2, the other terminal of the horn, the other side of the red warning light, and one side of relay coil S. A 5 is placed on the other side of relay coil S, the other side of the normally open reset button, and on the common terminal of relay contact S which is used as a normally open contact.

Notice that the numbers used to label the components of the wiring diagram are the same as

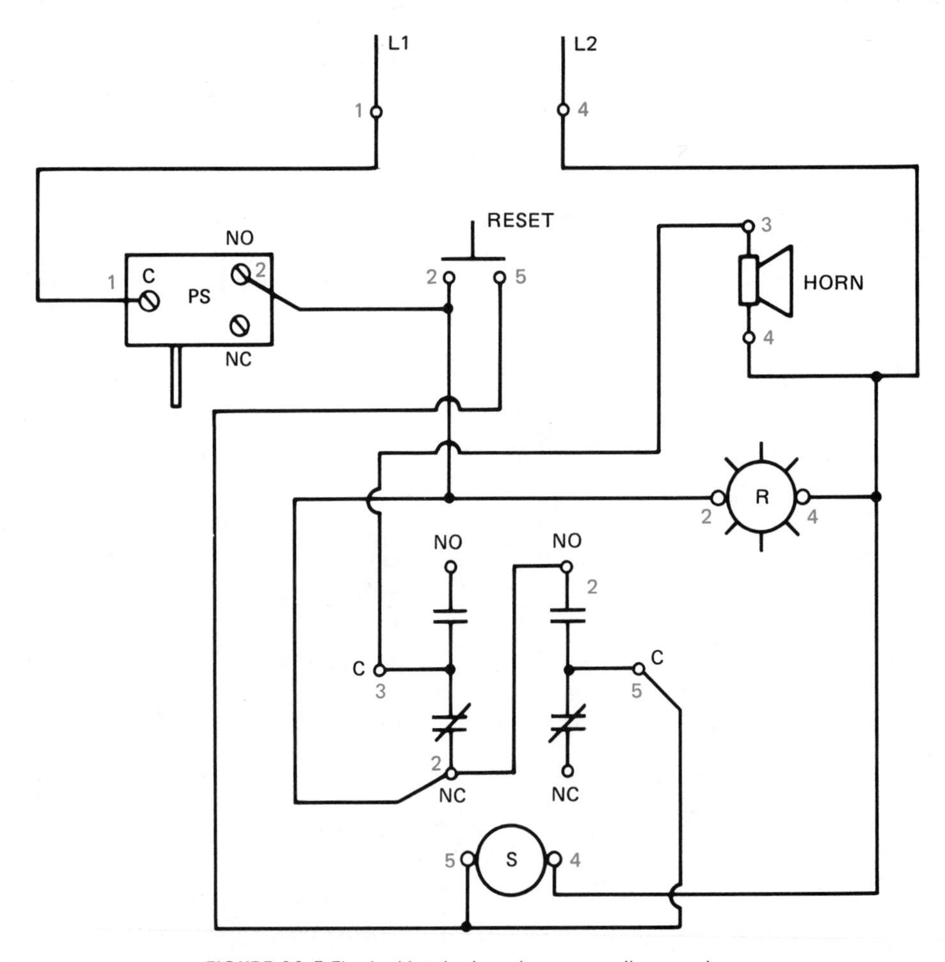

FIGURE 33-5 Final wiring is done by connecting numbers

the numbers used to label the components of the schematic. For instance, the pressure switch in the schematic is shown as being normally open and is labeled with a 1 and a 2. The pressure switch in the wiring diagram is labeled with a 1 beside the common terminal and a 2 beside the NO terminal. The normally closed S contact in the schematic is labeled with a 2 and a 3. Relay S in the wiring diagram has a normally closed contact labeled with a 2 and a 3. The numbers used to label the components in the wiring diagram correspond to the number used to label the same components in the schematic.

After labeling the components in the wiring diagram with the proper numbers, it is simple to connect the circuit, figure 33-5. Connection of the circuit is made by connecting like numbers. For example, all of the components labeled with a 1 are connected, all of those labeled with a 2 are connected, all of the 3s are connected, all of the 4s are connected, and all of the 5s are connected.

REVIEW QUESTION

1. Why are numbers used when developing a wiring diagram from a schematic diagram?

Developing a Wiring Diagram (Circuit #2)

Objectives ***After studying this unit, the student will be able to:***

- **Develop a wiring diagram using this schematic**
- **Connect this circuit**

The circuit shown in figure 34-1 is the same as the schematic shown in figure 31-1, except the schematic in 34-1 has been labeled with numbers. Figure 34-2 shows the components of the wiring diagram. The numbers used to label the components in the wiring diagram correspond to the numbers in the schematic. For instance, the schematic shows the numbers 1 and 8 beside normally

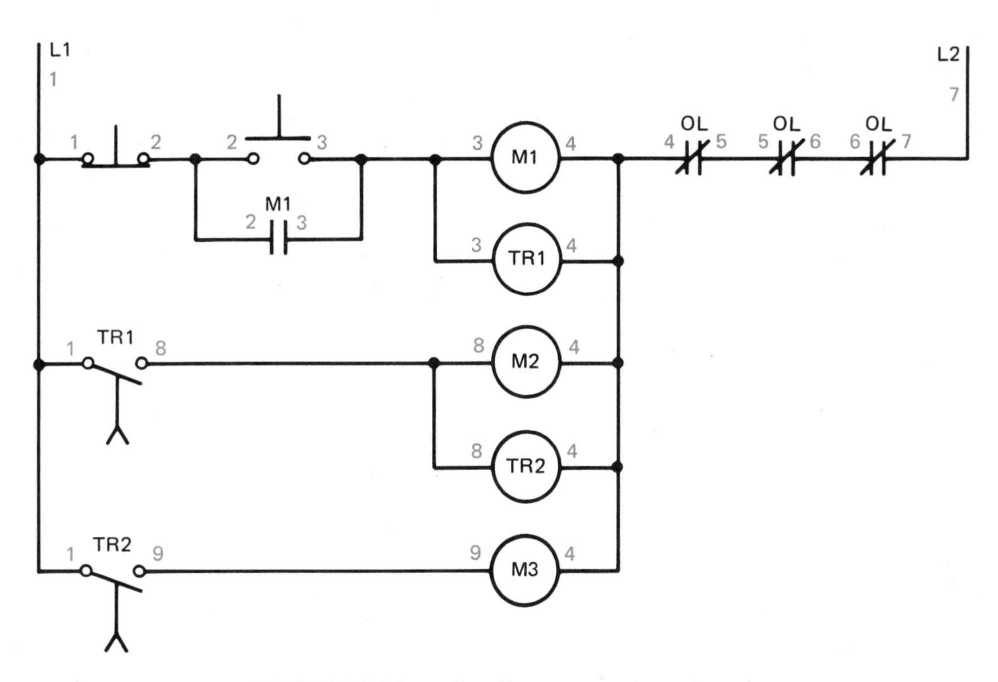

FIGURE 34-1 Schematic with components numbered

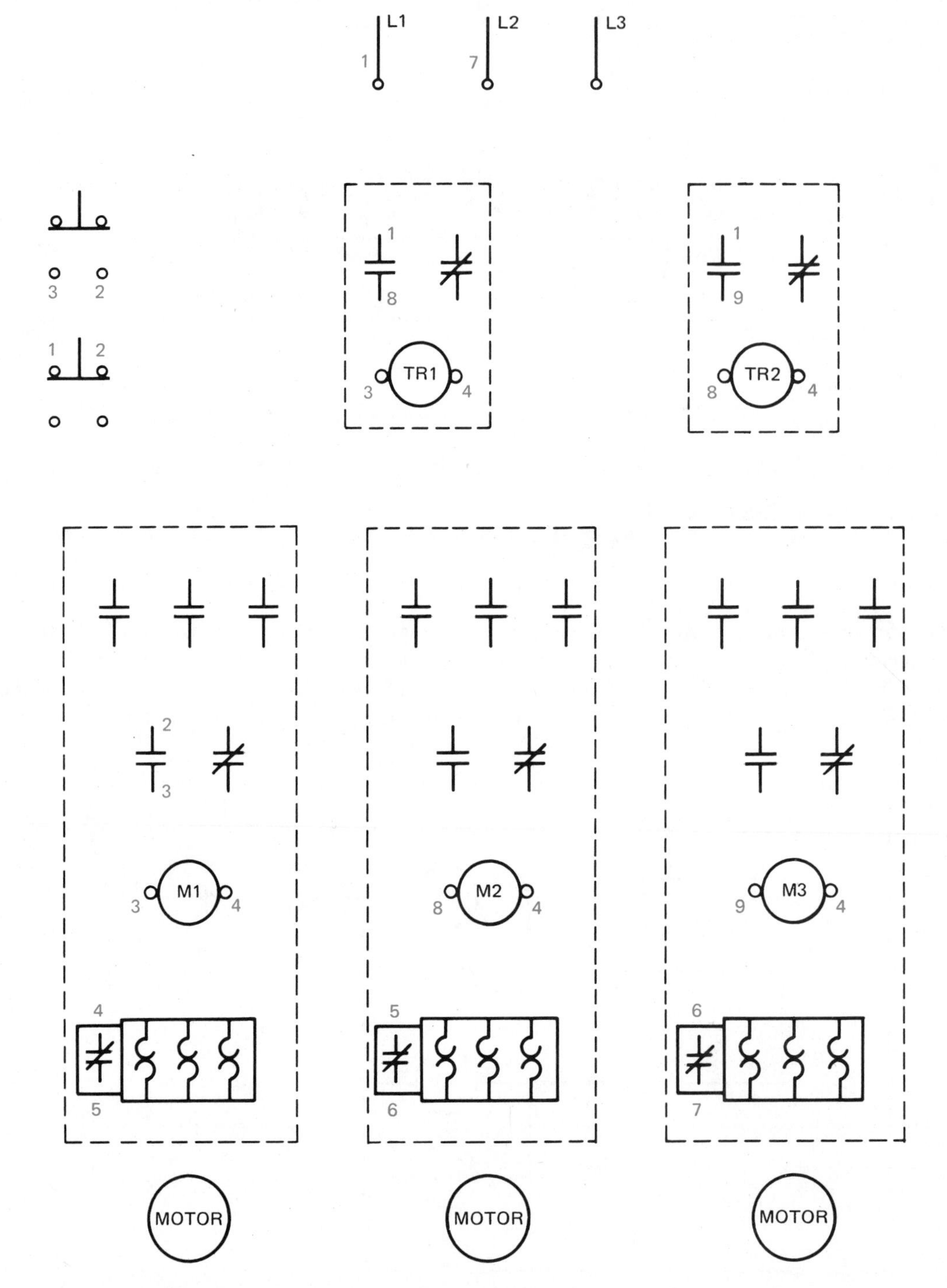

FIGURE 34-2 Components have been numbered to match the schematic

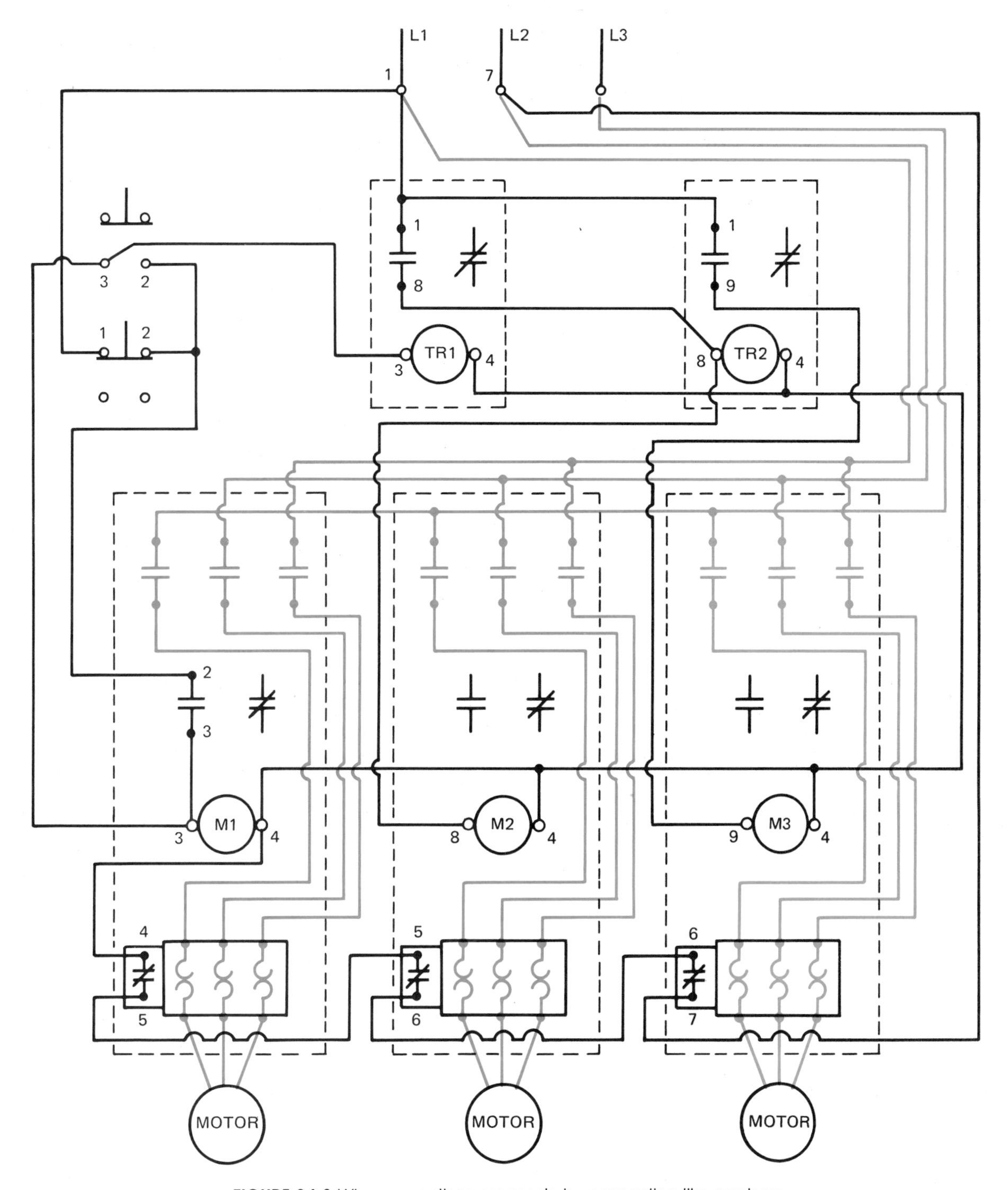

FIGURE 34-3 Wire connections are made by connecting like numbers

open contact TR1. The wiring diagram also shows the numbers 1 and 8 beside normally open contact TR1. The numbers used with each component shown on the schematic have been placed beside the proper component shown in the wiring diagram.

Figure 34-3 shows the wiring diagram with connected wires. Notice that the wiring diagram shows motor connections while the schematic does not. Although it is a common practice to omit motor connections in control schematics, wiring diagrams do show the motor connections.

REVIEW QUESTION

1. Referring to circuit 34-1, would it be possible to change the components that have been numbered with an 8 to a number 9, and the components that have been numbered with a 9 to a number 8 without affecting the operation of the circuit?

UNIT 35

Developing a Wiring Diagram (Circuit #3)

Objectives *After studying this unit, the student will be able to:*

- **Develop a wiring diagram using this schematic**
- **Connect this circuit**

Figure 35-1 shows the same schematic as figure 32-1, except that figure 35-1 has been labeled with numbers in the same manner as the two previous circuits. Figure 35-2 shows the components of the wiring diagram labeled with numbers that correspond to the numbered components shown in the schematic. Figure 35-3 shows the wiring diagram with connected wires.

The same method has been used to number the circuits in the last few units. Although most control schematics are numbered to aid the electrician in troubleshooting, several methods are used. Regardless of the method used, all numbering systems use the same principles. An electrician who learns this method of numbering a schematic will have little difficulty understanding a different method.

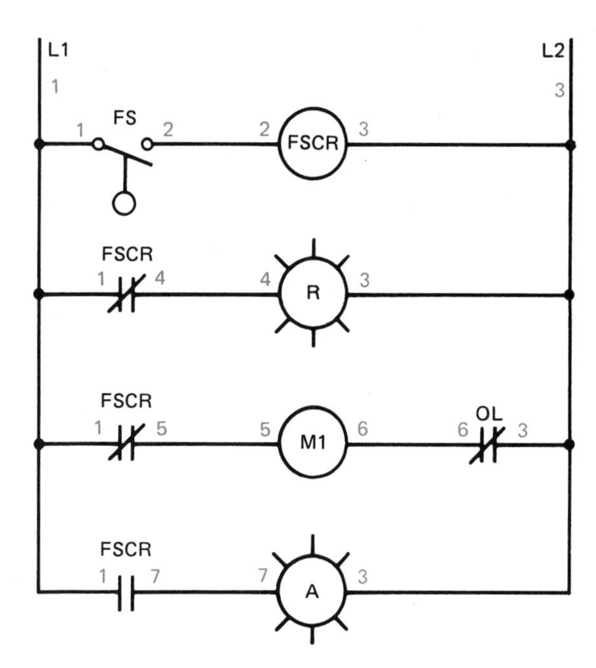

FIGURE 35-1 Schematic with components numbered

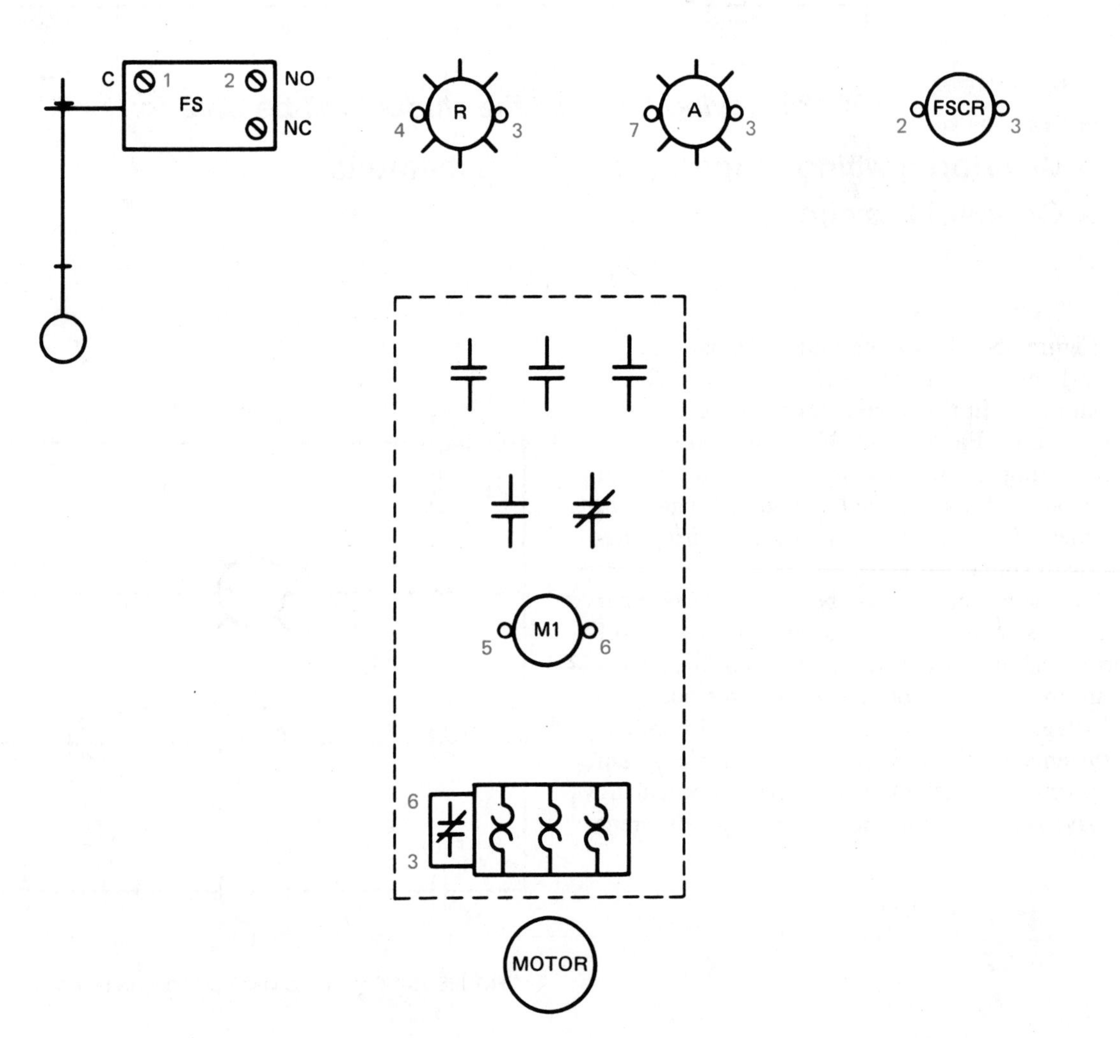

FIGURE 35-2 Components have been numbered to correspond with the schematic

REVIEW QUESTION

1. Are numbering systems other than the one described in this text used to develop wiring diagrams from schematic diagrams?

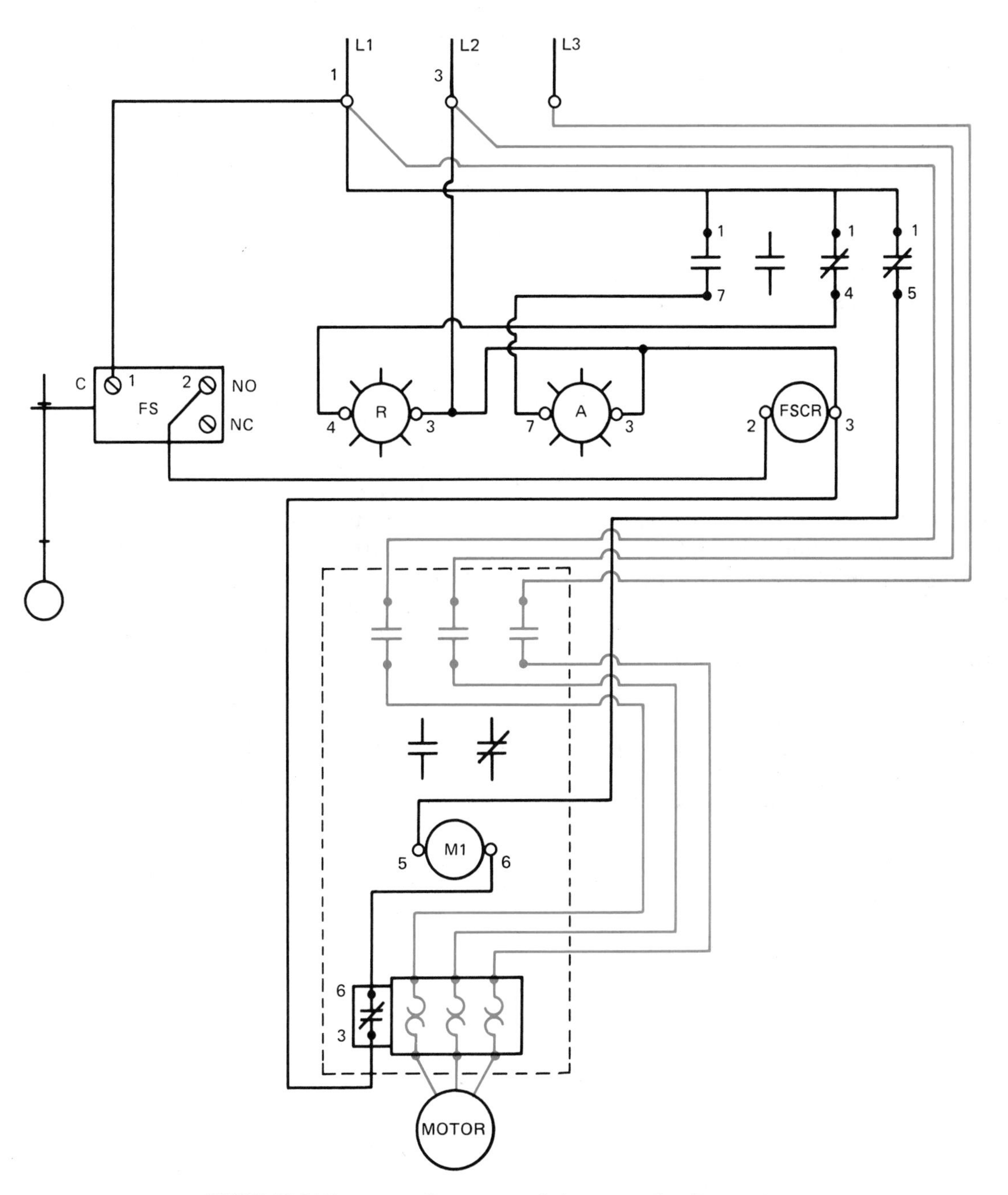

FIGURE 35-3 Wire connections are made by connecting like numbers

Installing Control Systems

Objectives ***After studying this unit, the student will be able to:***

- **Discuss the installation of control circuits**
- **Use the number system on a schematic diagram to troubleshoot a circuit**

Wiring diagrams can be misleading in that they show all components grouped together. In actual practice, the control relays and motor starters may be located in one cabinet, the push buttons and pilot lights in a control panel, and pilot devices, such as limit switches, and pressure switches, may be located on the machine itself, figure 36-1. Most control systems use a relay cabinet to house the control relays and motor starters. Wiring is brought to the relay cabinet from the push buttons and pilot lights located in the control panel, and from the pilot devices located on the machine. All of the connections are made inside the relay cabinet.

Relay cabinets generally contain rows of terminal strips. These terminal strips are used as connection points between the control wiring inside the cabinet and inputs or outputs to the machine or control panel. Most terminal strips are designed so that connection points can be num-

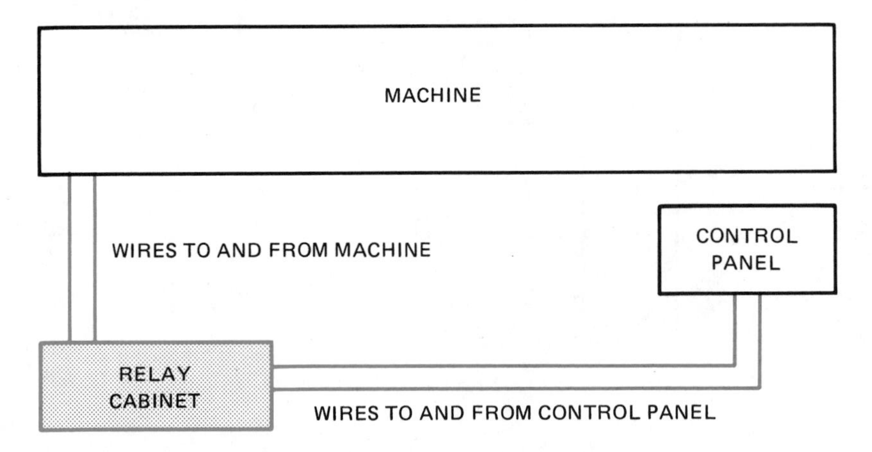

FIGURE 36-1 Most wire connections are made inside the relay cabinet

bered. This type of system can be more costly to install, but it will more than pay for the extra cost in the time saved when troubleshooting. For example, assume that an electrician desires to check limit switch LS15 to see if its contacts are open or closed. Limit switch LS15 is located on the machine, but assume that in the schematic one side of the limit switch is number 25 and the other side of the limit switch is number 26. If numbers 25 and 26 can be located on the terminal strip, the electrician can check the limit switch from the relay cabinet without having to go to the machine and remove the cover from the limit switch. The electrician will only have to connect the probes of his voltmeter across terminals 25 and 26. If the voltmeter indicates the control voltage of the circuit, the limit switch contacts are open. If the voltmeter indicates 0 volts, the limit switch contacts are closed.

REVIEW QUESTIONS

1. Define a schematic diagram.
2. Define a wiring diagram.
3. Are symbols in a schematic shown in their energized or de-energized condition?
4. Draw the standard NEMA symbol for the following components.
   A. Float switch (NO)
   B. Limit switch (NC)
   C. Normally open push button
   D. Relay coil
   E. Overload contact

SECTION 4

Basic Control Circuits

UNIT 37

Hand-off Automatic Controls

Objectives *After studying this unit, the student will be able to:*

- **State the purpose of hand-off-automatic controls**
- **Connect hand-off-automatic controls**
- **Read and draw diagrams using hand-off-automatic controls**

Hand-off-automatic switches are used to select the function of a motor controller either manually or automatically. This selector switch may be a separate unit or built into the starter enclosure cover. A typical control circuit using a single-break selector switch is shown in figure 37-1.

With the switch turned to the HAND (manual) position, coil (M) is energized all the time and the motor runs continuously. In the OFF position, the motor does not run at all. In the AUTOMATIC position, the motor runs whenever the two-wire control device is closed. An operator does not need to be present. The control device may be a pressure switch, limit switch, thermostat, or other two-wire control pilot device.

The heavy-duty, three-position double-break selector switch shown in figure 37-2 is also used for manual and automatic control. When the switch is turned to "hand," the coil is energized, by-passing the automatic control device in the "auto" position.

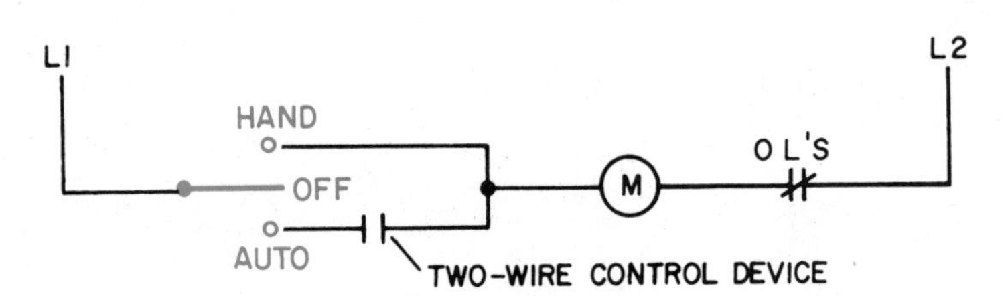

FIGURE 37-1 Standard duty, three-position selector switch in control circuit

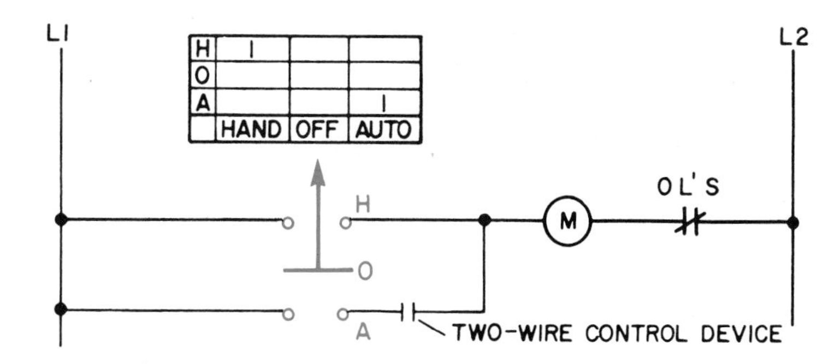

FIGURE 37-2 Diagram of a heavy duty, three-position selector switch in control circuit

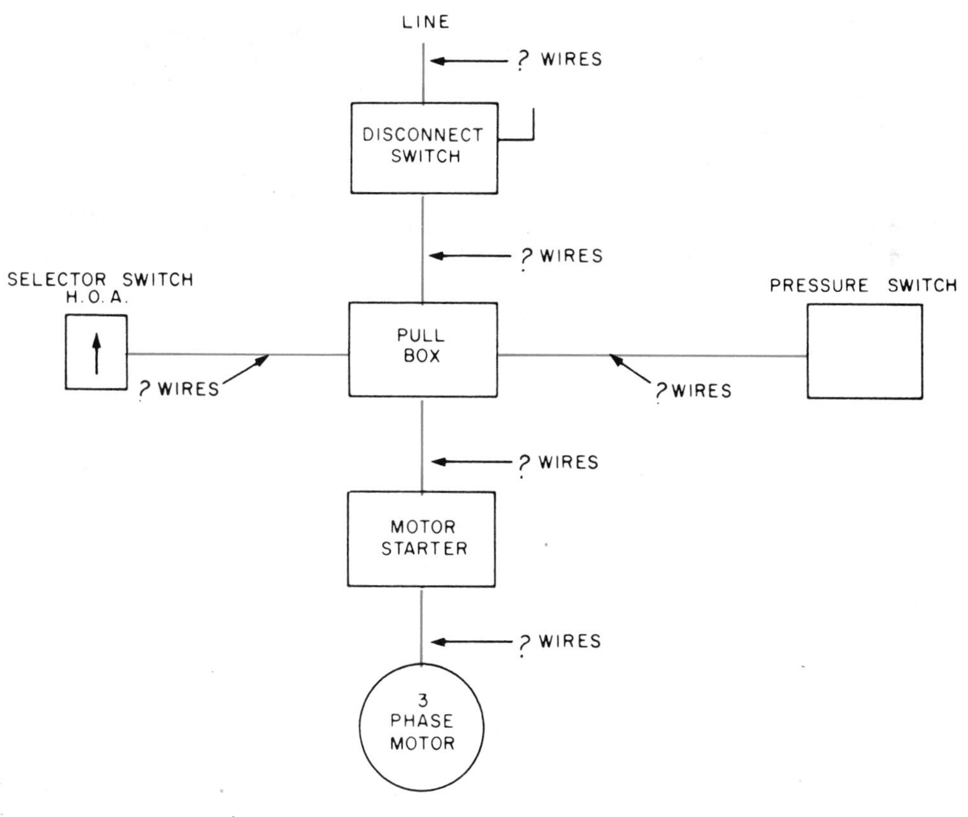

FIGURE 37-3

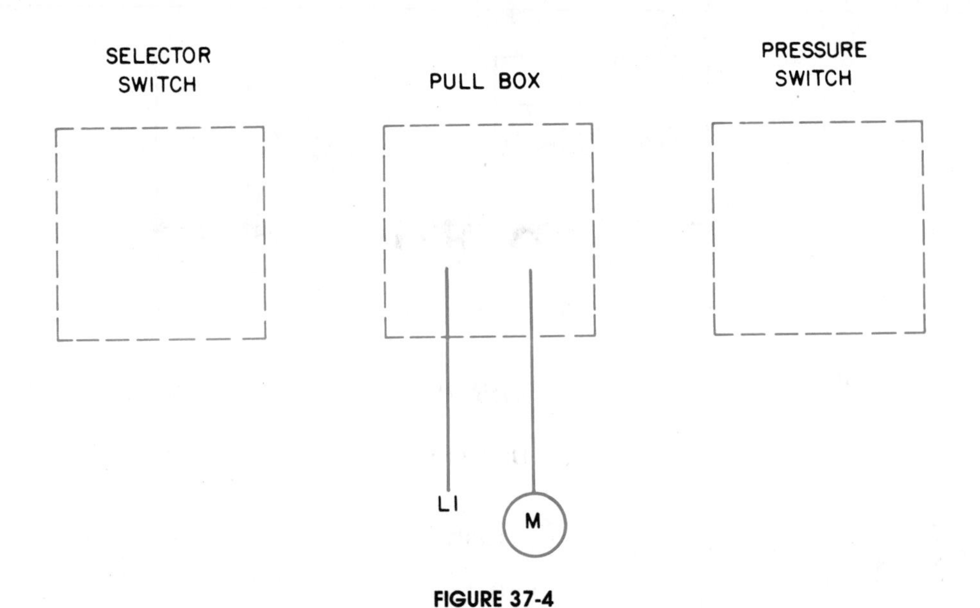

FIGURE 37-4

REVIEW QUESTIONS

1. A selector switch and two-wire pilot device cannot be used to control a large motor directly, but rather must be connected to a magnetic starter. Explain why this is true.
2. Determine the minimum number of wires in each conduit shown in figure 37-3.
3. Complete figure 37-4 to show the wiring diagram of the hand-off-automatic selector switch, pull box, pressure switch and L1 and coil M.

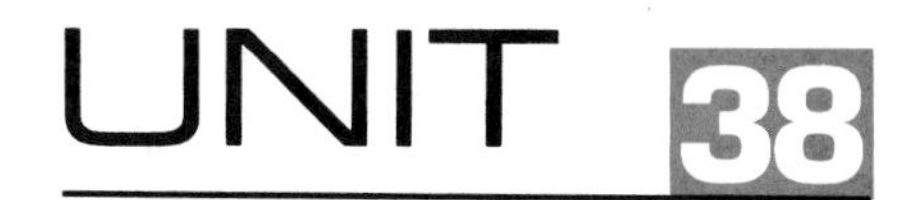

Multiple Push-Button Stations

Objectives ***After studying this unit, the student will be able to:***

- **Read and interpret diagrams using multiple push-button stations**
- **Draw diagrams using multiple push-button stations**
- **Connect multiple push-button stations**

The conventional three-wire, push-button control circuit may be extended by the addition of one or more push-button control stations. The motor may be started or stopped from a number of separate stations by connecting all start buttons in parallel and all stop buttons in series. The operation of each station is the same as that of the single push-button control in the basic three-wire circuit covered in unit 29. Note in figure 38-1 that pressing any stop button de-energizes the coil. Pressing any start button energizes the coil.

When a motor must be started and stopped from more than one location, any number of start and stop buttons may be connected. Another possible arrangement is to use only one start-stop station and several stop buttons at different locations to serve as emergency stops.

Multiple push-button stations are used to control conveyor motors on large shipping and receiving freight docks.

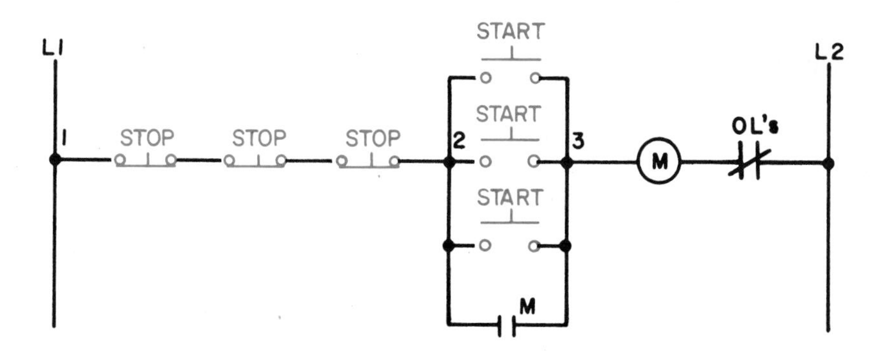

FIGURE 38-1 Three-wire control using a momentary contact, multiple push-button station

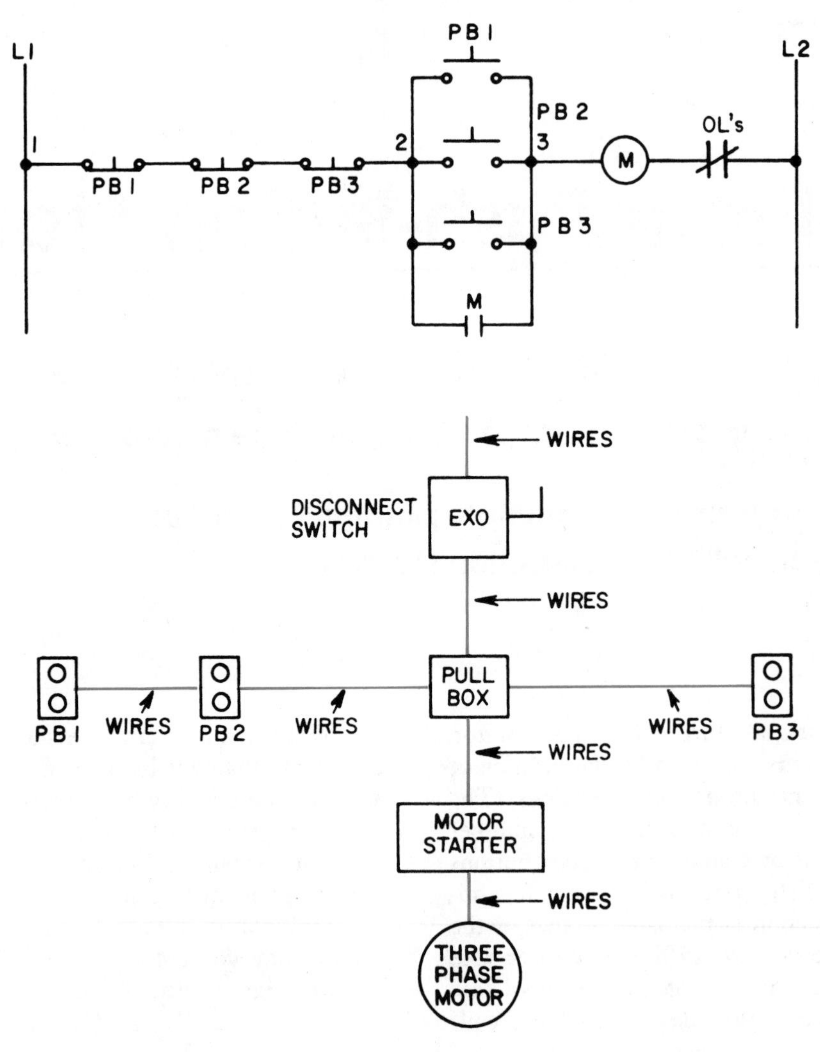

FIGURE 38-2

REVIEW QUESTIONS

1. Using the line diagram in figure 38-2, determine the number of wires controlling the three-phase motor. The conduit layout plan indicates the sections of conduit for which the quantity of wires is to be determined.
2. Referring to figure 38-2, select the minimum size conduit for each run. Use the appropriate locally enforced electrical code—national, state, county, or city.

UNIT 39

Interlocking Methods for Reversing Control

Objectives ***After studying this unit, the student will be able to:***

- **Explain the purpose of the various interlocking methods**
- **Read and interpret wiring and line diagrams of reversing controls**
- **Read and interpret wiring and line diagrams of interlocking controls**
- **Wire and troubleshoot reversing and interlocking controls**

The direction of rotation of three-phase motors can be reversed by interchanging any two motor leads to the line. If magnetic control devices are to be used, then reversing starters accomplish the reversal of the motor direction, figure 39-1A. Reversing starters wired to NEMA standards interchange lines L1 and L3, figure 39-1B. To do this, two contactors for the starter assembly are required—one for the forward direction and one for the reverse direction, figure 39-1C. A technique called *interlocking* is used to prevent the contactors from being energized simultaneously or closing together and causing a short circuit. There are three basic methods of interlocking.

MECHANICAL INTERLOCK

A mechanical interlocking device is assembled at the factory between the forward and reverse contactors. This interlock locks out one contactor at the beginning of the stroke of either contactor to prevent short circuits and burnouts.

The mechanical interlock between the contactors is represented in the elementary diagram of figure 39-2 by the broken line between the coils. The broken line indicates that coils F and R cannot close contacts simultaneously because of the mechanical interlocking action of the device.

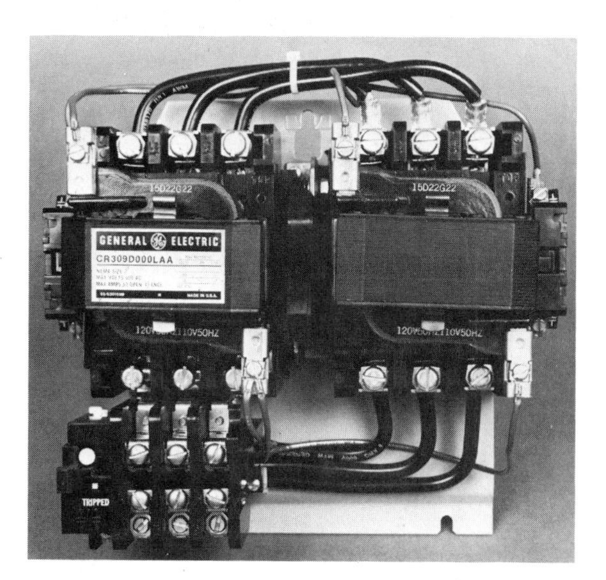

FIGURE 39-1A Reversing starter (Courtesy General Electric Co.)

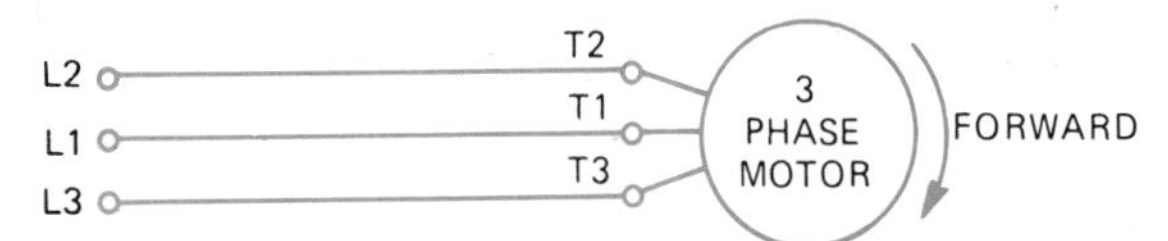

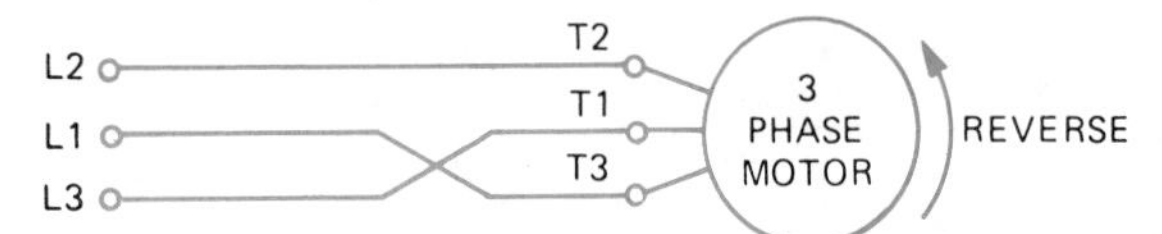

FIGURE 39-1B Reversing rotation of an induction motor

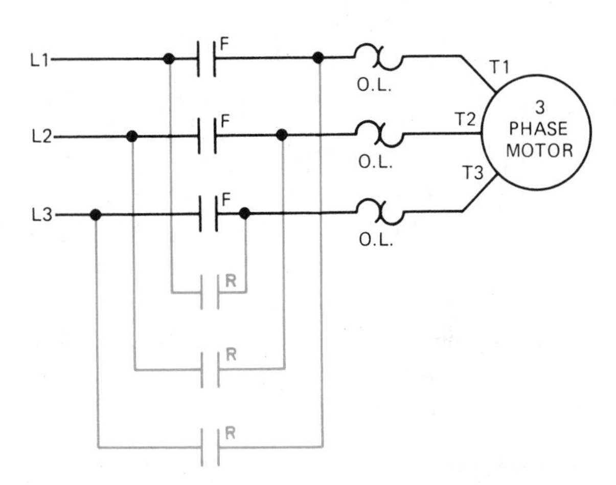

FIGURE 39-1C Elementary diagram of a reversing starter power circuit

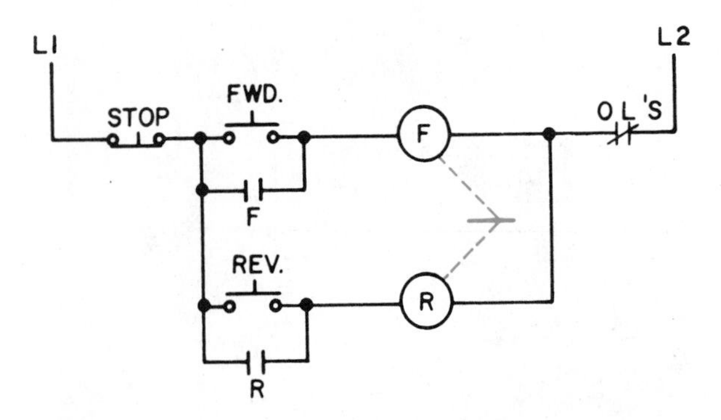

FIGURE 39-2 Mechanical interlock between the coils prevents the starter from closing all contacts simultaneously. Only one contactor can close at a time.

When the forward contactor coil (F) is energized and closed through the forward push button, the mechanical interlock prevents the accidental closing of coil R. Starter F is blocked by coil R in the same manner. The first coil to close moves a lever to a position that prevents the other coil from closing its contacts when it is energized. If an oversight allows the second coil to remain energized without closing its contacts, the excess current in the coil due to the lack of the proper inductive reactance will damage the coil.

Note in the elementary diagram of figure 39-2 that the stop button must be pushed before the motor can be reversed.

Reversing starters are available in horizontal and vertical construction. A vertical starter is shown in figure 39-3A.

A mechanical interlock is installed on the majority of reversing starters in addition to the use of one or both of the following electrical methods: push-button interlock and auxiliary contact interlock.

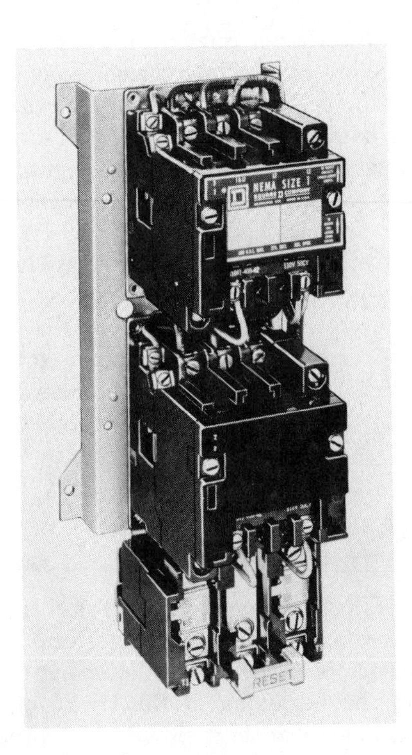

FIGURE 39-3A Vertical reversing motor starter (Courtesy Square D Co.)

PUSH-BUTTON INTERLOCK

Push-button interlocking is an *electrical method* of preventing both starter coils from being energized simultaneously.

When the forward button in figure 39-3B is pressed, coil F is energized and the normally open (NO) contact F closes to hold in the forward contactor. Because the normally closed (NC) contacts are used in the forward and reverse push-button units, there is no need to press the stop button before changing the direction of rotation. If the reverse button is pressed while the motor is running in the forward direction, the forward control circuit is de-energized and the reverse contactor is energized and held closed.

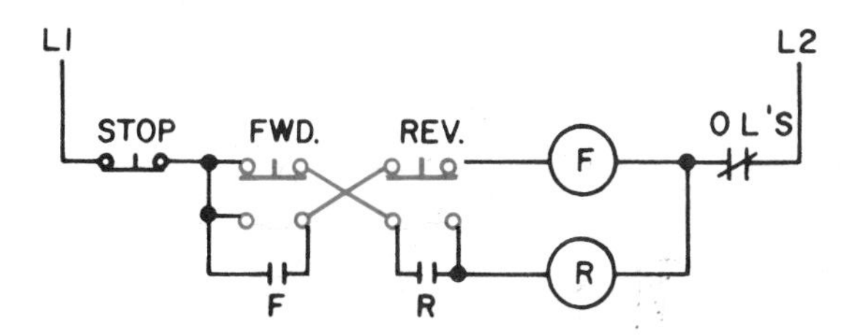

FIGURE 39-3B Double-circuit push buttons are used for push-button interlocking

Repeated reversals of the direction of motor rotation are not recommended. Such reversals may cause the overload relays and starting fuses to overheat; this disconnects the motor from the circuit. The driven machine may be damaged also. It may be necessary to wait until the motor has coasted to a standstill.

NEMA specifications call for a starter to be derated. That is, the next size larger starter must be selected when it is to be used for "plugging" to stop, or "reversing" at a rate of more than five times per minute.

Reversing starters consisting of mechanical and electrical interlocked devices are preferred for maximum safety.

AUXILIARY CONTACT INTERLOCK

Another method of electrical interlock consists of normally closed auxiliary contacts on the forward and reverse contactors of a reversing starter, figure 39-4.

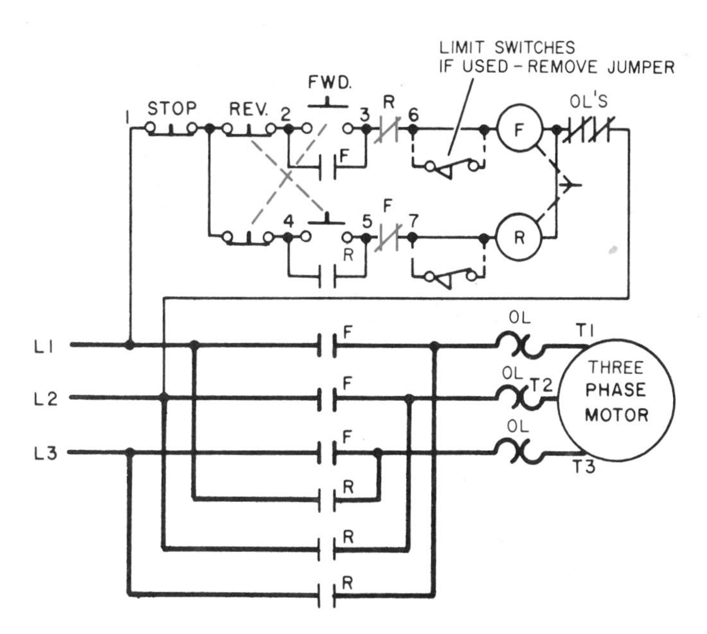

FIGURE 39-4 Elementary diagram of the reversing starter shown in figure 39-1A. The mechanical push button and auxiliary contact interlocks are indicated.

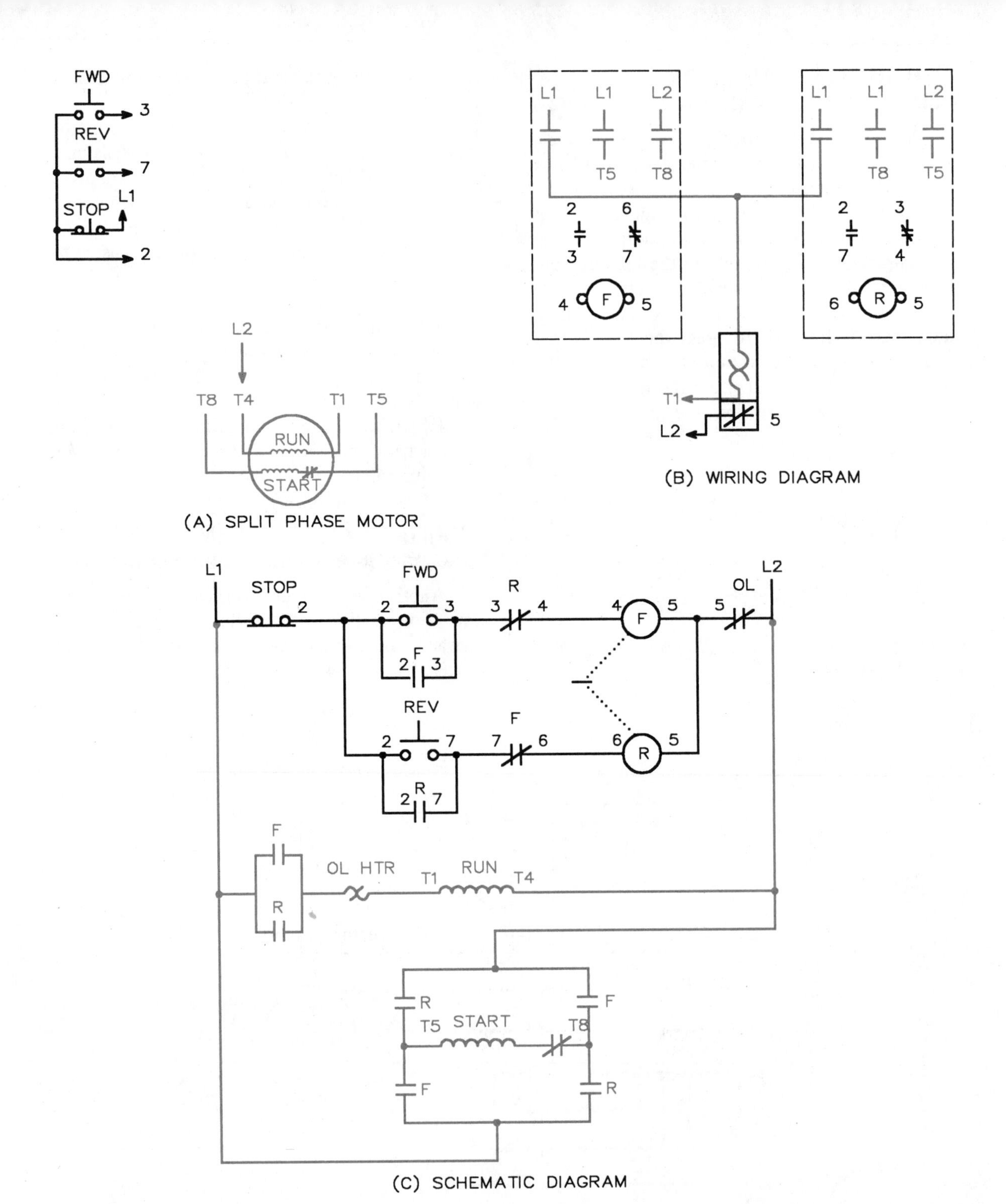

FIGURE 39-5 Sizes 0 and 1 reversing starters used with single split-phase induction motors

When the motor is running forward, an NC contact (F) on the forward contactor opens and prevents the reverse contactor from being energized by mistake and by closing. The same operation occurs if the motor is running in reverse.

The term, *interlocking*, is also used generally when referring to motor controllers and control stations that are interconnected to provide control of production operations.

To reverse the direction of rotation of single-phase motors, *either* the starting *or* running winding motor leads are interchanged, but not both. Figure 39-5A completes the wiring diagram for the single-phase, four-wire, split phase induction motor; figure 39-5B is a wiring diagram for a single-phase vertical starter; and figure 39-5C is a line diagram of the connections.

REVIEW QUESTIONS

1. How is a change in the direction of rotation of a three-phase motor accomplished?
2. What is the purpose of interlocking?
3. What will happen if both start buttons are pushed in a control with push-button interlocking? Why?
4. How is auxiliary contact interlocking obtained on a reversing starter?
5. When the forward coil is energized, in what position is the forward interlock (F)?
6. If a mechanical interlock is the only means of interlocking used, describe the operation that must be followed to reverse the direction of rotation of the motor while running.
7. If pilot lights are to indicate the direction of rotation of a motor, where should the devices be connected so as not to add any contacts?
8. What is the sequence of the operations if limit switches are used in figure 39-4?
9. What will happen in figure 39-4 if limit switches are installed and the jumpers from terminals 6 and 7 to the coils are not removed?
10. In place of the push buttons in figure 39-2, draw a selector switch for forward-reverse-stop control. Show the target table for this selector switch.

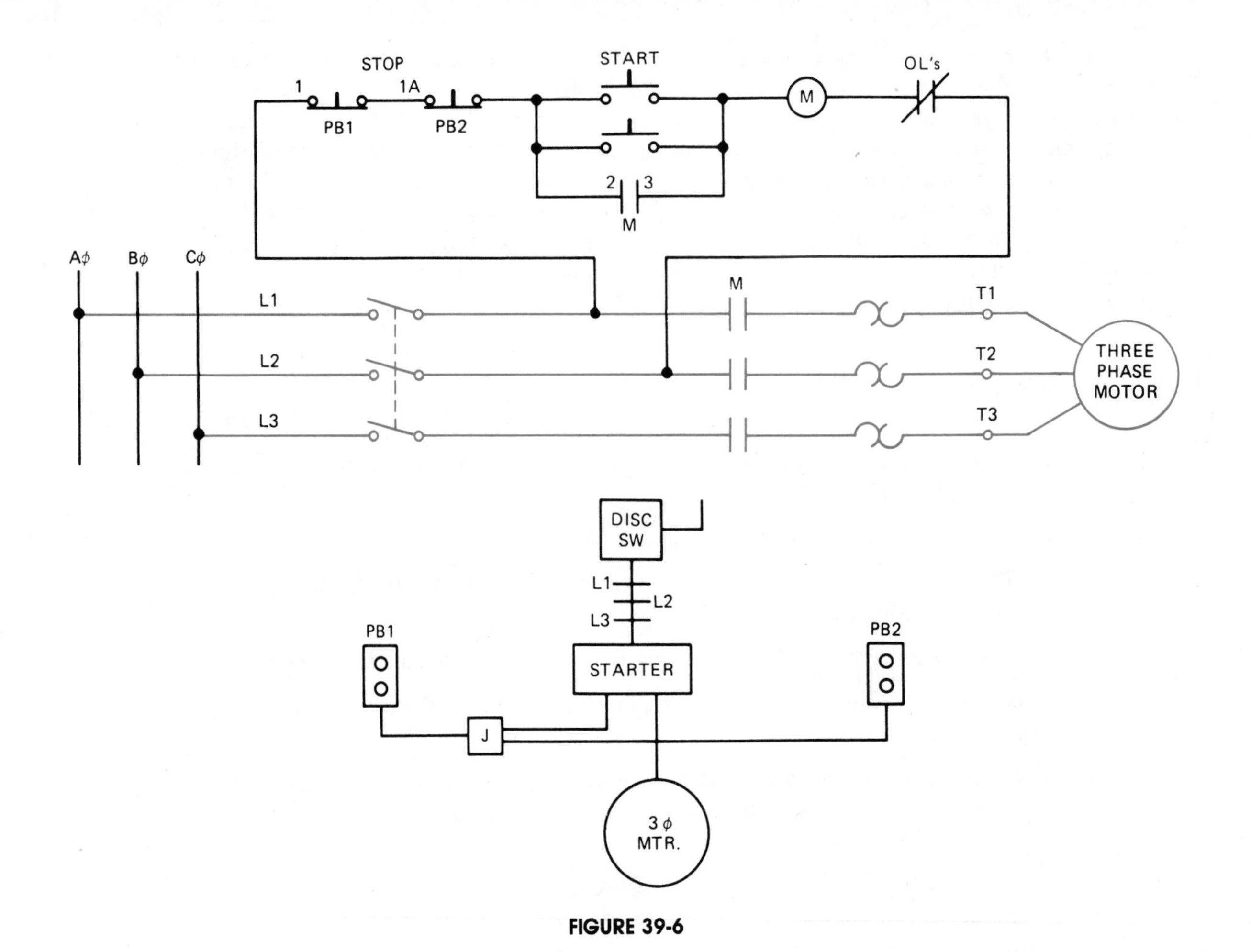

FIGURE 39-6

11. From the elementary drawing in figure 39-6, determine the number and terminal identification of the wiring in each conduit in the conduit layout. Indicate your solutions in the same manner as the example given below the disconnect switch.
12. Convert the control circuit only, figure 39-7, from the wiring diagram to an elementary diagram. Include the limit switches (RLS, FLS) as operating in the control circuit.

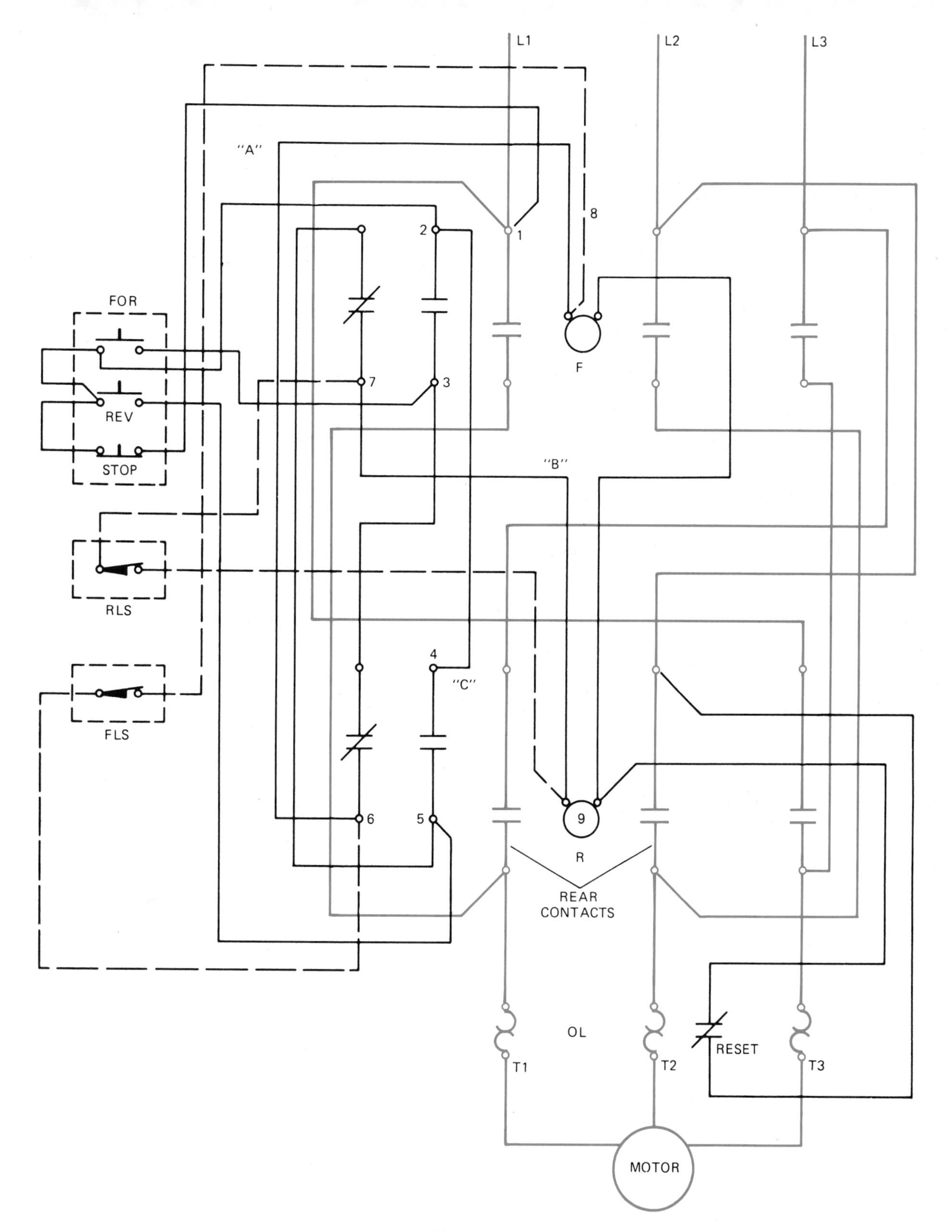
L1
L2
L3
"A"
8
2
1
FOR
F
7
3
REV
STOP
"B"
RLS
4
"C"
FLS
6
5
9
R
REAR
CONTACTS
OL
RESET
T1
T2
T3
MOTOR

FIGURE 39-7

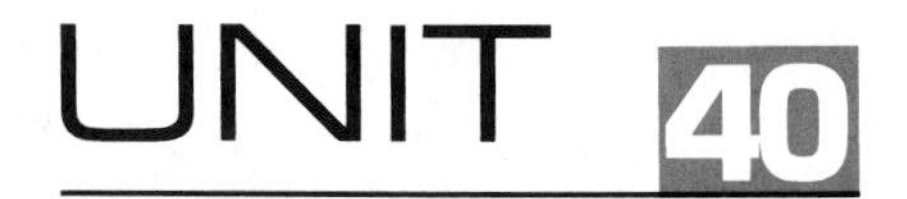

Sequence Control

Objectives ***After studying this unit, the student will be able to:***

- **Describe the purpose of starting motors in sequence**
- **Read and interpret sequence control diagrams**
- **Make the proper connections to operate motors in sequence**
- **Troubleshoot sequence motor control circuits**

Sequence control is the method by which starters are connected so that one cannot be started until the other is energized. This type of control is required whenever the auxiliary equipment associated with a machine, such as high-pressure lubricating and hydraulic pumps, must be operating before the machine itself can be operated safely. Another application of sequence control is in main or subassembly line conveyors.

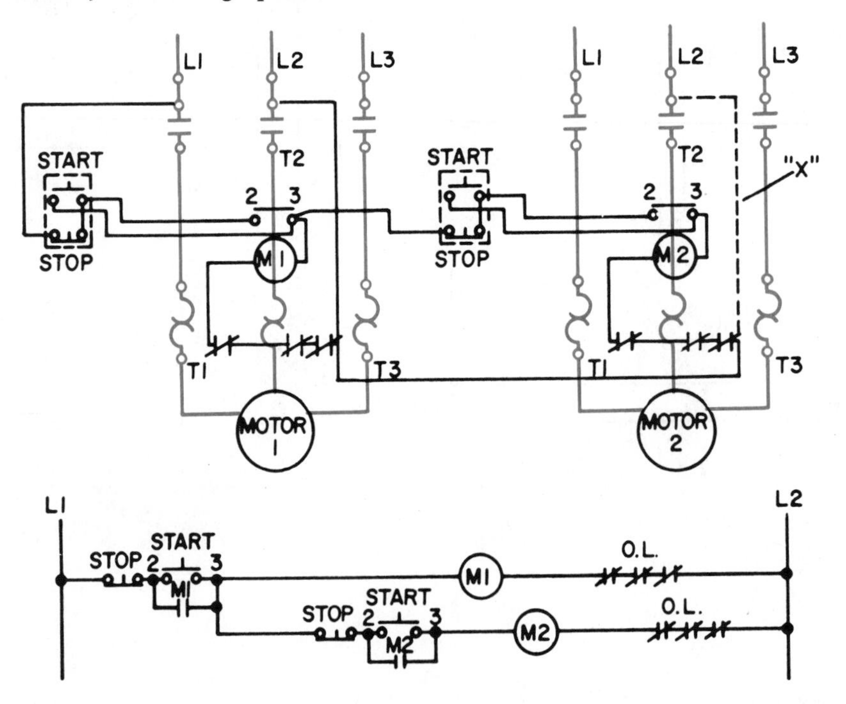

FIGURE 40-1 Standard starters wired for sequence control (Courtesy Allen-Bradley Co.)

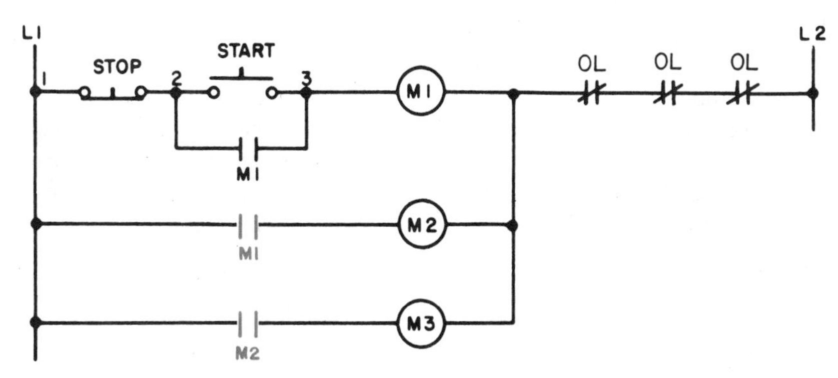

FIGURE 40-2 Auxiliary contacts (or interlocks) used for automatic sequence control. Contact M1 energizes coil M2; contact M2 energizes coil M3.

The proper push-button station connections for sequence control are shown in figure 40-1. Note that the control circuit of the second starter is wired through the maintaining contacts of the first starter. As a result, the second starter is prevented from starting until after the first starter is energized. If standard starters are used, the connection wire (X) must be removed from one of the starters.

If sequence control is to be provided for a series of motors, the control circuits of the additional starters can be connected in the manner shown in figure 40-2. That is, M3 will be connected to M2 in the same step arrangement by which M2 is connected to M1.

The stop button or an overload on any motor will stop all motors with this method.

AUTOMATIC SEQUENCE CONTROL

A series of motors can be started automatically with only one start-stop control station as shown in figures 40-2 and 40-3. When the lube oil pump, (M1) in figure 40-3, is started by pressing the start button, the pressure must be built up

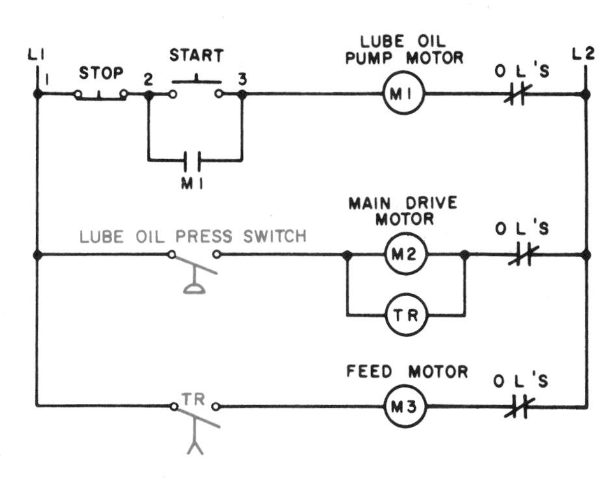

FIGURE 40-3 Pilot devices used in an automatic sequence control scheme

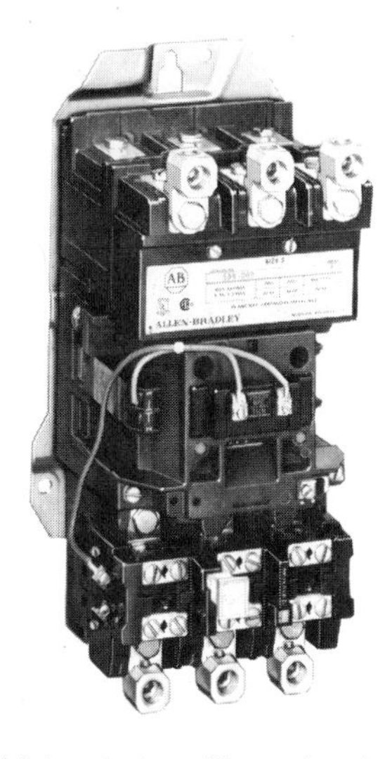

FIGURE 40-4 Motor starter with overload relay (Courtesy Allen-Bradley Co.)

enough to close the pressure switch before the main drive motor (M2) will start. The pressure switch also energizes a timing relay (TR). After a preset time delay, the contact (TR) will close and energize the feed motor starter coil (M3).

If the main drive motor (M2) becomes overloaded, the starter and timing relay (TR) will open. As a result, the feed motor circuit (M3) will be de-energized due to the opening of the contact (TR). If the lube oil pump motor (M1) becomes overloaded, all of the motors will stop. Practically any desired overload control arrangement is possible. A motor starter with an overload relay is shown in figure 40-4.

REVIEW QUESTIONS

1. Describe what is meant by sequence control.
2. Referring to the diagram in figure 40-2, explain what will happen if the motor on the coil (M1) becomes overloaded.
3. In figure 40-2, what will happen if there is an overload on the motor starter (M2)?
4. What is the sequence of operation in figure 40-2?
5. Redraw figure 40-3 so that if the feed motor (M3) is tripped out because of overload, it will also stop the main drive motor (M2).

UNIT 41

Jogging (Inching) Control Circuits

Objectives *After studying this unit, the student will be able to:*

- **Define the process of jogging control**
- **State the purpose of jogging controllers**
- **Describe the operation of a jogging control circuit using a control relay**
- **Describe the operation of a jogging control using a control relay on a reversing starter**
- **Describe the operation of a jogging control using a selector switch**
- **Connect jogging controllers and circuits**
- **Recommend solutions for troubleshooting jogging controllers**

Jogging, or *inching*, is defined by the National Electrical Manufacturers' Association (NEMA) as "the quickly repeated closure of a circuit to start a motor from rest for the purpose of accomplishing small movements of the driven machine." The term *jogging* is often used when referring to across-the-line starters; the term *inching* can be used to refer to reduced voltage starters. Generally, the terms are used interchangeably because they both prevent a holding circuit.

JOGGING CONTROL CIRCUITS

The control circuits covered in this unit are representative of the various methods that are used to obtain jogging.

Figure 41-1 is a line diagram of a very simple jogging control circuit. The stop button is held open mechanically, figure 41-2. With the stop button held open, maintaining contact M cannot hold the coil energized after the start button is closed. The disadvantage of a circuit connected in this manner is the loss of the lock-stop safety feature. This circuit can be mistaken for a conventional three-wire control circuit, locked off for safety reasons, such as to keep a circuit or machine from being energized. A padlock should be installed for safety purposes.

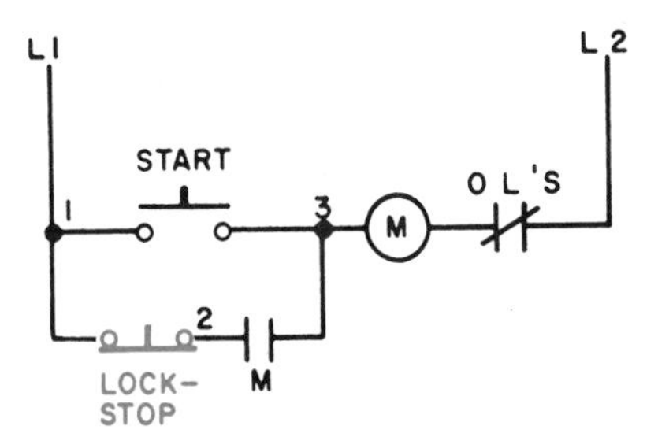

FIGURE 41-1 Lock-stop push button in jogging circuit

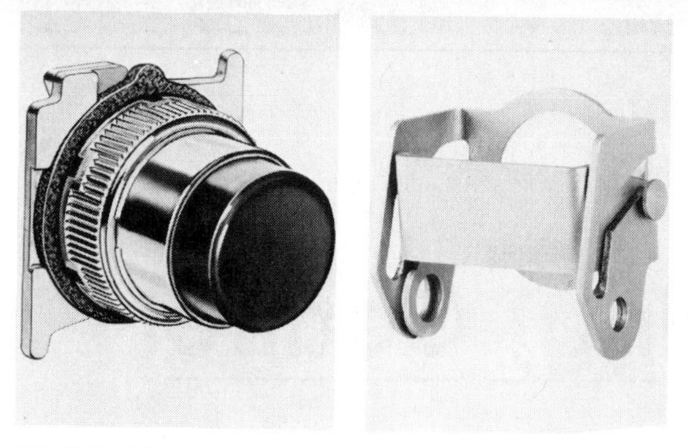

FIGURE 41-2 Oiltight push-button operator with extra long button to accept padlock attachment (on right) to provide lockout-on-stop feature (Courtesy EATON Corp., Cutler-Hammer Products)

If the lock-stop push button is used for jogging, it should be clearly marked for this purpose.

Figure 41-3 illustrates other simple schemes for jogging circuits. The normally closed push-button contacts on the jog button in figure 41-3(B) are connected in series with the holding circuit contact on the magnetic starter. When the jog button is pressed, the normally open contacts energize the starter magnet. At the same time, the normally closed contacts disconnect the holding circuit. When the button is released, therefore, the starter immediately opens to disconnect the motor from the line. The action is similar in figure 41-3(A). A *jogging attachment can be used to prevent the reclosing of the normally closed contacts of the jog button.* This device assures that the starter holding circuit is not re-established if the jog button is released too rapidly. Jogging can be repeated by reclosing the jog button; it can be continued until the jogging attachment is removed.

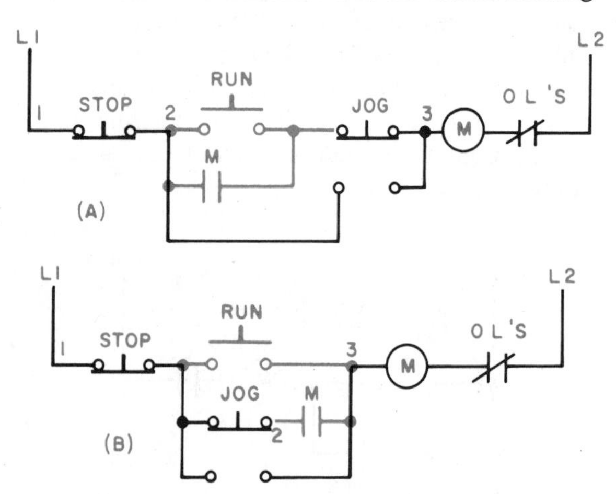

FIGURE 41-3 Line diagrams of simple jogging control circuits

CAUTION: If the circuits shown in figure 41-3 are used without the jogging attachment mentioned, they are hazardous. A control station using such a circuit, less a jogging attachment, can maintain the circuit when the operator's finger is quickly removed from the button. This could injure production workers, equipment, and machinery. This circuit should not be used by responsible people committed to safety in the electrical industry.

JOGGING USING A CONTROL RELAY

When a jogging circuit is used, the starter can be energized only as long as the jog button is depressed. This means the machine operator has instantaneous control of the motor drive.

The addition of a control relay to a jogging circuit provides even greater control of the motor drive. A control relay jogging circuit is shown in figure 41-4. When the start button is pressed, the control relay is energized and a holding circuit is formed for the control relay and the starter magnet. The motor will now run. The jog button is connected to form a circuit to the starter magnet. This circuit is independent of the control relay. As a result, the jog button can be pressed to obtain the jogging or inching action.

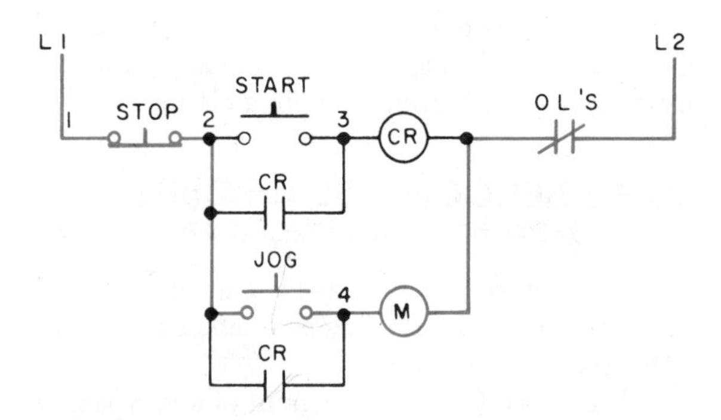

FIGURE 41-4 Jogging is achieved with added use of control relay

Other typical jogging circuits using control relays are shown in figure 41-5. In figure 41-5(A), pressing the start button energizes the control relay. In turn, the relay energizes the starter coil. The normally open starter interlock and relay contact then form a holding circuit around the start button. When the jog button is pressed, the starter coil is energized independently of the relay and a holding circuit does not form. As a result, a jogging action can be obtained.

Jogging With a Control Relay on a Reversing Starter

The control circuit shown in figure 41-6 permits the motor to be jogged in either the forward or the reverse direction while the motor is at standstill or is rotating in either direction. Pressing either the start-forward button or the start-reverse button causes the corresponding starter coil to be energized. The coil then closes the circuit to the control relay which picks up and completes the holding circuit around the start button. While the relay is energized, either the forward or the reverse starter will also remain energized. If either jog button is pressed, the relay is de-energized and the closed starter is released. Continued pressing of either jog button results in a jogging action in the desired direction.

Jogging with a Selector Switch

The use of a selector switch in the control circuit to obtain jogging requires a three-element control station with start and stop controls and a selector switch. A standard duty, two-position selector switch is shown connected in the circuit in figure 41-7. The starter maintaining circuit is disconnected when the selector switch is placed in the jog position. The motor is then inched with the start button. Figure 41-8 is the same circuit as that shown in figure 41-7 with the substitution of a heavy-duty, two-position selector switch.

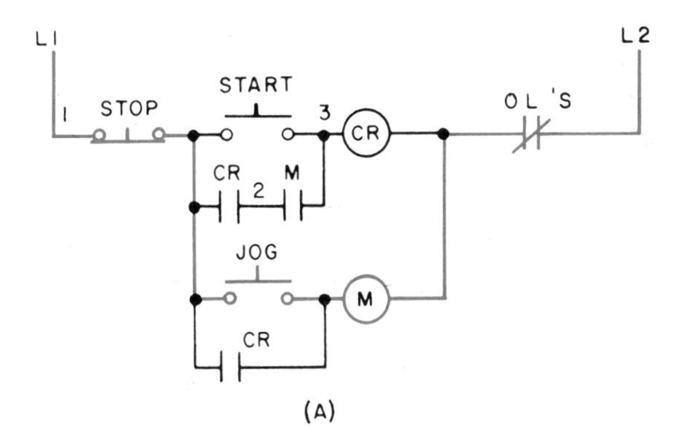

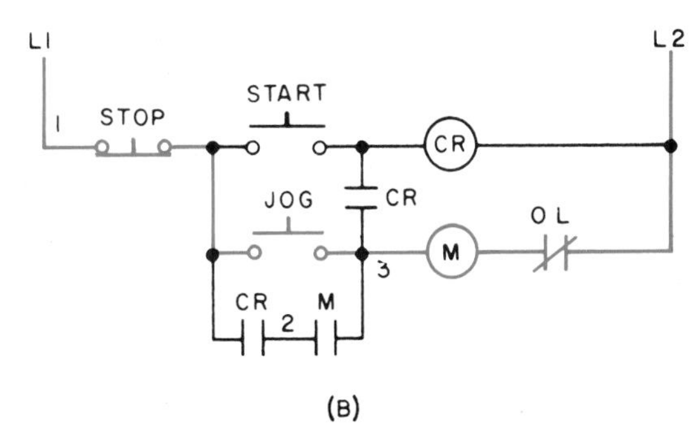

FIGURE 41-5 Line diagrams using control relays in typical installations

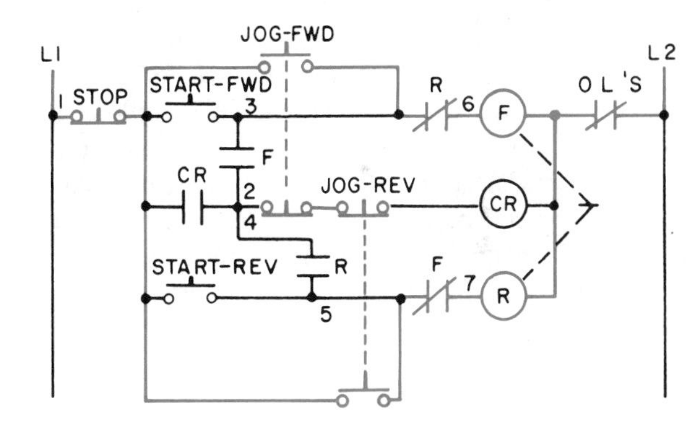

FIGURE 41-6 Jogging using control relay on a reversing starter

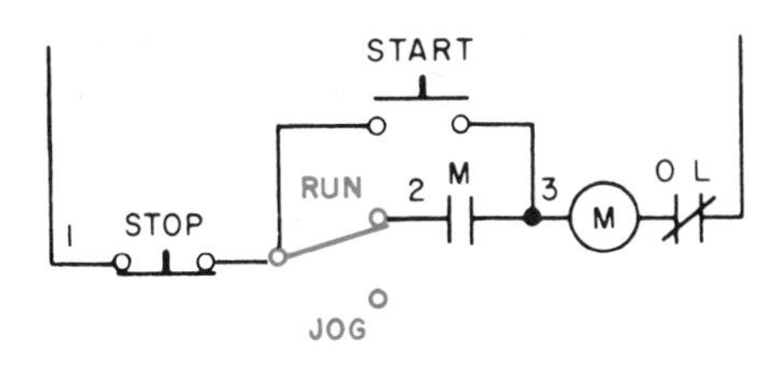

FIGURE 41-7 Jogging using a standard duty, two-position selector switch

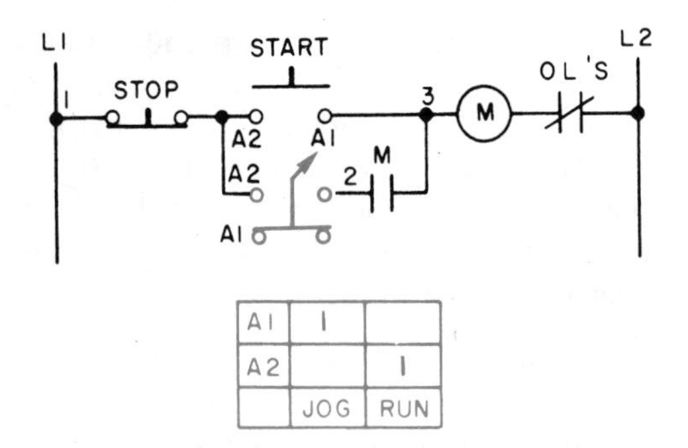

A1	1	
A2		1
	JOG	RUN

FIGURE 41-8 Jogging using a selector-switch jog with start button

The use of a selector push button to obtain jogging is shown in figure 41-9. In the jog position, the holding circuit is broken, and jogging is accomplished by depressing the push button.

Jogging with a Push-Pull Operator

Another type of jog-run control can be connected using a push-pull operator. The push-pull operator used in this circuit contains two normally open momentary contacts, figure 41-10. When the control is pulled outward, contact A completes a circuit to coil CR. When coil CR energizes, both CR contacts close. Contact CR1 completes a circuit to the coil of motor starter M. When motor starter M energizes, contact M closes. Since contacts CR2 and M are closed, a circuit is main-

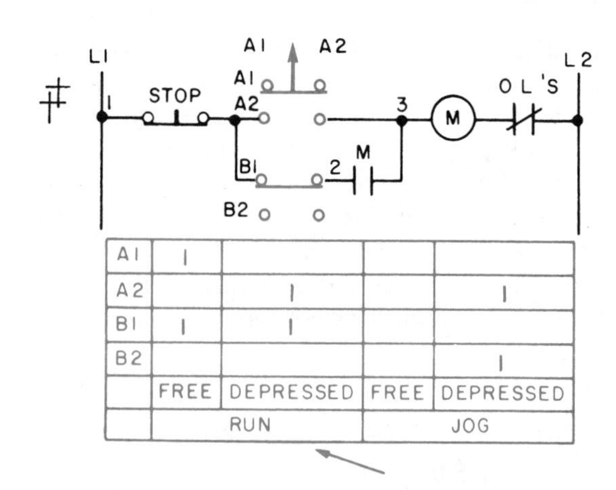

A1	1			
A2		1		1
B1	1	1		
B2				1
	FREE	DEPRESSED	FREE	DEPRESSED
	RUN		JOG	

FIGURE 41-9 Jogging using selector push button

tained to coil M when the push-pull operator is released and movable contact A returns to its normally open position. The circuit will remain in this condition until the stop button is pushed and coils CR and M de-energize.

When the push-pull operator is pressed, movable contact B completes a circuit to coil M. When coil M energizes, auxiliary contact M closes. Contact CR2, however, is open and there is no complete circuit to maintain current flow to coil M. When the push button is released and movable contact B returns to its open position, the circuit to coil M is broken and the motor starter de-energizes.

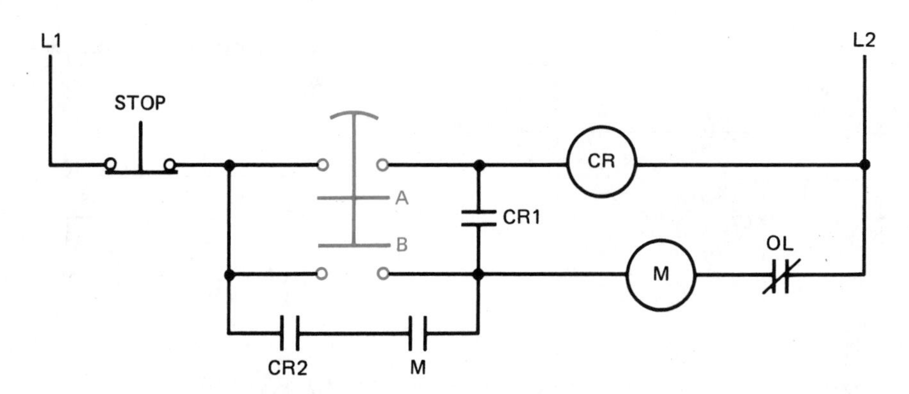

FIGURE 41-10 Jog-run control using a push-pull operator

REVIEW QUESTIONS

1. Why is jogging (inching) included in this section on "Methods of Deceleration"?
2. What is the safety feature of a lock-stop push button?
3. In (A) of figure 41-3, what will happen if both the run and jog push buttons are closed?
4. What will happen if the start and jog push buttons of the circuit shown in figure 41-4 are pushed at the same time?
5. In figure 41-6, what will happen if both jog push buttons are pushed momentarily?
6. Draw an elementary control diagram of a reversing starter. Use a standard duty selector switch with forward, reverse, and stop push buttons with three methods of interlocking.
7. Describe the contact arrangement of the push-pull operator used for a jog-run control.

UNIT 42

Plugging

Objectives *After studying this unit, the student will be able to:*

- **Define what is meant by the plugging of a motor**
- **Describe how a control circuit using a zero-speed switch operates to stop a motor**
- **Describe the action of a time-delay relay in a plugging circuit**
- **Describe briefly the action of the several alternate circuits which use the zero-speed switch**
- **Connect plugging control circuits**
- **Recommend troubleshooting solutions for plugging problems**

Plugging is defined by NEMA as a system of braking, in which the motor connections are reversed so that the motor develops a counter torque which acts as a retarding force. Plugging controls provide for the rapid stop and quick reversal of motor rotation.

Motor connections can be reversed while the motor is running unless the control circuits are designed to prevent this type of connection. Any standard reversing controller can be plugged, either manually or with electromagnetic controls. Before the plugging operation is attempted, however, several factors must be considered including:

1. the need to determine if methods of limiting the maximum permissible currents are necessary, especially with repeated operations and dc motors.

2. the need to examine the driven machine to insure that repeated plugging will not cause damage to the machine.

PLUGGING SWITCHES AND APPLICATIONS

Plugging switches, or zero-speed switches, are designed to be added to control circuits as pilot devices to provide quick, automatic stopping of machines. In most cases, the machines will be driven by squirrel cage motors. If the switches are adjusted properly, they will prevent the direction reversal of rotation of the controlled drive after it reaches a standstill following the reversal of the motor connections. One typical use of plugging switches is for machine tools which must stop suddenly at some point in their cycle of operation to prevent inaccuracies in the work or damage to the machine. Another use is for processes in which the machine must stop completely before the next step of work begins. In this case, the reduced stopping time means that more time can be applied to production to achieve a greater total output.

Typical plugging switches are shown in figure 42-1. The shaft of a plugging switch is connected mechanically to the motor shaft or to a shaft on

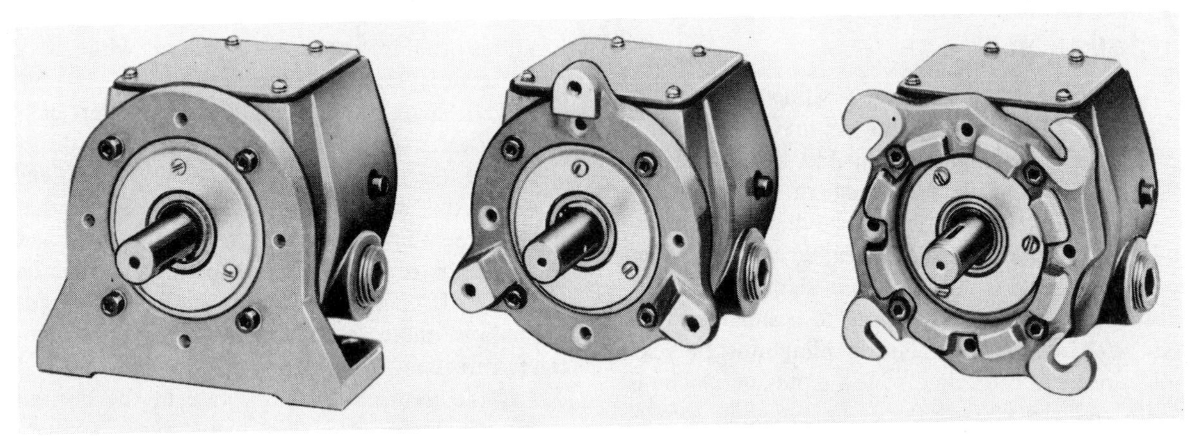

FIGURE 42-1 Plugging (zero-speed) switches. Note mounting methods. (Courtesy Allen-Bradley Co.)

the driven machine. The rotating motion of the motor is transmitted to the plugging switch contacts either by a centrifugal mechanism or by a magnetic induction arrangement (eddy current disc) within the switch. The switch contacts are wired to the reversing starter which controls the motor. The switch acts as a link between the motor and the reversing starter. The starter applies just enough power in the reverse direction to bring the motor to a quick stop.

Plugging a Motor to a Stop From One Direction Only

The forward rotation of the motor in figure 42-2 closes the normally open plugging switch contact. When the stop button is pushed, the forward contactor drops out. At the same time, the reverse contactor is energized through the plugging switch and the normally closed forward interlock. Thus, the motor connections are reversed and the motor is braked to a stop. When the motor is stopped, the plugging switch opens to disconnect the reverse contactor. This contactor is used only to stop the motor using the plugging operation; it is not used to run the motor in reverse.

Adjustment

The torque that operates the plugging switch contacts will vary according to the speed of the motor. An adjustable contact spring is used to oppose the torque to insure that the contacts open and close at the proper time regardless of the motor speed. To operate the contacts, the motor must produce a torque that will overcome the spring pressure. The spring adjustment is generally made with screws that are readily accessible when the switch cover is removed.

Care must be exercised to prevent the entry of chips, filings, and hardware into the housing when it is opened. Such material may be attracted to the magnets or hamper spring action. The housing must be carefully cleaned before the cover is removed for maintenance or inspection.

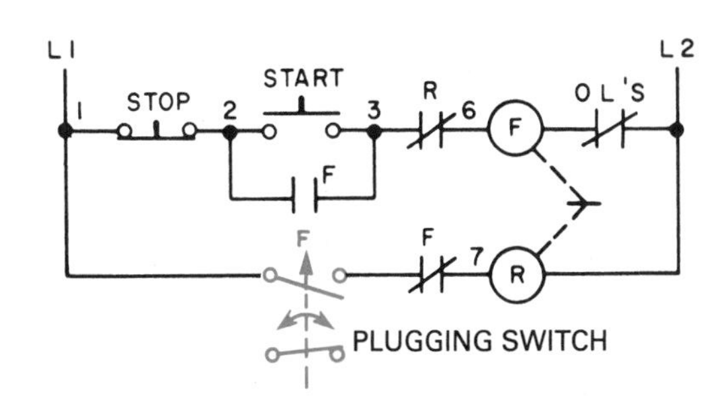

FIGURE 42-2 Plugging motor to stop from one direction only

Installation

To obtain the greatest possible accuracy in braking, the switch should be driven from the shaft with the highest available speed that is within the operating speed range of the switch.

The plugging switch may be driven by gears, by a chain drive, or a direct flexible coupling. The preferred method of driving the switch is to connect a direct flexible coupling to a suitable shaft on the driven machine. The coupling must be flexible since the centerline of the motor or machine shaft and the centerline of the plugging switch shaft are difficult to align accurately enough to use a rigid coupling. The switch must be driven by a positive means. Thus, a belt drive should not be used. In addition, a positive drive must be used between the various parts of the machine being controlled, especially where these parts have large amounts of inertia.

The starter used for this type of circuit is a reversing starter that interchanges two of the three motor leads for a three-phase motor, reverses the direction of current through the armature for a dc motor, and reverses the relationship of the running and starting windings for a single-phase motor.

Motor Rotation

Experience shows that there is little way of predetermining the direction of the rotation of motors when the phases are connected externally in proper sequence. This is an important consideration for the electrician and the electrical contractor when the applicable electrical code or specifications require that each phase wire of a distribution system be color coded.

If the shaft end of a motor runs counterclockwise rather than in the desired clockwise direction, the electrician must reconnect the motor leads at the motor. For example, assume that many three-phase motors are to be connected and the direction of rotation of all the motors must be the same. If counterclockwise rotation is desired, the supply phase should be connected to the motor terminals in the proper sequence, T_1, T_2, and T_3. If the motor does not rotate in the desired counterclockwise direction using these connections, the leads may be interchanged at the motor. Once the proper direction of rotation is established, the remaining motors can be connected in a similar manner if they are from the same manufacturer. If the motors are from different manufacturers, they may rotate in different directions even when all the connections are similar and the supply lines have been phased out for the proper phase sequence and color coded. The process of correcting the rotation may be difficult if the motors are located in a place that is difficult to reach.

Lockout Relay

The zero-speed switch can be equipped with a lockout relay or a safety latch relay. This type of relay provides a mechanical means of preventing the switch contacts from closing unless the motor starting circuit is energized. The safety feature insures that if the motor shaft is turned accidentally, the plugging switch contacts do not close and start the motor. The relay coil generally is connected to the T1 and T2 terminals of the motor. The lock-

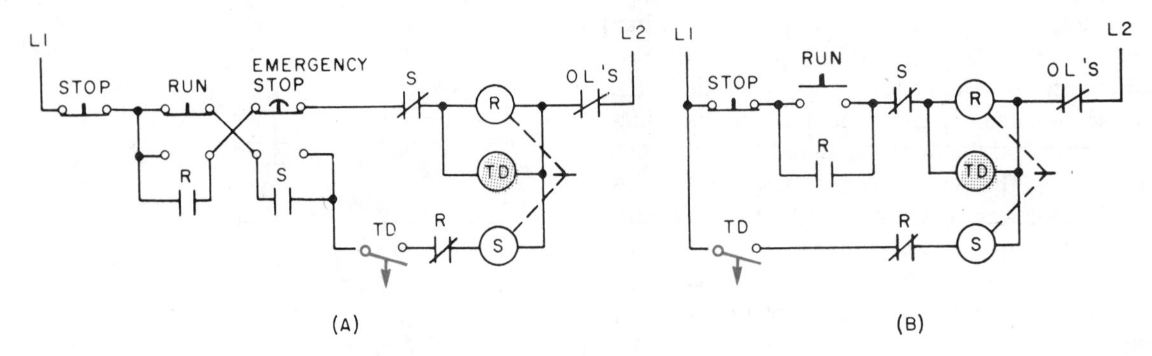

FIGURE 42-3 Plugging with time-delay relay

out relay should be a standard requirement for circuits to protect people, machines, and production processes.

PLUGGING WITH THE USE OF A TIMING RELAY

A time-delay relay may be used in a motor plugging circuit, figure 42-3. Unlike the zero-speed switch, this control circuit does not compensate for a change in the load conditions on the motor. The circuit shown in figure 42-3 can be used for a constant load condition once the timer is preset. If the emergency stop button in figure 42-3(A) is pushed *momentarily* and the normally open circuit is *not* completed, the motor will coast to a standstill. (This action is also true of the normal double contact stop button.) If the emergency stop button is pushed to *complete* the normally open circuit of the push button, contactor S is energized through the closed contacts (TD and R). Contactor S closes and reconnects the motor leads, causing a reverse torque to be applied. When the relay coil is de-energized, the opening of contact TD can be retarded. The time lag is set so that contact TD opens at or near the point at which motor shaft speed reaches 0 r/min.

ALTERNATE CIRCUITS FOR PLUGGING SWITCH

The circuit in figure 42-4 is used for operation in one direction only. When the stop push button is pressed and immediately released, the

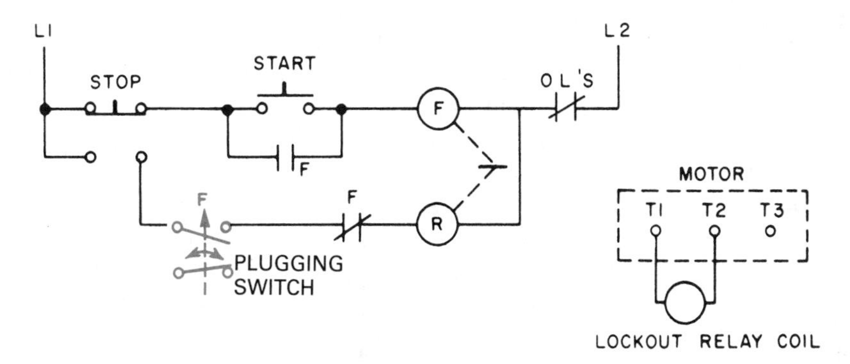

FIGURE 42-4 Holding stop button will stop motor in one direction

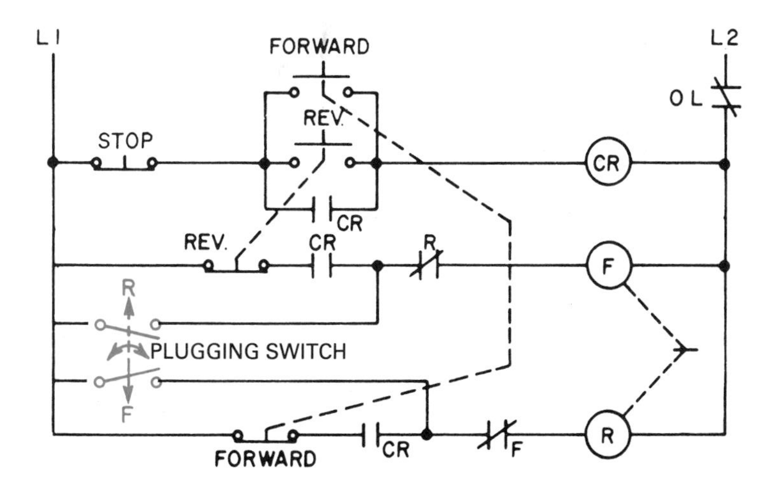

FIGURE 42-5 In this circuit, pressing stop button stops motor in either direction

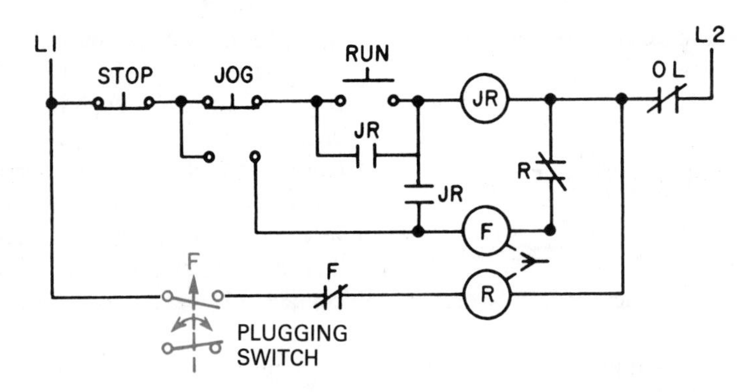

FIGURE 42-6 Pressing stop button will stop motor in one direction

motor and the driven machine coast to a standstill. If the stop button is held down, the motor is plugged to a stop.

Using the circuit shown in figure 42-5, the motor may be started in either direction. When the stop button is pressed, the motor can be plugged to a stop from either direction.

The circuit shown in figure 42-6 provides operation in one direction. The motor is plugged to a stop when the stop button is pressed. Jogging is possible with the use of a control relay.

Figure 42-7 shows a circuit for controlling the direction of rotation of a motor in either direction. Jogging in either the forward or reverse direction is possible if control jogging relays are used. The motor can be plugged to a stop from either direction by pressing the stop button.

The circuit in figure 42-8 provides control in either direction using a maintained contact selector switch with forward, off, and reverse positions. The plugging action is available from either direction of rotation when the switch is turned to the off position. Low-voltage protection is not provided with this circuit.

The circuit in figure 42-9 allows motor operation in one direction. The plugging switch is used as a speed interlock. The solenoid, or coil F, will not operate until the main motor reaches its running speed. A typical application of this circuit is to provide an interlock for a conveyor system. The

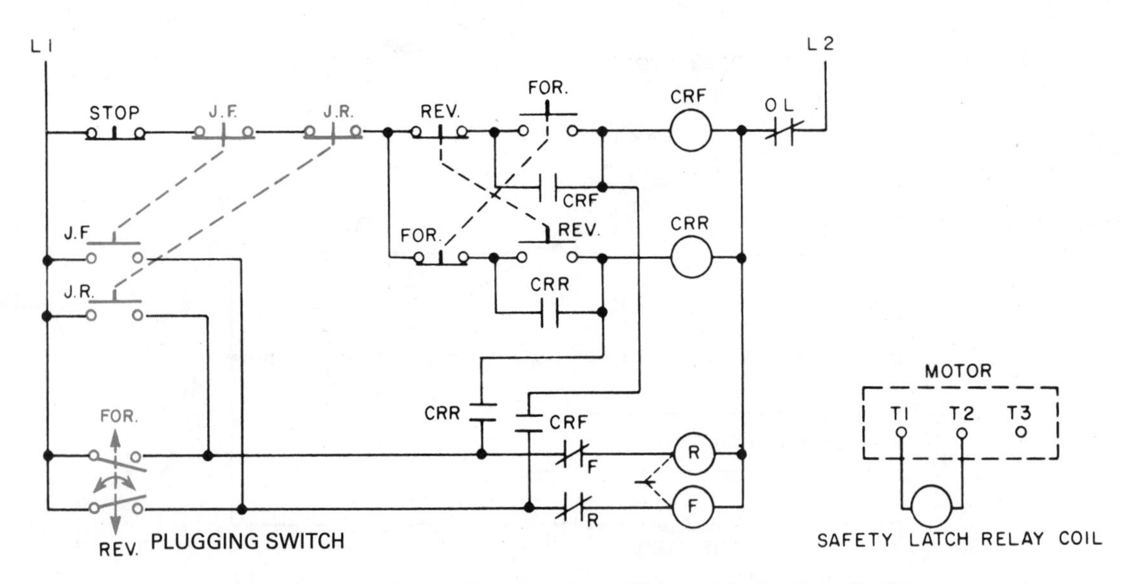

FIGURE 42-7 Use of control jogging relays will stop motor in either direction

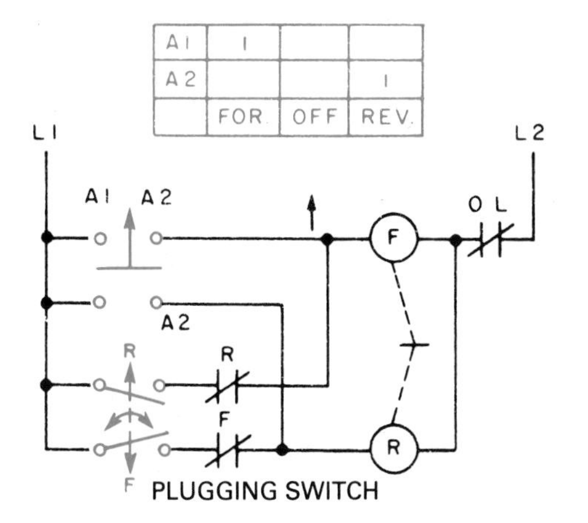

FIGURE 42-8 Using the maintained contact selector switch

feeder conveyor motor cannot be started until the main conveyor is operating.

ANTIPLUGGING PROTECTION

Antiplugging protection, according to NEMA, is obtained when a device prevents the application of a counter torque until the motor speed is reduced to an acceptable value. An antiplugging circuit is shown in figure 42-10. With the motor operating in one direction, a contact on the antiplugging switch opens the control circuit of the contactor used to achieve rotation in the opposite direction. This contact will not close until the motor speed is reduced. Then the other contactors can be energized.

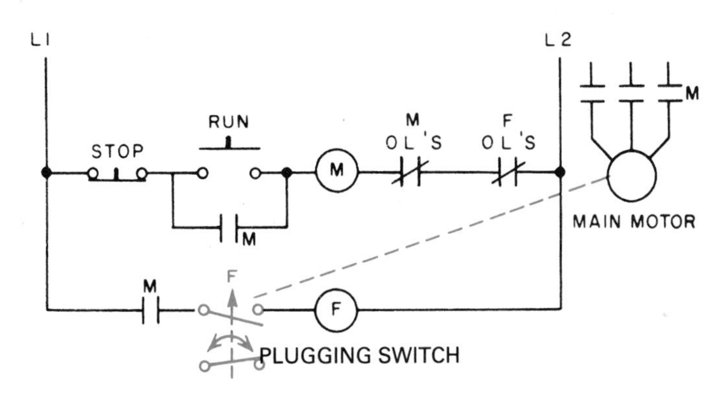

FIGURE 42-9 Using plugging switch as speed interlock

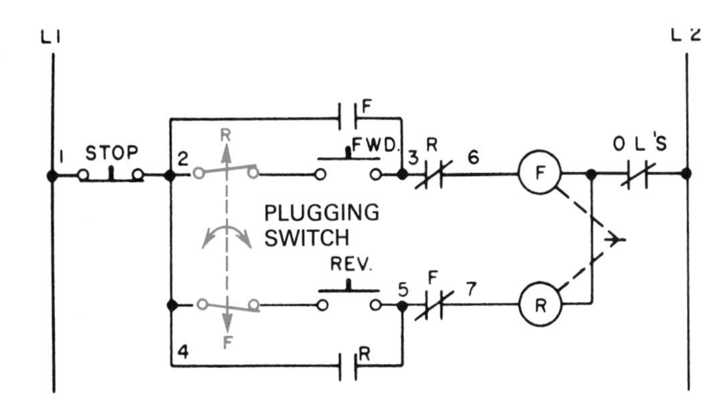

FIGURE 42-10 Antiplugging protection; the motor is to be reversed but not plugged

Alternate Antiplugging Circuits

The direction of rotation of the motor is controlled by the motor starter selector switch, figure 42-11. The antiplugging switch completes the reverse circuit only when the motor slows to a safe, preset speed. Undervoltage protection is not available.

In figure 42-12, the direction of rotation of the motor is selected by using the maintained contact, two-position selector switch. The motor is started with the push button. The direction of rotation cannot be reversed until the motor slows to a safe, preset speed. Low-voltage protection is provided by a three-wire, start-stop, push-button station.

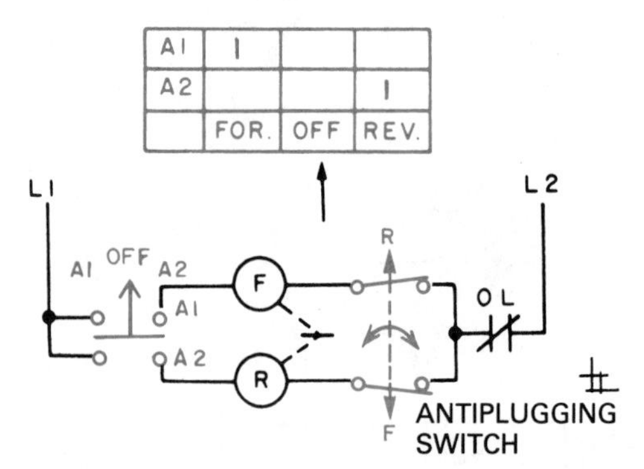

FIGURE 42-11 Antiplugging with rotation direction selector switch

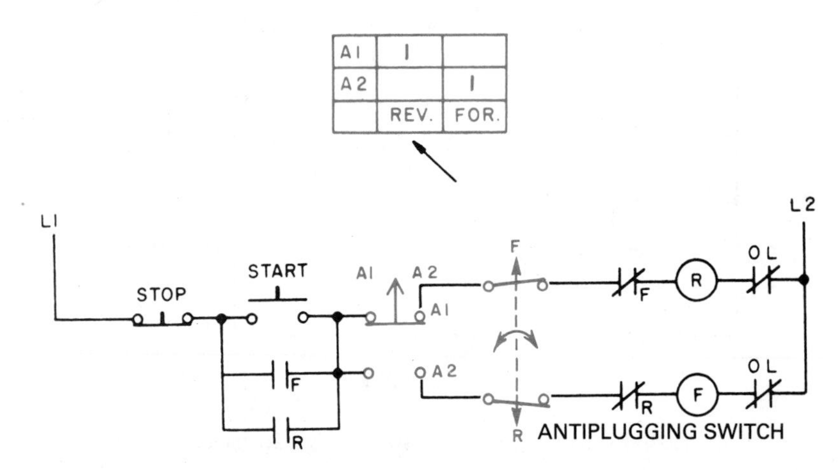

FIGURE 42-12 Antiplugging circuit using a selector switch and providing low-voltage protection

REVIEW QUESTIONS

1. In figure 42-2, what is the purpose of normally closed contact (F)?
2. Can a time-delay relay be used satisfactorily in a plugging circuit? Explain.
3. In what position must a plugging switch be mounted? Explain.
4. What is the preferred method of connecting the plugging switch to the motor or the driven machine?
5. What happens if the zero-speed switch contacts are adjusted to open too late?
6. What is the purpose of the lockout relay or safety latch relay?
7. What happens if the reverse push button is closed when the motor is running in the forward direction, as in figure 42-5?
8. What alternate methods of stopping are provided by the circuit described in figure 42-4?

9. If the motor described in figure 42-6 is plugging to a stop and the operator suddenly wants it to inch ahead or run, what action must be taken?
10. In figure 42-7, is it necessary to push the stop button when changing the direction of rotation? Explain your answer.
11. In figure 42-8, what happens to the motor running in the forward direction if the power supply is lost for 10 minutes?
12. In figure 42-9, how is the feeder motor protected from an overload?
13. What is antiplugging protection?
14. During normal operation, when do the antiplugging switch contacts close?
15. If the supply lines are in the proper phase sequence (L1, L2, and L3) and are connected to their proper terminals on the motor (T1, T2, and T3), will all the motors rotate in the same direction? Why?

SECTION 5

Dc Motor Controls

UNIT 43

Dc Motors

Objectives *After studying this unit, the student will be able to:*

- List applications of dc motors
- Describe the electrical characteristics of dc motors
- Describe the field structure of a dc motor
- Change the direction of rotation of a dc motor
- Identify the series and shunt fields and the armature winding with an ohmmeter
- Connect motor leads to form a series, shunt, or compound motor
- Describe the difference between a differential and cumulative compound motor

APPLICATION

Dc motors are used in applications where variable speed and strong torque are required. They are used for cranes and hoists when loads must be started slowly and accelerated quickly. Dc motors are also used in printing presses, steel mills, pipe forming mills, and many other industrial applications where speed control is important.

SPEED CONTROL

The speed of a dc motor can be controlled by applying variable voltage to the armature or field. When full voltage is applied to both the armature and the field, the motor operates at its base or normal speed. When full voltage is applied to the field and reduced voltage is applied to the armature, the motor operates below normal speed. When full voltage is applied to the armature and reduced voltage is applied to the field, the motor operates above normal speed.

MOTOR CONSTRUCTION

The essential parts of a dc motor are the armature, field windings, brushes, and frame, figure 43-1.

The Armature

The armature is the rotating part of the motor. It is constructed from an iron cylinder that has slots cut into it. Wire is wound through the slots to form the windings. The ends of the windings are connected to the commutator which consists of insulated copper bars and is mounted on the same shaft as the windings. The windings and commutator together form the armature.

Carbon brushes, which press against the commutator segment, supply power to the armature

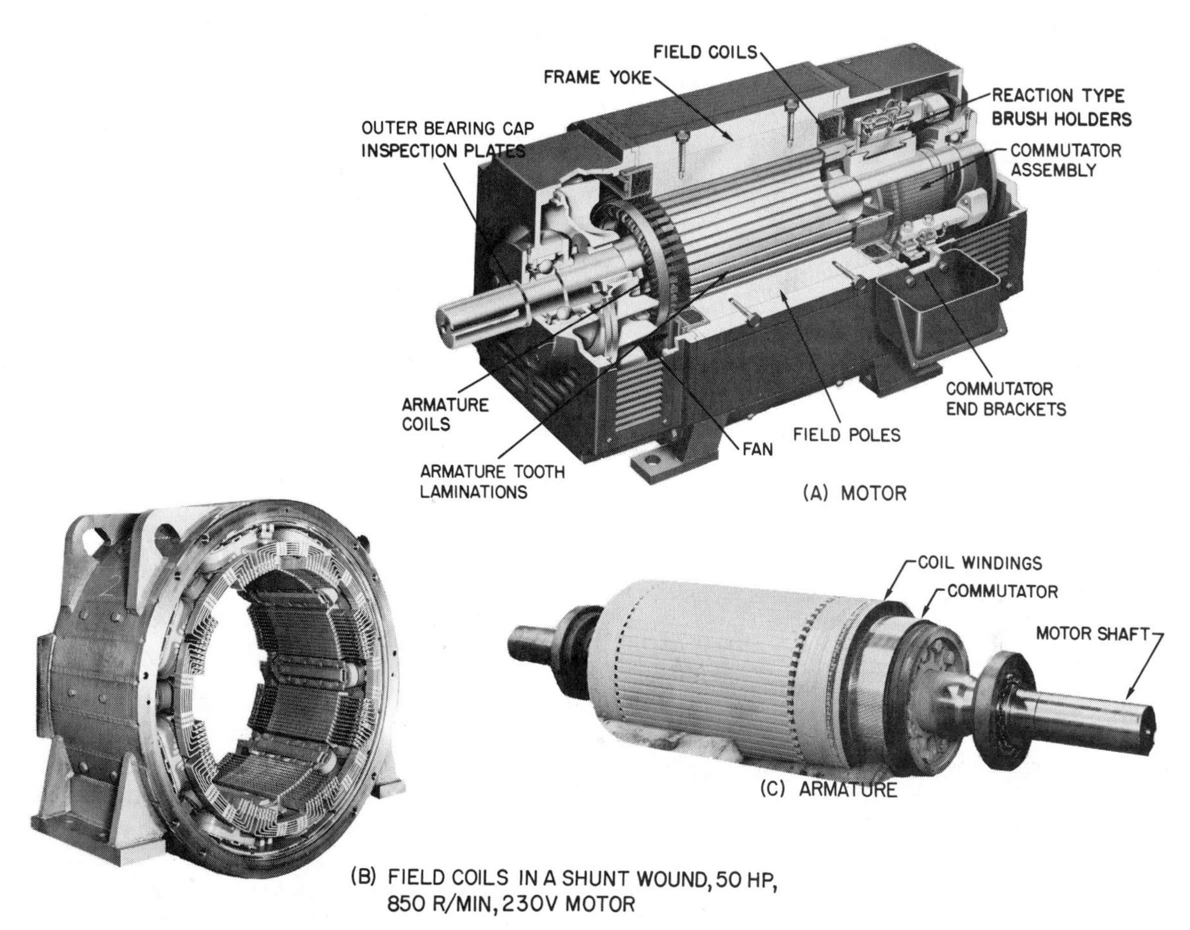

FIGURE 43-1 Dc motor, field structure, and armature assembly (Courtesy Reliance Electric Co.)

from the dc power line. The commutator is a mechanical switch which forces current to flow through the armature windings in the same direction. This enables the polarity of the magnetic field produced in the armature to remain constant as it turns.

Armature resistance is kept low, generally less than one ohm. This is because the speed regulation of the motor is proportional to the armature resistance. The lower the armature resistance, the better the speed regulation will be. Where the brush leads extend out of the motor at the terminal box, they are labeled A1 and A2.

Field Windings

There are two types of field windings used in dc motors: series, and shunt. The series field is made with a few turns of large wire. It has a low resistance and is designed to be connected in series with the armature. The terminal markings, S1 and S2, identify the series field windings.

The shunt field winding is made with many turns of small wire. It has a high resistance and is designed to be connected in parallel with the armature. Since the shunt field is connected in parallel with the armature, line voltage is connected across it. The current through the shunt field is, therefore, limited by its resistance. The terminal markings for the shunt field are F1 and F2.

IDENTIFYING WINDINGS

The windings of a dc motor can be identified with an ohmmeter. The shunt field winding can

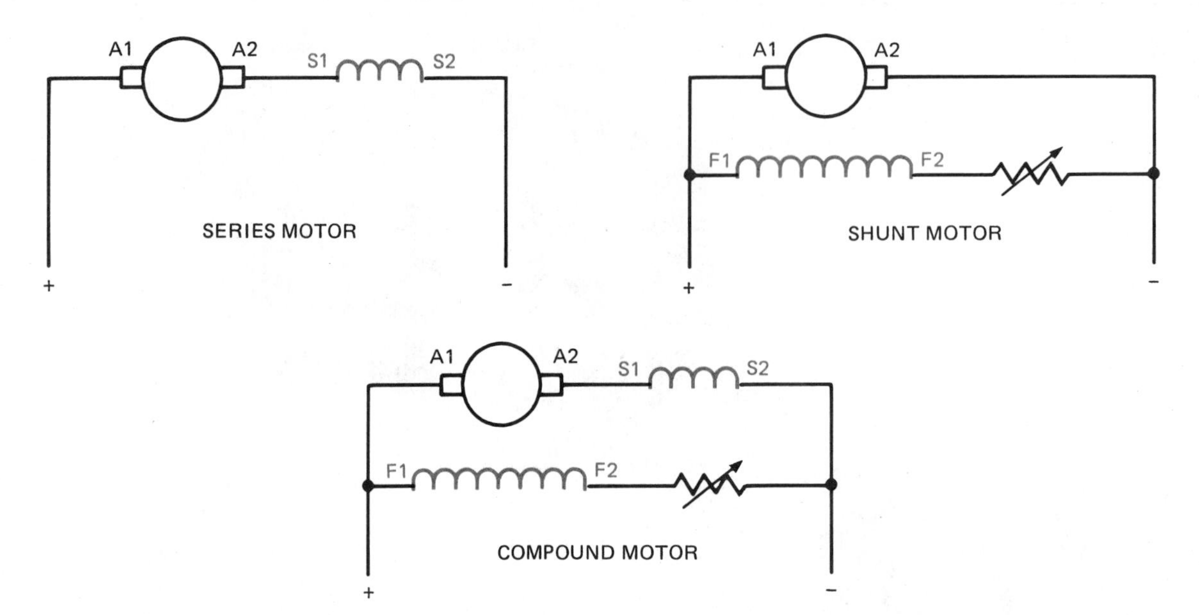

FIGURE 43-2 Dc motor connections

be identified by the fact that it has a high resistance as compared to the other two windings. The series field and armature windings have a very low resistance. They can be identified, however, by turning the motor shaft. When the ohmmeter is connected to the series field and the motor shaft is turned, the ohmmeter reading will not be affected. When the ohmmeter is connected to the armature winding and the motor shaft is turned, the reading will become erratic as the brushes make and break contact with different commutator segments.

TYPES OF DC MOTORS

There are three basic types of dc motors: the series, the shunt, and the compound. The type of motor used is determined by the requirements of the load. The series motor, for example, can produce very high starting torque, but its speed regulation is poor. The only thing that limits the speed of a series motor is the amount of load connected to it. A very common application of a series motor is the started motor used on automobiles. Shunt and compound motors are used in applications where speed control is essential.

Figure 43-2 shows the basic connections for series, shunt, and compound motors. Notice that the series motor contains only the series field connected in series with the armature. The shunt motor contains only the shunt field connected parallel to the armature. A rheostat is shown connected in series with the shunt field to provide above normal speed control.

The compound motor has both series and shunt field windings. Each pole piece in the motor will have both windings wound on it, figure 43-3. There are different ways of connecting compound motors. For instance, a motor can be connected as a long shunt compound or as a short shunt compound, figure 43-4. When a long shunt connection is made, the shunt field is connected parallel to

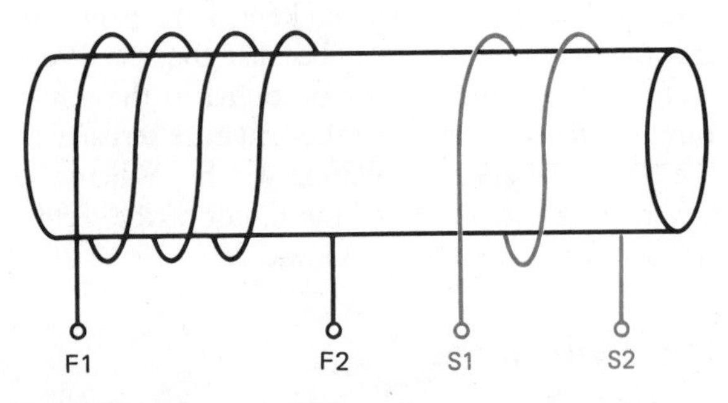

FIGURE 43-3 Series and shunt field windings are wound on one pole piece.

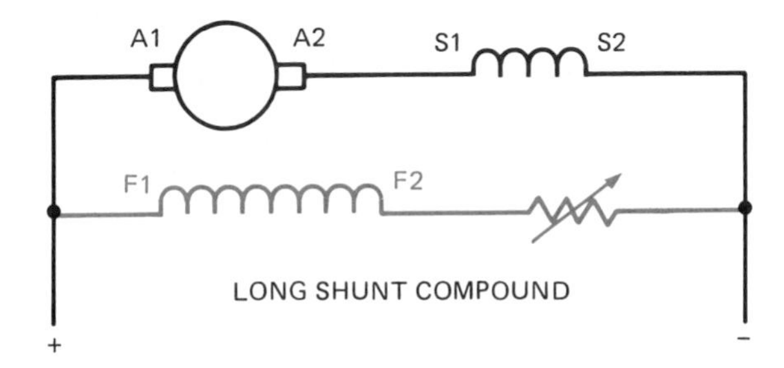

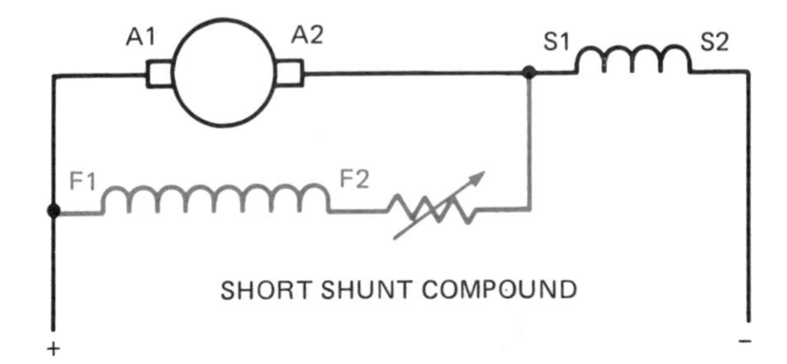

FIGURE 43-4 Compound motor connections

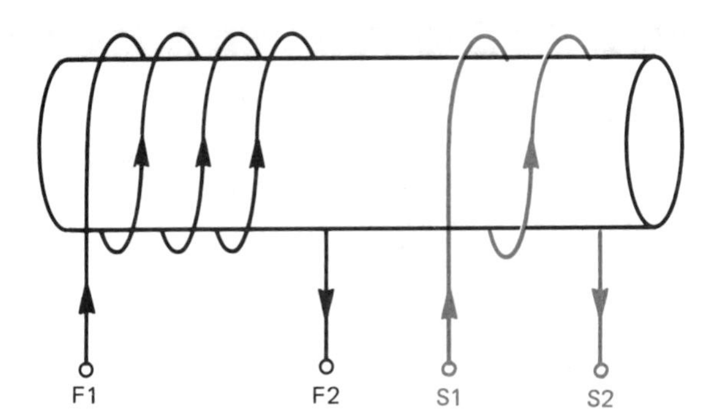

FIGURE 43-5 Cumulative compound connection

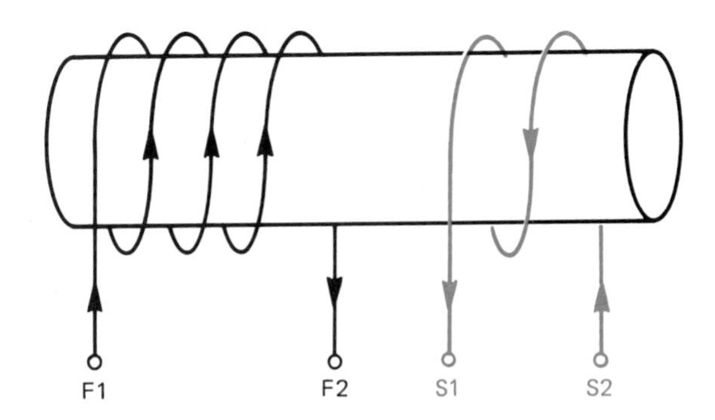

FIGURE 43-6 Differential compound connection

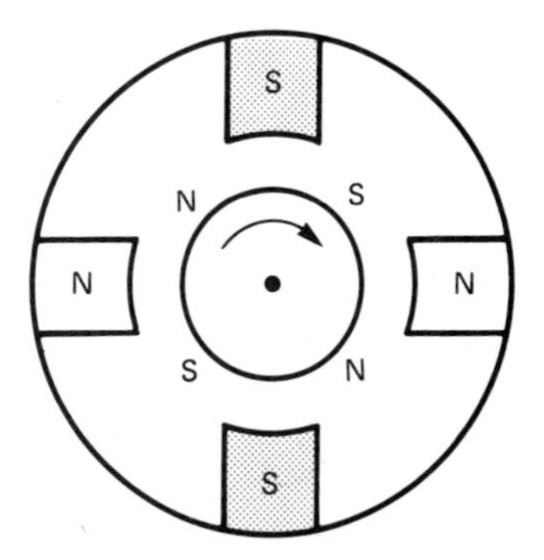

FIGURE 43-7 Armature rotates in a clockwise direction

both the armature and the series field. When a short shunt connection is made, the shunt field is connected parallel to the armature, but in series with the series field.

Compound motors can also be connected as cumulative or differential. When a motor is connected as a cumulative compound, the shunt and series fields are connected in such a manner that as current flows through the windings they aid each other in the production of magnetism, figure 43-5. When the motor is connected as a differential compound, the shunt and series field windings are connected in such a manner that as current flows through them they oppose each other in the production of magnetism, figure 43-6.

DIRECTION OF ROTATION

The direction of rotation of the armature is determined by the relationship of the polarity of the magnetic field of the armature to the polarity of the magnetic field of the pole pieces. Figure 43-7 shows a motor connected in such a manner that the armature will rotate in a clockwise direction due to the attraction and repulsion of magnetic fields. If the input lines to the motor are reversed, the magnetic polarity of both the pole pieces and the armature will be reversed and the motor will continue to operate in the same direction, figure 43-8.

To reverse the direction of rotation of the armature, the magnetic polarity of the armature and

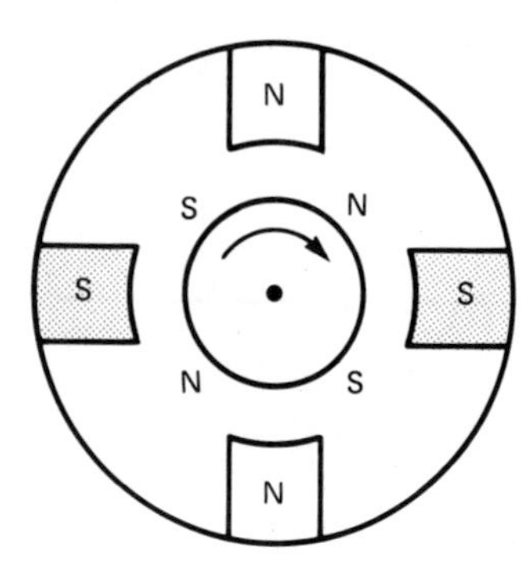

FIGURE 43-8 Changing input lines will not reverse the direction of rotation

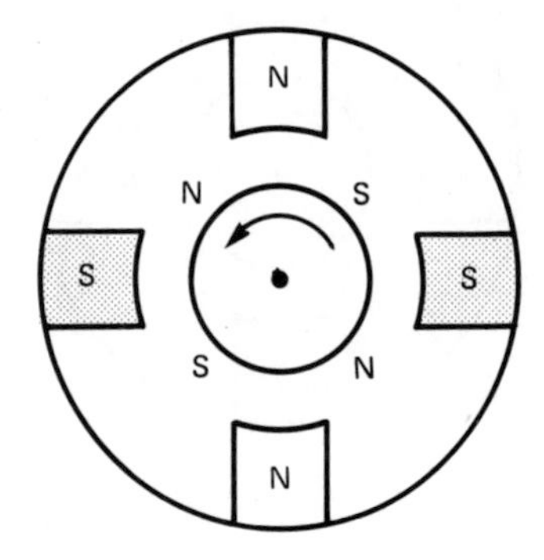

FIGURE 43-9 When the armature leads are reversed, only the direction of rotation is changed

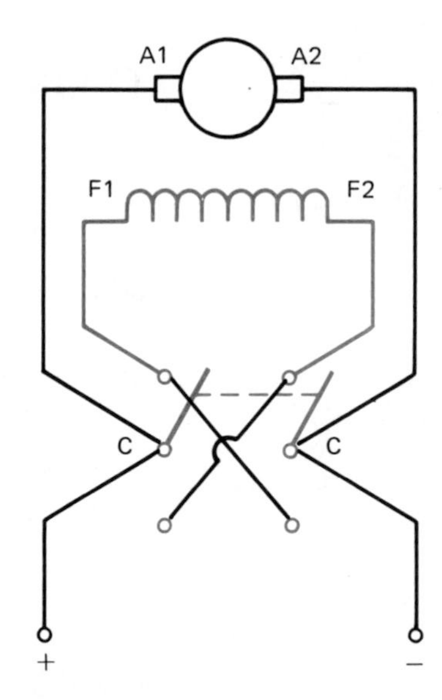

FIGURE 43-10 Double-pole, double-throw switch used to reverse the direction of rotation of a shunt motor

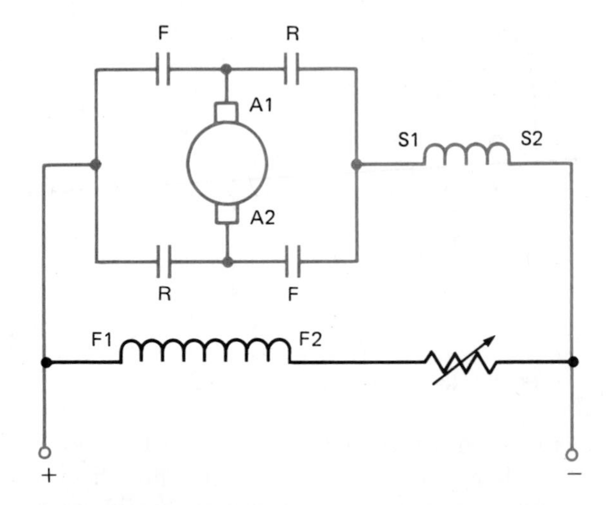

FIGURE 43-11 Contactors reverse the direction of current flow through the armature

the field must be changed in relation to each other. In figure 43-9, the armature leads have been changed, but the field leads have not. Notice that the attraction and repulsion of the magnetic fields now cause the armature to turn in a counterclockwise direction.

When the direction of rotation of a series or shunt motor is to be changed, either the field or the armature leads can be reversed. Many small dc shunt motors are reversed by reversing the connection of the shunt field leads. This is done because the current flow through the shunt field is much lower than the current flow through the armature. This permits a small switch, instead of a large solenoid switch, to be used as a reversing switch. Figure 43-10 shows a double-pole, double-throw (DPDT) switch used as a reversing switch. Power is connected to the common terminals of the switch and the stationary terminals are cross connected.

When a compound motor is to be reversed, only the armature leads are changed. If the motor is reversed by changing the shunt field leads, the motor will be changed from a cumulative compound motor to a differential compound motor. If this happens, the motor speed will drop sharply when load is added to the motor. Figure 43-11 shows a reversing circuit using magnetic contactors to change the direction of current flow through the armature. Notice that the direction of current flow through the series and shunt fields remains the same whether the F contacts or the R contacts are closed.

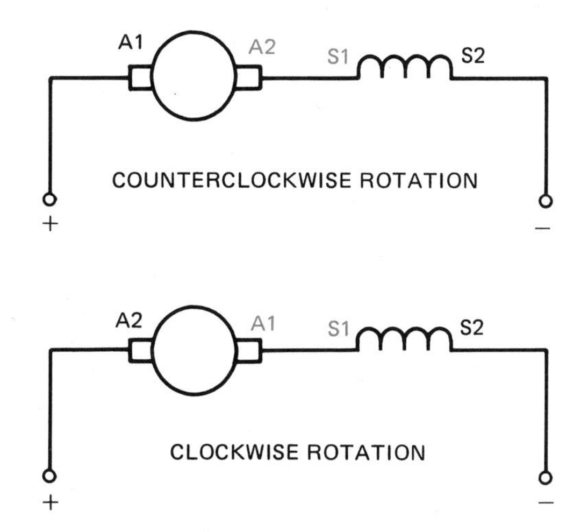

FIGURE 43-12 Standard connections for series motors

STANDARD CONNECTIONS

When dc motors are wound, the terminal leads are marked in a standard manner. This permits the direction of rotation to be determined when the motor windings are connected. The direction of rotation is determined by facing the commutator end of the motor, which is generally located on the rear of the motor, but not always. Figure 43-12 shows the standard connections for a series motor, figure 43-13 shows the standard connections for a shunt motor, and figure 43-14 shows the standard connections for a cumulative compound motor.

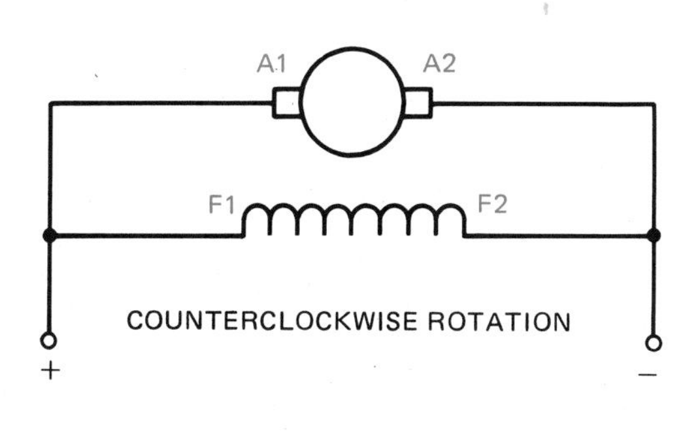

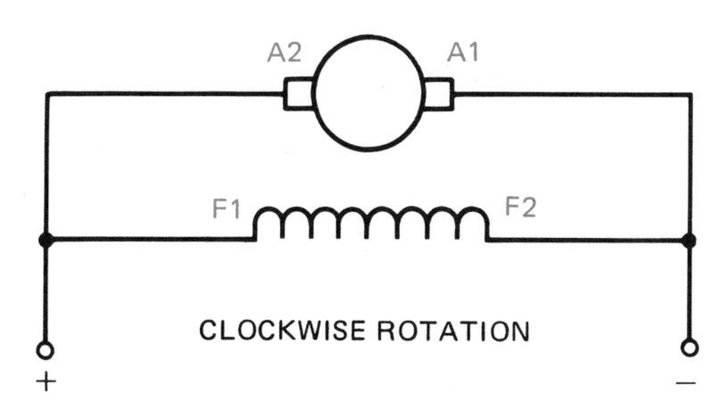

FIGURE 43-13 Standard connections for shunt motors

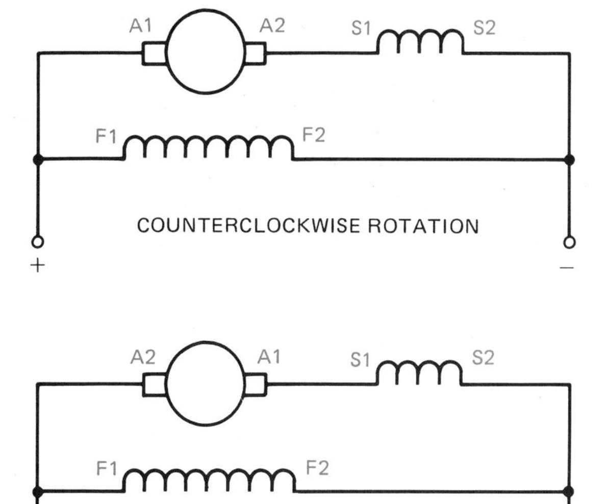

FIGURE 43-14 Standard connections for compound motors

REVIEW QUESTIONS

1. How can a dc motor be made to operate below its normal speed?
2. Name the three basic types of dc motors.
3. Explain the physical difference between series field windings and shunt field windings.
4. The speed regulation of a dc motor is proportional to what?
5. What connection is made to form a long shunt compound motor?
6. Explain the difference between the connection of a cumulative compound and a differential compound motor.
7. How is the direction of rotation of a dc motor changed?
8. Why is it important to reverse only the armature leads when changing the rotation of a compound motor?

Across-the-line Starting

Objectives

After studying this unit, the student will be able to:

- **Describe across-the-line starting for small dc motors**
- **State why a current-limiting resistor may be used in the starting circuit for a dc motor**
- **Connect across-the-line starters used with small dc motors**
- **Recommend troubleshooting solutions for across-the-line starters**
- **Draw diagrams for three motor starter control circuits**

Small dc motors can be connected directly across the line for starting because a small amount of friction and inertia is overcome quickly in gaining full speed and developing a counter emf. Fractional horsepower manual starters (discussed in unit 12) or magnetic contactors and starters (unit 13) are used for across-the-line starting of small dc motors, figure 44-1.

Magnetic across-the-line control of small dc motors is similar to ac control or to two- or three-wire control. Somc dc across-the-line starter coils have dual windings because of the added load of multiple break contacts and the fact that the dc circuit lacks the inductive reactance which is present with ac electromagnets. Both windings are used to lift and close the contacts, but only one winding remains in the holding position. The starting (or lifting) winding of the coil is designed for momentary duty only. In figure 44-2, assume that coil M is energized momentarily by the start button. When the starter is closed, it maintains itself through the normally open maintaining contact (M) and the upper winding of the coil since the normally closed contact (M) is now open. Power contacts M close, and the motor starts across the full line voltage. The double-break power contacts are designed to minimize the effects of arcing. (Dc arcs are greater than those due to alternating current.)

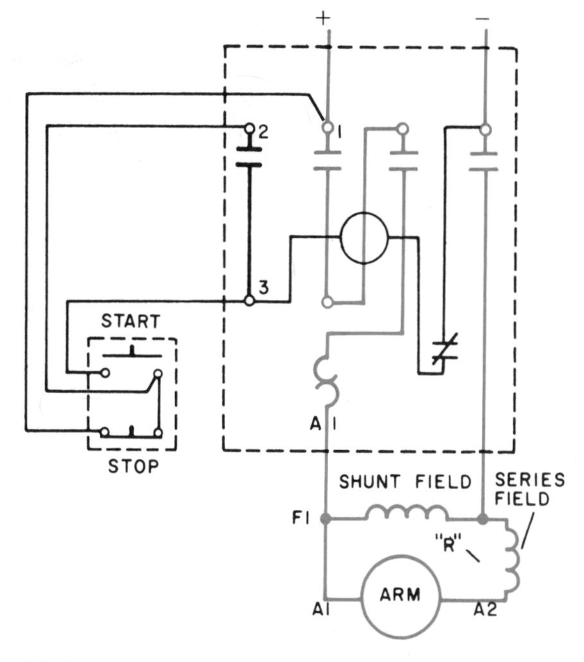

FIGURE 44-1 Dc full voltage starter wiring diagram. Connection ("R") is removed with use of series field.

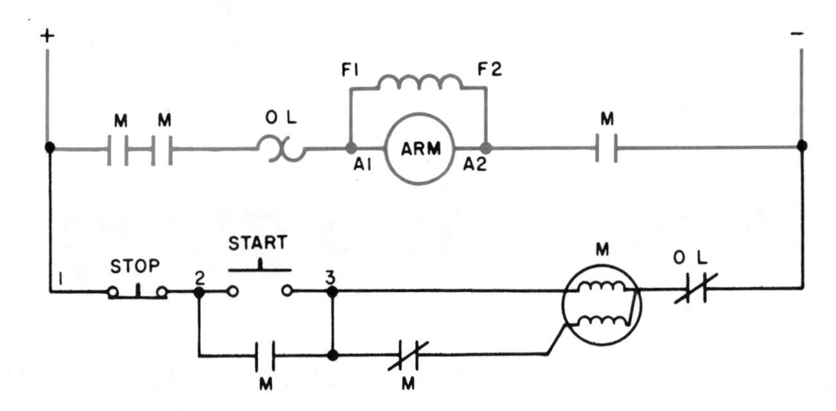

FIGURE 44-2 Line diagram of dc motor starter with dual winding coil

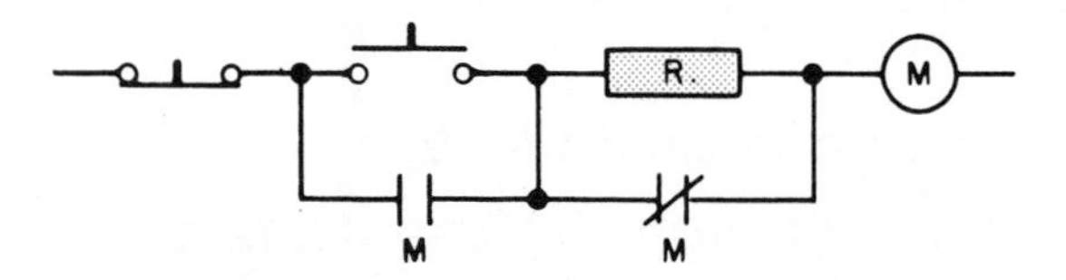

FIGURE 44-3 Dc starting circuit using current limiting resistor

Figure 44-3 shows another control method used to start a dc motor. In this method, a current-limiting resistor is provided to prevent coil burnout. It is used to limit a continuous duty current flow to some coils or when the coils are overheating.

The coil first receives the maximum current required to close the starter. It then receives the minimum current necessary to hold in the contacts and for continuous duty through the current-limiting resistor.

REVIEW QUESTIONS

1. Why may small dc motors be started directly across the line?
2. When using a coil that is not designed for continuous duty, what may happen if a resistor is not added to the circuit?
3. What is the purpose of a double-break power contact?

UNIT 45

Definite Time Starting Control

Objectives

After studying this unit, the student will be able to:

- Describe field current protection for a dc motor
- Describe the use of series resistance for limiting the armature current when starting a dc motor
- Connect a timed starting control for a dc motor

When large dc motors are to be started, current inrush to the armature must be limited. One method of limiting this current is to connect resistors in series with the armature. When the armature begins to turn, counter-emf is developed in the armature. As counter-emf increases, resistance can be shunted out of the armature circuit, permitting the armature to turn at a higher speed. When armature speed increases, counter-emf also increases. Resistance can be shunted out of the circuit in steps until the armature is connected directly to the power line.

Limiting the starting current of the armature is not the only factor that should be considered in a dc control circuit. Most dc motor control circuits use a *field current relay (FCR)* connected in series with the shunt field of the motor. The field current relay insures that current is flowing through the shunt field before voltage can be connected to the armature.

If the motor is running and the shunt field opens, the motor will become a series motor and begin to increase rapidly in speed. If this happens, both the motor and the equipment it is operating can be destroyed. For this reason, the shunt field relay must disconnect the armature from the line if shunt field current stops flowing.

The circuit shown in figure 45-1 is a dc motor control with two steps of resistance connected in series with the armature. When the motor is started, both resistors limit current flow to the armature. Time-delay relays are used to shunt the starting resistors out of the circuit in time intervals of five seconds each until the armature is connected directly to the line.

The circuit operates as follows: When the start button is pushed, current is supplied to relay coil F and all F contacts change position. One F contact is connected parallel to the start button and acts as a holding contact. Another F contact connects the field current relay and the shunt field to the line.

When shunt field current begins to flow, contact FCR closes. When contact FCR closes, a circuit is completed to motor starter coil M and coil TR1. When starter M energizes, contact M closes and connects the armature circuit to the dc line. Five seconds after coil TR1 energizes, contact TR1 closes. This permits current to flow to relay coils S1 and TR2. When contact S1 closes, resistor

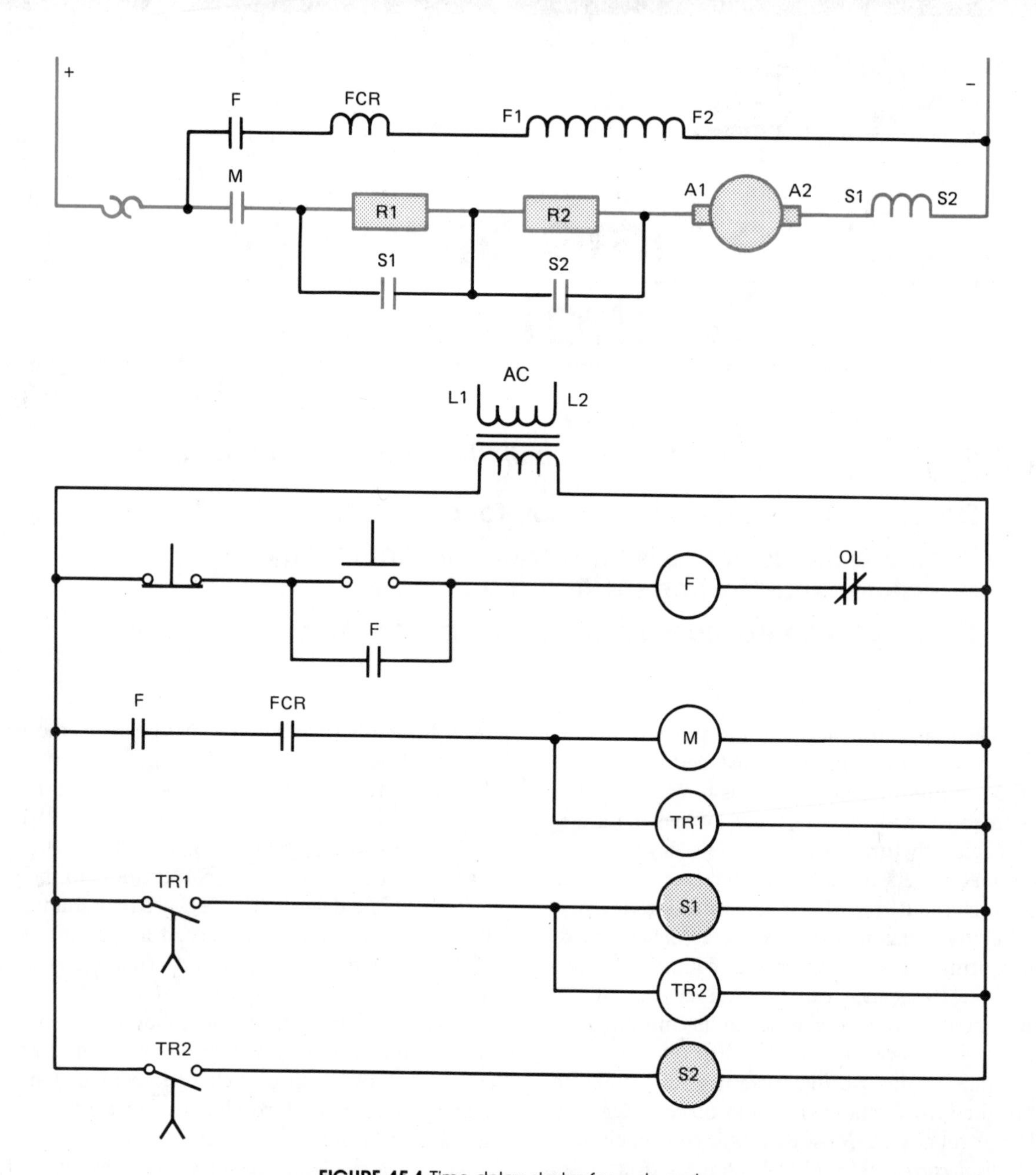

FIGURE 45-1 Time-delay starter for a dc motor

R1 is shunted out of the circuit. Five seconds after relay coil TR2 energizes, contact TR2 closes. When contact TR2 closes, current can flow to coil S2. When contact S2 closes, resistor R2 is shunted out of the circuit and the armature is connected directly to the dc power line.

When the stop button is pushed, relay F de-energizes and opens all F contacts. This breaks the circuit to starter coil M which causes contact M to open and disconnect the armature from the line. When coil TR1 de-energizes, contact TR1 opens immediately and de-energizes coils S1 and

TR2. When coil TR2 de-energizes, contact TR2 opens immediately and de-energizes coil S2. All contacts in the circuit are back in their original positions, and the circuit is ready to be started again.

REVIEW QUESTIONS

1. Why is resistance sometimes connected in series with large dc motors?
2. Is the field current relay a current relay or a voltage relay?
3. What happens if the motor is operating and the shunt field current stops flowing?
4. Describe the action of this circuit if TR1 and TR2 are replaced with off-delay relays.

UNIT 46

Solid-State Dc Motor Controls

Objectives *After studying this unit, the student will be able to:*

- **Describe armature control**
- **Discuss dc voltage control with a three-phase bridge rectifier**
- **Describe methods of current limit control**
- **Discuss feedback for constant speed control**

Direct current motors are used throughout much of industry because of their ability to produce high torque at low speed, and because of their variable speed characteristics. Dc motors are generally operated at or below *normal speed*. Normal speed for a dc motor is obtained by operating the motor with full rated voltage applied to the field and armature. The motor can be operated at below normal speed by applying rated voltage to the field and reduced voltage to the armature.

In unit 45, resistance was connected in series with the armature to limit current and, therefore, speed. Although this method does work and was used in industry for many years, it is seldom used today. When resistance is used for speed control, much of the power applied to the circuit is wasted in heating the resistors, and the speed control of the motor is not smooth because resistance is taken out of the circuit in steps.

Speed control of a dc motor is much smoother if two separate *power supplies*, which convert the ac voltage to dc voltage, are used to control the motor instead of resistors connected in series with the armature, figure 46-1. Notice that one power supply is used to supply a constant voltage to the shunt field of the motor, and the other power supply is variable and supplies voltage to the armature only.

THE SHUNT FIELD POWER SUPPLY

Most solid-state dc motor controllers provide a separate dc power supply which is used to fur-

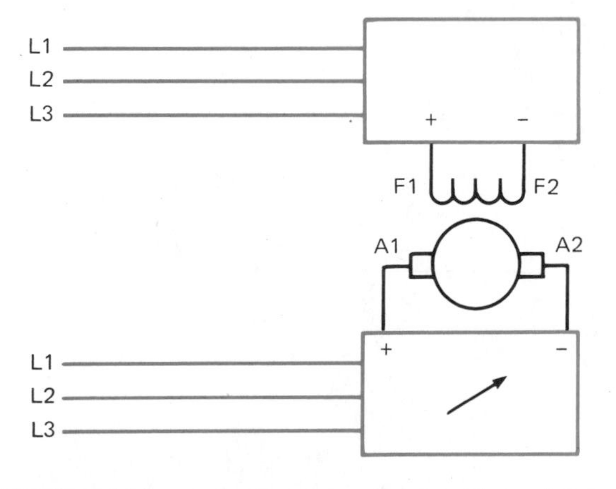

FIGURE 46-1 Separate power supplies used to control armature and field

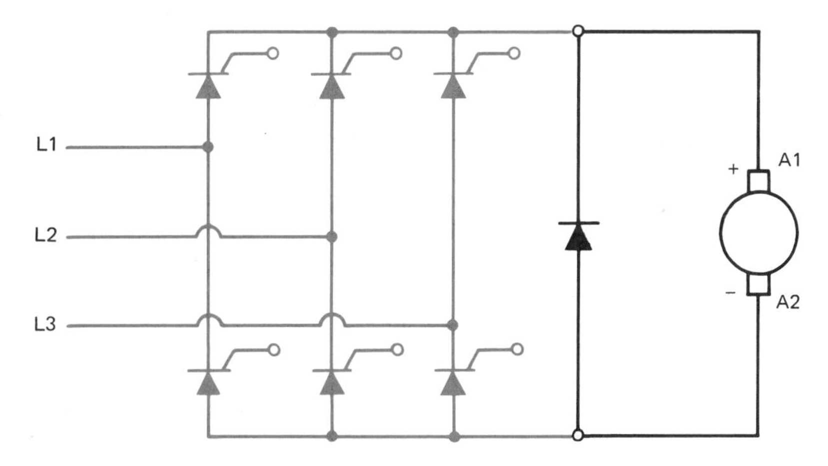

FIGURE 46-2 Three-phase bridge rectifier

nish excitation current to the shunt field. The shunt field of most industrial motors requires a current of only a few amps to excite the field magnets; therefore, a small power supply can be used to fulfill this need. The shunt field power supply is generally designed to remain turned on even when the main (armature) power supply is turned off. If power is connected to the shunt field even when the motor is not operating, the shunt field will act as a small resistance heater for the motor. This heat helps prevent moisture from forming in the motor due to condensation.

THE ARMATURE POWER SUPPLY

The armature power supply is used to provide variable dc voltage to the armature of the motor. This power supply is the heart of the solid-state motor controller. Depending on the size and power rating of the controller, armature power supplies can be designed to produce from a few amps to hundreds of amps. Most of the solid-state motor controllers intended to provide the dc power needed to operate large dc motors convert three-phase ac voltage directly into dc voltage with a three-phase bridge rectifier. (See figure 3-6.)

The diodes of the rectifier, however, are replaced with SCRs to provide control of the output voltage, figure 46-2. Figure 46-3 shows SCRs used for this type of dc motor controller. A large diode is often connected across the output of the bridge. This diode is known as a *freewheeling* or *kickback* diode and is used to kill inductive spike voltages produced in the armature. If armature power is suddenly interrupted, the collapsing magnetic

FIGURE 46-3 This unit is designed to control a 150-hp dc motor. The fuses shown protect the three-phase input lines. The large SCRs rectify the ac voltage into dc voltage. (Courtesy EATON Corp., Cutler-Hammer Products)

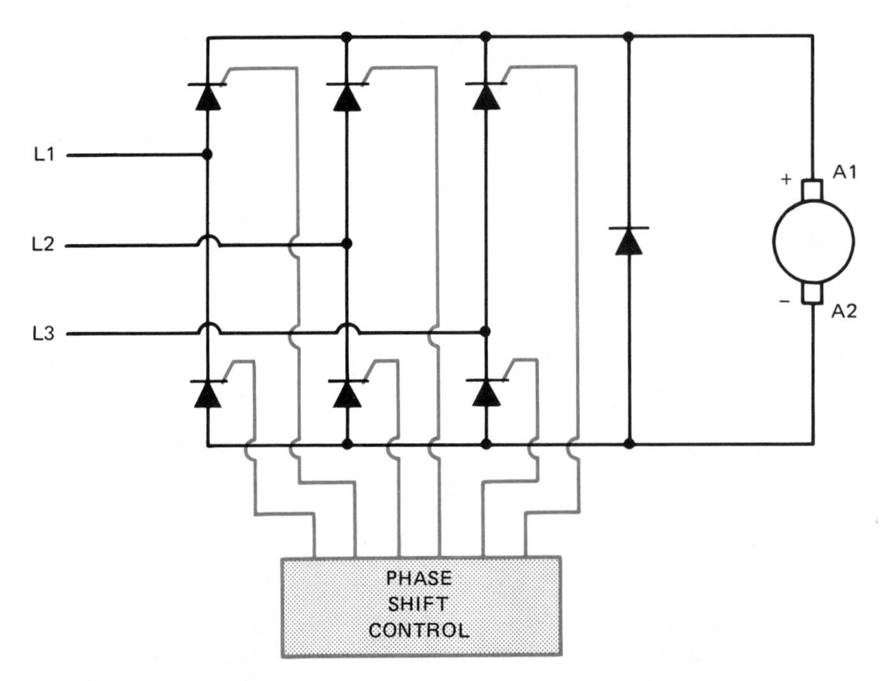

FIGURE 46-4 Phase shift controls output voltage

field induces a high voltage into the armature windings. The diode is reverse biased when the power supply is operating under normal conditions, but an induced voltage is opposite in polarity to the applied voltage. This means the kickback diode will be forward biased to any voltage induced into the armature. Since a silicon diode has a voltage drop of .6 to .7 volts in the forward direction, a high voltage spike cannot be produced in the armature.

VOLTAGE CONTROL

Output voltage control is achieved by phase shifting the SCRs. The phase shift control unit determines the output voltage of the rectifier, figure 46-4. Since the phase shift unit is the real controller of the circuit, other sections of the circuit provide information to the phase shift control unit. Figure 46-5 shows a typical phase shift control unit.

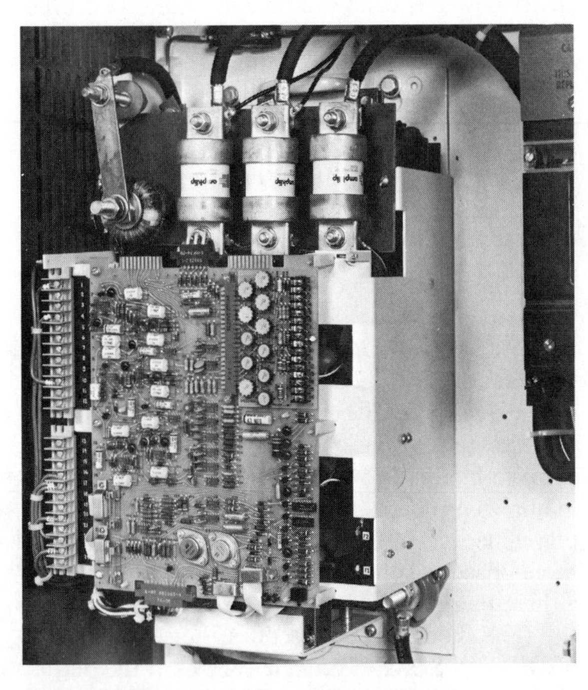

FIGURE 46-5 Phase shift control for the SCRs (Courtesy EATON Corp., Cutler-Hammer Products)

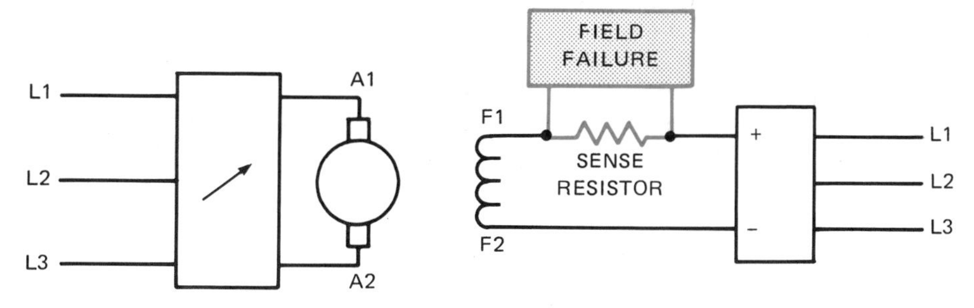

FIGURE 46-6 Resistor used to sense current flow through field

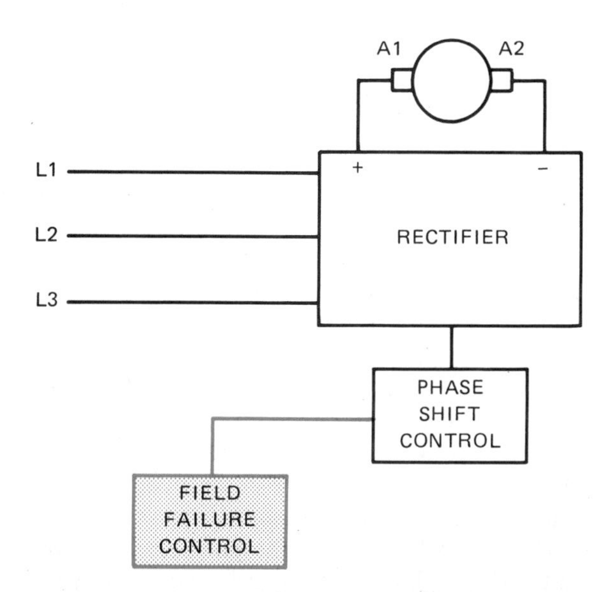

FIGURE 46-7 Field failure control signals the phase shift control

FIELD FAILURE CONTROL

As stated previously, if current flow through the shunt field is interrupted, a compound wound, dc motor will become a series motor and race to high speeds. Some method must be provided to disconnect the armature from the circuit in case current flow through the shunt field stops. Several methods can be used to sense current flow through the shunt field. In unit 45, a current relay was connected in series with the shunt field. A contact of the current relay was connected in series with the coil of a motor starter used to connect the armature to the power line. If current flow were stopped, the contact of the current relay would open causing the circuit of the motor starter coil to open.

Another method used to sense current flow is to connect a low value of resistance in series with the shunt field, figure 46-6. The voltage drop across the sense resistor is proportional to the current flowing through the resistor ($E = I \times R$). Since the sense resistor is connected in series with the shunt field, the current flow through the sense resistor must be the same as the current flow through the shunt field. A circuit can be designed to measure the voltage drop across the sense resistor. If this voltage falls below a certain level, a signal is sent to the phase shift control unit and the SCRs are turned off, figure 46-7.

CURRENT LIMIT CONTROL

The armature of a large dc motor has a very low resistance, typically less than one ohm. If the controller is turned on with full voltage applied to the armature, or if the motor stalls while full voltage is applied to the armature, a very large current will flow. This current can damage the armature of the motor or the electronic components of the controller. For this reason, most solid-state, dc motor controls use some method to limit the current to a safe value.

One method of sensing the current is to insert a low value of resistance in series with the armature circuit. The amount of voltage dropped across the sense resistor is proportional to the cur-

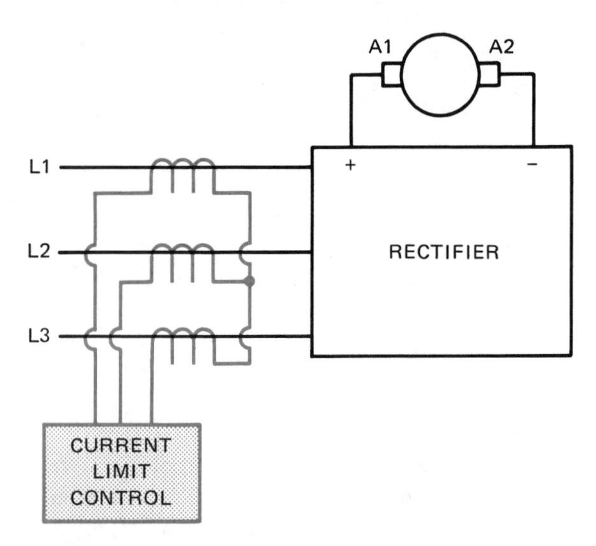

FIGURE 46-8 Current transformers measure ac line current

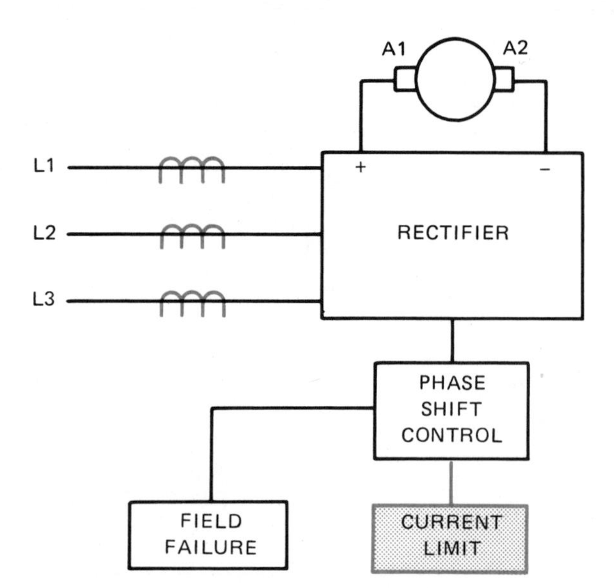

FIGURE 46-9 Current flow to armature is limited

rent flow through the resistor. When the voltage drop reaches a certain level, a signal is sent to the phase shift control telling it not to permit any more voltage to be applied to the armature.

When dc motors of about 25 hp or larger are to be controlled, resistance connected in series with the armature can cause problems. Therefore, another method of sensing armature current can be used, figure 46-8. In this circuit, current transformers are connected to the ac input lines. The current supplied to the rectifier will be proportional to the current supplied to the armature. When a predetermined amount of current is detected by the current transformers, a signal is sent to the phase shift control telling it not to permit the voltage applied to the armature to increase, figure 46-9. This method of sensing the armature current has the advantage of not adding resistance to the armature circuit. Regardless of the method used, the current limit control signals the phase shift control, and the phase shift control limits the voltage applied to the armature.

SPEED CONTROL

The greatest advantage of using direct current motors is their variable speed characteristic. Although the ability to change motor speed is often desirable, it is generally necessary that the motor maintain a constant speed once it has been set. For example, assume that a dc motor can be adjusted to operate at any speed from 0 to 1800 rpm. Now assume that the operator has adjusted the motor to operate at 1200 rpm. The operator controls are connected to the phase shift control unit, figure 46-10. If the operator desires to change speed, a signal is sent to the phase shift control unit and the phase shift control permits the voltage applied to the armature to increase or decrease.

Dc motors, like many other motors, will change speed if the load is changed. If the voltage connected to the armature remains constant, an increase in load will cause the motor speed to decrease, or a decrease in load will cause the motor speed to increase. Since the phase shift unit controls the voltage applied to the armature, it can be used to control motor speed. If the motor speed is to be held constant, some means must be used to detect the speed of the motor. A very common method of detecting motor speed is with the use of an *electrotachometer,* figure 46-11. An electrotachometer is a small, permanent, magnet generator connected to the motor shaft. The output voltage of the generator is proportional to its speed. The

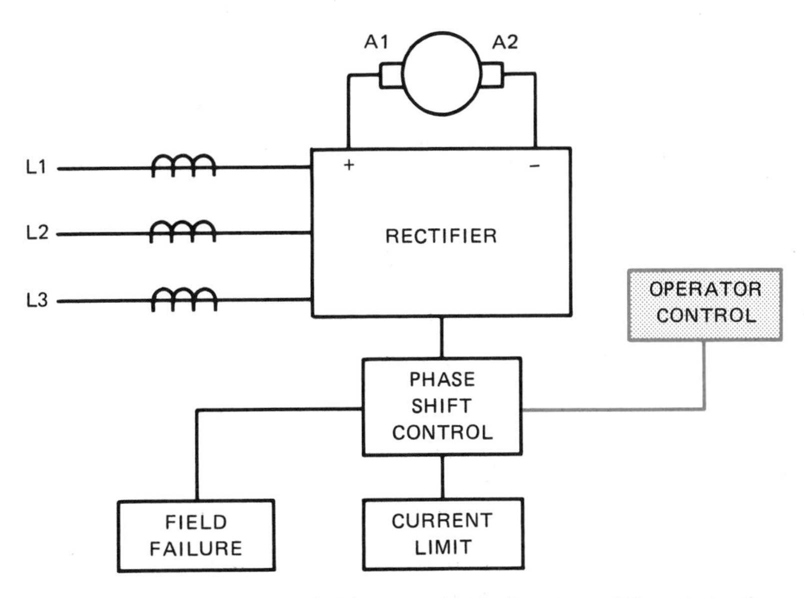

FIGURE 46-10 Operator control is connected to the phase shift control unit

FIGURE 46-11 Dc motor with tachometer attached (Courtesy Allen-Bradley Co., Drives Division)

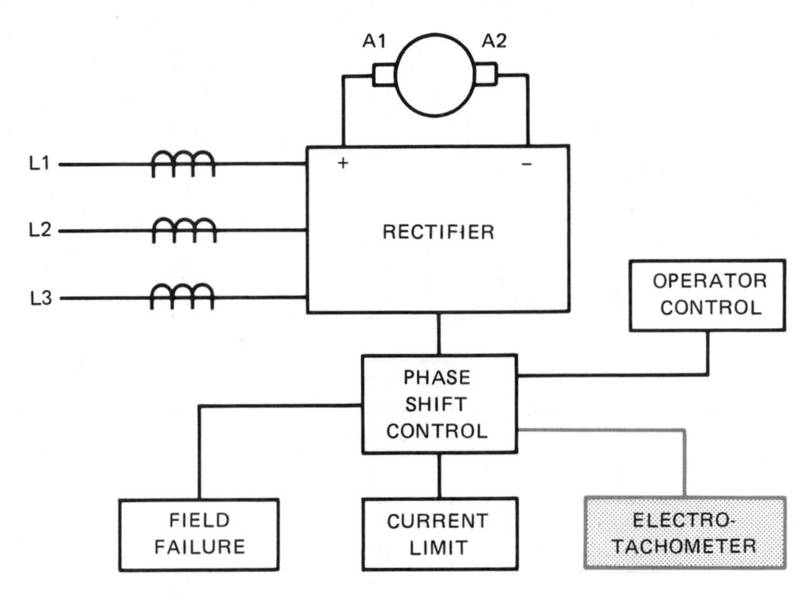

FIGURE 46-12 Electrotachometer measures motor speed

output voltage of the generator is connected to the phase shift control unit, figure 46-12. If load is added to the motor, the motor speed will decrease. When the motor speed decreases, the output voltage of the electrotachometer drops. The phase shift unit detects the voltage drop of the tachometer and increases the armature voltage until the tachometer voltage returns to the proper value.

If the load is removed, the motor speed will increase. An increase in motor speed causes an increase in the output voltage of the tachometer. The phase shift unit detects the increase of tachometer voltage and causes a decrease in the voltage applied to the armature. Electronic components respond so fast that there is almost no noticeable change in motor speed when load is added or removed. An SCR motor control unit is shown in figure 46-13.

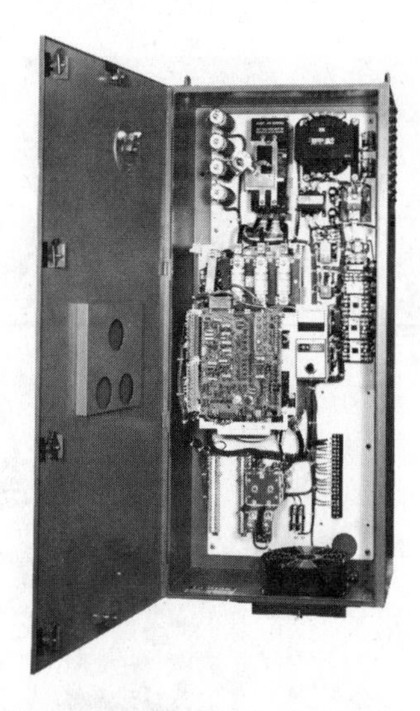

FIGURE 46-13 An SCR motor control unit mounted in a cabinet (Courtesy EATON Corp., Cutler-Hammer Products)

REVIEW QUESTIONS

1. What electronic component is generally used to change the ac voltage into dc voltage in large dc motor controllers?
2. Why is this component used instead of a diode?
3. What is a "freewheeling" or "kickback" diode?

4. Name two methods of sensing the current flow through the shunt field.
5. Name two methods of sensing armature current.
6. What unit controls the voltage applied to the armature?
7. What device is often used to sense motor speed?
8. If the motor speed decreases, does the output voltage of the electrotachometer increase or decrease?

SECTION 6

Ac Motor Control

UNIT 47

Stepping Motors

Objectives

After studying this unit, the student will be able to:

- **Describe the operation of a dc stepping motor**
- **Describe the operation of a stepping motor when connected to ac power**
- **Discuss the differences between stepping motors and other types of motors**
- **Discuss the differences between four-step and eight-step switching**

Stepping motors are devices that convert electrical impulses into mechanical movement. Stepping motors differ from other types of dc or ac motors in that their output shaft moves through a specific angular rotation each time the motor receives a pulse. Each time a pulse is received, the motor shaft moves a precise amount. The stepping motor allows a load to be controlled with regard to speed, distance, or position. These motors are very accurate in their control performance. Generally, less than 5% error per angle of rotation exists, and this error is not cumulative regardless of the number of rotations. Stepping motors are operated on dc power, but can be used as a two-phase synchronous motor when connected to ac power.

THEORY OF OPERATION

Stepping motors operate on the theory that like magnetic poles repel and unlike magnetic poles attract. Consider the circuit shown in figure 47-1. In this illustration, the rotor is a permanent magnet and the stator winding consists of two electromagnetics. If current flows through the winding of stator pole A in such a direction that it creates a north magnetic pole, and through B in such a direction that it creates a south magnetic pole, it would be impossible to determine the direction of rotation. In this condition, the rotor could turn in either direction.

Now consider the circuit shown in figure 47-2. In this circuit, the motor contains four stator

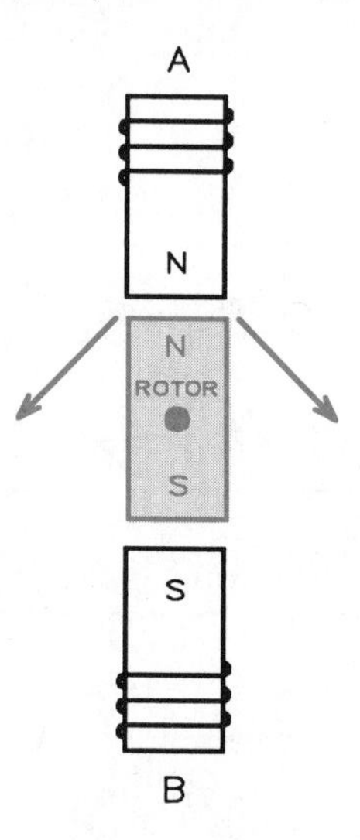

FIGURE 47-1 The rotor could turn in either direction

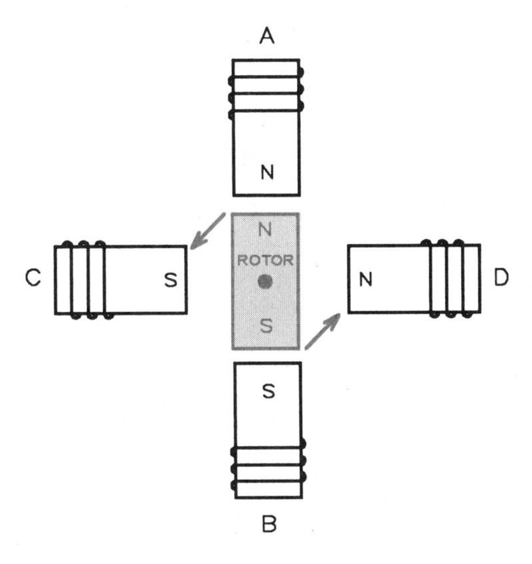

FIGURE 47-2 Direction of rotation is known

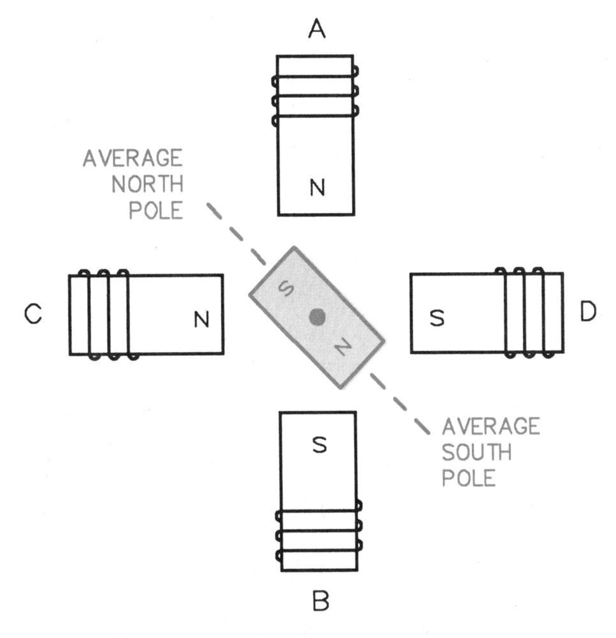

FIGURE 47-3

poles instead of two. The direction of current flow through stator pole A is still in such a direction as to produce a north magnetic field; the current flow through pole B produces a south magnetic field. The current flow through stator pole C, however, produces a south magnetic field, and the current flow through pole D produces a north magnetic field. As illustrated, there is no doubt regarding the direction or angle of rotation. In this example, the rotor shaft will turn 90° in a counter-clockwise direction.

Figure 47-3 shows yet another condition. In this example, the current flow through poles A and C is in such a direction as to form a north magnetic pole, and the direction of current flow through poles B and D forms south magnetic poles. In this illustration, the permanent magnetic rotor has rotated to a position between the actual pole pieces.

To allow for better stepping resolution, most stepping motors have eight stator poles, and the pole pieces and rotor have teeth machined into them as shown in figure 47-4. In practice, the number of teeth machined in the stator and rotor determines the angular rotation achieved each time the motor is stepped. The stator-rotor tooth configuration shown in figure 47-4 produces an angular rotation of 1.8° per step.

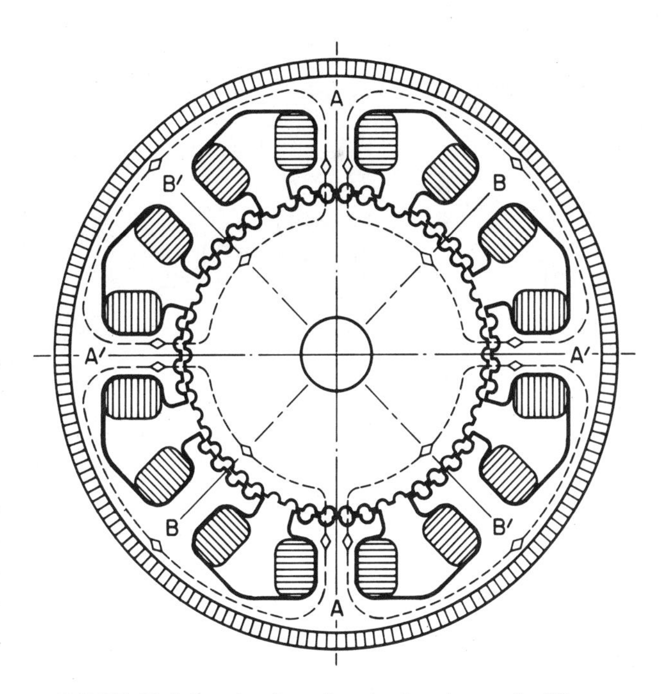

FIGURE 47-4 Construction of a dc stepping motor (Courtesy The Superior Electric Company)

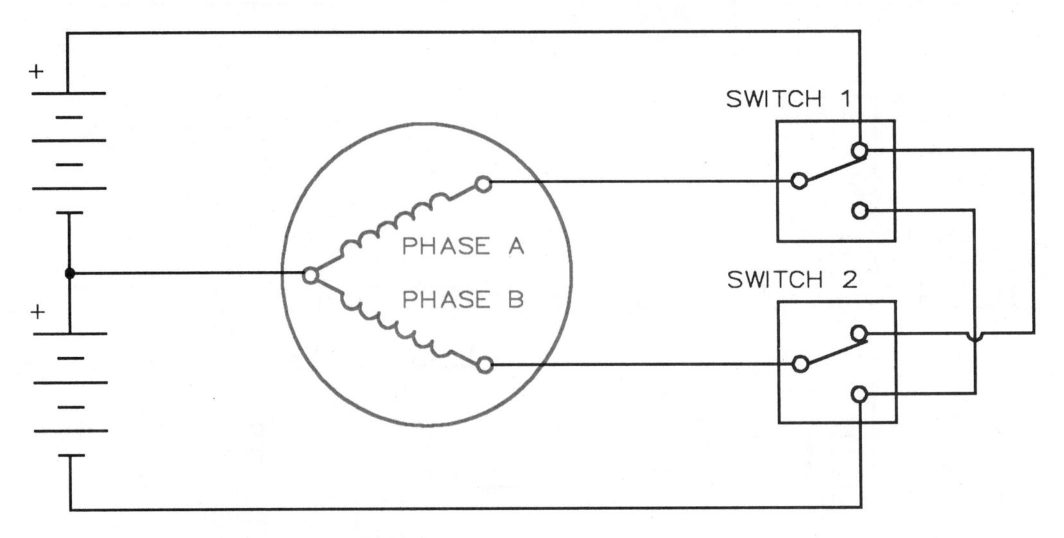

FIGURE 47-5 Standard three-lead motor

WINDINGS

There are different methods for winding stepper motors. A standard three-lead motor is shown in figure 47-5. The common terminal of the two windings is connected to ground of an above- and below-ground power supply. Terminal 1 is connected to the common of a single-pole double-throw switch (switch #1) and terminal 3 is connected to the common of another single-pole double-throw switch (switch #2). One of the stationary contacts of each switch is connected to the positive or above-ground voltage, and the other stationary contact is connected to the negative or below-ground voltage. The polarity of each

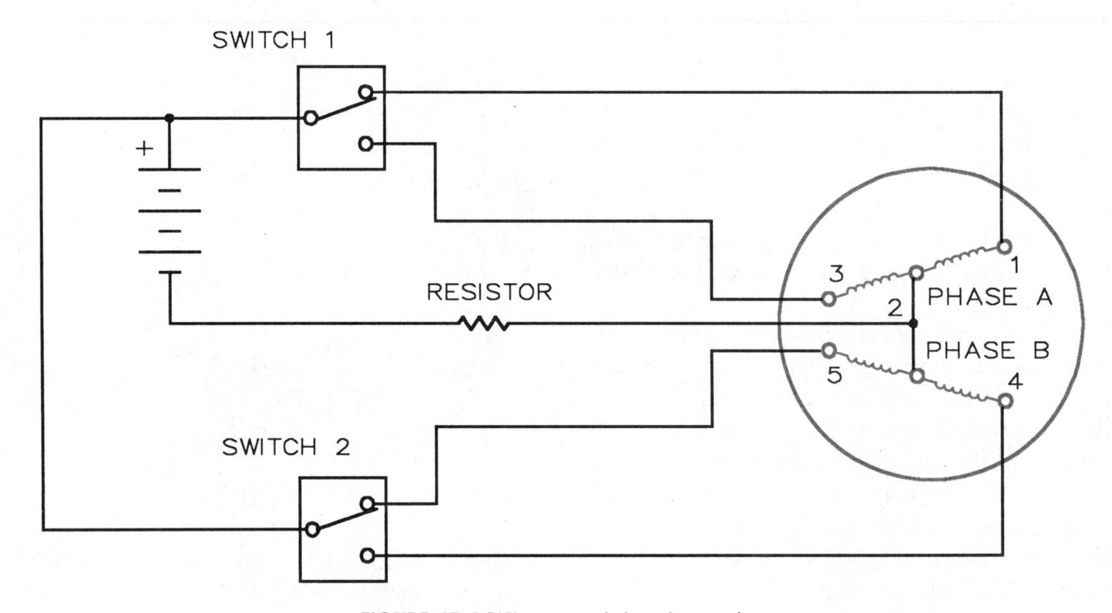

FIGURE 47-6 Bifilar wound stepping motor

winding is determined by the position setting of its control switch.

Stepping motors can also be wound *bifilar* as shown in figure 47-6. The term *bifilar* means that there are two windings wound together. This is similar to a transformer winding with a center tap lead. Bifilar stepping motors have twice as many windings as the three-lead type, which makes it necessary to use smaller wire in the windings. This results in higher wire resistance in the winding, producing a better inductive-resistive (L/R) time constant for the bifilar wound motor. The increased L/R time constant results in better motor performance. The use of a bifilar stepper motor also simplifies the drive circuitry requirements. Notice that the bifilar motor does not require an above- and below-ground power supply. As a general rule, the power supply voltage should be about five times greater than the motor voltage. A current-limiting resistance is used in the common lead of the motor. This current-limiting resistor also helps to improve the L/R time constant.

FOUR-STEP SWITCHING (Full Stepping)

The switching arrangement shown in figure 47-6 can be used for a four-step sequence. Each time one of the switches changes position, the rotor will advance one-fourth of a tooth. After four steps, the rotor has turned the angular rotation of one "full" tooth. If the rotor and stator have fifty teeth, it will require 200 steps for the motor to rotate one full revolution. This corresponds to an angular rotation of 1.8° per step. (360°/200 steps = 1.8° per step.) Figure 47-7 illustrates the switch positions for each step.

STEP	SWITCH #1	SWITCH #2
1	1	5
2	1	4
3	3	4
4	3	5
1	1	5

FIGURE 47-7 Four-step switching sequence

EIGHT-STEP SWITCHING (Half Stepping)

Figure 47-8 illustrates the connections for an eight-stepping sequence. In this arrangement, the center tap leads for phases A and B are connected through their own separate current limiting resistors back to the negative of the power supply. This circuit contains four separate single pole switches instead of two switches. The advantage of this arrangement is that each step causes the motor to rotate one-eighth of a tooth instead of one-fourth of a tooth. The motor now requires 400 steps to produce one revolution which produces an angular rotation of 0.9° per step. This results in better stepping resolution and greater speed capability. The chart in figure 47-9 illustrates the switch position for each step. Figure 47-10 depicts a solid-state switching circuit for an eight-step switching arrangement. A stepping motor is shown in figure 47-11.

AC OPERATION

Stepping motors can be operated on ac voltage. In this mode of operation, they become two-phase ac synchronous constant speed motors and are classified as a *permanent magnet induction motor*. Refer to the exploded diagram of a stepping motor in figure 47-12. Notice that this motor has no brushes, slip rings, commutator, gears, or belts. Bearings maintain a constant air gap between the permanent magnet rotor and the stator windings. A typical eight-stator pole stepping motor will have a synchronous speed of 72 rpm when connected to a 60-hertz two-phase ac power line.

A resistive-capacitive network can be used to provide the 90° phase shift needed to change single-phase ac into two-phase ac. A simple forward-off-reverse switch can be added to provide direc-

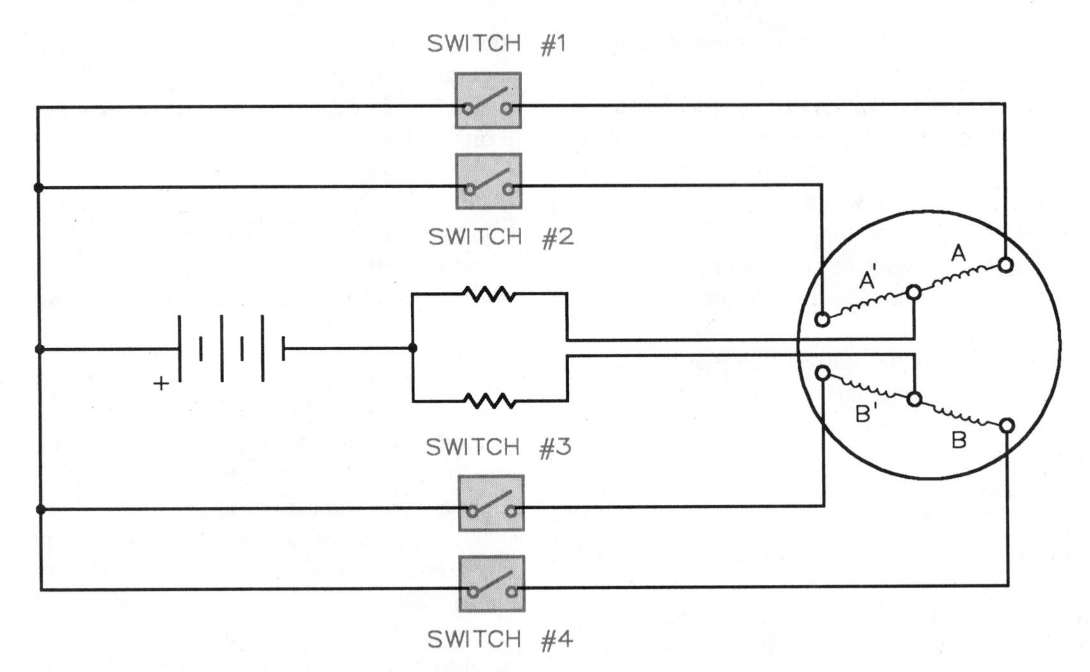

FIGURE 47-8 Eight-step switching

tional control. A sample circuit of this type is shown in figure 47-13. The correct values of resistance and capacitance are necessary for proper operation. Incorrect values can result in random direction of rotation when the motor is started,

STEP	SW #1	SW #2	SW #3	SW #4
1	ON	OFF	ON	OFF
2	ON	OFF	OFF	OFF
3	ON	OFF	OFF	ON
4	OFF	OFF	OFF	ON
5	OFF	ON	OFF	ON
6	OFF	ON	OFF	OFF
7	OFF	ON	ON	OFF
8	OFF	OFF	ON	OFF
1	ON	OFF	ON	OFF

FIGURE 47-9 Eight-step switching sequence

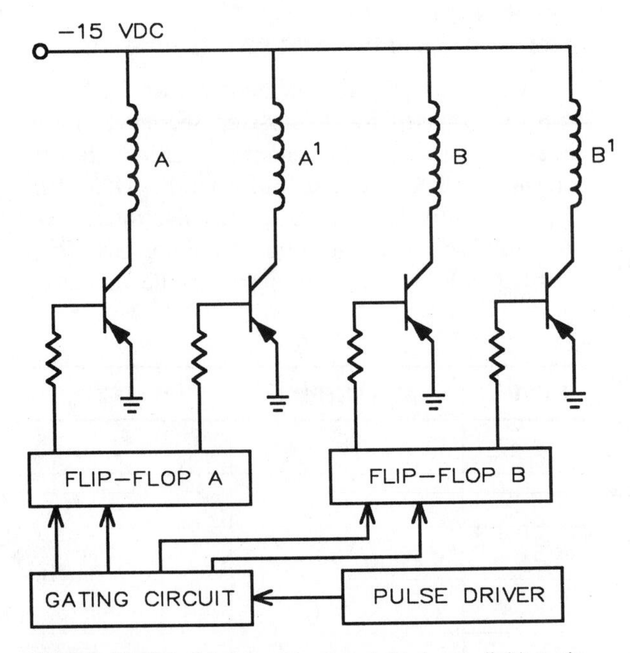

FIGURE 47-10 Solid state drive for eight-step switching circuit

FIGURE 47-11 Dc stepping motor (Courtesy The Superior Electric Company)

change of direction when the load is varied, erratic and unstable operation, as well as failure of the motor to start. The correct values of resistance and capacitance will be different with different stepping motors. The manufacturer's recommendations should be followed for the particular type of stepping motor used.

MOTOR CHARACTERISTICS

When stepping motors are used as two-phase synchronous motors, they have the ability to virtually start, stop, or reverse direction of rotation instantly. The motor will start within about $1\frac{1}{2}$ cycles of the applied voltage and stop within 5 to 25 milliseconds. The motor can maintain a stalled condition without harm to it. Because the rotor is a permanent magnet, no induced current is in the rotor, and no high inrush of current occurs when the motor is started. The starting and running currents are the same. This simplifies the power requirements of the circuit used to supply the motor. Due to the permanent magnetic structure of the rotor, the motor does provide holding torque when turned off. If more holding torque is needed, dc voltage can be applied to one or both windings when the motor is turned off. An example circuit of this type is shown in figure 47-14. If dc voltage is applied to one winding, the holding torque will be approximately 20% greater than the *rated* torque of the motor. If dc voltage is applied to both windings, the holding torque will be about $1\frac{1}{2}$ times greater than the rated torque.

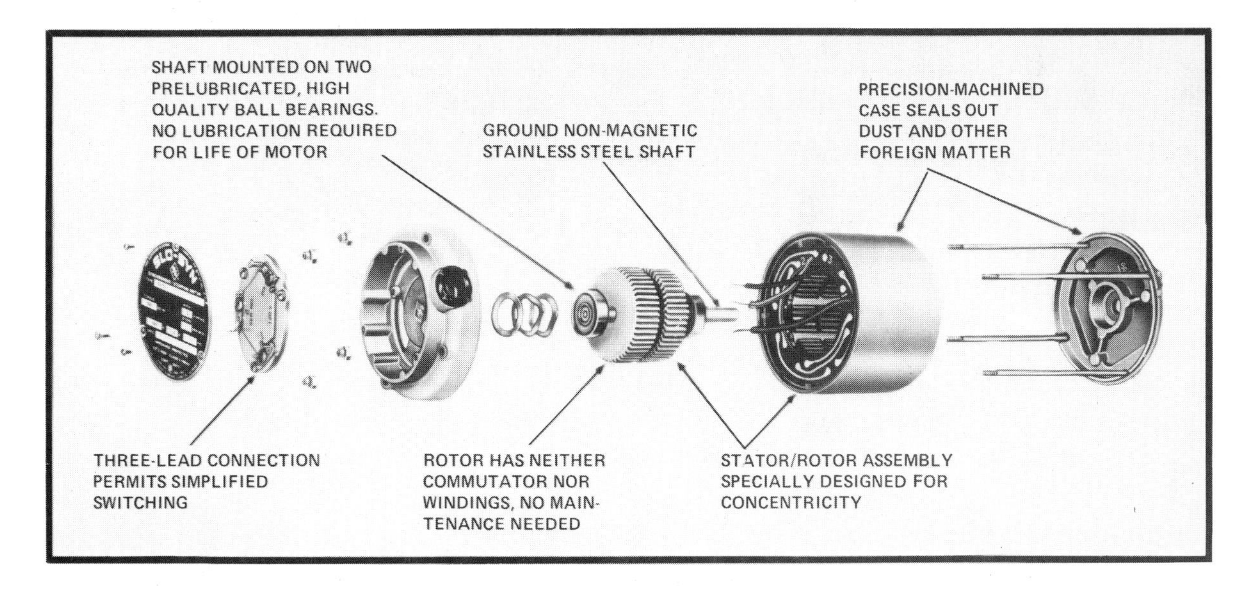

FIGURE 47-12 Exploded diagram of a stepping motor (Courtesy The Superior Electric Company)

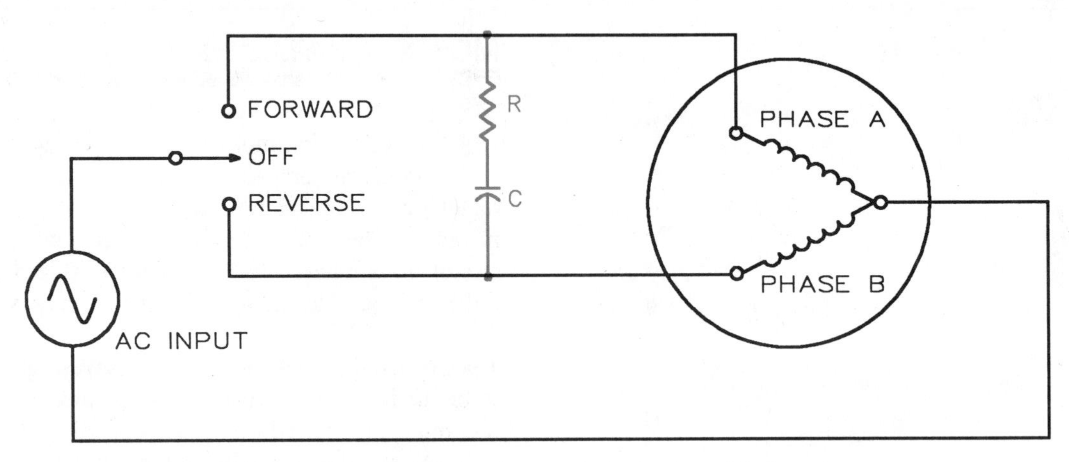

FIGURE 47-13 Phase shift circuit converts single-phase into two-phase

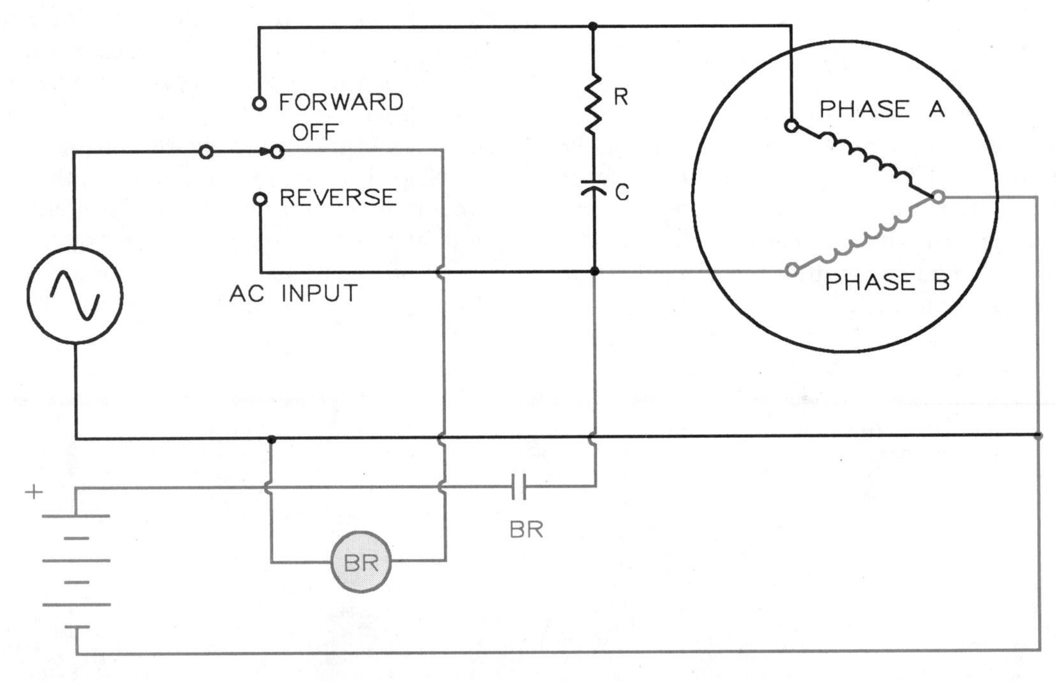

FIGURE 47-14 Applying dc voltage to increase holding torque

REVIEW QUESTIONS

1. Explain the difference in operation between a stepping motor and a common dc motor.
2. What is the principle of operation of a stepping motor?
3. What does the term bifilar mean?
4. Why do stepping motors have teeth machined in the stator poles and rotor?

5. When a stepping motor is connected to ac power, how many phases must be applied to the motor?
6. How many degrees out of phase are the voltages of a two-phase system?
7. What is the synchronous speed of an eight-pole stepping motor when connected to a two-phase 60-hertz ac line?
8. How can the holding torque of a stepping motor be increased?

UNIT 48

The Motor and Starting Methods

Objectives *After studying this unit, the student will be able to:*

- **Describe the most important factors to consider when selecting motor starting equipment**
- **State why reduced current starting is important**
- **Describe typical starting methods**
- **Identify squirrel cage induction motors**
- **Describe how a squirrel cage motor functions**

There are two reasons for the use of reduced voltage starting:

1. To reduce the high starting current drawn by the motor.
2. To reduce the starting torque provided by the motor.

The simplicity, ruggedness, and reliability of squirrel cage induction motors have made them the standard choice for alternating-current, all-purpose, constant speed motor applications. Several types of motors are available; therefore, various kinds of starting methods and control equipment are also obtainable.

THE MOTOR

The Revolving Field

The squirrel cage motor consists of a frame, a stator, and a rotor. The *stator,* or stationary portion carries the stator windings, figure 48-1 (center). The *rotor* is a rotating member, figure 48-1 (bottom) which is constructed of steel laminations mounted rigidly on the motor shaft. The rotor winding consists of many copper, or aluminum, bars fitted into slots in the rotor. The bars are connected at each end by a closed continuous ring. The assembly of the rotor bars and end rings resembles a squirrel cage. This similarity gives the motor its name, *squirrel cage motor.*

For a three-phase motor, the stator frame has three windings. The stator for a squirrel cage motor never has fewer than two poles. The stator windings are connected to the power source. When a 60-hertz current flows in the stator winding, a magnetic field is produced. Because of the three-phase alternating current and the displacement of each phase winding, this field circles the rotor at a speed equal to the number of revolutions per minute (r/min or rpm) divided by the number of pairs of stator poles. Therefore, on 60 hertz, a motor having one pair of two poles will run at 3600 r/min; a four-pole motor (two pairs of two

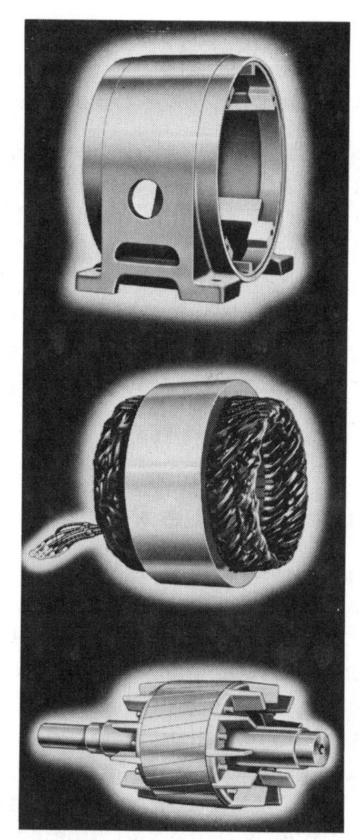

FIGURE 48-1 Squirrel cage induction motor frame, stator, and rotor. Note the coiling blades on rotor. (Courtesy U.S. Electrical Motors)

FIGURE 48-2 Cutaway view of a squirrel cage motor (Courtesy U.S. Electrical Motors)

poles each) will run at 1800 r/min (3600/2). The revolving stator magnetic field induces current in the short-circuited rotor bars. The induced current in the squirrel cage then has a magnetic field of its own. The two fields interact, with the rotor field following the stator rotating field, thereby establishing a torque on the motor shaft. The current induced has its largest value when the rotor is at a standstill. The current then decreases as the motor comes up to speed. In designing motors for specific applications, changing the resistance and reactance of the rotor will alter the characteristics of the motor. For any one rotor design, however, the characteristics are fixed. There are no external connections to the rotor. A cutaway view of an assembled squirrel cage motor is shown in figure 48-2.

Locked Rotor Currents

The locked rotor current and the resulting torque are factors which determine if the motor can be connected across the line or if the current must be reduced to obtain the required performance. Locked rotor currents for different motor types vary from 2 1/2 to 10 times the full-load current of the motor. Some motors, however, have even higher inrush currents.

The Induction Motor At Start

Figure 48-3 illustrates the behavior of the current taken by an induction motor at various speeds. First, note that the starting current is high compared to the running current. In addition, the starting current remains fairly constant at this high value as the motor speed increases. The current then drops sharply as the motor approaches its full rated speed. Since the motor heating rate is a function of I^2R (copper loss), this rate is high during acceleration. For most of the acceleration period, the motor can be considered to be in the locked condition.

No-Load Rotor Speed

The induced current in the rotor gives rise to magnetic forces which cause the rotor to turn in

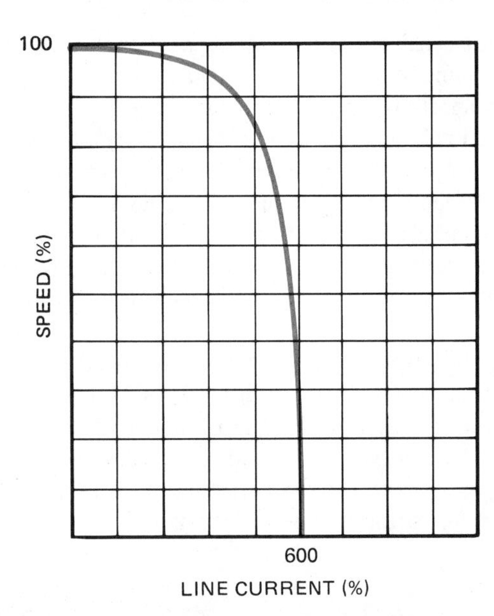

FIGURE 48-3 Induction motor current at various speeds.

the direction of rotation of the stator field. The motor accelerates until the necessary speed is reached to overcome windage and friction losses. This speed is referred to as the *no-load speed*. The motor never reaches synchronous speed since a current will not be induced in the rotor under these conditions and thus the motor will not produce a torque. *Torque* refers to the twisting or turning of the motor shaft. (The rotor bars of the squirrel cage must be cut by the rotating magnetic ac field to produce a torque.)

Speed Under Load

As the rotor slows down under load, the speed adjusts itself to the point where the forces exerted by the magnetic field on the rotor are sufficient to overcome the torque required by the load. *Slip* is the term given to the difference between the speed of the magnetic field and the speed of the rotor.

The slip necessary to carry the full load depends on the characteristics of the motor. In general, the following situations are true:

1. The higher the inrush current, the lower the slip at which the motor can carry full load, and the higher the efficiency.
2. The lower the value of inrush current, the higher the slip at which the motor can carry full load, and the lower the efficiency.

An increase in line voltage causes a decrease in the slip, and a decrease in line voltage causes an increase in the slip. In either case, sufficient current is induced in the rotor to carry the load. A decrease in the line voltage causes an increase in the heating of the motor. An increase in the line voltage decreases the heating. In other words, the motor can carry a larger load. The slip at rated load may vary from 3 percent to 20 percent for different types of motors.

Variation of Torque Requirements

Different loads have different torque requirements. This must be kept in mind when considering the starting torque required and the rate of acceleration most desirable for the load. In general, a number of motors will satisfy the load requirements of an installation under normal running conditions. However, it may be more difficult to use a motor that will perform satisfactorily during both starting and running. Often, it is necessary to decide which is the more important factor to consider for a particular application. For example, a motor may be selected to give the best starting performance, but there may be a sacrifice in the running efficiency. In another case, to obtain a high running efficiency, it may be necessary to use a motor with a high current inrush. For these and other examples, the selection of the proper starter to overcome the objectionable features is an important consideration in view of the high cost of energy today.

The machine to which the motor is connected may be started at no load, normal load, or overload conditions. Many industrial applications require that the machine be started when it is not loaded. Thus, the only torque required is that necessary to overcome the inertia of the machine. Other applications may require that the motor be started while the machine it is driving is subjected to the same load it handles during normal running. In this instance, the starting requirements include the ability to overcome both the normal load and the starting inertia.

Important factors in providing the proper starting equipment include using a starter that satisfies the horsepower rating of the motor, and connecting the motor directly across the line. Another factor is that the motor itself must meet the torque requirements of the industrial application. Actually, the starting equipment may be selected to provide adequate control of the torque after the motor is selected.

Controlling Torque

The most common method of starting a polyphase, squirrel cage induction motor is to connect the motor directly to the plant distribution system at full voltage. In this case a manual or a magnetic starter is used. From the standpoint of the motor itself, this is a perfectly acceptable practice. As a matter of fact, it is probably the most desirable method of starting this type of motor.

Overload protective devices have reached such a degree of reliability that the motor is given every opportunity to make a safe start. The application of a reduced voltage to a motor in an attempt to prevent overheating during acceleration is generally wasted effort. The accelerating time will increase and correctly sized overload elements may still trip.

Reduced voltage starting minimizes the shock on the driven machine by reducing the starting torque of the motor. A high torque, applied suddenly, with full voltage starting, may cause belts to slip and wear or may damage gears, chains, or couplings. The material being processed, or conveyed, may be damaged by the suddenly applied jerk of high torque. By reducing the starting voltage, or current, at the motor terminals, the starting torque is decreased.

Reduced Voltage, Reduced Current, Reduced Torque

The category of reduced voltage methods generally includes all starting methods which deviate from standard, line voltage starting. Not all of these starting schemes reduce the voltage at the motor terminals. Even reduced voltage starters reduce the voltage only to achieve either the reduction of line current or the reduction of starting torque. The reduction of line current is the most commonly desired result.

You should note one important point: *When the voltage is reduced to start a motor, the current is also reduced, and so is the torque that the machine can deliver.* Regardless of the desired result (either reduced current or reduced torque), remember that the other will always follow.

If this fact is kept in mind, it is apparent that a motor which will not start a load on full voltage cannot start that same load at reduced voltage or reduced current conditions. Any attempt to use a reduced voltage or current scheme will not be successful in accelerating troublesome loads. The very process of reducing the voltage and current will further reduce the available starting torque.

Need for Reduced Current Starting

The most common function of reduced voltage starting devices is to reduce, or in some way modify, the starting current of an induction motor. In other words, the rate of change of the starting current is confined to prescribed limits, or there is a predetermined current-time picture that the motor presents to the supply wiring network.

A current-time picture for an entire area is maintained and regulated by the public power utility serving the area. The power company attempts to maintain a reasonably constant voltage at the points of supply so that lamp flicker will not be noticeable. The success of the power company in this attempt depends on the generating capacity to the area; transformer and line loading conditions and adequacies; and the automatic voltage regulating equipment in use. Voltage regulation also depends on the sudden demands imposed on the supply facilities by residential, commercial, and industrial customers. Transient overloading of the power supply may be caused by: (1) sudden high surges of reactive current from large motors on starting, (2) pulsations in current taken by electrical machinery driving reciprocating compressors and similar machinery, (3) the impulse demands of industrial x-ray equipment, and (4) the variable power factor of electric furnaces. All of these demands are capable of producing voltage fluctuations.

Therefore, each of these particularly difficult loads is regulated in some way by the power utility. The utility requires the use of some form of reduced voltage, reduced current method and helps its customers determine the best method.

Power company rules and regulations vary between individual companies and the areas served. The following list gives some commonly applied regulations. (All of the possible restrictions on energy usage are not given.) An installation may be governed by just one of these restrictions, or two or more rules may be combined. Regulations include:

1. A maximum number of starting amperes, either per horsepower or per motor.
2. A maximum horsepower for line starting. A limit in percent of full load current is set for anything above this value.
3. A maximum current in amperes for a particular feeder size. It is up to the user to determine if the motor will conform to the power company requirements.
4. A maximum rate of change of line current taken by the motor; for example, 200 amperes per half-second.

It should be apparent that it is very important for the electrician to understand the behavior of an induction motor during the startup and acceleration periods. Such an understanding enables you to select the proper starting method to conform to local power company regulations. Although several starting methods may appear to be appropriate, a careful examination of the specific application will usually indicate the one best method for motor starting.

TYPICAL STARTING METHODS

The most common methods of starting polyphase squirrel cage motors include:

- *Full voltage starting:* a hand-operated or automatic starting switch throws the motor directly across the line.

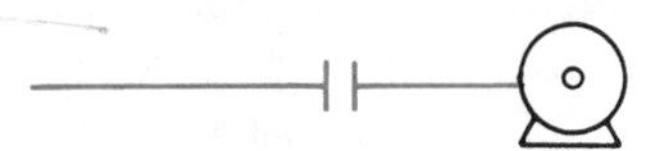

- *Primary resistance starting:* a resistance unit connected in series with the stator reduces the starting current.

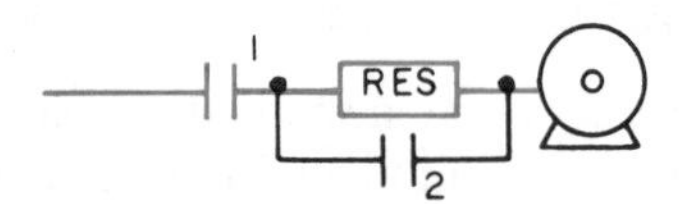

- *Autotransformer or compensator starting:* manual or automatic switching between the taps of the autotransformer gives reduced voltage starting.

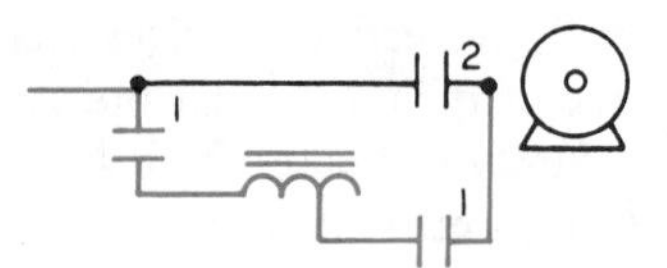

- *Impedance starting:* reactors are used in series with the motor.

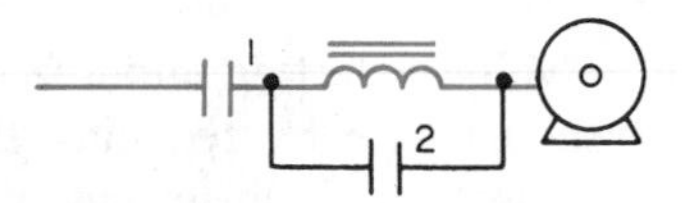

- *Star-delta starting:* the stator of the motor is star connected for starting and delta connected for running.

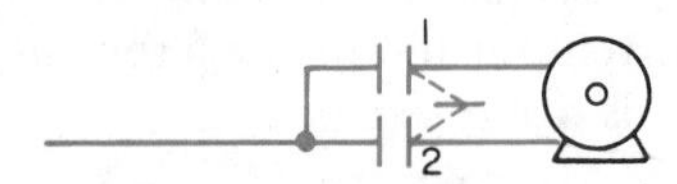

- *Part winding starting:* the stator windings of the motor are made up of two or more circuits; the individual circuits are connected to the line in series for starting and in parallel for normal operation.

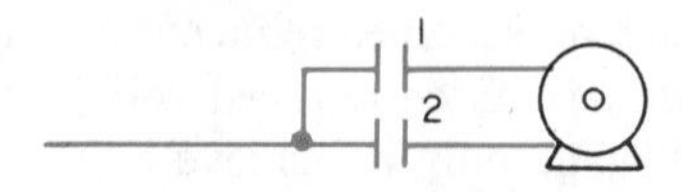

Of these methods, the two most fundamental ways of starting squirrel cage motors are full voltage starting and reduced voltage starting. Once again, full voltage starting can be used where the driven load can stand the shock of starting and objectionable line disturbances are not created. Reduced voltage starting may be required if the starting torque must be applied gradually or if the starting current produces objectionable line disturbances.

REVIEW QUESTIONS

1. List five commonly used starting methods.
2. When can full voltage starting be used?
3. Name the simplest, most rugged and reliable ac motor. Describe how it operates.
4. Describe the term, *slip*.
5. When the voltage is reduced to start a motor, what happens to the
   a. torque?
   b. current?
6. Why is it more advisable to start some machines at reduced torque?

UNIT 49

Primary Resistor-type Starters

Objectives *After studying this unit, the student will be able to:*

- **State why reduced current starting is important**
- **Describe the construction and operation of primary resistor starters**
- **Interpret and draw diagrams for primary resistor starters**
- **Connect squirrel cage motors to primary resistor starters**
- **Troubleshoot electrical problems on primary resistor starters**

PRIMARY RESISTOR-TYPE STARTERS

A simple and common method of starting a motor at reduced voltage is used in primary resistor-type starters. In this method, a resistor is connected in series in the lines to the motor, figure 49-1. Thus, there is a voltage drop across the resistors and the voltage is reduced at the motor terminals. Reduced motor starting speed and current are the result. As the motor accelerates, the current through the resistor decreases, reducing the voltage drop (E = IR) and increasing the voltage across the motor terminals. A smooth acceleration is obtained with gradually increasing torque and voltage.

The resistance is disconnected when the motor reaches a certain speed. The motor is then connected to run on full line voltage, figure 49-2. The introduction and removal of resistance in the motor starting circuit may be accomplished manually or automatically.

Primary resistor starters are used to start squirrel cage motors in situations where limited torque is required to prevent damage to driven machinery. These starters are also used with limited current inrush to prevent excessive power line disturbances.

It is desirable to limit the starting current in the following cases:

1. When the power system does not have the capacity for full voltage starting.
2. When full voltage starting may cause serious line disturbances, such as in lighting circuits, electronic circuits, the simultaneous starting of many motors, or the motor is distant from the incoming power supply.

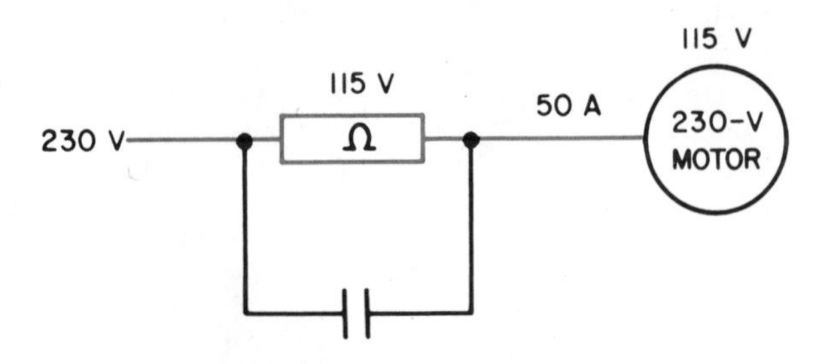

FIGURE 49-1 Reduced voltage starting

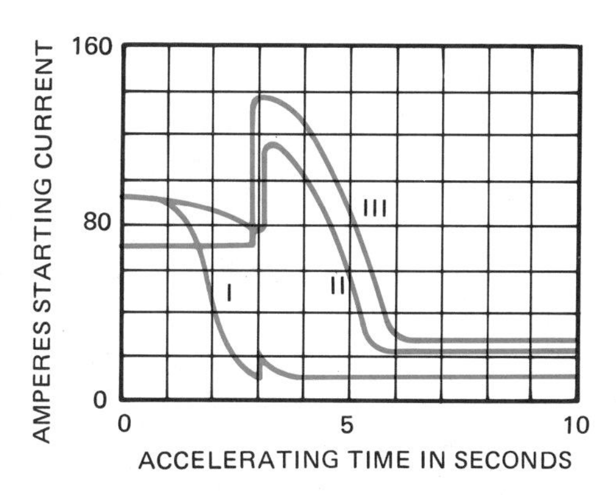

FIGURE 49-2 The curves illustrate how a primary resistance starter reduces the starting current of a 10-hp, 230-volt motor under three different load conditions. Curve 1 is at a light load and Curve 11 is at a heavy load. In either case, the running switch closes after three seconds. With a heavy load, an increase in the starting time reduces the second current inrush; the motor will reach a higher speed before it is connected to full line voltage. In Curve 111 the motor cannot start on the current allowed through the resistance; it comes up to speed only after connection to full line voltage.

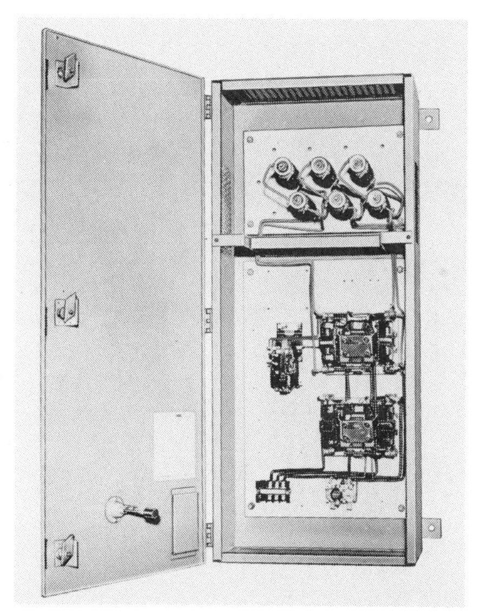

FIGURE 49-3 Reduced voltage, primary resistance-type starter. Resistors are in the top, ventilated system. (Courtesy EATON Corp., Cutler-Hammer Products)

In these situations, reduced voltage starters may be recommended for motors with ratings as small as five horsepower.

Reduced voltage starting must be used for driving machinery that must not be subjected to a sudden high starting torque and the shock of sudden acceleration. Among typical applications are those where belt drives may slip or where large gears, fan blades, or couplings may be damaged by sudden starts.

Automatic primary resistor starters may use one or more than one step of acceleration, depending upon the size of the motor being controlled. These starters provide smooth acceleration without the line current surges normally experienced when switching autotransformer types of reduced voltage starters.

Primary resistor starters provide closed transition starting. This means the motor is never disconnected from the line from the moment it is first connected until the motor is operating at full line voltage. This feature may be important in wiring systems sensitive to voltage changes. Primary resistor starters do consume energy, with the energy being dissipated as heat. However, the motor starts at a much higher power factor than with other starting methods.

Special starters are required for very high inertia loads with long acceleration periods or where power companies require that current surges be limited to specific increments at stated intervals.

Primary Resistor-type Reduced Voltage Starter

Figure 49-3 and 49-4 illustrate an automatic, primary resistor-type, reduced voltage starter. Figure 49-4 shows the starter with two-point acceleration connected to a three-phase, squirrel cage induction motor.

When the start button is pressed, a complete circuit is established from L1 through the stop button, start button, coil M, and the overload relay contacts to L2. When coil (M) is energized, the main power contacts (M) and the control circuit maintaining contact (M) are closed. The motor is energized through the overload heaters and the starting resistors. Because the resistors are connected in series with the motor terminals, a voltage drop occurs in the resistors and the motor starts on reduced voltage.

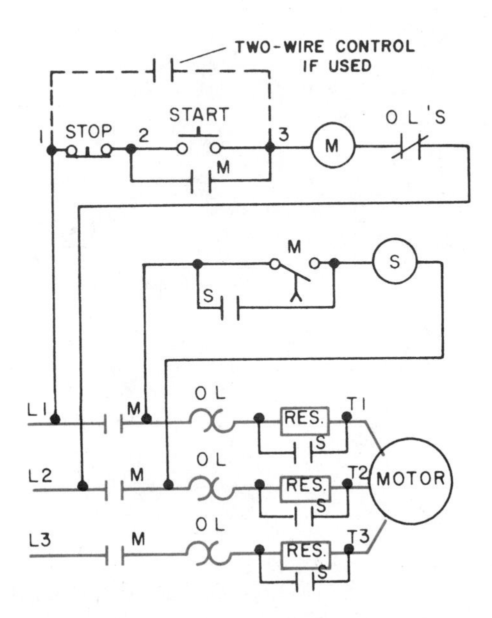

FIGURE 49-4 Line diagram of a primary resistor starter with two-step acceleration

As the motor accelerates, the voltage drop across the resistors decreases gradually because of a reduction in the starting current. At the same time, the motor terminal voltage increases.

After a predetermined acceleration time, delay contact M closes the circuit to contactor coil S. Coil S, in turn, closes contacts S, the resistances are shunted out, and the motor is connected across the full line voltage.

Note that the stop button controls coil M directly. When the main power contacts M open, coil S drops out. After coil M is energized, a pneumatic timing unit attached to starter unit M retards the closing of time contact M. This scheme uses starter M for a dual purpose and eliminates one coil, a timing relay. Additional control contact (S) connected parallel to the on-delay contact (M) provides additional electrical security to insure coil S normally open is maintained.

For maximum operating efficiency, push buttons or other pilot devices are usually mounted on the driven machinery within easy reach of the operator. The starter is located near the motor to keep the heavy power circuit wiring as short as possible. Only two or three small connecting wires are necessary between the starter and pilot device.

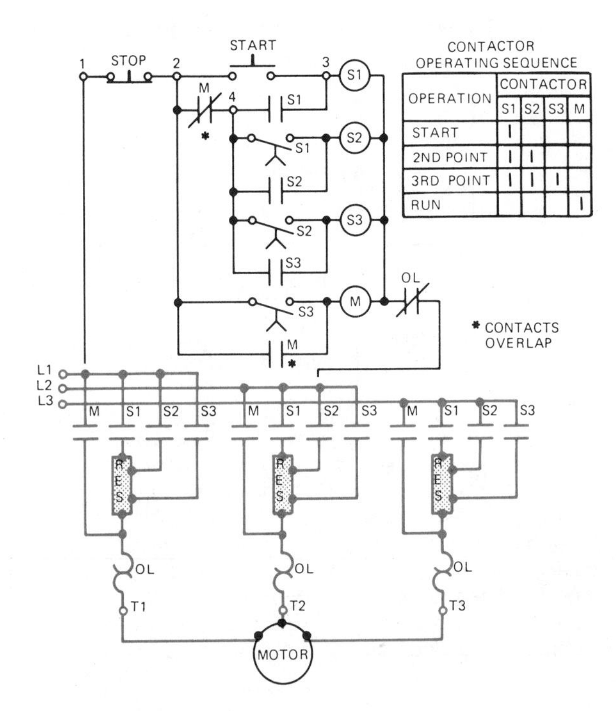

OPERATION	CONTACTOR S1	S2	S3	M
START	I			
2ND POINT	I	I		
3RD POINT	I	I	I	
RUN				I

FIGURE 49-5 Schematic diagram of a four-point primary resistor starter

A motor can be operated from any of several remote locations if a number of push buttons or pilot switches are used with one magnetic starter, such as on a conveyor system.

Ac primary resistor starters are available for use on single-phase and three-phase reversing operations. They are also available with multiple points of acceleration.

Four-point Resistor Starter

The elementary diagram in figure 49-5 illustrates a four-point resistor starter. Operating the start button energizes contactor S1, connecting the motor to the line. The total resistance is in each line. The mechanically operated timer S1 closes to energize contactor S2. The closing of contacts S2 shunts out a portion of the starting resistor. The mechanically operated timer S2 closes to energize contactor S3. The closing of contacts S3 shunts out additional resistance in the resistor bank. The timer S3 closes to energize contactor M. The closing of contacts M connects the motor at full voltage. An electrical interlock on M then opens to drop out contactors S1, S2, and S3. The rotor may not start turning until the second or third point, but reduced voltage will have been accomplished.

The timers or timing relays used are of the preset time type such as pneumatic, dashpot or solid-state timers. Compensating-type current relays are also used.

Figure 49-6 shows two types of resistor banks used to start small motors. The resistors (shown in A of figure 49-6) consist of resistance wire wound around porcelain bases and imbedded in refractory cement. The other resistor (B of figure 49-6) has a ribbon-type construction and is used on larger motor starters. This type of resistor consists of an alloy ribbon formed in a zig-zag shape. The formed

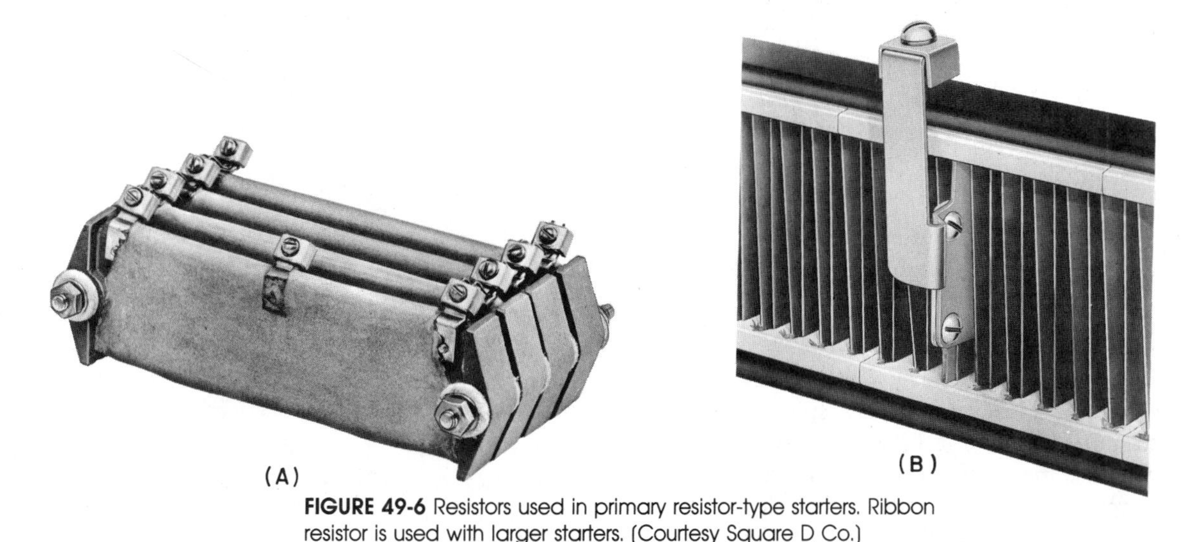

FIGURE 49-6 Resistors used in primary resistor-type starters. Ribbon resistor is used with larger starters. (Courtesy Square D Co.)

FIGURE 49-7 Manual resistance starter with "carbon pile" resistors (Courtesy Allen-Bradley Co.)

ribbon is supported between porcelain blocks which have recesses for each bend of the ribbon. The ribbons are not stressed and do not have a reduced cross section, since they are bent on the flat and not on the edge. Any number of units can be combined vertically or horizontally, in series or in parallel.

Some starters have graphite compression disk resistors and are used for starting polyphase squirrel cage motors, figure 49-7.

A primary resistance-type starter has the following features:

- simple construction
- low initial cost
- low maintenance
- smooth acceleration in operation
- continuous connection of the motor to the line during the starting period
- a high power factor.

These starters should *not* be used for starting *very heavy* loads because of their low starting torque. These starters are said to have a low starting economy because the starting resistors dissipate electrical energy. A solid-state, reduced voltage starter is shown in figure 49-8. Motor and remote control terminal connections are made at a marked terminal block.

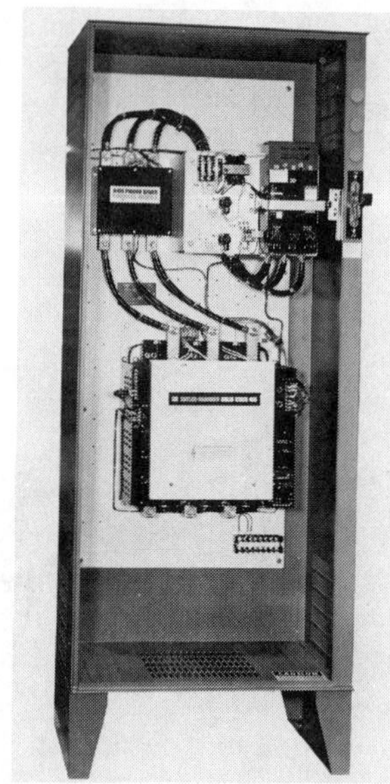

FIGURE 49-8 Solid-state, three-phase, reduced voltage motor starter (Courtesy EATON Corp., Cutler-Hammer Products)

REVIEW QUESTIONS

1. What is the purpose of inserting resistance in the stator circuit during starting?
2. Why is the power not interrupted when the motor makes the transition from start to run?
3. If the starter is to function properly and a timing relay is used, where must the coil of the relay be connected?
4. How many additional steps of acceleration be added?
5. What is meant by the low or poor starting economy of a primary resistor starter?
6. How does this starter provide smooth acceleration and a gradual increase of torque?

UNIT 50

Autotransformer Starters

Objectives *After studying this unit, the student will be able to:*

- **Describe the construction and operation of autotransformer starters**
- **Draw and interpret diagrams for autotransformer starters**
- **Connect squirrel cage motors to autotransformer starters**
- **Define what is meant by open transition and closed transistion starting**
- **Troubleshoot electrical problems on autotransformer starters**

Autotransformer reduced voltage starters are similar to primary resistor starters in that they are used primarily with ac squirrel cage motors to limit the inrush current or to lessen the starting strain on driven machinery, figure 50-1. This type of starter uses autotransformers between the motor and the supply lines to reduce the motor starting voltage. Taps are provided on the autotransformer to permit the user to start the motor at approximately 50%, 65%, or 80% of line voltage.

Most motors are successfully started at 65% of line voltage. In situations where this value of voltage does not provide sufficient starting torque, the 80% tap is available. If the 50% starting voltage creates excessive line drop to the motor, the 65% taps are available. This way of changing the starting voltage is not usually available with other types of starters. The starting transformers are inductive loads; therefore, they momentarily affect the power factor. They are suitable for long starting periods, however.

FIGURE 50-1 Reduced voltage autotransformer starter, size 3 (Courtesy EATON Corp., Cutler-Hammer Products)

Autotransformer Starters

To reduce the voltage across the motor terminals during the accelerating period, an autotransformer-type starter generally has two autotransformers connected in open delta. During the reduced voltage starting period, the motor is connected to the taps on the autotransformer. With the lower starting voltage, the motor draws less current and develops less torque than if it were connected to the line voltage, figure 50-2.

An adjustable time-delay relay controls the transfer from the reduced voltage condition to full voltage. A current-sensitive relay may be used to control the transfer to obtain current-limiting acceleration.

Figure 50-2(A) shows the power circuit for starting the motor with two autotransformers. Figure 50-2(B) shows the circuit for starting a motor with three autotransformers.

To understand the operation of the autotransformer starter more clearly, refer to the line diagram in figure 50-3. When the start button is closed momentarily, the timing relay (TR) is energized. The relay maintains the circuit across the start button with the normally open instantaneous contact (TR) which now closes. Starting coil (S) is energized from terminal four through the normally closed "time delay in opening" contact (TR), through the normally closed interlock (R), through coil S, and through the overload contact to L2, completing the circuit. The running starter cannot be closed at this point because the normally closed interlock S is open and the mechanical interlock is operating.

After a preset timing period (TR), the normally closed contacts open and the normally open (TR) contacts close. When coil S is de-energized, normally closed interlock S closes and energizes the running starter R.

The contact switching arrangement for a typical power circuit is shown in figure 50-3. When two transformers are used, there will be an imbalance in the motor voltage during starting. This imbalance will produce a torque variation of approximately 10 percent. In the running position, the motor is connected directly across the line and the autotransformers are disconnected from the line. As a result, only three contacts are shown.

Full line voltage is applied to the outside terminals of the autotransformer on starting. Reduced voltage for starting the motor is obtained from the autotransformer taps. The current taken by a motor varies directly with the applied voltage.

Starting compensators (autotransformer starters) using a five-pole starting contactor are classified as open transition starters. The motor is disconnected momentarily from the line during the transfer from the start to the run conditions.

Closed transition connections are usually found on standard size 6 and larger starters. For

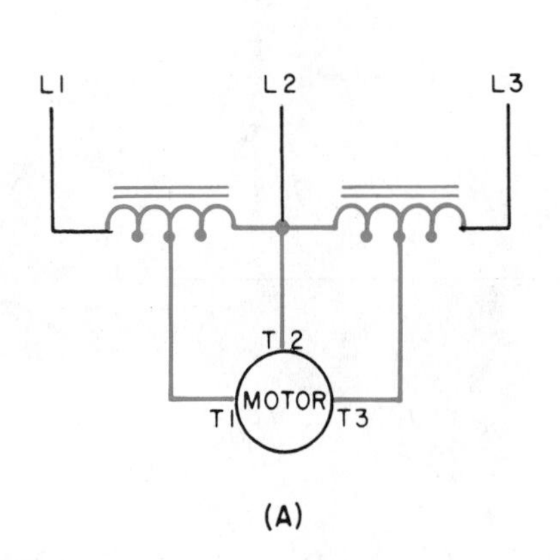

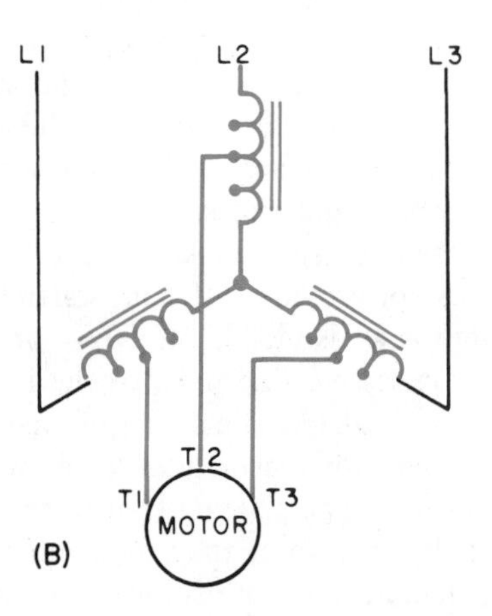

FIGURE 50-2 Power circuit connections showing two and three autotransformers used for reduced voltage starting

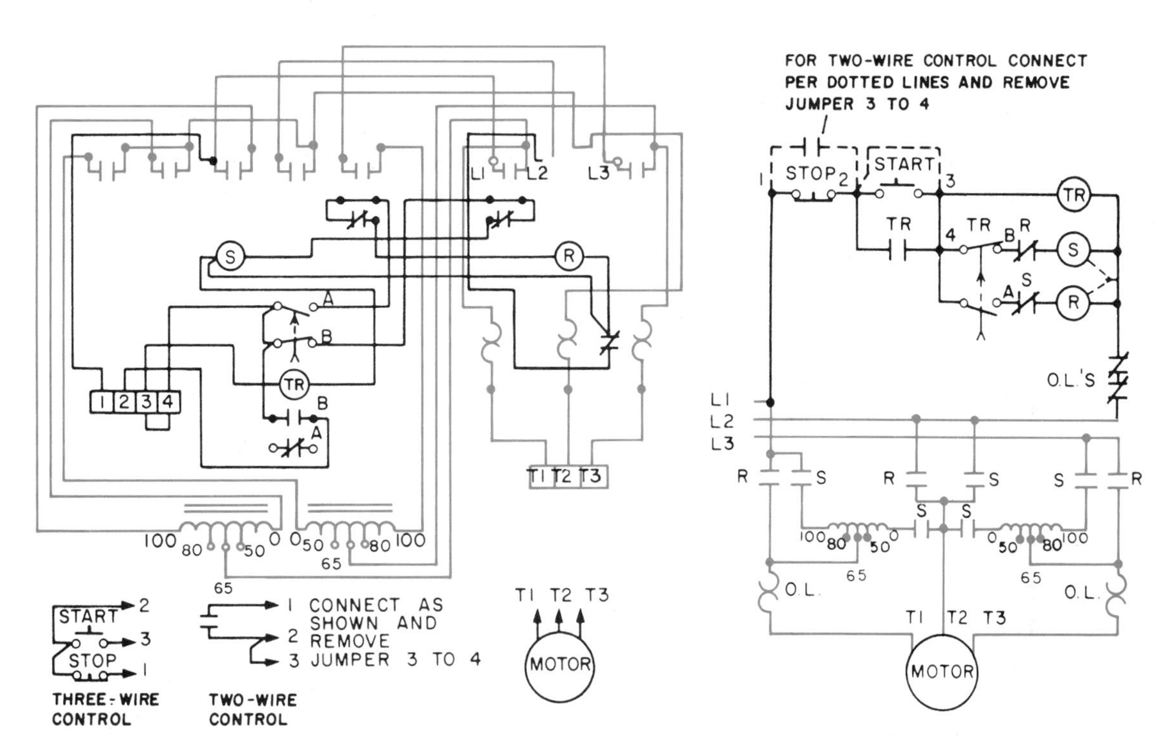

FIGURE 50-3 Autotransformer starters provide greater starting torque per ampere drawn from the line than any other type of reduced voltage starter. A typical wiring diagram, similar to that for the device shown in figure 50-1, is shown on the left. The line diagram is shown on the right.

the closed transition starter, figure 50-4, the starting contactors consist of a three-pole (S2) and a two-pole (S1) contactor operating independently of each other. During the transfer from start to run, the two-pole contactor is open and the three-pole contactor remains closed. The motor continues to accelerate with the autotransformer serving as a reactor. With this type of starter, the motor is not disconnected from the line during the transfer period. Thus, there is less line disturbance and a smoother acceleration.

The transformers are de-energized while in the running position. This is done to conserve electrical energy and to extend their life.

The (CT) designations in figure 50-4 indicate current transformers. These transformers are used on large motor starters to step down the current so that a conventional overload relay and heater size may be used. Magnetic overload relays are used on large reduced voltage starters also.

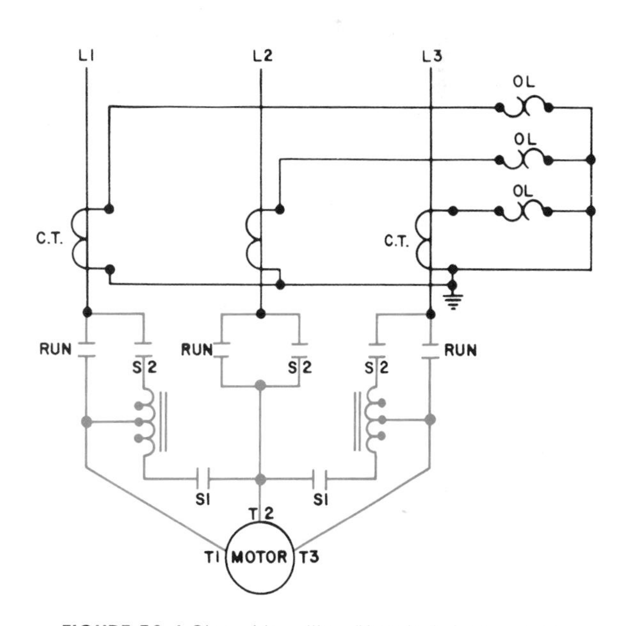

FIGURE 50-4 Closed transition (Korndorfer) connection

REVIEW QUESTIONS

1. Why is it desirable to remove the autotransformers from the line when the motor reaches its rated speed?
2. What is meant by an "open transition" from start to run? Why is this condition objectionable at times when used with large horsepower motors?
3. Which of the following applies to an autotransformer starter with a five-pole starting contactor: open transition or closed transition? Locate in figure 50-3.
4. How are reduced voltages obtained from autotransformer starters?
5. Assume the motor is running. What happens when the stop button is pressed and then the start button is pressed immediately?
6. What is a disadvantage of starting with autotransformer coils rather than with resistors?
7. What is one advantage of using an autotransformer starter?

UNIT 51

Automatic Starters for Star-Delta Motors

Objectives *After studying this unit, the student will be able to:*

- **Identify terminal markings for a star-delta motor and motor starter**
- **Describe the purpose and function of star-delta starting**
- **Troubleshoot star-delta motor starters**
- **Connect star-delta motors and starters**

A commonly used means of reducing inrush currents without the need of external devices is star-delta motor starting (sometimes called wye-delta starting). Figure 51-1 shows a typical star-delta starter.

Star-delta motors are similar in construction to standard squirrel cage motors. However, in star-delta motors, both ends of each of the three windings are brought out to the terminals. If the starter used has the required number of properly wired contacts, the motor can be started in star and run in delta.

The motor must be wound in such a manner that it will run with its stator windings connected in delta. The leads of all of the windings must be brought out to the motor terminals for their proper connection in the field.

FIGURE 51-1 Reduced current star-delta starter, Size 2, NEMA 1 (Courtesy EATON Corp., Cutler-Hammer Products)

APPLICATIONS

The primary applications of star-delta motors are for driving centrifugal chillers of large, central air conditioning units for loads such as fans, blowers, pumps, or centrifuges, and for situations where a reduced starting torque is necessary. Star-delta motors also may be used where a reduced starting current is required. Since all of the stator winding is used and there are no limiting devices such as resistors or autotransformers, star-delta motors are widely used on loads having high inertia and a long acceleration period.

The speed of a star-delta or wye-delta squirrel cage induction motor depends on the frequency of the applied voltage and the number of stator poles. Since both these values are the same for the wye or delta connection, the motor will run at approximately the same speed regardless of how the windings are connected. The inrush line current is much less when the windings are connected in wye, however. Assume that a motor is to be connected directly to a 480 volt line during the starting period. Also assume that each winding exhibits an impedance of 0.4 Ω during the starting period. If the windings are connected in delta when power is first applied to the motor, 480 volts will be connected directly across the phase windings, figure 51-2. This will produce a phase current of 1200 amps.

$$I_{PHASE} = \frac{E_{PHASE}}{Z_{PHASE}}$$

$$I_{PHASE} = \frac{480}{0.4}$$

$$I_{PHASE} = 1200 \text{ amps}$$

Since the line current supplying a delta connection is 1.732 times greater than the phase current, the line current will be 2078.4 amps (1200 × 1.732 = 2078.4).

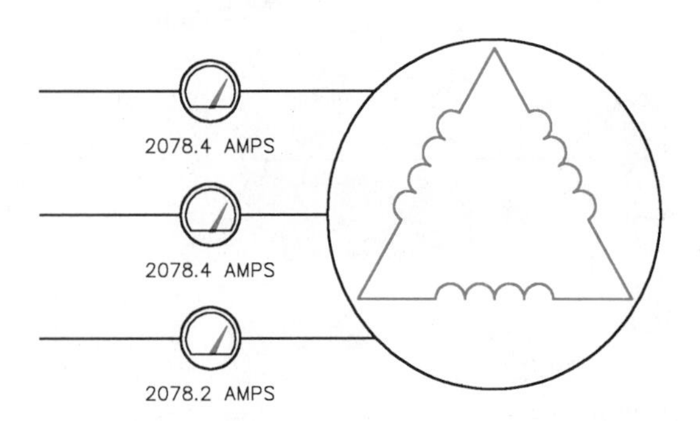

FIGURE 51-2 Delta inrush current is high

If the stator windings are connected in a wye configuration during the starting period, figure 51-3, the inrush line current will be only one third the value of the delta connection. Since the windings are now connected in a wye or star, the voltage applied across each phase winding will be less than the line voltage by a factor of 1.732 or 277 volts (480 ÷ 1.732 = 277). This will produce a phase current of 692.8 amps when power is first connected to the motor.

$$I_{PHASE} = \frac{E_{PHASE}}{Z_{PHASE}}$$

$$I_{PHASE} = \frac{277}{0.4}$$

$$I_{PHASE} = 692.8 \text{ amps}$$

In a wye-connected system, the line current and phase current are the same. Therefore, the line current has been reduced from 2078.2 amps to 692.8 amps during the initial starting period.

OVERLOAD PROTECTION

Three overload relays are connected in the phase windings during both the starting and running period, figure 51-4. This means that the overload heaters must be selected on the basis of the winding or phase current, not on the full load line current indicated on the motor nameplate. To determine the proper current for the overload heaters, divide the line current by 1.732. A diagram of the entire connection for the motor windings,

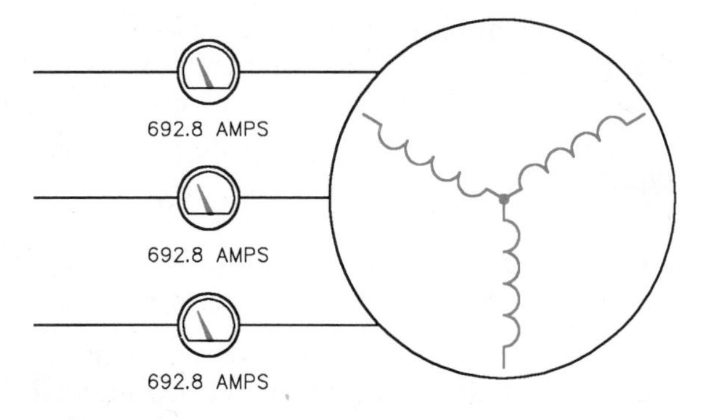

FIGURE 51-3 A Wye-connected winding draws one third the starting current of a delta-connected winding

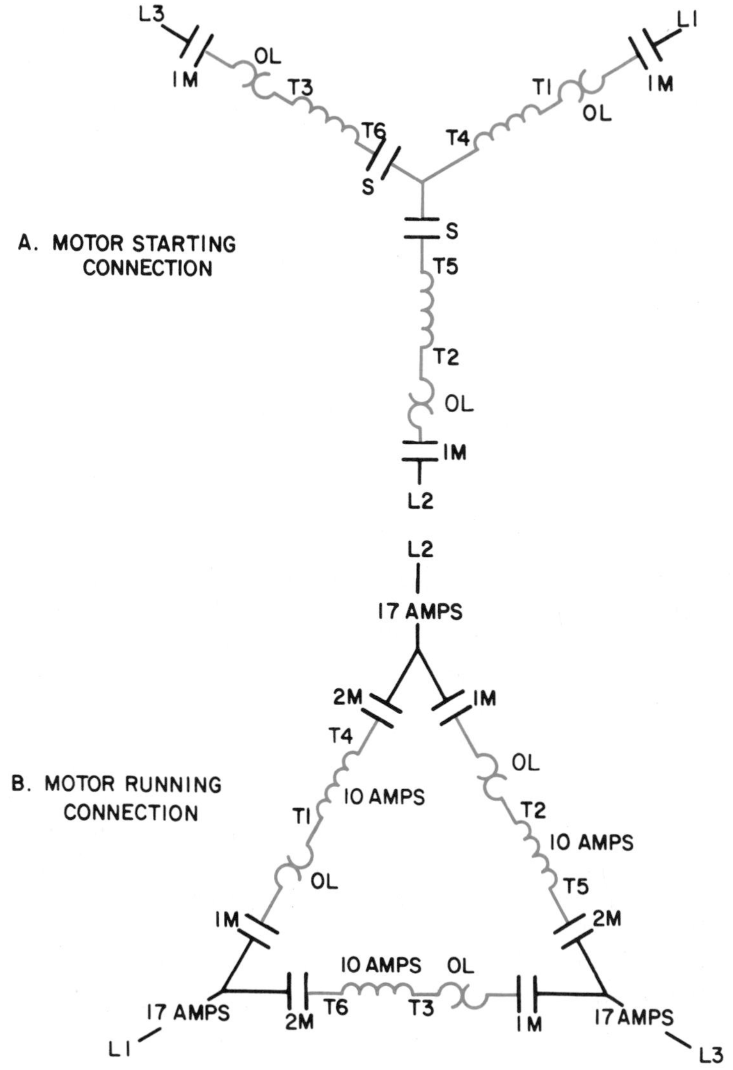

FIGURE 51-4 Elementary diagrams of motor power circuits of figure 51-3. Controller connects motor in wye on start and in delta for run. Note that the overload relays are connected in the motor winding circuit, not in the line. Note also that the line current is higher than the phase winding current in the diagram for the delta connection (B). Winding current is the same as the line current in diagram A.

overload heaters, and load contacts is shown in figure 51-5.

OPEN TRANSITION STARTING

Probably the most common method for wye-delta starting is *open transition* starting. This method receives its name from the fact that the motor windings are open during the transition period of changing the windings from a wye connection to a delta connection. A control circuit for performing this transition is shown in figure 51-6. In this circuit, an on delay timer is used to change the motor windings from a wye connection to a delta connection. When the start button is pressed, coils 1M, TR, and S energize immediately. Coil 1M closes all 1M contacts to supply

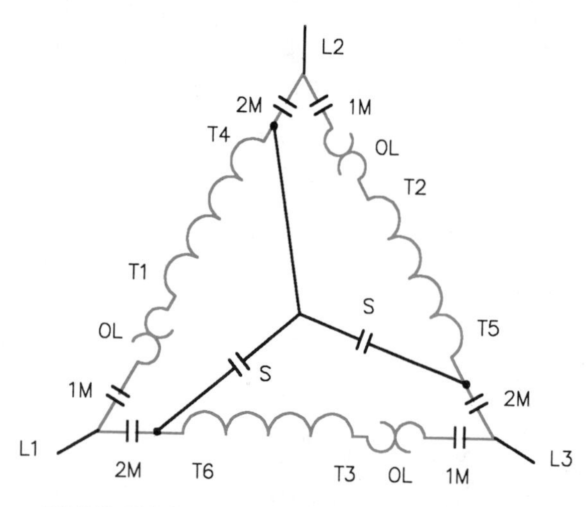

FIGURE 51-5 Stator connection for Wye-Delta starting

power to the motor windings. Coil S closes both of the S contacts that connect the motor windings in a wye.

After some period of time, both TR contacts change position. The normally closed contact connected in a series with an S coil opens and de-energizes the S coil. The S load contacts open and disconnect the motor windings. When the normally open TR contact closes, coil 2M energizes and reconnects the motor windings in a delta configuration. It is the transition period between S contacts opening and 2M contacts closing that this starting method receives its name. The normally closed S and 2M contacts act as interlocks to prevent the possibility of coils S and 2M being energized at the same time.

Care should be exercised when connecting the stator windings to the load contacts. If the circuit is not connected properly, it will generally result in the motor reversing direction of rotation when the windings are changed from wye to delta. Figure 51-7 illustrates a schematic diagram of the stator winding connection and a wiring diagram of the motor connection to the load contacts. Notice that the wire numbers have been added to the stator schematic which corresponds to the components shown in the wiring diagram.

CLOSED TRANSITION STARTING

In figure 51-8, resistors maintain continuity to the motor to avoid the difficulties associated with the open circuit form of transition between start and run.

With closed transition starting, the transfer from the star to delta connections is made without disconnecting the motor from the line. When the transfer from star to delta is made in open transition starting, the starter momentarily disconnects the motor and then reconnects it in delta. While an open transition is satisfactory in many cases, some installations may require closed transition starting to prevent power line disturbances. Closed transition starting is achieved by adding a three-pole contactor and three resistors to the

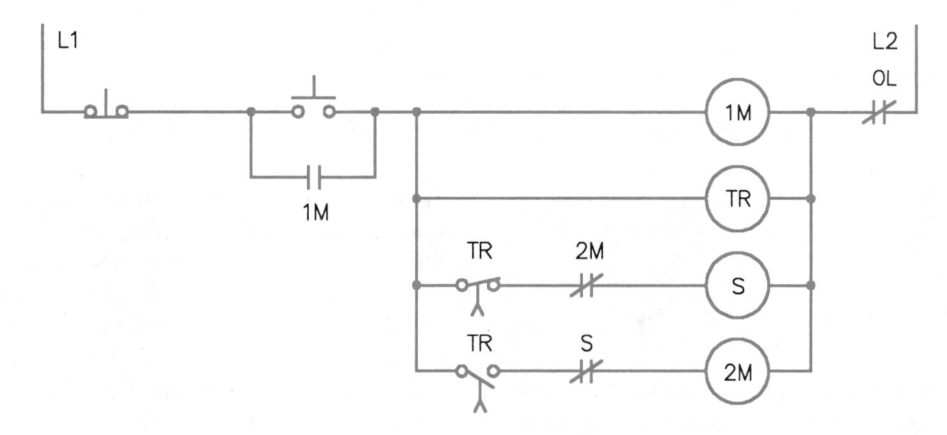

FIGURE 51-6 Basic control circuit for a Wye-Delta starter

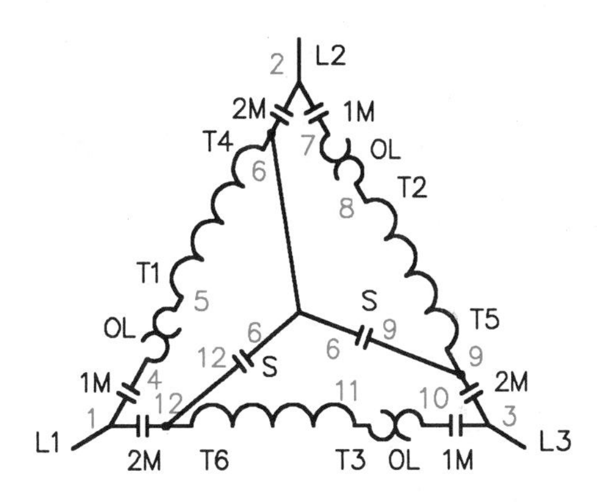

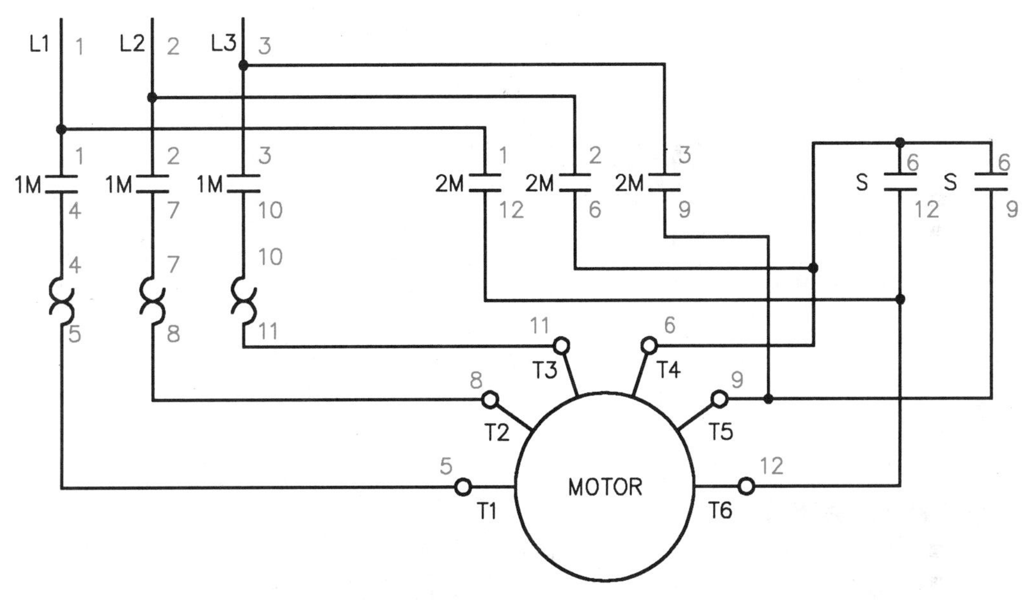

FIGURE 51-7 Load circuit connection for Wye-Delta starter

starter circuit. The connections are made as shown in the closed transition schematic diagram, figure 51-8. The contactor is energized only during the transition from star to delta. It keeps the motor connected to the power source through the resistors during the transition period, figure 51-9. There is a reduction in the incremental current surge which results from the transition. The balance of the operating sequence of the closed transition starter is similar to that of the open transition star-delta motor starter.

A single method of reduced current starting may not achieve the desired results because the motor starting requirements are so involved, the restrictions so stringent, and the needs so conflicting. It may be necessary to use a combination of starting methods before satisfactory performance is realized. For special installations, it may be necessary to design a starting system to fit the particular conditions.

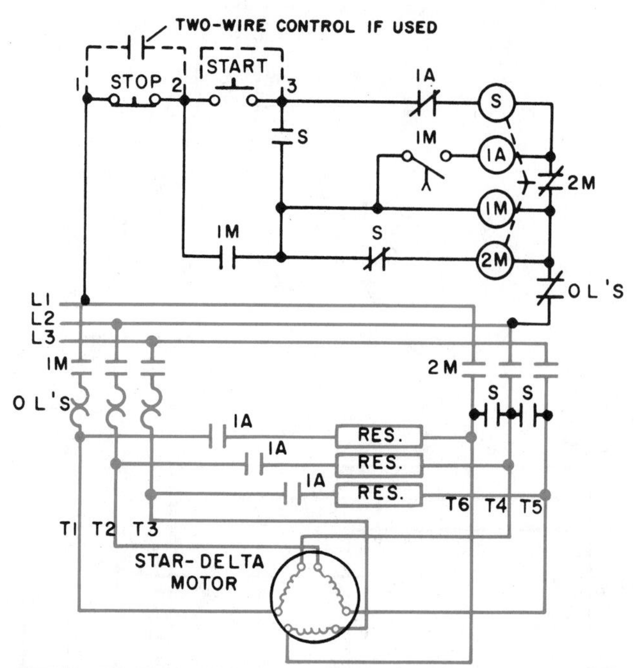

FIGURE 51-8 Elementary diagram of Sizes 1, 2, 3, 4, and 5 star-delta starters with transition starting

REVIEW QUESTIONS

1. Indicate the correct terminal markings for a star-delta motor on the diagram below.

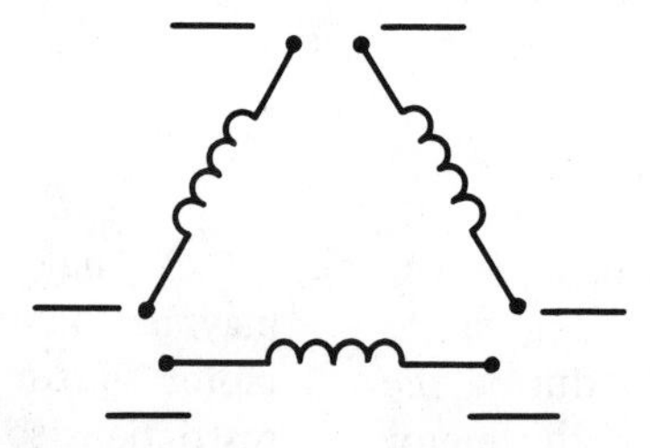

2. What is the principal reason for using star-delta motors?
3. In figure 51-6, which contactor closes the transition?
4. The closed transition contactor is energized only on transfer from star to delta. How is this accomplished?
5. If a delta-connected, six-lead motor nameplate reads "Full Load Current 170 Amperes," upon what current rating should the overload relay setting, or the selection of the heater elements be based?

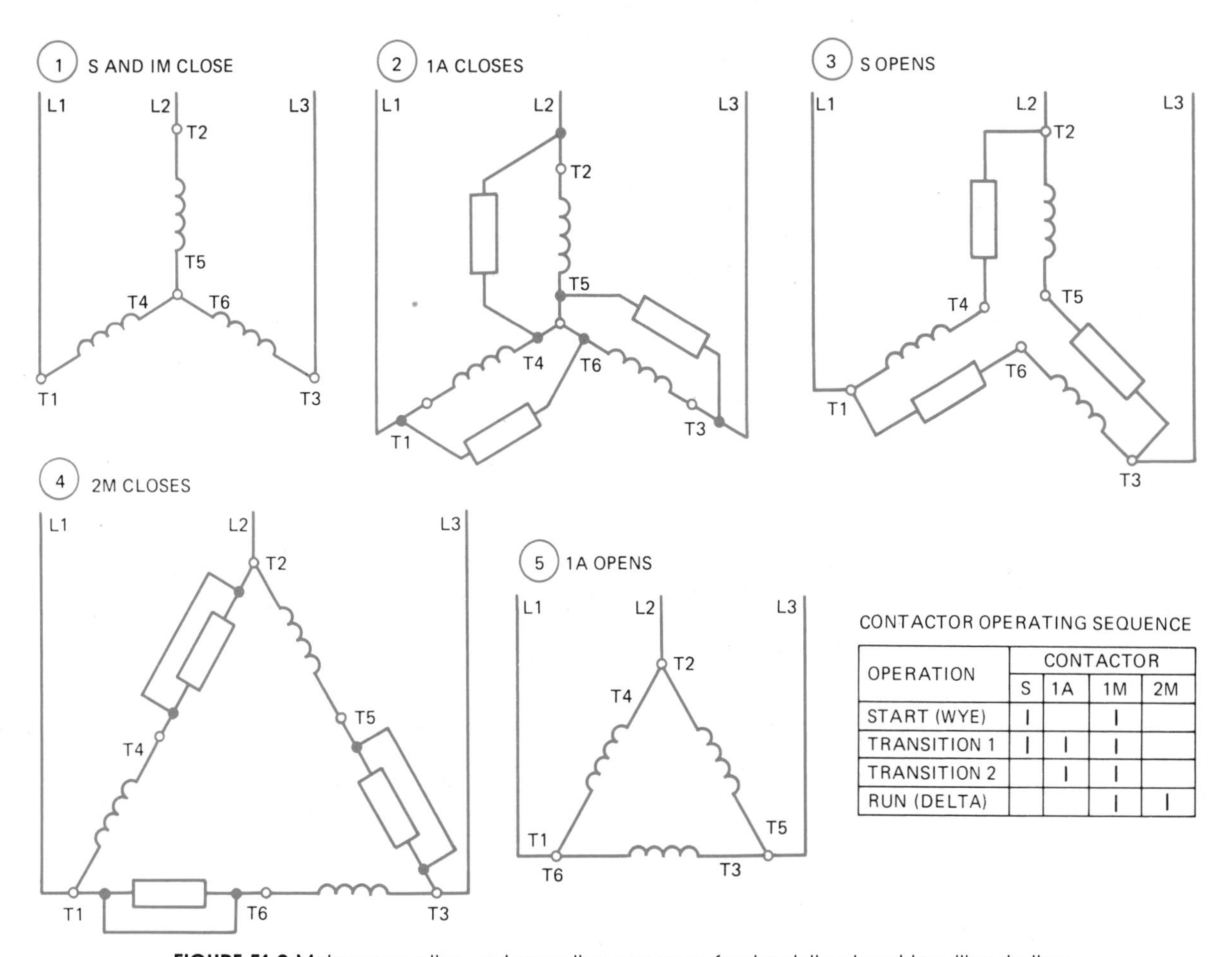

OPERATION	CONTACTOR			
	S	1A	1M	2M
START (WYE)	I		I	
TRANSITION 1	I	I	I	
TRANSITION 2		I	I	
RUN (DELTA)			I	I

FIGURE 51-9 Motor connection and operating sequence for star-delta closed transition starting

UNIT 52

Two-speed, One-winding (Consequent Pole) Motor Controller

Objectives *After studying this unit, the student will be able to:*

- **Identify terminal markings for two-speed, one-winding motors and controllers**
- **Describe the purpose and function of two-speed, one-winding motor starters and motors**
- **Connect two-speed, one-winding controllers and motors**
- **Connect a two-speed starter with reversing controls**
- **Recommend solutions for troubleshooting these motors and controllers**

Certain applications require the use of a squirrel cage motor having a winding arranged so that the number of poles can be changed by reversing some of the currents. If the number of poles is doubled, the speed of the motor is cut approximately in half.

The number of poles can be cut in half by changing the polarity of alternate pairs of poles, see figure 52-1. The polarity of half the poles can be changed by reversing the current in half the coils, figure 52-2.

If a stator field is laid flat (as in figure 52-1) the established stator field must move the rotor twice as far in B as in A and in the same amount of time. As a result, the rotor must travel faster. The fewer the number of poles established in the stator, the greater is the speed in rpm of the rotor.

A three-phase squirrel cage motor can be wound so that six leads are brought out, figure 52-3. By making suitable connections with these leads, the windings can be connected in series delta or parallel wye, figure 52-4. If the winding is such that the series delta connection gives the high speed and the parallel wye connection gives the low speed, the horsepower rating is the same at both speeds. If the winding is such that the series delta connection gives the low speed and the parallel wye connection gives the high speed, the torque rating is the same at both speeds.

Consequent pole motors have a single winding for two speeds. Extra taps can be brought from the winding to permit reconnection for a different number of stator poles. The speed range is limited to a 1:2 ratio, such as 600-1200 rpm or 900-1800 rpm.

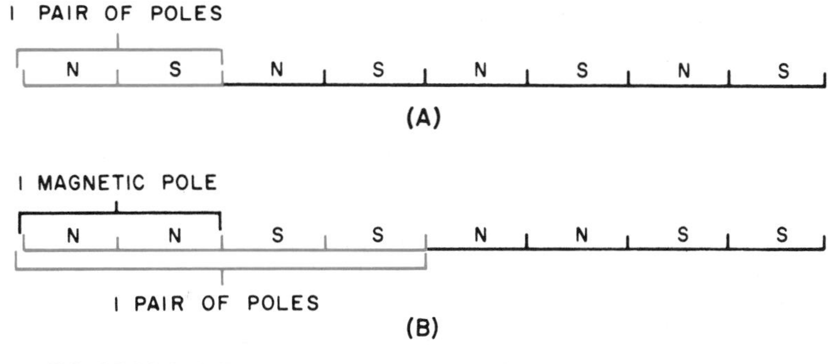

FIGURE 52-1 (A) Eight poles for low speed (B) Four poles for high speed

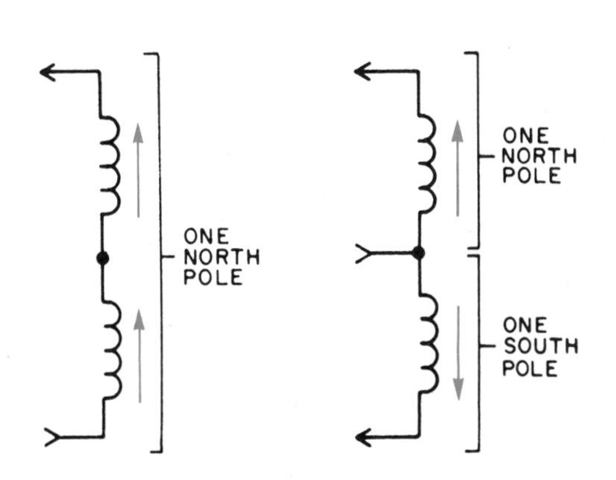

FIGURE 52-2 The number of poles is doubled by reversing current through half a phase. Two speeds are obtained by producing twice as many consequent poles for low-speed operation as for high speed.

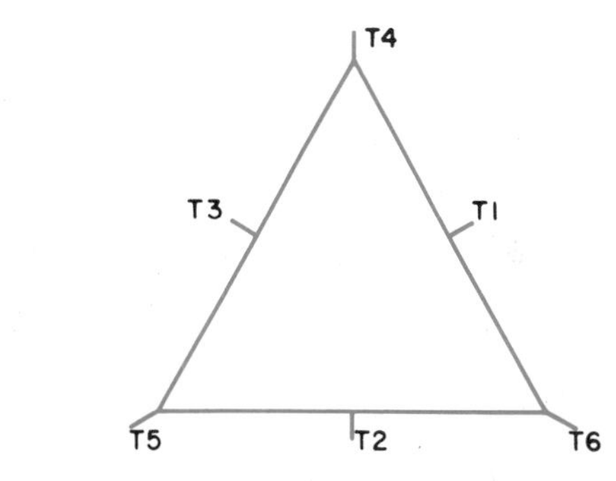

SPEED	L1	L2	L3	OPEN	TOGETHER
LOW	T1	T2	T3	———	T4, T5, T6
HIGH	T6	T4	T5	ALL OTHERS	———

FIGURE 52-3 Connection table for a three-phase, two-speed, one-winding, constant horsepower motor

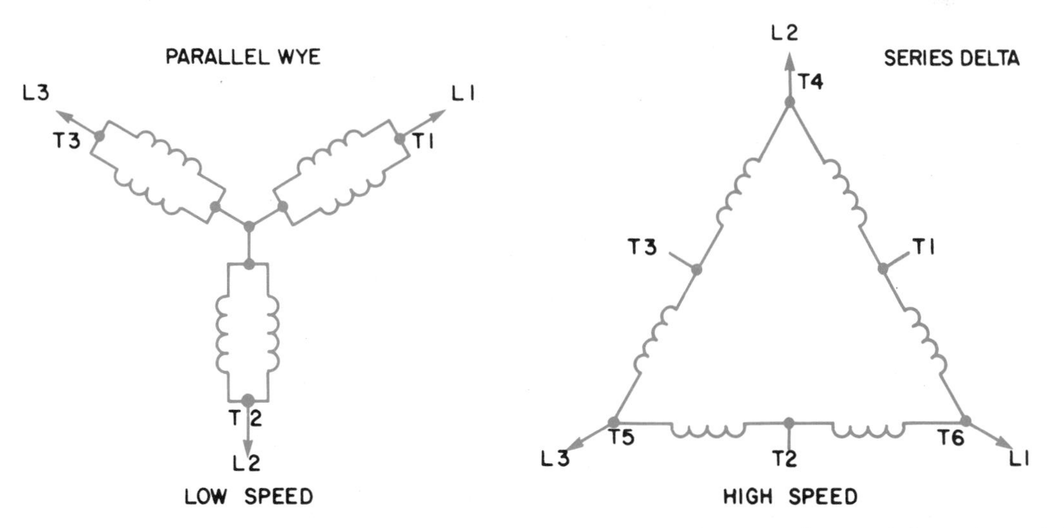

FIGURE 52-4 Three-phase, two-speed, one-winding, constant horsepower motor connections made by motor controller

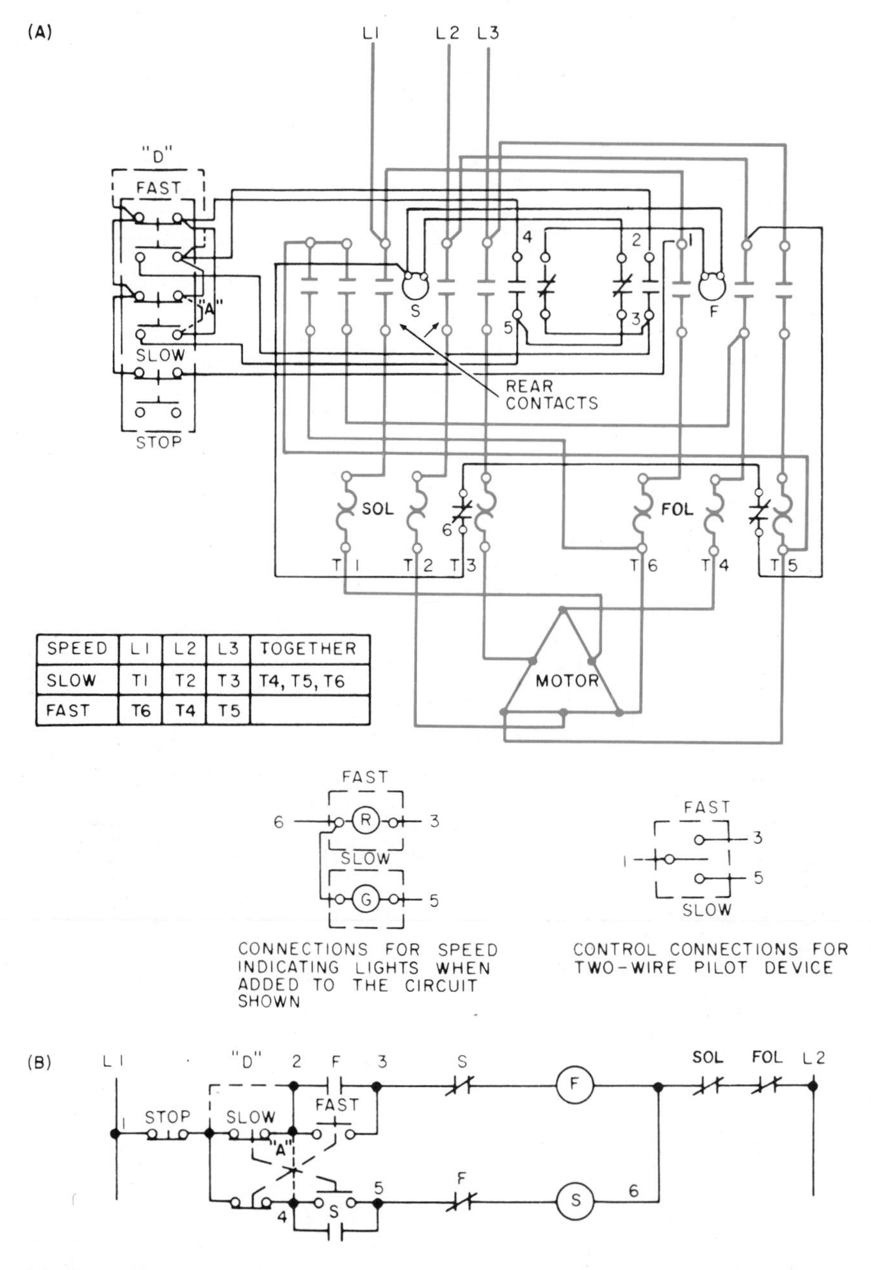

SPEED	L1	L2	L3	TOGETHER
SLOW	T1	T2	T3	T4, T5, T6
FAST	T6	T4	T5	

FIGURE 52-5 Wiring diagram (A) and line diagram (B) of an ac, full voltage, two-speed magnetic starter for single-winding (reconnectable pole) motors

Two-speed, consequent pole motors have one reconnectable winding. However, three-speed, consequent pole motors have two windings, one of which is reconnectable. Four-speed consequent pole motors have two reconnectable windings.

Referring back to the motor connection table in figure 52-3, note that for low speed operation, T1 is connected to L1; T2 to L2; T3 to L3; and T4, T5, and T6 are connected together. For high

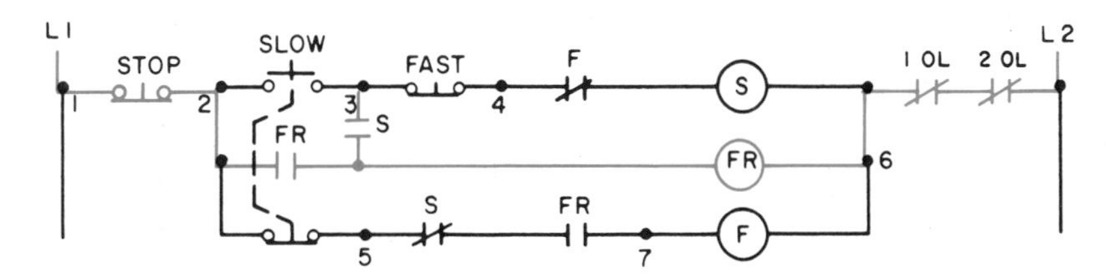

FIGURE 52-6 Two-speed control circuit using a compelling relay

speed operation, T6 is connected to L1; T4 to L2; T5 to L3; and all other motor leads are open.

Figure 52-5 illustrates the circuit for a Size 1, selective multispeed starter connected for operation with a reconnectable, constant horsepower motor. The control station is a three-element, fast-slow-stop station connected for starting at either the fast or slow speed. The speed can be changed from fast to slow or slow to fast without pressing the stop button between changes. If equipment considerations make it desirable to stop the motor before changing speeds, this feature can be added to the control circuit by making connections "D" and "A", shown in figure 52-5 by the dashed lines. Adding these jumper wires eliminates push-button interlocking in favor of stopping the motor between speed changes. This feature may be desirable for some applications.

Connections for the addition of indicating lights or a two-wire pilot device instead of the control shown are also given.

A compelling-type control scheme is shown in figure 52-6. These connections mean that the operator must start the motor at the slow speed. This controller cannot be switched to the fast speed until after the motor is running.

When the slow button is pressed, the slow-speed starter S and the control relay FR are energized. Once the motor is running, pressing the fast button causes the slow-speed starter to drop out. The high-speed starter is picked up through the normally closed interlock contacts S of the slow-speed starter and through the normally open contacts of control relay FR. The normally open contacts of control relay FR will now close.

If the fast button is pressed in an attempt to start the motor, nothing will happen because the normally open contacts of control relay FR will prevent the high-speed starter from energizing. When the fast button is pressed, it breaks a circuit but does not make a circuit. This scheme is another form of sequence starting.

MISTAKEN REVERSAL CAUTION

When multispeed controllers are installed, the electrician should check carefully to insure that the phases between the high- and low-speed windings are not accidentally reversed. Such a phase reverse will also reverse the direction of motor rotation. The driven machine should remain disconnected from the motor until an operational inspection is completed. The upper oil inspection plugs (pressure plugs) in large gear reduction boxes should be removed. Failure to remove these plugs may create broken casings if the motor is reversed accidentally.

Machines can be damaged if the direction of rotation is changed from that for which they are designed. In general, the correct rotational direction is indicated by arrows on the driven machine.

TWO-SPEED STARTER WITH REVERSING CONTROLS

Figure 52-7 is an elementary diagram of a two-speed, reversing controller. The desired speed, either high or low, is determined with a two-position selector switch. The direction of rotation is selected with either the forward or reverse push buttons. When the power contacts (F) or (R) are closed, current is supplied for the "high-low" controls. Assuming that the selector switch is in the high position, contactor L is energized to start

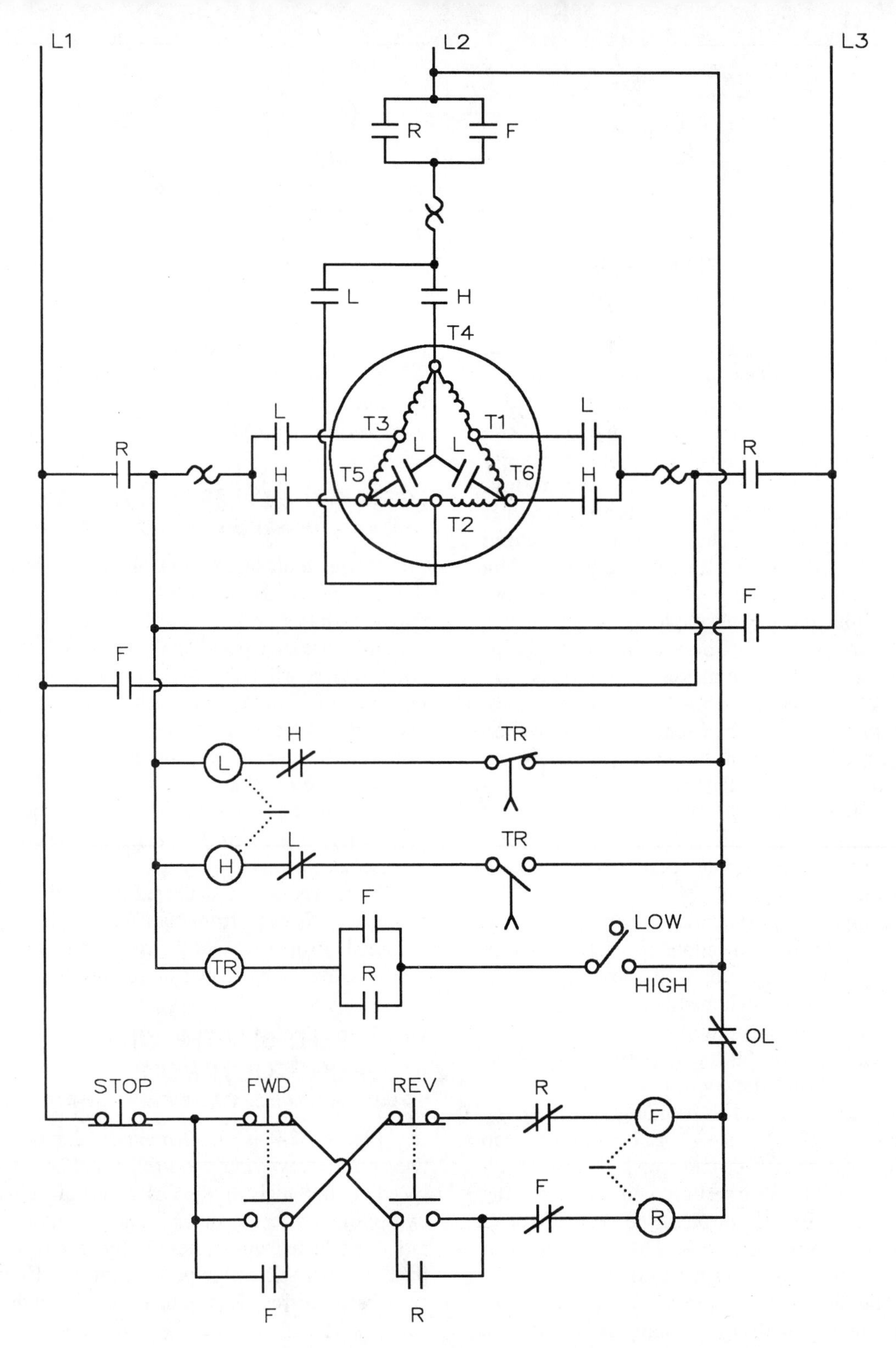

FIGURE 52-7 Two-speed reversing controller

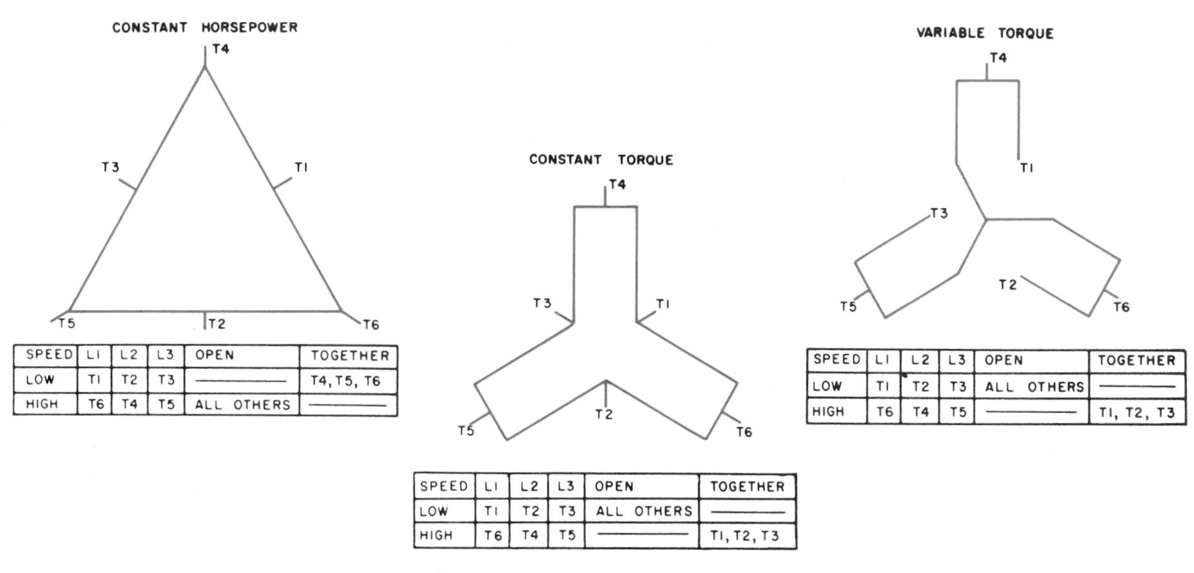

SPEED	L1	L2	L3	OPEN	TOGETHER
LOW	T1	T2	T3	————	T4, T5, T6
HIGH	T6	T4	T5	ALL OTHERS	————

SPEED	L1	L2	L3	OPEN	TOGETHER
LOW	T1	T2	T3	ALL OTHERS	————
HIGH	T6	T4	T5	————	T1, T2, T3

SPEED	L1	L2	L3	OPEN	TOGETHER
LOW	T1	T2	T3	ALL OTHERS	————
HIGH	T6	T4	T5	————	T1, T2, T3

FIGURE 52-8 Typical motor connection arrangement for three-phase, two-speed, one-winding motors. These connections conform to NEMA and ASA standards. (All possible arrangements are not shown.)

the motor in low speed. At the same moment, timing relay coil TR is energized. When the normally closed delay-in-opening contact TR opens after a preset time delay, normally open contact (TR) closes at the same time. This operation drops out contactor L, and the normally closed interlock L energizes contactor H. The motor is now at high speed.

The motor may be started in low speed in either direction before it is transferred to high speed. If the low speed is to be maintained, the selector switch is turned to *low* in order to open the circuit supplying the timing relay coil TR. (This coil would normally transfer the control to high-speed operation.) Typical motor terminal connections are shown in figure 52-8.

REVIEW QUESTIONS

1. The rotating magnetic field of a two-pole motor travels 360 electrical degrees. How many mechanical degrees does the rotor travel?
2. How can two speeds be obtained from a one-winding motor?
3. What will happen if terminals T5 and T6 in figure 52-3 are interchanged on high speed?
4. Is the motor operating at high speed or low speed when it is connected in series delta? In parallel wye?
5. Describe the operation of the line diagram in figure 52-5(B) by adding jumper "D" only.
6. Describe the operation of figure 52-5(B) by adding jumper wire "A" to the original circuit.
7. In figure 52-6, the motor cannot be started in high speed. Why?
8. Draw a schematic diagram of a two-speed control circuit, using a standard duty selector switch for the high and low speeds. Add a red pilot light for fast speed and a green pilot light for low speed. Omit "A" and "D" jumper wires.

UNIT 53

Wound Rotor Motors and Manual Speed Control

Objectives

After studying this unit, the student will be able to:

- **Identify terminal markings for wound rotor (slip ring) motors and controllers**
- **Describe the purpose and function of manual speed control and wound rotor motor applications**
- **Explain the difference between two-wire and three-wire control for wound rotor motors**
- **Connect wound rotor motors with manual speed controllers**
- **Recommend solutions to troubleshoot problems with these motors**

The ac three-phase wound rotor, or slip ring, induction motor was the first alternating-current motor that successfully provided speed control characteristics. This type of motor was an important factor in successfully adapting alternating current for industrial power applications. Because of their flexibility in specialized applications, wound rotor motors and controls are widely used throughout industry to drive conveyors for moving materials, hoists, grinders, mixers, pumps, variable speed fans, saws, and crushers. Advantages of this type of motor include maximum utilization of driven equipment, better coordination with the overall power system, and reduced wear on mechanical equipment. The wound rotor motor has the added features of high starting torque and low starting current. These features give the motor better operating characteristics for applications requiring a large motor or where the motor must start under load. This motor is especially desirable where its size is large with respect to the capacity of the transformers or power lines.

The phrase "wound rotor" actually describes the construction of the rotor. In other words, it is wound with wire. When the rotor is installed in a motor, three leads are brought out from the rotor winding to solid conducting slip rings, figure 53-1. Carbon brushes ride on these rings and carry the rotor winding circuit out of the motor to a controller. Unlike the squirrel cage motor, the induced current can be varied in the wound rotor motor. As a result, the motor speed can also be varied, figure 53-2. (Wound rotor motors have stator windings identical to those used in squirrel cage motors.) The controller varies the resistance (and thus the current), in the rotor circuit to control the acceleration and speed of the rotor once it is operating.

Resistance is introduced into the rotor circuit when the motor is started or when it is operating at slow speed. As the external resistance is eliminated by the controller, the motor accelerates.

Generally wound rotor motors under load are not suitably starting with the rings shunted. If such a motor is started in this matter, the rotor resistance is so low that starting currents are too high to be acceptable. In addition, the starting torque is less than if suitable resistors are inserted in the slip-ring circuit. By inserting resistance in the ring circuit, starting currents are decreased and starting torque is increased.

FIGURE 53-1 Rotor of an 800-hp, wound rotor induction motor (Courtesy Electric Machinery Mfg. Co.)

A control for a wound rotor motor consists of two separate elements: (1) a means of connecting the primary or stator winding to the power lines, and (2) a mechanism for controlling the resistance in the secondary or rotor circuit. For this reason, wound rotor motor controllers are often called *secondary resistor starters*.

Basically, there are two types of manual controllers, *starters* and *regulators*. The resistors used in *starters* are designed for starting duty only. This means that the operating lever must be moved to the full *ON* position. The lever must *not* be left in any intermediate position. The resistors used in *regulators* are designed for continuous duty. As a result, the operating lever in regulators can be left in any speed position.

When wound rotor motors are used as adjustable speed drives, they are operated on a continuous basis with resistance in the rotor circuit. In this case, the speed regulation of the motor is changed, and the motor operates at less than the full-load speed.

The use of three-wire control for starting (normally closed control contact), figure 53-3, means that low-voltage protection is provided. The motor is disconnected from the line in the event of voltage failure. To restart the motor when the voltage returns to its normal value, the normal starting procedure must be followed. To reverse the motor, any two of the three-phase motor leads may be interchanged.

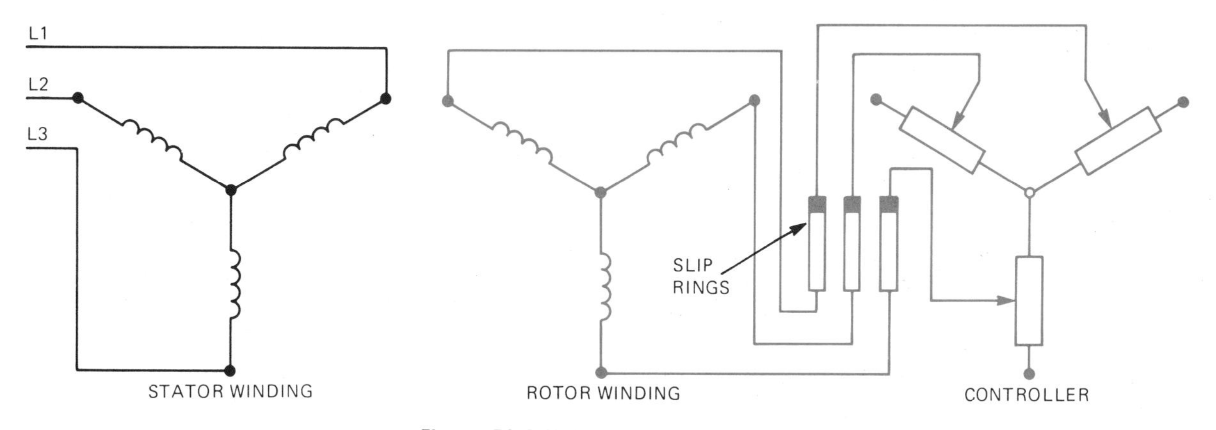

Figure 53-2 Stator, rotor, and controller

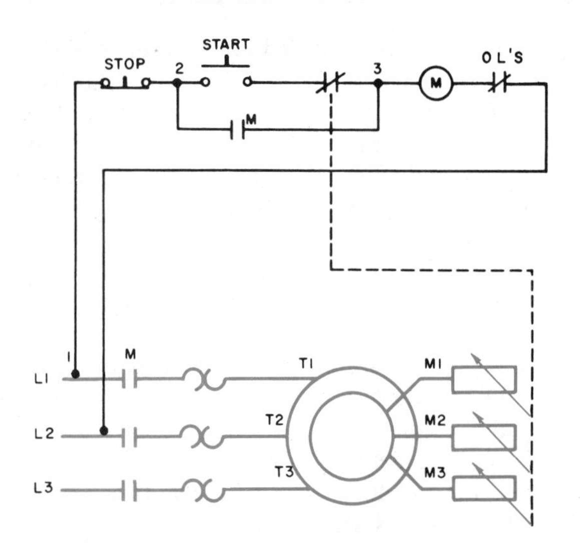

FIGURE 53-3 Elementary diagram of manual speed regulator interlocked with magnetic starter for control of slip ring motor (three-wire control)

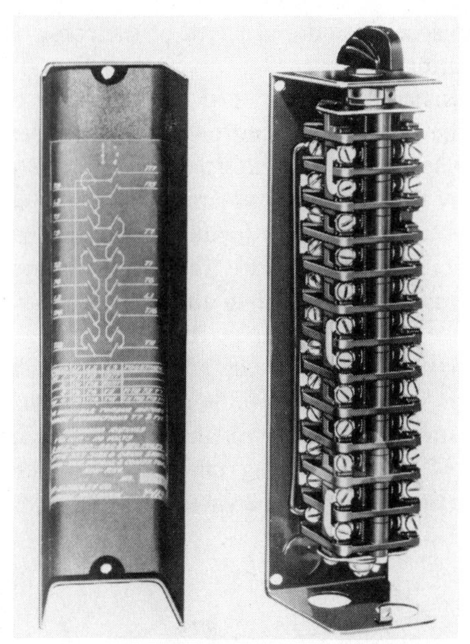

FIGURE 53-4 Drum controller (Courtesy Allen-Bradley Co.)

Drum controllers, figure 53-4, may be used for manual starting. The drum controller, however, is an independent component; it is separate from the resistors. A starting contact controls a line voltage, across-the-line starter. Generally, the electrician must make the connections on the job site.

REVIEW QUESTIONS

1. What characteristic of wound rotor motors led to their wide use for industrial applications?
2. By what means is the rotor coupled to the stator?
3. What other name is given to the wound rotor motor?
4. Does increased resistance in the rotor circuit produce low or high speed?
5. What two separate elements are used to control a wound rotor motor?
6. What is the difference between a manual starter and a manual regulator?

Automatic Acceleration for Wound Rotor Motors

Objectives *After studying this unit, the student will be able to:*

- State the advantages of using controllers to provide automatic acceleration of wound rotor motors
- Identify terminal markings for automatic acceleration controllers used with wound rotor motors
- Describe the process of automatic acceleration using reversing control
- Describe the process of automatic acceleration using frequency relays
- Connect wound rotor motor and automatic accleration and reversal controllers using push buttons and limit switches
- Recommend troubleshooting solutions for these problems

Secondary resistor starters used for the automatic acceleration of wound rotor motors consist of (1) an across-the-line starter for connecting the primary circuit to the line and (2) one or more accelerating contactors to shunt out resistance in the secondary circuit as the rotor speed increases. The secondary resistance consists of banks of three uniform wye sections. Each section is to be connected to the slip rings of the motor. The wiring of the accelerating starters and the design of the resistor sections are meant for starting duty only. This type of controller cannot be used for speed regulation. The current inrush on starters with two steps of acceleration is limited by the secondary resistors to a value of approximately 250 percent at the point of the initial acceleration. Resistors on starters with three or more steps of acceleration limit the current inrush to 150 percent at the point of initial acceleration. Resistors for acceleration generally are designed to withstand one 10-second accelerating period in each 80 seconds of elapsed time, for a duration of one hour without damage.

The operation of accelerating contactors is controlled by a timing device. This device provides timed acceleration in a manner similar to the operation of primary resistor starters. Normally, the timing of the steps of acceleration is controlled by adjustable accelerating relays. When these timing relays are properly adjusted, all starting periods are the same regardless of variations in the starting load. This automatic timing feature eliminates the danger of an improper startup sequence by an inexperienced machine operator.

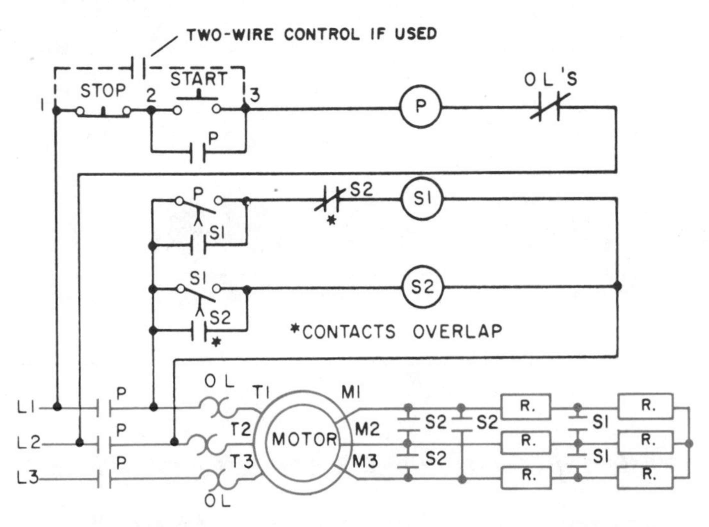

FIGURE 54-1 Typical elementary diagram of wound rotor motor starter with three points of acceleration

The primary circuit (stator) in figure 54-1 is energized with the start button. The motor starts with a full value of resistance in the secondary circuit. Coil P actuates the normally open, delay-in-closing contact P. After a preset timing period, contact P closes, energizes contactor S1, and maintains itself through the maintaining contact S1. When contacts S1 in the secondary resistor circuit close, the motor continues to accelerate. After the normally open, delay-in-closing contact S1 times out and closes, contactor S2 is energized and closes the resistor circuit contacts S2. The motor then is accelerated to its maximum speed. The normally closed interlock S2 opens the contactor (S1) circuit. The closing of S2 is assured by staggered, or overlapping, control contacts S2.

AUTOMATIC ACCELERATION WITH REVERSING CONTROL

Automatic acceleration can be obtained in either direction of rotation with the addition to the circuit of reversing contactors and push buttons. The wiring of these devices is shown in figure 54-2.

As described for a squirrel cage motor, a wound rotor motor can be reversed by interchanging any two stator leads.

The motor may be started in either direction of rotation—at low speed with the full secondary resistance inserted in the circuit. For either direction of rotation, the timing relay TR is energized by the normally open auxiliary contacts (F or R). Coil TR activates the normally open, delay-in-

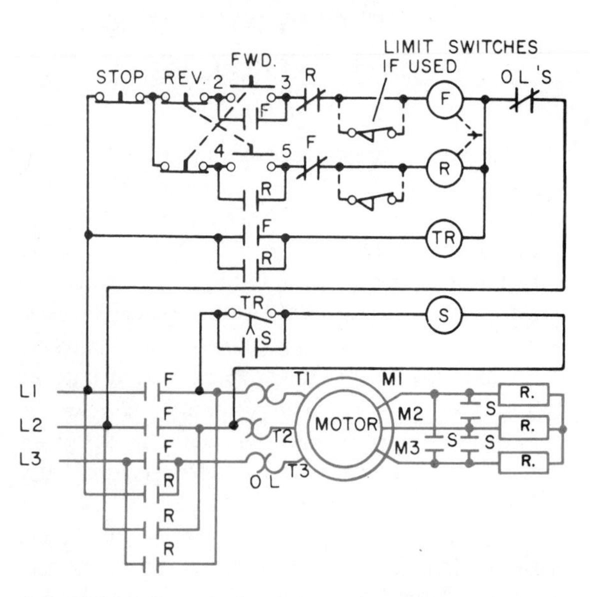

FIGURE 54-2 Typical elementary diagram of a starter with two points of acceleration for a reversing wound rotor motor

closing contact TR. Coil S is energized when contact TR times out and closes, and removes all of the resistors from the circuit to achieve maximum motor speed. The primary contactors are interlocked with the push buttons, normally closed interlocking contacts R and F, and the mechanical devices. The connections that occur if a limit switch is used are shown by the dashed lines in figure 54-2. The motor will stop when the limit switch is struck and opened. In this situation, it is necessary to restart the motor in the opposite direction with the push button. As a result, lines 1 and 3 on the primary side will be interchanged. Of course if limit switches are used, the jumper wire must be removed to avoid shunting out the operation of the limit switches.

AUTOMATIC ACCELERATION USING FREQUENCY RELAYS

Definite timers or compensated timers may be used to control the acceleration of wound rotor motors. Definite timers, which usually consist of pneumatic or dashpot relays, are set for the highest load current and remain at the same setting regardless of the load. The operation of a compensated timer is based on the applied load. In other words, the motor will be allowed to accelerate faster for a light load and slower for a heavy load. The frequency relay is one type of compensating timer. This relay uses the principle of electrical resonance in its operation.

When a 60-hertz ac, wound rotor motor is accelerated, the frequency induced in the secondary circuit decreases from 60 hertz at zero speed to 2 or 3 hertz at full speed, see figure 54-3. The voltage between the phases of the secondary circuit decreases in the same proportion from zero speed to full-speed operation. At zero speed, the voltage induced in the rotor is determined by the ratio of the stator and rotor turns. This action is similar to the operation of a transformer. The frequency, however, is the same as that of the line supply. As the rotor accelerates, the magnetic fields induced in it almost match the rotating magnetic field of the stator. As a result, the number of lines of force cut by the rotor is decreased, causing a decrease in the frequency and voltage of the rotor. The rotor never becomes fully synchronized with the rotating field. This is due to the slip necessary to achieve the relative motion required for induction and the operation of the rotor. The percentage of slip determines the value of the secondary frequency and voltage. If the slip is five percent, then the secondary frequency and voltage are five percent of normal.

Figure 54-4 illustrates a simplified frequency relay system operated by push-button starting. This system has two contactor coils connected in parallel (A and B) and a capacitor connected in series with coil B. A three-step automatic acceleration results from this arrangement. When the motor starts, full voltage is produced across coils A and B, causing normally closed contacts A and B to open. The full resistance is connected across the secondary of the motor. As the motor accelerates, the secondary frequency decreases. This means that coil B drops out and contacts B close to decrease the resistance in the rotor circuit, resulting in continued acceleration of the motor. The capacitor depends upon the frequency of an alternating current. As the motor continues to accelerate, coil A drops out and closes contacts A, accelerating the motor further. Because the normally closed contacts are used, the secondary of the motor cannot be shunted out completely. If the secondary could be completely removed from the circuit, the electron flow would take the path of least resistance, resulting in no energy being delivered to coils A and B upon starting.

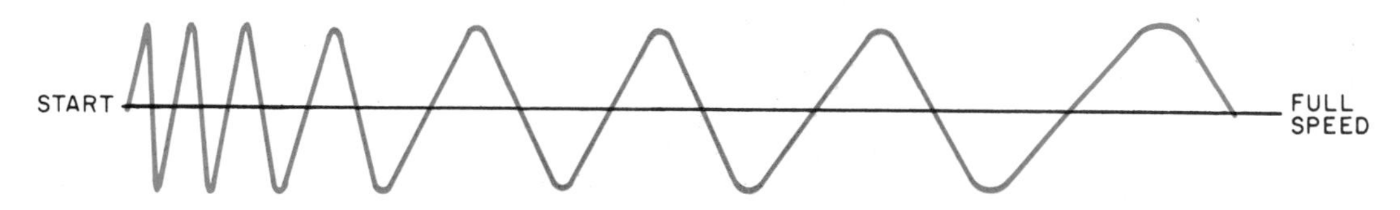

FIGURE 54-3 Rotor frequency decreases as the motor approaches full speed

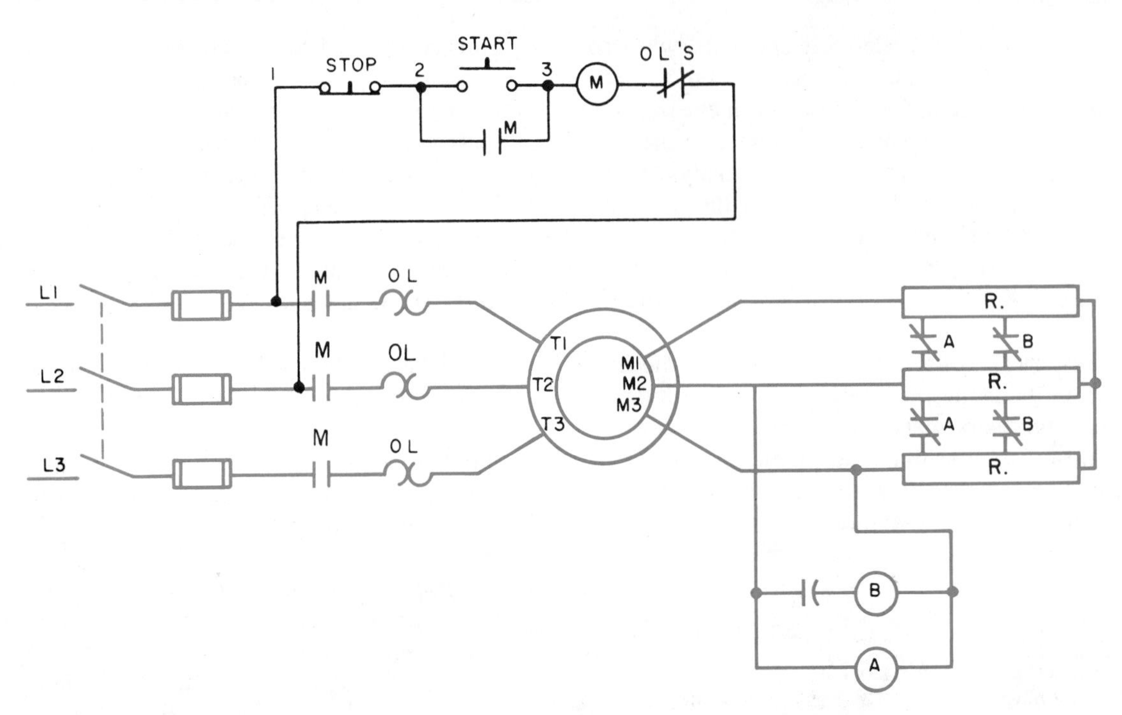

FIGURE 54-4 Automatic acceleration of wound rotor motor using simplified frequency relay system

The controllers for large crane hoists have a resistance, capacitance, and inductance control circuit network that is independent of the secondary rotor resistors.

Frequency relays have a number of advantages, including:

(1) positive response

(2) operating current drops sharply as the frequency drops below the point of resonance

(3) accuracy is maintained because this type of relay operates in a resonant circuit

(4) simple circuit

(5) changes in temperature and variations in line voltage do not affect the relay

(6) an increase in motor load prolongs the starting time.

REVIEW QUESTIONS

1. Are the secondary resistors connected in three uniform wye or delta sections?
2. Do secondary resistors on starters with three or more steps of acceleration have more or less current inrush than those with two steps of acceleration?
3. Does reversing the secondary rotor leads mean that the direction of rotation will reverse?
4. If one of the secondary resistor contacts (S2) fails in figure 54-1, what will happen?
5. In figure 54-2, how many different interlocking conditions exist? Name them.
6. Referring to figure 54-4, why isn't it possible to remove all of the resistance from the secondary circuit?

7. If frequency relays are used on starting, why is the starting cycle prolonged with an increase in motor load?
8. If there is a locked rotor in the secondary circuit, what will be the value of the frequency?
9. Why is it necessary to remove the jumpers in figure 54-2 if limit switches are used?
10. Referring to figure 54-2, why must the push buttons be used to restart if the limit switches are used?

UNIT 55

Synchronous Motor Operation

Objectives

After studying this unit, the student will be able to:

- Describe the operation and applications of a synchronous motor
- Describe lagging and leading power factor and the causes of each
- Describe how the use of a synchronous motor improves the efficiency of an electrical system having a lagging power factor
- Identify a brushless synchronous motor

One of the distinguishing features of the synchronous motor is that it runs without slip at a speed determined by the frequency of the connected power source and the number of poles it contains. This type of motor sets up a rotating field through stator coils energized by alternating current. (This action is similar to the principle of an induction motor.) An independent field is established by a rotor energized by direct current through slip rings mounted on the shaft. The rotor has the same number of coils as the stator. At running speed, these fields (north and south) lock into one another magnetically so that the speed of the rotor is in step with the rotating magnetic field of the stator. In other words, the rotor turns at the synchronous speed. Variations in the connected

FIGURE 55-1 Rotor of a 300-hp, 720-rpm synchronous motor (Courtesy Electric Machinery Mfg. Co.)

load do not cause a corresponding change in speed, as they would with the induction motor.

The rotor, figure 55-1, is excited by a source of direct current so that it produces alternate north and south poles. These poles are then attracted by the rotating magnetic field in the stator. The rotor must have the same number of poles as the stator winding. Every rotor pole, north and south, has an alternate stator pole, south and north, with which it can synchronize.

The rotor has dc field windings to which direct current is supplied through collector rings (slip rings). The current is provided from either an external source or a small dc generator connected to the end of the rotor shaft.

The magnetic fields of the rotor poles are locked into step with—and pulled around by—the revolving field of the stator. Assuming that the rotor and stator have the same number of poles, the rotor moves at the stator frequency (in hertz) actually produced by the generator supplying the motor.

Synchronous motors are constructed almost exactly like alternators. They differ only in those features of design that may make the motor better adapted to its particular purpose.

A synchronous motor cannot start without help because the dc rotor poles at rest are alternately attracted and repelled by the revolving stator field. Therefore, an induction squirrel cage, or starting, winding is embedded in the pole faces of the rotor. This is called an *amortisseur* (ah-more-ti-sir) winding.

This starting winding resembles a squirrel cage winding. The induction effect of the starting winding provides the starting, accelerating, and pull-up torques required. The winding is designed to be used only for starting and for damping oscillations during running. It cannot be used like the winding of the conventional squirrel cage motor. It has a relatively small cross-sectional area and will overheat if the motor is used as a squirrel cage induction motor.

The slip is equal to 100 percent at the moment of starting. Thus, when the ac rotating magnetic field of the stator cuts the rotor windings, which are stationary at startup, the induced voltages produced may be high enough to damage the insulation if precautions are not taken.

If the dc rotor field is either connected as a closed circuit or connected to a discharge resistor during the starting period, the resulting current produces a voltage drop that is opposed to the generated voltage. Thus, the induced voltage at the field terminals is reduced. The squirrel cage winding is used to *start* the synchronous motor in the same way it is used in the squirrel cage induction motor. When the rotor reaches the maximum speed to which it can be accelerated as a squirrel cage motor (about 95 percent or more of the synchronous speed), direct current is applied to the rotor field coils to establish north and south rotor

FIGURE 55-2 A 2000-hp, 300-rpm synchronous motor driving compressor

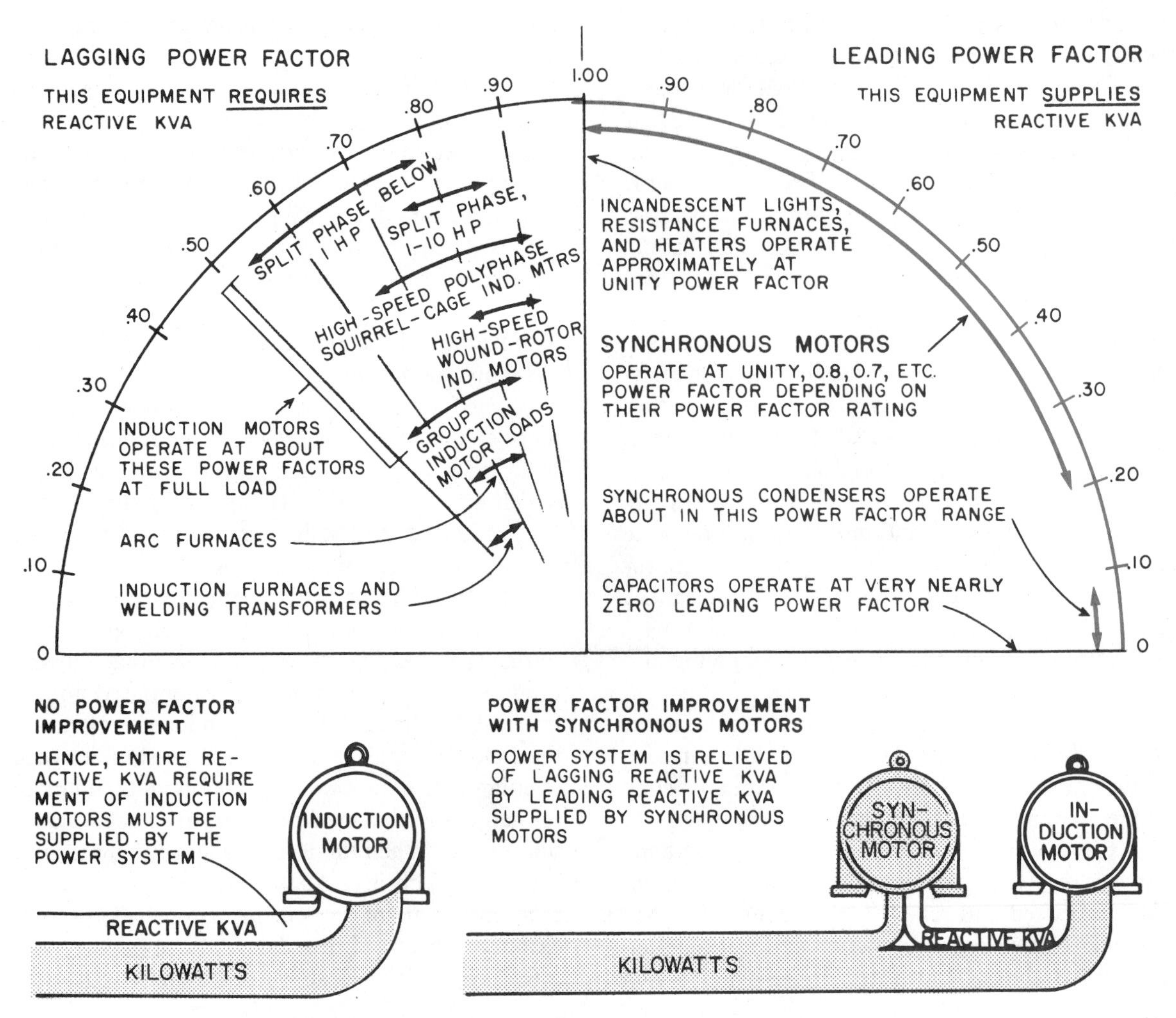

FIGURE 55-3 Power factor operation of various devices may be improved through the use of a synchronous motor (Courtesy Electric Machinery Mfg. Co.)

poles. These poles then are attracted by the poles on the stator. The rotor then accelerates until it locks into synchronous motion with the stator field.

Synchronous motors are used for applications involving large, slow-speed machines with steady loads and constant speeds. Such applications include compressors, fans and pumps, many types of crushers and grinders, and pulp, paper, rubber, chemical, flour, and metal rolling mills, figure 55-2.

POWER FACTOR CORRECTION BY SYNCHRONOUS MOTOR

A synchronous motor converts alternating-current electrical energy into mechanical power. In addition, it also provides power factor correction. It can operate at a leading power factor or at unity. In very rare occasions, it can operate at a lagging power factor.

The power factor is of great concern to industrial users of electricity with respect to energy conservation. Power factor is the ratio of the actual power being used in a circuit, expressed in watts or kilowatts, to the power apparently being drawn from the line, expressed in volt-amperes or kilovolt-amperes (kVA). The kVA value is obtained by multiplying a voltmeter reading and an ammeter reading of the same circuit or equipment. Inductance within the circuit will cause the current to lag the voltage.

When the values of the apparent and actual power are equal or in phase, the ratio of these values is 1:1. In other words, when the voltage and amperage are in phase, the ratio of these values is 1:1. This is the case of pure resistive loads. The unity power factor value is the highest power factor that can be obtained. The higher the power factor, the greater is the efficiency of the electrical equipment.

Ac loads generally have a lagging power factor. As a result, these loads burden the power system with a large reactive load. Refer to the induction motor in figure 55-3. A synchronous motor with an overexcited dc field may be used to offset the low power factor of the other loads on the same electrical system. An overexcited synchronous motor means that it is operating at more than the unity power factor. Therefore, it is working to improve the power factor of the power system.

BRUSHLESS SYNCHRONOUS MOTORS

Solid-state technology has brought about the use of brushless synchronous motors. The dc field excitation for such a motor is provided by a special ac generator mounted on the main motor shaft. The excitation is converted to direct current by a rotating rectifier assembly.

The operating characteristics are the same as those of synchronous motors with brushes. However, elimination of the collector rings, brushes, commutator and some control contactors gives the brushless motor several outstanding advantages:

- Brush sparking is eliminated, reducing safety hazards in some areas.
- Field control and excitation are provided by a static system, requiring much less maintenance.
- Field excitation is automatically removed whenever the motor is out of step. Automatic resynchronization can be achieved whenever it is practical.

REVIEW QUESTIONS

1. Why is it necessary that the rotor and stator have an equal number of poles?
2. What is the effect of the starting winding of the synchronous motor on the running speed?
3. What are typical applications of synchronous motors?
4. Explain how the rotor magnetic field is established.
5. A loaded synchronous motor cannot operate continuously without dc excitation on the rotor. Why?
6. Why must a discharge resistor be connected in the field circuit for starting?
7. What is meant when a synchronous motor is called overexcited?
8. Depending on their power factor ratings, what is the range of the leading power factor at which synchronous motors operate?
9. At what power factor do incandescent lights operate?
10. At what power factor do high-speed, wound rotor motors operate?

Select the *best* answer for the following items.

11. The speed of a synchronous motor is fixed by the
    a. rotor winding
    b. amortisseur winding
    c. supply voltage
    d. frequency of power supply and number of poles
12. Varying the dc voltage to the rotor field changes the
    a. motor speed
    b. power factor
    c. phase excitation
    d. slip
13. Amortisseur windings are located
    a. in the stator pole faces
    b. in the rotor pole faces
    c. in the controller
    d. leading the power factor
14. Dc excitation is applied to the
    a. starting winding
    b. stator winding
    c. rotor winding
    d. amortisseur winding
15. Induction motors and welding transformers require magnetizing current which causes
    a. lagging power factor
    b. leading power factor
    c. unity power factor
    d. zero power factor
16. A synchronous motor can be used to increase the power factor of an electrical system by
    a. reducing the speed
    b. overexciting the stator field
    c. overexciting the rotor field
    d. applying direct current to the stator field

Synchronous Automatic Motor Starter

Objectives *After studying this unit, the student will be able to:*

- **Describe how an out-of-step relay protects the starting winding of a synchronous motor**
- **Describe the action of a polarized field frequency relay in applying and removing dc field excitation on a synchronous motor**
- **Connect synchronous motors and controllers which use out-of-step relays and polarized field frequency relays to achieve automatic motor synchronization**
- **Recommend troubleshooting solutions for problems**

An automatic synchronous motor starter can be used with a synchronous motor to provide automatic control of the startup sequence. That is, the controller automatically sequences the operation of the motor so that the rotor field is synchronized with the revolving magnetic field of the stator.

There are two basic methods of starting synchronous motors automatically. In the first method, full voltage is applied to the stator winding. In the second method, the starting voltage is reduced. A commonly used method of starting synchronous motors is the across-the-line connection. In this method, the stator of the synchronous motor is connected directly to the plant distribution system at full voltage. A magnetic starter is used in this method of starting.

A polarized field frequency relay can be used for the automatic application of field excitation to a synchronous motor.

ROTOR CONTROL EQUIPMENT

Field Contactor

The field contactor opens both lines to the source of excitation, figure 56-1. During starting, the contactor also provides a closed field circuit through a discharge resistor. A solenoid-operated field contactor is similar in appearance to the standard dc contactor. However, for this dc operated contactor, the center pole is normally closed. It is designed to provide a positive overlap between the normally closed contact and the two normally open contacts. This overlap is an important feature because it means that the field winding is never open. The field winding of the motor must always be short-circuited through a discharge resistor or connected to the dc line. The coil of the field contactor is operated from the same direct-

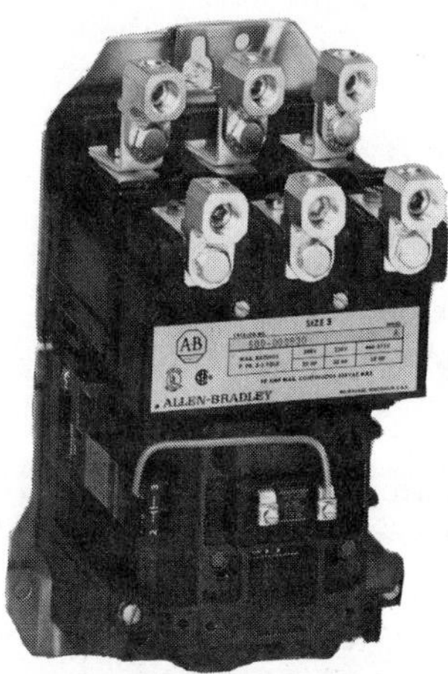

FIGURE 56-1 Magnetic contactor used on synchronous starters for field control (Courtesy Allen-Bradley Co.)

current source that provides excitation for the synchronous motor field.

Out-of-Step Relay

The squirrel cage winding, or starting (amortisseur) winding will not overheat if a synchronous motor starts, accelerates, and reaches synchronous speed within a time interval determined to be normal for the motor. In addition, the motor must continue to operate at synchronous speed. Under these conditions, adequate protection for the entire motor is provided by three overload relays in the stator winding. The squirrel cage winding, however, is designed for starting only. If the motor operates at subsynchronous speed, the squirrel cage winding may overheat and be damaged. It is not unusual for some synchronous motors to withstand a maximum locked rotor interval of only five to seven seconds.

An out-of-step relay (OSR), figure 56-2, is provided on automatic synchronous starters to protect the starting winding. The normally closed contacts of the relay will open to de-energize the line contactor under the following conditions:

1. the motor does not accelerate and reach the synchronizing point after a preset time delay.

FIGURE 56-2 Out-of-step relay used on synchronous starters (Courtesy Allen-Bradley Co.)

2. the motor does not return to a synchronized state after leaving it.
3. the amount of current induced in the field winding exceeds a value determined by the core setting of the out-of-step relay.

As a result, power is removed from the stator circuit before the motor overheats.

Polarized Field Frequency Relay

A synchronous motor is started by accelerating the motor to as high a speed as possible from the squirrel cage winding and then applying the dc field excitation. The components responsible for correctly and dependably applying and removing the field excitation are a *polarized field frequency relay* (PFR) and a reactor, figure 56-3.

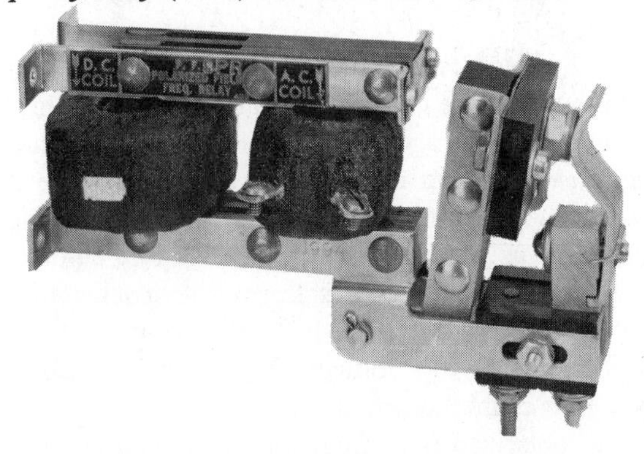

FIGURE 56-3 Polarized field frequency relay with contacts in normally closed position (Courtesy Allen-Bradley Co.)

The operation of the frequency relay is shown in figure 56-4. The magnetic core of the relay has a direct-current coil (C), an induced field-current coil (B), and a pivoted armature (A) to which contact (S) is attached. Coil C is connected to the source of dc excitation. This coil establishes a constant magnetic flux in the relay core. This flux causes the relay to be polarized. Superimposed on this magnetic flux in the relay core is the alternating magnetic flux produced by the alternating induced rotor field current flowing in coil B. The flux through armature A depends on the flux produced by ac coil B and dc coil C. Coil B produces an alternating flux of equal positive and negative magnitude each half-cycle. Thus, the combined flux flowing through armature (A) is much larger when the flux from coil B opposes that from coil C. In figure 56-4(A), the flux from coil B opposes the flux from dc coil C, resulting in a strong flux being forced through armature A of the relay. This condition is shown by the lower shaded loops of figure 56-4(C). One-half cycle later, the flux produced by coil B reverses and less flux flows through armature A. This is due to the fact that the flux from coil B no longer forces as much flux from coil C to take the longer path through armature A. The resultant flux is weak and is illustrated by the small, upper shaded loops of figure 56-4(C). The relay armature opens only during the period of the induced field current wave, which is represented by the small, upper loops of the relay armature flux.

As the motor reaches synchronous speed, the induced rotor field current in relay coil B decreases in amplitude. A value of relay armature flux (upper shaded loop) is reached at which the relay armature A no longer stays closed. The relay then opens to establish contact S. Dc excitation is then applied at the point indicated on the induced field current wave.

Excitation is applied in the direction shown by the arrow. The excitation is opposite in polarity to that of the induced field current at the point of application. This requirement is necessary to compensate for the time needed to build up excitation. The time interval results from the magnetic inertia of the motor field winding. Because of the inertia, the dc excitation does not become effective until the induced current reverses (point O on the wave) to the same polarity as the direct current.

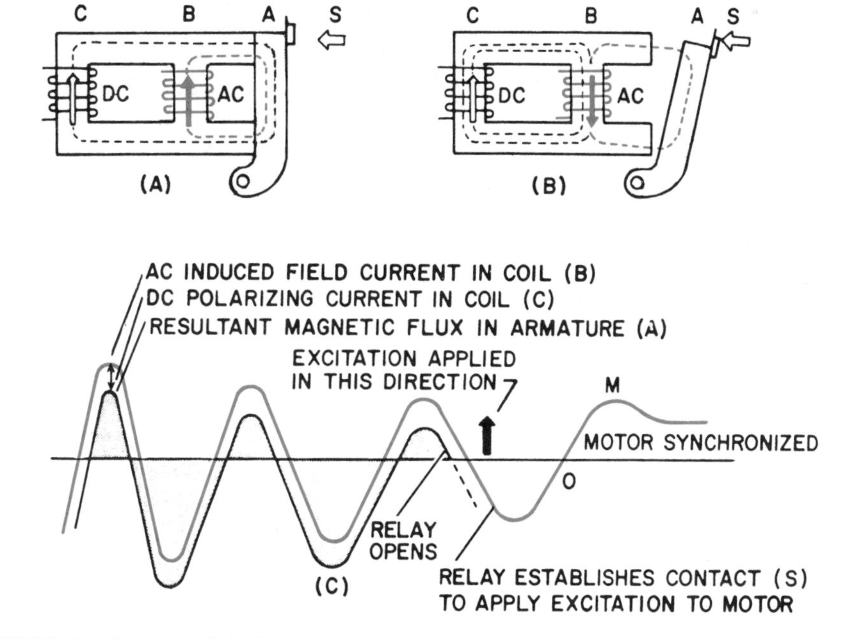

FIGURE 56-4 Polarized field frequency relay operation (Courtesy Electric Machinery Mfg. Co.)

The excitation continues to build up until the motor is synchronized as shown by point M on the curve.

Figure 56-5 indicates the normal operation of the frequency relay. Dc excitation is applied to the coil of the relay at the instant the synchronous motor is started. When the stator winding is energized, using either full voltage or reduced voltage methods, line current is allowed to flow through the three overload relays and the stator winding. Line frequency currents are induced in the two electrically independent circuits of the rotor: (1) the squirrel cage or starting windings and (2) the field windings. The current induced in the field windings flows through the reactor. This device shunts part of the current through the ac coil of the frequency relay, the coil of the out-of-step relay, the field discharge resistor, and finally to the normally closed contact of the field contactor. The flux established in the frequency relay core pulls the armature against the spacer and opens the normally closed relay contacts, figure 56-5. As the motor accelerates to the synchronous speed, the frequency of the induced currents in the field windings diminishes. There is, however, sufficient magnetic flux in the relay core to hold the armature against the core. This flux is due to a considerable amount of induced current forced through the ac coil of the frequency relay by the impedance of the reactor at high slip frequency.

At the point where the motor reaches its synchronizing speed (usually 92 to 97 percent of the synchronous speed) the frequency of the induced field current is at a very low value. The reactor impedance also is greatly reduced at this low frequency. Thus, the amount of current shunted to the ac coil is reduced to the point where the resultant core flux is no longer strong enough to hold the armature against the spacer. At the moment that the rotor speed and the frequency and polarity of the induced currents are most favorable for synchronization, the armature is released, the re-

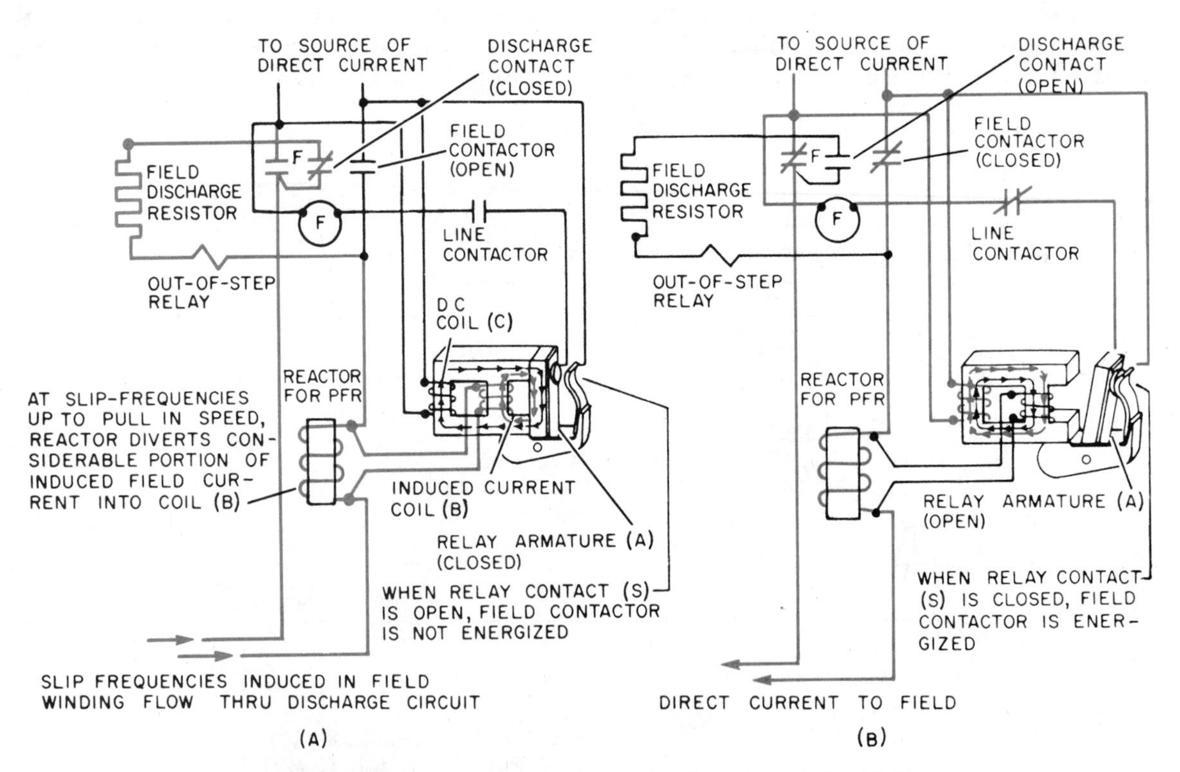

FIGURE 56-5 Wiring connections and operation of a polarized field frequency relay (Courtesy Electric Machinery Mfg. Co.)

lay contacts close, and the control circuit is completed to the operating coil of the field contactor. Dc excitation is applied to the motor field winding, figure 56-5(B). At the same time, the out-of-step relay and discharge resistor are de-energized by the normally closed contacts of the field contactor.

An overload or voltage fluctuation may cause the motor to pull out of synchronism. In this case, a current at the slip frequency is induced in the field windings. Part of this current flows through the ac coil of the polarized field frequency relay, opens the relay contact, and removes the dc field excitation. The motor automatically resynchronizes if the line voltage and load conditions return to normal within a preset time interval, and the motor has enough pull-in torque. However, if the overload and low-voltage conditions continue so

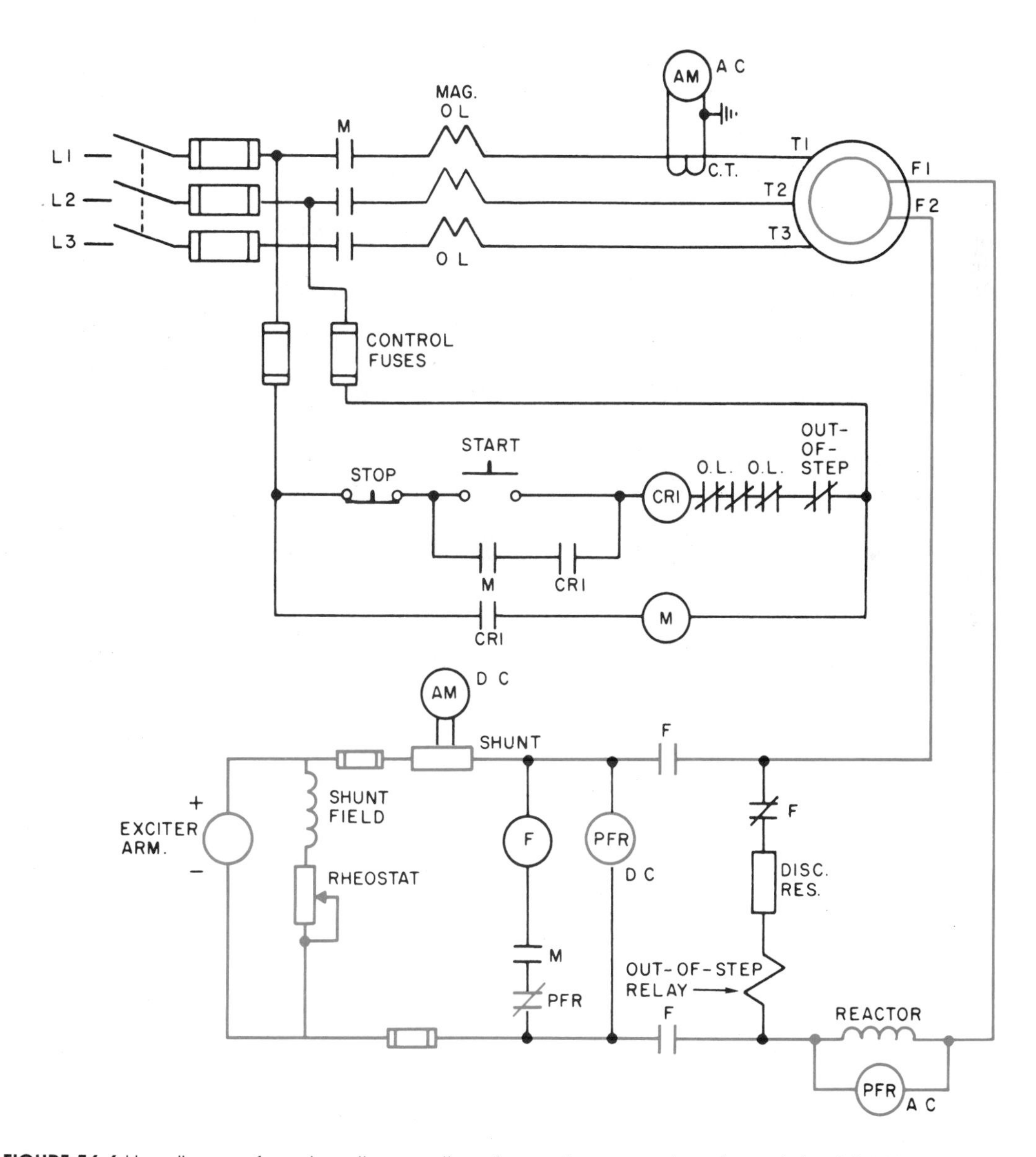

FIGURE 56-6 Line diagram for automatic operation of a synchronous motor using polarized field frequency relay

that the motor cannot resynchronize, then either the out-of-step relay or the overload relays activate to protect the motor from overheating.

SUMMARY OF AUTOMATIC STARTER OPERATION

The line diagram in figure 56-6 shows the automatic operation of a synchronous motor. For starting, the motor field winding is connected through the normally closed power contact of the field contactor (F), the discharge resistor, the coil of the out-of-step relay, and the reactor. When the start button is pressed, the circuit is completed to the control relay coil (CR1) through the control fuses, the stop button, and contacts of the overload and out-of-step relays. The closing of CR1 energizes the line contactor M which applies full voltage at the motor terminals with the overload relays in the circuit. A normally open contact on CR1 and a normally open interlock on line contactor M provide the hold-in, or maintaining circuit. The starting and running current drawn by the motor is indicated by an ammeter with a current transformer.

At the moment the motor starts, the polarized field frequency relay (PFR) opens its normally closed contact and maintains an open circuit to the field contactor (F) until the motor accelerates to the proper speed for synchronizing. When the motor reaches a speed equal to 92 to 97 percent of its synchronous speed, and the rotor is in the correct position, the contact of the polarized field frequency relay closes to energize field contactor F through an interlock on line contactor M. The closing of field contactor F applies the dc excitation to the field winding and causes the motor to synchronize. After the rotor field circuit is established through the normally open power contacts of the field contactor, the normally closed contact on this contactor opens the discharge circuit. The motor is now operating at the synchronous speed. If the stop button is pressed, or if either magnetic overload relay is tripped, the starter is de-energized and disconnects the motor from the line.

REVIEW QUESTIONS

1. What are the two basic methods of automatically starting a synchronous motor?
2. What is an out-of-step relay?
3. Why is an out-of-step relay used on automatic synchronous starters?
4. Under what conditions will the out-of-step relay trip out the control circuit?
5. What is the last control contact which closes on a starting and synchronizing operation?
6. What influence do both of the polarized field frequency relay (PFR) coils exert on the normally closed contact?
7. Why is the PFR polarized with a dc coil?
8. Approximately how much time (in terms of electrical cycles) elapses from the moment the PFR opens to the moment the motor actually synchronizes?
9. How does the ac coil of the PFR receive the induced field current without receiving the full field current strength?
10. Why is a control relay (CRI) used in figure 56-6?

UNIT 57

Variable Speed Ac Motor Control

Objectives *After studying this unit, the student will be able to:*

- **Discuss different methods of changing the speed of ac induction motors**
- **Discuss how an alternator is used to provide variable frequency motor control**
- **Discuss electronic methods of variable frequency control**

Two factors determine the speed of the rotating magnetic field of an ac induction motor:

1. Number of stator poles
2. Frequency

If either of these factors is changed, the speed of the motor can be changed.

In Unit 52, the operation of the consequent pole motor was discussed. The speed of the consequent pole motor can be changed by changing the number of stator poles. This method of speed control causes the speed to change in steps and does not permit control over a wide range of speed. For example, the synchronous speed of the rotating magnetic field in a two-pole motor is 3600 rpm when connected to a 60 Hz line. If this motor is changed to a four-pole motor, the synchronous speed will change to 1800 rpm. Notice that this method of speed control permits the motor to operate with a synchronous field speed of 3600 rpm or 1800 rpm. The motor cannot be operated with a synchronous speed between 3600 rpm and 1800 rpm.

VARIABLE VOLTAGE SPEED CONTROL

Another method of controlling the speed of some ac induction motors is by reducing the applied voltage to the stator. This method does not change the synchronous speed of the rotating magnetic field, but it does cause the magnetic field of the stator to become weaker. As the magnetic field of the stator becomes weaker, the rotor slip becomes greater and, therefore, causes a reduction in rotor speed.

Variable voltage speed control is used with fractional horsepower motors that operate light loads such as fans or blowers. Motors that are intended to be operated with variable voltage are designed with high resistance rotors, such as the type "A" rotor, to help limit the amount of current induced into the rotor at low speed. Induction motors that use a centrifugal switch cannot be used with variable voltage control. This limits the types of induction motors that can be used to shaded pole motors or capacitor start-capacitor run motors.

There are several methods used to control the voltage supplied to the motor. One method is to use a triac with a phase shift network similar to the circuit shown in figure 57-1. When triac circuits are used with inductive loads, care must be taken to insure that both halves of the ac waveform are conducted. Only triac controllers intended to be used with inductive loads should be used for motor control. Triac circuits intended to control incandescent lamp loads will often begin conducting on one half of the ac cycle and not the other. For example, assume the triac in figure 57-1 begins conducting on the positive half cycle of voltage before it conducts the negative half cycle. A waveform similar to the one shown in figure 57-2 could be produced across the load. Since only positive voltage pulses are being conducted, the voltage applied to the load is dc. A dc voltage applied to a resistive load such as an incandescent lamp will not cause any harm, but a great deal of harm can be done if a dc voltage is applied to an inductive load such as the stator winding of a motor. A variable speed control using a triac is shown in figure 57-3.

Another device used to control the voltage applied to a small induction motor is the autotransformer, figure 57-4. In this circuit, a rotary switch is used to connect the motor to different taps on the transformer winding. This permits the motor to be operated at any one of several different speeds.

A tapped inductor can also be used to control motor speed, figures 57-5 and 57-6. In this circuit, the inductor is not used to control the voltage applied to the motor, but is used to control the impedance of the circuit. If more of the inductor

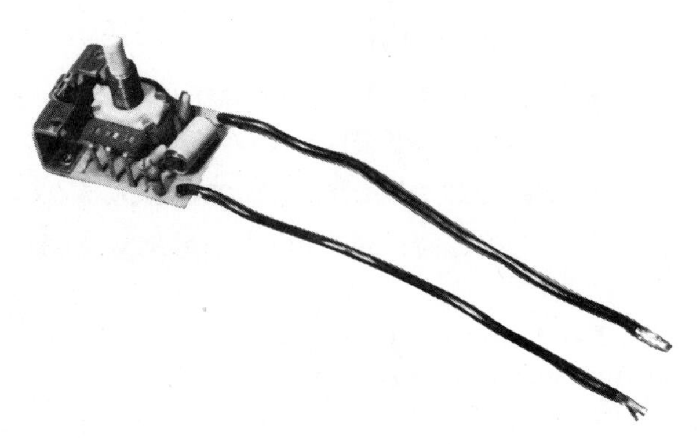

FIGURE 57-3 Variable speed control using a triac to control the voltage applied to the motor

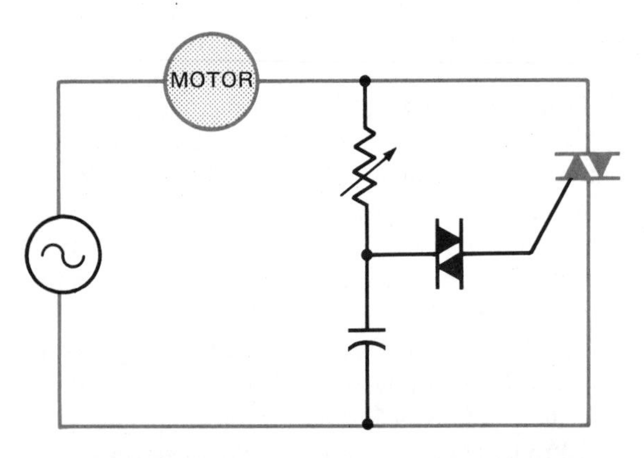

FIGURE 57-1 Triac used to control motor speed

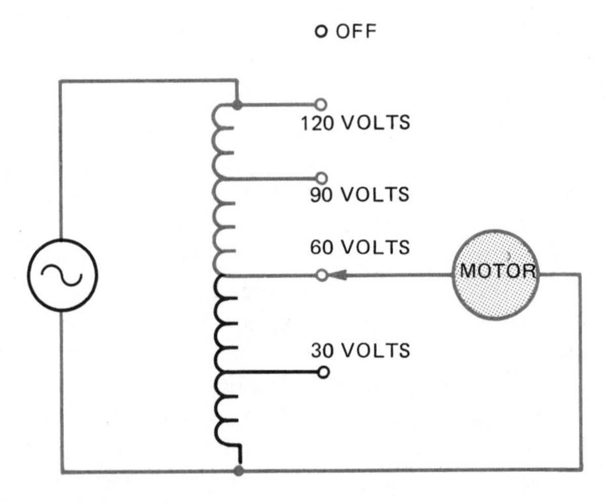

FIGURE 57-4 Autotransformer controls motor voltage

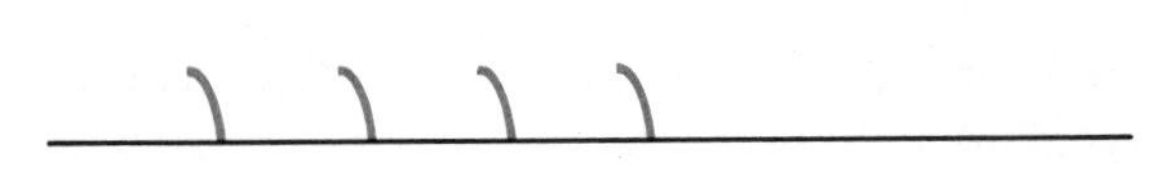

FIGURE 57-2 Triac conducts only the positive half of the waveform

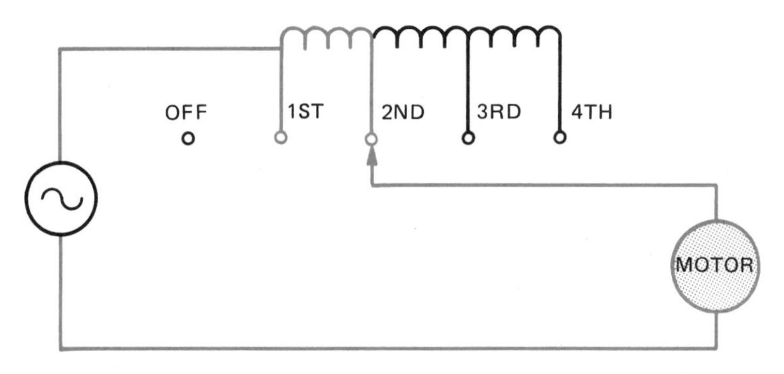

FIGURE 57-5 Series inductor changes impedance of circuit

FIGURE 57-6 Tapped inductor used to control motor speed by connecting it in series with the motor winding

is connected in series with the motor, the total impedance of the circuit is increased and, therefore, the current flow through the motor is decreased. As current flow is decreased, the magnetic field of the stator becomes weaker and the increase in rotor slip causes the motor speed to slow down. The motor speed can be adjusted by changing the tap with the rotary switch to insert more or less of the inductor winding in series with the motor.

VARIABLE FREQUENCY CONTROL

One of the factors that determines the speed of the rotating magnetic field of an induction motor is the frequency of the applied voltage. If the frequency is changed, the speed of the rotating magnetic field changes also. For example, a four-pole stator connected to a 60 Hz line will have a synchronous speed of 1800 rpm. If the frequency is lowered to 30 Hz, the synchronous field speed falls to 900 rpm.

When the frequency is lowered, care must be taken not to damage the stator winding. The current flow through the winding is limited to a great extent by inductive reactance. When the frequency is lowered, inductive reactance is lowered also ($X_L = 2\pi FL$). For this reason, variable frequency motor controllers must have some method of reducing the applied voltage to the stator as frequency is reduced.

Alternator Control

One method of producing variable frequencies for operating induction motors is with the use of an alternator, figure 57-7. In this arrangement some type of variable speed drive, such as dc motor, is used to turn the shaft of the alternator. The speed of the alternator determines the frequency of the voltage applied to the induction motors. The alternator can furnish power to as many induction motors as desired provided the power rating of the alternator is not exceeded.

Since the output voltage of the alternator is controlled by the amount of dc excitation current applied to the rotor, a variable voltage dc power supply is used to determine the output voltage of the alternator. As frequency is reduced, the output voltage must also be reduced to prevent excessive current flow in the windings of the induction

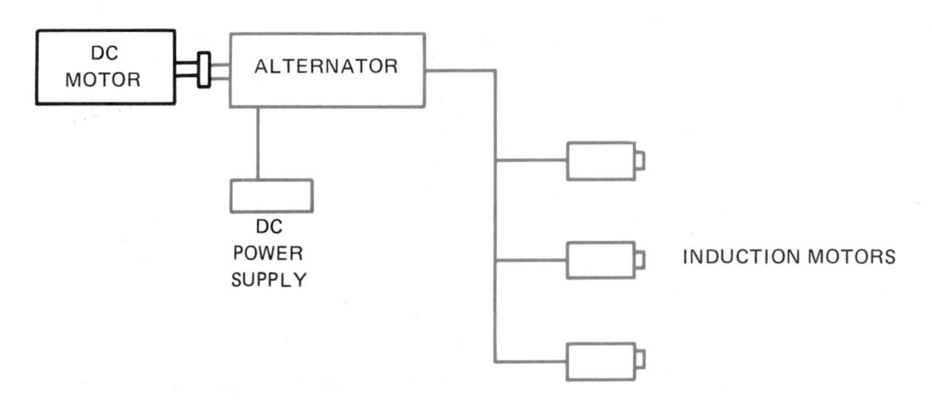

FIGURE 57-7 Alternator controls speed of induction motors

motors. This method of speed control is frequently used on conveyor systems where it is desirable to have a large number of motors controlled from one source.

Solid-State Control

Most solid-state variable frequency drives operate by first changing the ac voltage to dc, and then changing the dc voltage back to ac at the desired frequency. A variable frequency controller is shown in figure 57-8. The circuit shown in figure 57-9 uses a three-phase bridge rectifier to convert three-phase ac voltage into dc voltage. A phase shift unit controls the output voltage of the rectifier. This permits the voltage applied to the motor to be decreased as the frequency is decreased.

A choke coil and capacitor bank are used to filter the output voltage of the rectifier before transistors Q1 through Q6 change the dc voltage back to ac. An electronic control unit is connected to the bases of transistors Q1 through Q6. The electronic control unit converts the dc voltage back into three-phase alternating current by turning transistors on or off at the proper time and in the proper sequence. For example, assume that transistors Q1 and Q4 are switched on at the same time. This permits T1 to be positive at the same time T2 is negative. If conventional current flow is assumed, current will flow through transistor Q1 to T1, from T1 through the motor winding to T2, and then through transistor Q4 to negative. Now assume that transistors Q1 and Q4 have been

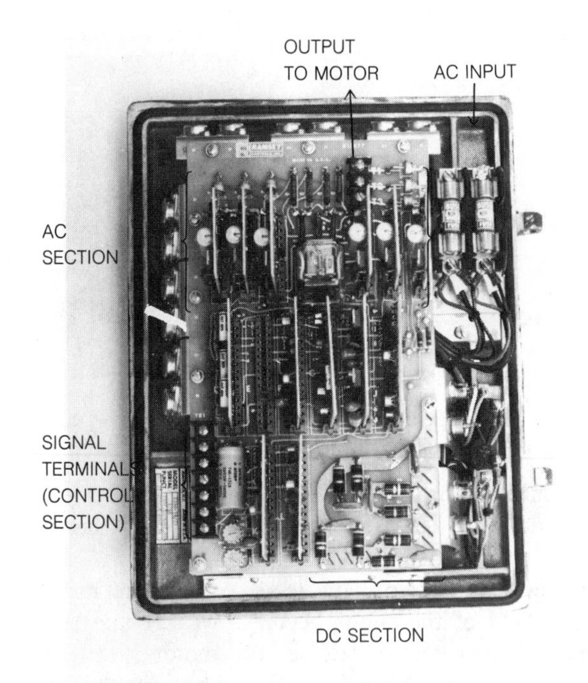

FIGURE 57-8 A 2-hp, variable frequency motor controller (Courtesy Ramsey Controls Inc.)

turned off, and transistors Q3 and Q6 have been turned on. Current can now flow through transistor Q3 to T2, from T2 through the motor to T3, and through transistor Q6 to negative.

Since the transistors are turned completely on or completely off, the waveform produced is a square wave instead of a sine wave, figure 57-10.

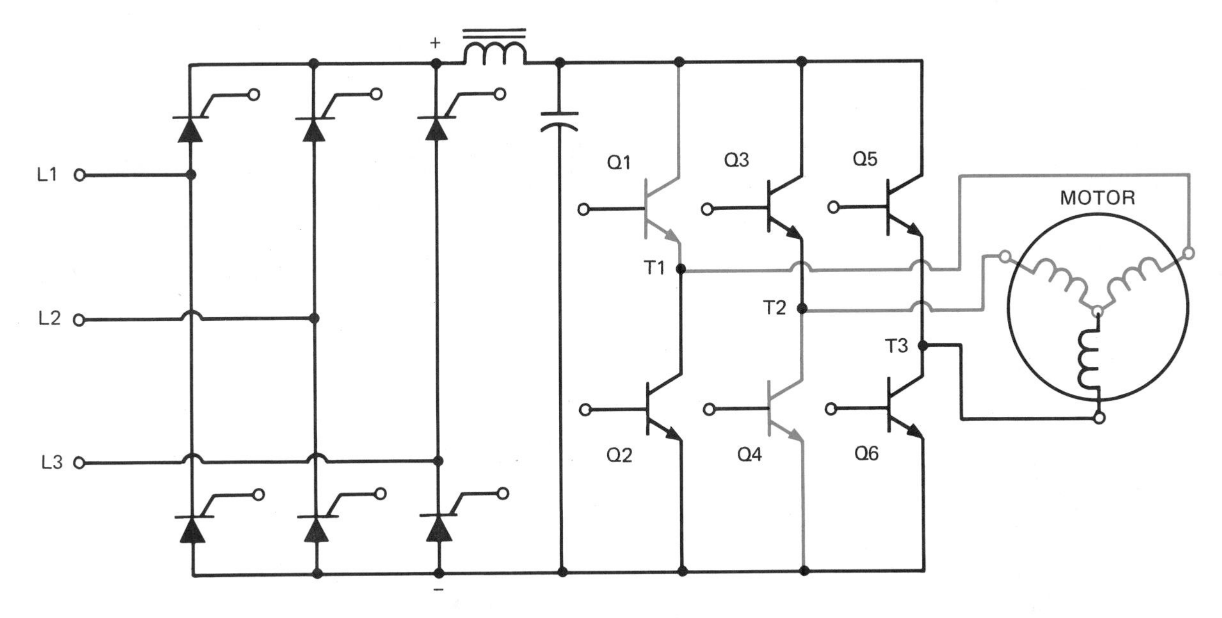

FIGURE 57-9 Solid-state variable frequency control

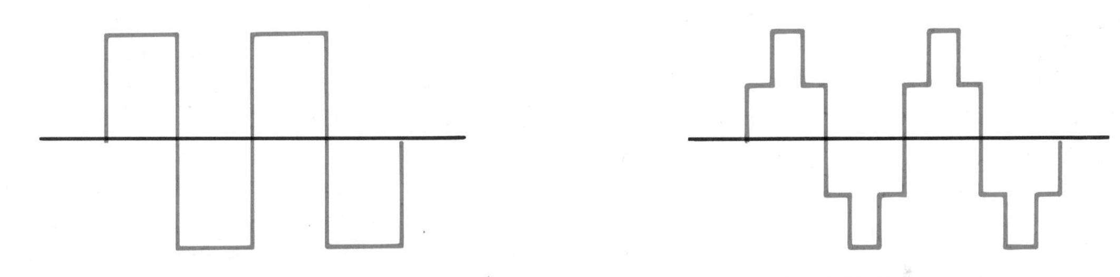

FIGURE 57-10 Square wave

FIGURE 57-11 Stepped wave

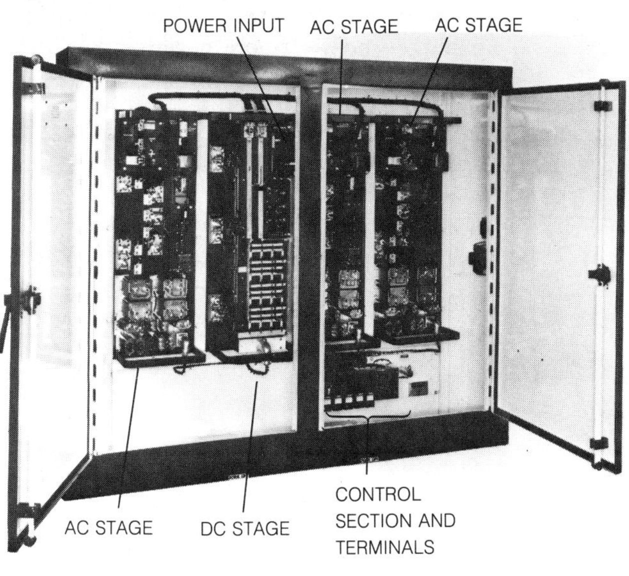

FIGURE 57-12 A 125-hp, variable frequency, ac motor controller (Courtesy Ramsey Controls Inc.)

Induction motors will operate on a square waveform without any problem. Some manufacturers design units that will produce a stepped waveform as shown in figure 57-11. This stepped waveform is used because it is similar to a sine wave. Another type of variable frequency ac motor controller is shown in figure 57-12.

VARIABLE FREQUENCY CONTROL USING SCRs

Because of their ability to handle large amounts of power, SCRs are often used for converting direct current into alternating current. An example of this type of circuit is shown in figure 57-13. In this circuit, the SCRs are connected to a control unit which controls the sequence and rate at which the SCRs are gated on. The circuit is constructed so that SCRs A and A′ are gated on at the same time and SCRs B and B′ are gated on at the same time. Inductors L_1 and L_2 are used for filtering and wave shaping. Diodes D_1 through D_4 are clamping diodes and are used to prevent the output voltage from becoming excessive. Capacitor C_1 is used to turn one set of SCRs off when the other set is gated on. This capacitor must be a true ac capacitor because it will be charged to the alternate polarity each half cycle. In a converter intended to handle large amounts of power, capacitor C_1 will be a bank of capacitors. To understand the operation of this circuit, assume that SCRs A and A′ are gated on at the same time. Current will flow through the circuit as shown in figure 57-14. Notice the direction of current flow through the load and that capacitor C_1 has been charged to the polarity shown. Recall that when an SCR has been turned on by the gate, it can only be turned off by permitting the current flow through the anode-cathode section to drop below the holding current level. Now assume that SCRs B and B′ are gated on. Because SCRs A and A′ are still turned on, two separate current paths now exist through the circuit. The negative charge on capacitor C_1, however, causes the positive current to see a path more negative than the one through SCRs A and

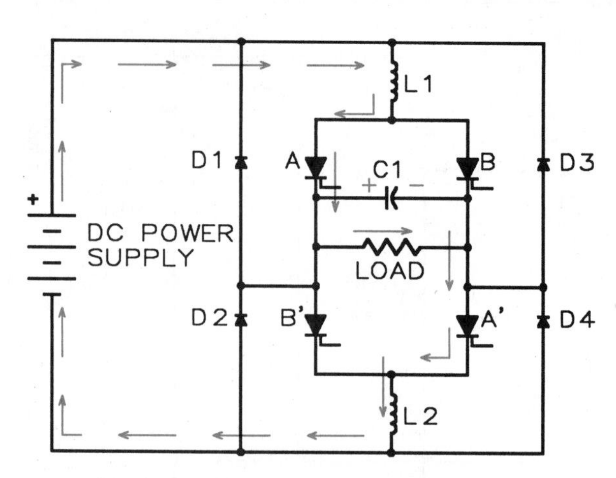

FIGURE 57-14 Current flows through SCRs A and A′

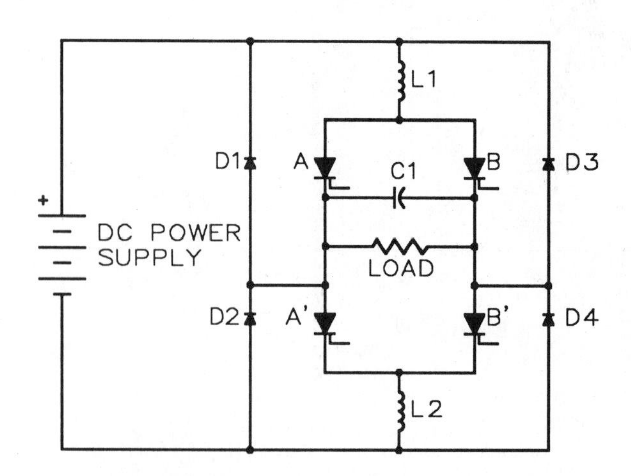

FIGURE 57-13 Changing dc into ac using SCRs

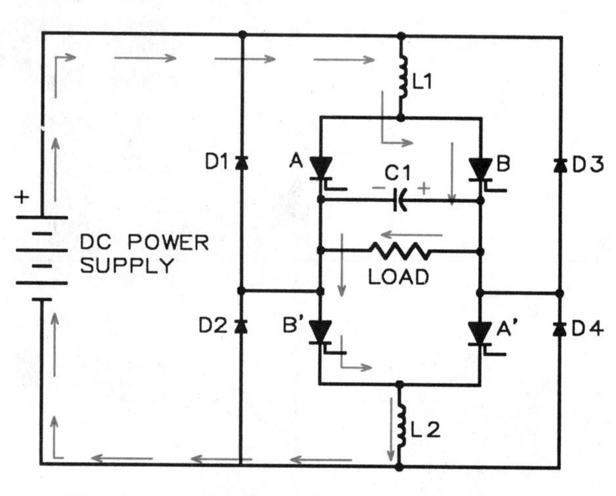

FIGURE 57-15 Current flows through SCRs B and B′

A′. The current now flows through SCRs B and B′ to charge capacitor C_1 to the opposite polarity as shown in figure 57-15. Because the current now flows through SCRs B and B′, SCRs A and A′ turn off. Notice that the current flows through the load in the opposite direction, which produces alternating current through the load, and that capacitor C_1 has been charged to the opposite polarity.

To produce the next half cycle of ac current, SCRs A and A′ are gated on again. The negatively charged side of capacitor C_1 will now cause the current to stop flowing through SCRs B and B′ and begin flowing through SCRs A and A′ as shown in figure 57-14. The frequency of the circuit is determined by the rate at which the SCRs are gated on.

REVIEW QUESTIONS

1. What two factors determine the synchronous speed of the rotating magnetic field of an induction motor?
2. What method of speed control that is often used with small motors does not involve changing the speed of the rotating magnetic field?
3. Name two devices commonly used to control the voltage applied to the motor.
4. Name the two types of motors most often used when speed control is accomplished by reducing the voltage applied to the motor.
5. Why are these two motors generally used for this type of speed control?
6. What type of rotor is generally used with motors designed to be controlled by variable voltage?
7. What determines the output frequency of an alternator?
8. What determines the output voltage of an alternator?
9. In the circuit shown in figure 57-9, why are SCRs used to form the three-phase bridge rectifier instead of diodes?
10. Why must the voltage applied to an induction motor be reduced when frequency is reduced?

UNIT 58

Magnetic Clutch and Magnetic Drive

Objectives *After studying this unit, the student will be able to:*

- **State several advantages of the use of a clutch in a drive**
- **Describe the operating principles of magnetic clutches and drives**
- **Distinguish between single and multiple-face clutches**
- **Connect magnetic clutch and magnetic drive controls**
- **Recommend troubleshooting solutions for magnetic clutch and drive problems**

ELECTRICALLY CONTROLLED MAGNETIC CLUTCHES

Machinery clutches were originally designed to engage very large motors to their loads after the motors had reached running speeds, figure 58-1. Clutches provide smooth starts for operations in which the material being processed might be damaged by abrupt starts, figure 58-2. Clutches are also used to start high-inertia loads, since the starting may be difficult for a motor that is sized to handle the running load. When starting conditions are severe, a clutch inserted between the motor and the load means that the motor can run within its load capacity. The motor will take longer to bring the load up to speed, but the motor and load will not be damaged.

As more automatic cycling and faster cycling rates are being required in industrial production, electrically controlled clutches are being used more often.

Single-face Clutch

The single-face clutch consists of two discs: one is the field member (electromagnet) and the other is the armature member. The operation of the clutch is similar to that of the electromagnet in a motor starter, figure 58-3. When current is applied to the field winding disc through collector (slip) rings, the two discs are drawn together magnetically. The friction face of the field disc is held tightly against the armature disc to provide positive engagement between the rotating drives. When the current is removed, a spring action separates the faces to provide a definite clearance between the discs. In this manner, the motor is mechanically disconnected from the load.

Multiple-face Clutch

Multiple-face clutches are also available. In a double-face clutch, both the armature and field discs are mounted on a single hub with a double-

FIGURE 58-1 Magnetic clutches in cement mill service. Note slip rings for clutch supply.

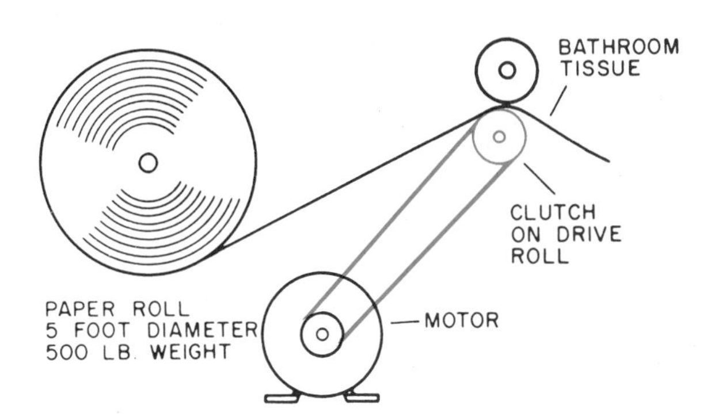

FIGURE 58-2 To prevent tearing, a cushioned start is required on drive roll that winds bathroom tissue off large roll. Roll is five feet in diameter and weighs 500 pounds when full. Pickup of thin tissue must be very gradual to avoid tearing. Application also can be used for filmstrip processing machine.

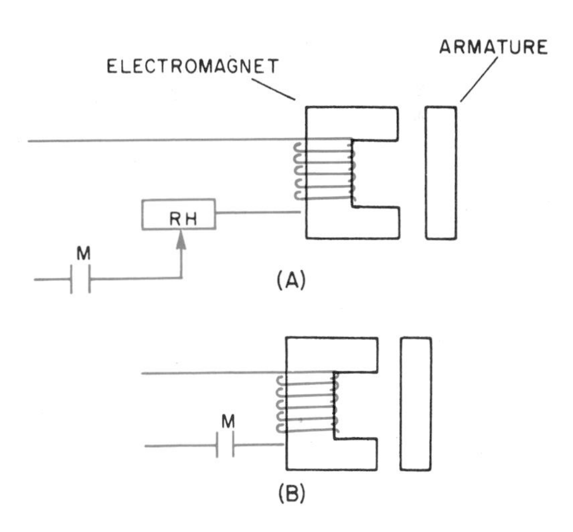

FIGURE 58-3 Principle of operation of electrically controlled clutches: (A) gradual clutch engagement; (B) more rapid clutch engagement.

faced friction lining supported between them. When the magnet of the field member is energized, the armature and field members are drawn together. They grip the lining between them to provide the driving torque. When the magnet is

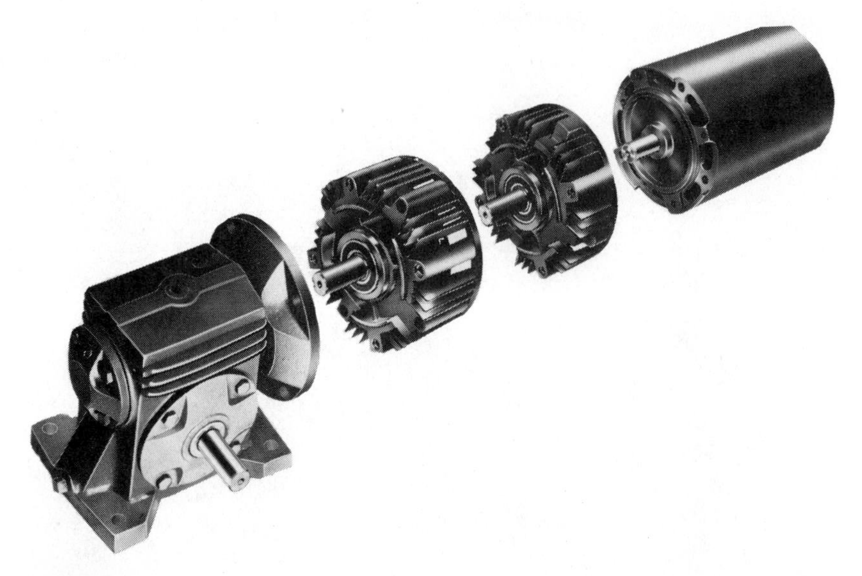

FIGURE 58-4 Electric brake and electric clutch modules. Brake is shown on the center left on the driven machine side of the clutch. (Courtesy Warner Electric Brake and Clutch Co.)

de-energized, a spring separates the two members and they rotate independently of each other. Double-face clutches are available in sizes up to 78 inches in diameter.

A water-cooled magnetic clutch is available for applications that require a high degree of slippage between the input and output rotating members. Uses for this type of clutch include tension control (wind-up and payoff) and cycling (starting and stopping) operations in which large differences between the input and the output speeds are required. Flowing water removes the heat generated by the continued slippage within the clutch. A rotary water union mounted in the end of the rotor shaft means that the water-cooled clutch cannot be end-coupled directly to the prime mover. Chains or gears must be used.

A combination clutch and magnetic brake disconnects the load from the drive and simultaneously applies a brake to the load side of the drive. Magnetic clutches and brakes are often used as mechanical power-switching devices in module form. Figure 58-4 shows a drive line with an electric clutch (center right) and an electric brake. An application of this arrangement is shown in figure 58-5. Remember that the quicker the start or stop, the shorter the life of this equipment.

Magnetic clutches are used on automatic machines for starting, running, cycling, and torque-limiting. The combinations and variations of these functions are practically limitless.

MAGNETIC DRIVES

The magnetic drive couples the motor to the load magnetically. The magnetic drive can be used

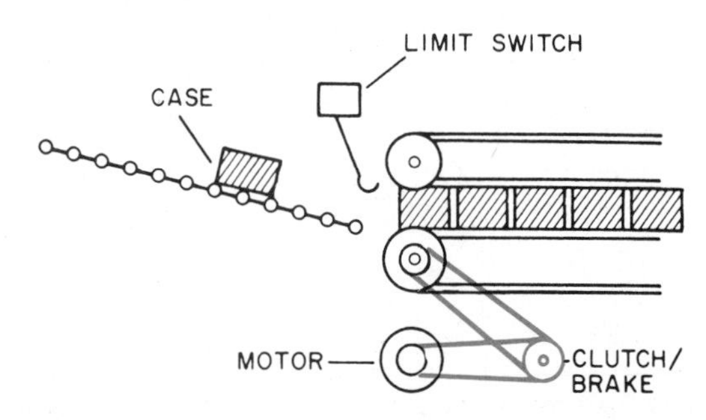

FIGURE 58-5 Case sealer is used to hold top of carton while glue is drying. In this application, cartons come down gravity conveyor and hit switch in front of sealer. Clutch on sealer drive is engaged, moving all cartons in sealer forward. When new carton passes trip switch, brake is engaged and clutch disengaged. Positioning provides even spacing of cartons, insuring that they are in the sealer for as long a time as possible.

as a clutch and can be adapted to an adjustable speed drive.

The electromagnetic (or eddy current) coupling is one of the simpler ways to obtain an adjustable output speed from the constant input speed of squirrel cage motors.

There is no mechanical contact between the rotating members of the magnetic drive. Thus, there is no wear. Torque is transmitted between the two rotating units by an electromagnetic reaction created by an energized coil winding. The slip between the motor and load can be controlled continuously, with more precision, and over a wider range than is possible with the mechanical friction clutch.

As shown in figure 58-6, the magnet rotates within the steel ring or drum. There is an air gap between the ring and the magnet. The magnetic flux crosses the air gap and penetrates the iron ring. The rotation of the ring with relation to the magnet generates eddy currents and magnetic fields in the ring. Magnetic interaction between the two units transmits torque from the motor to the load. This torque is controlled with a rheostat which manually or automatically adjusts the direct current supplied to the electromagnet through the slip rings.

When the electromagnetic drive responds to an input or command voltage, a further refinement can be obtained in automatic control to regulate and maintain the output speed. The magnetic drive can be used with any type of actuating device or transducer that can provide an electrical signal. For example, electronic controls and sensors that detect liquid level, air and fluid pressure, temperature, and frequency can provide the input required.

A tachometer generator provides feedback speed control in that it generates a voltage that is proportional to its speed. Any changes in load condition will change the speed. The resulting generator voltage fluctuations are fed to a control circuit which increases or decreases the magnetic drive field excitation to hold the speed constant.

For applications were a magnetic drive meets the requirements, an adjustable speed is frequently a desirable choice. Magnetic drives are used for applications requiring an adjustable speed, such as cranes, hoists, fans, compressors, and pumps, figure 58-7.

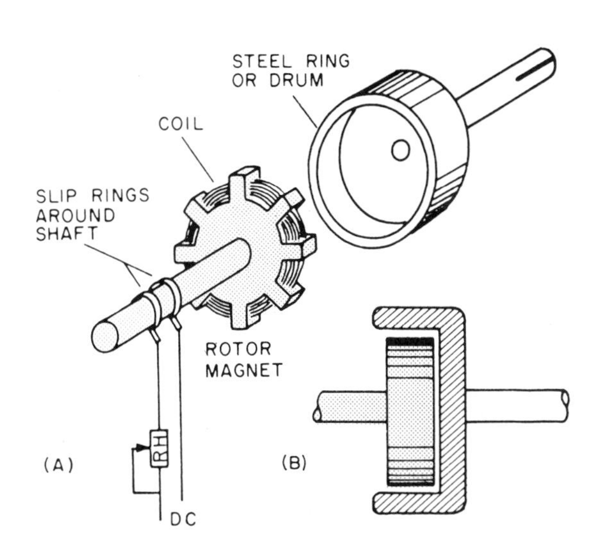

FIGURE 58-6 Diagram showing (A) open view of magnetic drive assembly; (B) spider rotor magnet rotates within ring.

FIGURE 58-7 Two magnetic drives driven by 100-hp, 1200-rpm induction motors mounted on top. Machines are used in typical sewage pumping plant. Pumps are mounted beneath floor of drives. (Courtesy Electric Machinery Mfg. Co.)

REVIEW QUESTIONS

1. How is the magnetic clutch engaged and disengaged?
2. What devices may be used to energize the magnetic clutch?
3. Which type of drive is best suited for maintaining large differences in the input and output speeds? Why?
4. What is meant by feedback speed control?
5. How is the magnetic drive used as an adjustable speed drive?

Motor Installation

Objectives ***After studying this unit, the student will be able to:***

- **Determine the full-load current rating of different types of motors using the *National Electrical Code (NEC)***
- **Determine the conductor size for installing motors**
- **Determine the short-circuit protection needed for different types of motors**
- **Determine the overload size for different types of motors**
- **Determine the conductor size and short-circuit protection for a multiple motor installation**

CALCULATION OF MOTOR CURRENT

There are different types of motors, such as direct current, single-phase ac, two-phase ac, and three-phase ac. Different tables are used to compute the running current for these different types of motors. Table 430-147, figure 59-1, is used to calculate the current for a direct-current motor; Table 430-148, figure 59-2, is used to determine the running current for single-phase ac motors; Table 430-149, figure 59-3, is used to determine full-load current for two-phase ac motors; and Table 430-150, figure 59-4, is used to determine the full-load current rating for three-phase motors. The table lists the amount of current the motor is expected to draw when it is under a full-load condition. The motor will have less current draw if it is not under a full load. These tables list the ampere rating of the motors according to horsepower (hp) and connected voltage. It should also be noted that according to 430-6a, these tables are to be used in determining *conductor size, fuse size, and ground fault protection* for a motor instead of the nameplate rating of the motor. The motor overload size, however, is to be determined by the nameplate rating of the motor.

Direct-Current Motors

Table 430-147 lists the current for direct-current motors. The horsepower rating of the motor is given in the far left hand column. Rated voltages are listed across the top of the table. The table shows that a one (1) hp motor will have a full-load current of 12.2 amperes when connected to 90 volts dc. If a 1 hp motor is designed to be connected to 240 volts, it will have a current draw of 4.7 amperes.

Single-Phase Ac Motors

The current ratings for single-phase ac motors are given in Table 430-148. Particular attention should be paid to the statements preceding the table. The statements assert that the values listed in this table are for motors which operate

Table 430-147. Full-Load Current in Amperes, Direct-Current Motors

The following values of full-load currents* are for motors running at base speed.

HP	Armature Voltage Rating*					
	90V	120V	180V	240V	500V	550V
1/4	4.0	3.1	2.0	1.6		
1/3	5.2	4.1	2.6	2.0		
1/2	6.8	5.4	3.4	2.7		
3/4	9.6	7.6	4.8	3.8		
1	12.2	9.5	6.1	4.7		
1 1/2		13.2	8.3	6.6		
2		17	10.8	8.5		
3		25	16	12.2		
5		40	27	20		
7 1/2		58		29	13.6	12.2
10		76		38	18	16
15				55	27	24
20				72	34	31
25				89	43	38
30				106	51	46
40				140	67	61
50				173	83	75
60				206	99	90
75				255	123	111
100				341	164	148
125				425	205	185
150				506	246	222
200				675	330	294

* These are average direct-current quantities.

FIGURE 59-1 Reprinted with permission from NFPA 70–1987, National Electrical Code, copyright © 1993, National Fire Protection Association, Quincy, MA 02269. This reprinted material is not the complete and official position of the NFPA on the referenced subject, which is represented only by the standard in its entirety.

under normal speeds and torques. Motors especially designed for low speed and high torque, or multispeed motors, shall have their running current determined from the nameplate rating of the motor.

The voltages listed in the table are 115, 200, 208, and 230. The last sentence of the preceding statement states that the currents listed shall be permitted for voltages of 110 to 120 volts and 220 to 240 volts. This means that if the motor is connected to a 120-volt line, it is permissible to use the currents listed in the 115-volt column. If a motor is connected to a 220-volt line, the current listed in the 230-volt column can be used.

EXAMPLE: A 3 hp single-phase ac motor is connected to a 208-volt line. What will be the full-load running current of this motor? Locate 3 horsepower in the far left-hand column. Follow across to the 208-volt column. The full-load current will be 18.7 amperes.

Two-Phase Motors

Although two-phase motors are seldom used, Table 430-149 lists rated full-load currents for

Table 430-148. Full-Load Currents in Amperes Single-Phase Alternating-Current Motors

The following values of full-load currents are for motors running at usual speeds and motors with normal torque characteristics. Motors built for especially low speeds or high torques may have higher full-load currents, and multispeed motors will have full-load current varying with speed, in which case the nameplate current ratings shall be used.

• The voltages listed are rated motor voltages. The currents listed shall be permitted for system voltage ranges of 110 to 120 and 220 to 240.

HP	115V	200V	208V	230V
1/6	4.4	2.5	2.4	2.2
1/4	5.8	3.3	3.2	2.9
1/3	7.2	4.1	4.0	3.6
1/2	9.8	5.6	5.4	4.9
3/4	13.8	7.9	7.6	6.9
1	16	9.2	8.8	8
1 1/2	20	11.5	11	10
2	24	13.8	13.2	12
3	34	19.6	18.7	17
5	56	32.2	30.8	28
7 1/2	80	46	44	40
10	100	57.5	55	50

FIGURE 59-2 Reprinted with permission from NFPA 70–1987, National Electrical Code, copyright © 1993, National Fire Protection Association, Quincy, MA 02269. This reprinted material is not the complete and official position of the NFPA on the referenced subject, which is represented only by the standard in its entirety.

these motors. Like single-phase motors, two-phase motors, which are especially designed for low speed, high torque applications and multispeed motors, use the nameplate rating instead of the values shown in the chart. When using a two-phase, three-wire system, the size of the neutral conductor must be increased by the square root of 2 or 1.41. The reason for this is that the voltages of a two-phase system are 90° out of phase with each other as shown in figure 59-5. The principle of two-phase power generation is shown in figure 59-6. In a two-phase alternator, the phase windings are arranged 90° apart. The magnet is the rotor of the alternator. When the rotor turns, it induces voltage into the phase windings which are 90° apart. When one end of each phase winding is joined to form a common terminal, or neutral, the current in the neutral conductor will be greater than the current in either of the two-phase conductors. An example of this is shown in figure 59-7. In this example, a two-phase alternator is connected to a two-phase motor. The current draw on each of the phase windings is 10 amperes. The current flow in the neutral, however, is 1.41 times greater than the current flow in the phase windings, or 14.1 amperes.

EXAMPLE: Compute the phase current and neutral current for a 60 hp, 460-volt two-phase motor. The phase current can be taken directly from Table 430-149.

Phase current = 67 amperes

The neutral current will be 1.41 times higher than the phase current.

Neutral current = 67 × 1.41

Neutral current = 94.5 amperes

Table 430-149. Full-Load Current
Two-Phase Alternating-Current Motors (4-Wire)

The following values of full-load current are for motors running at speeds usual for belted motors and motors with normal torque characteristics. Motors built for especially low speeds or high torques may require more running current, and multispeed motors will have full-load current varying with speed, in which case the nameplate current rating shall be used. Current in the common conductor of a 2-phase, 3-wire system will be 1.41 times the value given.

The voltages listed are rated motor voltages. The currents listed shall be permitted for system voltage ranges of 110 to 120, 220 to 240, 440 to 480, and 550 to 600 volts.

	Induction Type Squirrel-Cage and Wound-Rotor Amperes				
HP	115V	230V	460V	575V	2300V
½	4	2	1	.8	
¾	4.8	2.4	1.2	1.0	
1	6.4	3.2	1.6	1.3	
1½	9	4.5	2.3	1.8	
2	11.8	5.9	3	2.4	
3		8.3	4.2	3.3	
5		13.2	6.6	5.3	
7½		19	9	8	
10		24	12	10	
15		36	18	14	
20		47	23	19	
25		59	29	24	
30		69	35	28	
40		90	45	36	
50		113	56	45	
60		133	67	53	14
75		166	83	66	18
100		218	109	87	23
125		270	135	108	28
150		312	156	125	32
200		416	208	167	43

FIGURE 59-3 Reprinted with permission from NFPA 70–1987, National Electrical Code, copyright © 1993, National Fire Protection Association, Quincy, MA 02269. This reprinted material is not the complete and official position of the NFPA on the referenced subject, which is represented only by the standard in its entirety.

Three-Phase Motors

Table 430-150 is used to determine the full-load current of three-phase motors. The notes at the bottom of the table are very similar to the notes of Tables 430-148 and 430-149. The full-load current of low speed, high torque and multi-speed motors is to be determined from the name-plate rating instead of from the values listed in the tables. Table 430-150 has an extra note, which deals with synchronous motors. Notice that the

Table 430-150. Full-Load Current*
Three-Phase Alternating-Current Motors

	Induction Type Squirrel-Cage and Wound-Rotor Amperes							Synchronous Type †Unity Power Factor Amperes			
HP	115V	200V	208V	230V	460V	575V	2300V	230V	460V	575V	2300V
½	4	2.3	2.2	2	1	.8					
¾	5.6	3.2	3.1	2.8	1.4	1.1					
1	7.2	4.1	4.0	3.6	1.8	1.4					
1½	10.4	6.0	5.7	5.2	2.6	2.1					
2	13.6	7.8	7.5	6.8	3.4	2.7					
3		11.0	10.6	9.6	4.8	3.9					
5		17.5	16.7	15.2	7.6	6.1					
7½		25.3	24.2	22	11	9					
10		32.2	30.8	28	14	11					
15		48.3	46.2	42	21	17					
20		62.1	59.4	54	27	22					
25		78.2	74.8	68	34	27		53	26	21	
30		92	88	80	40	32		63	32	26	
40		119.6	114.4	104	52	41		83	41	33	
50		149.5	143.0	130	65	52		104	52	42	
60		177.1	169.4	154	77	62	16	123	61	49	12
75		220.8	211.2	192	96	77	20	155	78	62	15
100		285.2	272.8	248	124	99	26	202	101	81	20
125		358.8	343.2	312	156	125	31	253	126	101	25
150		414	396.0	360	180	144	37	302	151	121	30
200		552	528.0	480	240	192	49	400	201	161	40

* These values of full-load current are for motors running at speeds usual for belted motors and motors with normal torque characteristics. Motors built for especially low speeds or high torques may require more running current, and multispeed motors will have full-load current varying with speed, in which case the nameplate current rating shall be used.

† For 90 and 80 percent power factor the above figures shall be multipled by 1.1 and 1.25 respectively.

The voltages listed are rated motor voltages. The currents listed shall be permitted for system voltage ranges of 110 to 120, 220 to 240, 440 to 480, and 550 to 600 volts.

FIGURE 59-4 Reprinted with permission from NFPA 70–1987, National Electrical Code, copyright © 1993, National Fire Protection Association, Quincy, MA 02269. This reprinted material is not the complete and official position of the NFPA on the referenced subject, which is represented only by the standard in its entirety.

right side of Table 430-150 is devoted to the full-load currents of synchronous type motors. The currents listed are for synchronous type motors that are to be operated at unity or 100% power factor. Since synchronous motors are often made to have a leading power factor by over excitation of the rotor current, the full-load current rating must be increased when this is done. If the motor is operated at 90% power factor, the rated full-load current in the table is to be increased by 10%. If the motor is to be operated at 80% power factor, the full-load current is to be increased by 25%.

EXAMPLE: A 150 hp, 460-volt synchronous motor is to be operated at 80% power factor. What will be the full-load current rating of the motor?

$$151 \times 1.25 = 188.75 \text{ or } 189 \text{ amperes}$$

EXAMPLE: A 200 hp, 2300-volt synchronous

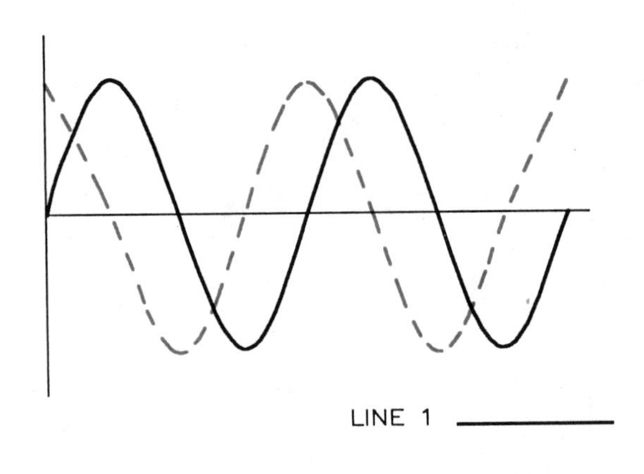

FIGURE 59-5 The voltages of a two-phase system are 90° out of phase with each other

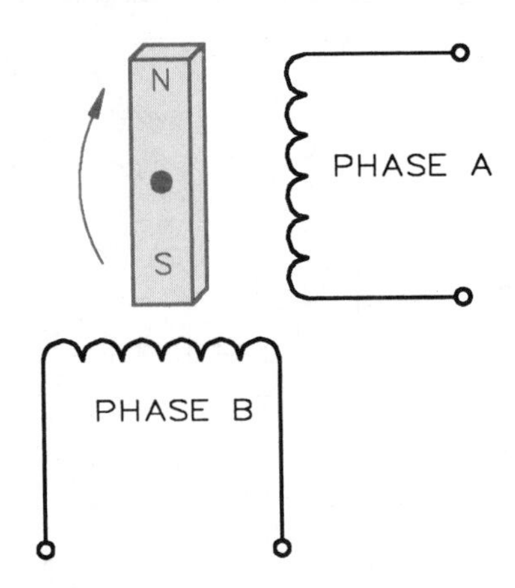

FIGURE 59-6 Two-phase alternator

motor is to be operated at 90% power factor. What will be the full-load current rating of this motor?

Locate 200 horsepower in the far left-hand column. Follow across to the 2300-volt column listed under synchronous type. Increase this value by 10%.

$$40 \times 1.10 = 44 \text{ amperes}$$

Calculation of Conductor Size

The size conductor needed for motor connection is determined by the full-load running current of the motor. Section 430-6a of the *NEC* states that the conductor size shall be determined by Tables 430-147, 430-148, 430-149, and 430-150, instead of motor nameplate current. Section 430-22 states that the conductors supplying a single motor shall have an ampacity of not less than 125% of the motor's full-load current. Section 310 of the NEC can be used to determine the conductor size after the ampacity has been determined. Table 310-16 is shown in figure 59-8.

EXAMPLE: A 30 hp, three-phase squirrel-cage induction motor is connected to a 480-volt line. The conductors are run in conduit to the motor. Copper conductors with THWN insulation are to be used for motor connection. What size conductor must be used?

The first step is to determine the full-load current of the motor. This is determined from Table 430-150. Find 30 hp in the far left-hand column and proceed across to the 460-volt column. The table indicates a current of 40 amperes for this motor. This current must be increased by 25% according to 430-22.

$$40 \times 1.25 = 50 \text{ amperes}$$

Table 310-16 can now be used to determine the conductor size. Locate the column which contains THWN insulation in the copper section of the table. Follow this column down to the nearest amperage that is not less than 50 amps. In this case, it will be 50. Follow this line across to the far left-hand side of the table. A #8 AWG copper conductor will be used.

Overload Size

When determining the overload size for a motor, the nameplate current rating of the motor is to be used instead of the current determined by the tables (430-6a). Other factors, such as the service factor (SF) or temperature rise of the motor, are also used to determine the overload size of a motor. Section 430-32, figure 59-9, of the NEC is used to determine the overload size for motors of 1 hp or more. The overload size is based on a percentage of the full-load current of the motor.

EXAMPLE: A 25 hp, three-phase induction motor has a nameplate rating of 32 amperes. The

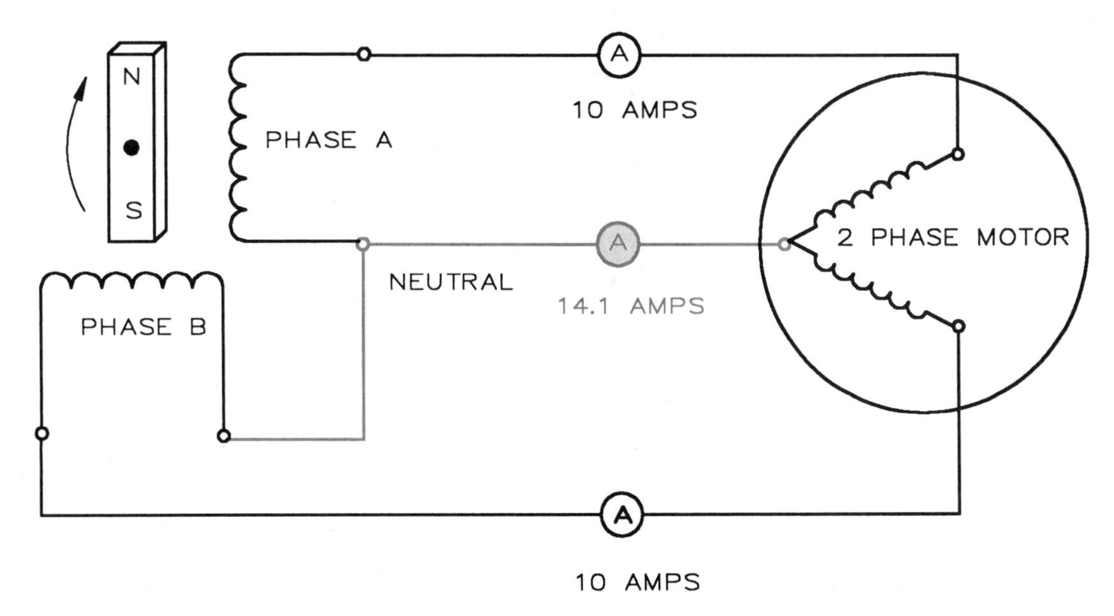

FIGURE 59-7 The neutral conductor must be larger than the phase conductors

nameplate also shows a temperature rise of 30°C. What is the overload size for this motor?

Table 430-32 indicates the overload size is to be 125% of the full-load current rating of the motor.

$$32 \times 1.25 = 40 \text{ amperes}$$

If, for some reason, this overload size does not permit the motor to start without tripping out, Section 430-34 permits the overload size to be increased to a maximum of 140% for this motor. If this increase in overload size does not solve the starting problem, the overload may be shunted out of the circuit during the starting period in accordance with Section 430-35a and b.

Locked-Rotor Current

Code letters located on the motor nameplate range from type A to type V. They indicate the type of squirrel-cage bars that are used when the rotor is made. Different types of bars are used to make motors designed for different applications. The type of bar largely determines the locked-rotor current of the motor. The locked-rotor current of the motor is used to determine the starting current of the motor. Table 430-7b, figure 59-10, lists the different code letters and gives the locked-rotor kilovolt-amperes rating per horsepower. The starting current can be determined by multiplying the KVA rating by the horsepower of the motor and then dividing by the applied voltage.

EXAMPLE: A 15 hp, three-phase squirrel-cage induction motor with a code letter of K is connected to a 240-volt line. Determine the locked-rotor current.

The table lists 8.0 to 8.99 KVA per horsepower for a code letter of K. An average value of 8.5 KVA will be used.

$$8.5 \times 15 = 127.5 \text{ KVA or } 127{,}500 \text{ volt amperes}$$

$$127{,}500 \div (240\sqrt{3}) = 306.7 \text{ amperes}$$

Short-Circuit Protection

The short-circuit protective device is determined from Table 430-152, figure 59-11. The far left-hand column lists the type of motor and, in the case of squirrel-cage motors, the type of code letter. The four columns to the right of the motor type column list the type of short-circuit protective device to be used and a percentage of the full-load current. The full-load motor current multi-

Table 310-16. Ampacities of Not More than Three Single Insulated Conductors, Rated 0 through 2000 Volts, in Raceway in Free Air and Ampacities of Cable Types AC, NM, NMC and SE

Based on Ambient Air Temperature of 30°C (86°F).

Size	Temperature Rating of Conductor. See Table 310-13.								Size
	60°C (140°F)	75°C (167°F)	85°C (185°F)	90°C (194°F)	60°C (140°F)	75°C (167°F)	85°C (185°F)	90°C (194°F)	
AWG MCM	TYPES †TW, †UF	TYPES †FEPW, †RH, †RHW, †THW, †THWN, †XHHW, †USE, †ZW	TYPE V	TYPES TA, TBS, SA, AVB, SIS, †FEP, †FEPB, †RHH, †THHN, †XHHW*	TYPES †TW, †UF	TYPES †RH, †RHW, †THW, †THWN, †XHHW †USE	TYPE V	TYPES TA, TBS, SA, AVB, SIS, †RHH, †THHN, †XHHW*	AWG MCM
	COPPER				ALUMINUM OR COPPER-CLAD ALUMINUM				
18	....	....	....	14	....	....	....	....	....
16	....	....	18	18	....	....	....	....	....
14	20†	20†	25	25†	....	....	....	....	....
12	25†	25†	30	30†	20†	20†	25	25†	12
10	30	35†	40	40†	25	30†	30	35†	10
8	40	50	55	55	30	40	40	45	8
6	55	65	70	75	40	50	55	60	6
4	70	85	95	95	55	65	75	75	4
3	85	100	110	110	65	75	85	85	3
2	95	115	125	130	75	90	100	100	2
1	110	130	145	150	85	100	110	115	1
1/0	125	150	165	170	100	120	130	135	1/0
2/0	145	175	190	195	115	135	145	150	2/0
3/0	165	200	215	225	130	155	170	175	3/0
4/0	195	230	250	260	150	180	195	205	4/0
250	215	255	275	290	170	205	220	230	250
300	240	285	310	320	190	230	250	255	300
350	260	310	340	350	210	250	270	280	350
400	280	335	365	380	225	270	295	305	400
500	320	380	415	430	260	310	335	350	500
600	355	420	460	475	285	340	370	385	600
700	385	460	500	520	310	375	405	420	700
750	400	475	515	535	320	385	420	435	750
800	410	490	535	555	330	395	430	450	800
900	435	520	565	585	355	425	465	480	900
1000	455	545	590	615	375	445	485	500	1000
1250	495	590	640	665	405	485	525	545	1250
1500	520	625	680	705	435	520	565	585	1500
1750	545	650	705	735	455	545	595	615	1750
2000	560	665	725	750	470	560	610	630	2000
	AMPACITY CORRECTION FACTORS								
Ambient Temp. °C	For ambient temperatures other than 30°C (86°F), multiply the ampacities shown above by the appropriate factor shown below.								Ambient Temp. °F
21-25	1.08	1.05	1.04	1.04	1.08	1.05	1.04	1.04	70-77
26-30	1.00	1.00	1.00	1.00	1.00	1.00	1.00	1.00	79-86
31-35	.91	.94	.95	.96	.91	.94	.95	.96	88-95
36-40	.82	.88	.90	.91	.82	.88	.90	.91	97-104
41-45	.71	.82	.85	.87	.71	.82	.85	.87	106-113
46-50	.58	.75	.80	.82	.58	.75	.80	.82	115-122
51-55	.41	.67	.74	.76	.41	.67	74	.76	124-131
56-60	....	.58	.67	.71	....	.58	.67	.71	133-140
61-70	....	.33	.52	.58	....	.33	.52	.58	142-158
71-80	....	....	.30	.41	....	....	.30	.41	160-176

† Unless otherwise specifically permitted elsewhere in this Code, the overcurrent protection for conductor types marked with an obelisk (†) shall not exceed 15 amperes for 14 AWG, 20 amperes for 12 AWG, and 30 amperes for 10 AWG copper; or 15 amperes for 12 AWG and 25 amperes for 10 AWG aluminum and copper-clad aluminum after any correction factors for ambient temperature and number of conductors have been applied.

* For dry and damp locations only. See 75°C column for wet locations.

FIGURE 59-8 Reprinted with permission from NFPA 70–1987, National Electrical Code, copyright © 1993, National Fire Protection Association, Quincy, MA 02269. This reprinted material is not the complete and official position of the NFPA on the referenced subject, which is represented only by the standard in its entirety.

430-32. Continuous-Duty Motors.

(a) More than 1 Horsepower. Each continuous-duty motor rated more than 1 horsepower shall be protected against overload by one of the following means:

(1) A separate overload device that is responsive to motor current. This device shall be selected to trip or rated at no more than the following percent of the motor nameplate full-load current rating.

Motors with a marked service factor not less than 1.15	125%
Motors with a marked temperature rise not over 40°C	125%
All other motors	115%

Modification of this value shall be permitted as provided in Section 430-34.

FIGURE 59-9 Reprinted with permission from NFPA 70–1987, National Electrical Code, copyright © 1993, National Fire Protection Association, Quincy, MA 02269. This reprinted material is not the complete and official position of the NFPA on the referenced subject, which is represented only by the standard in its entirety.

plied by the percentage factor determines the rating of the short-circuit protective device. The full-load current is to be determined from the motor amperage tables instead of the motor nameplate rating. Once the size of the device has been determined, the closest standard size of fuse or circuit breaker must be chosen from the standard sizes listed in 240-6, figure 59-12. The closest standard size without going *under* the computed size will be used.

Table 430-7(b). Locked-Rotor Indicating Code Letters

Code Letter	Kilovolt-Amperes per Horsepower with Locked Rotor
A	0 — 3.14
B	3.15 — 3.54
C	3.55 — 3.99
D	4.0 — 4.49
E	4.5 — 4.99
F	5.0 — 5.59
G	5.6 — 6.29
H	6.3 — 7.09
J	7.1 — 7.99
K	8.0 — 8.99
L	9.0 — 9.99
M	10.0 — 11.19
N	11.2 — 12.49
P	12.5 — 13.99
R	14.0 — 15.99
S	16.0 — 17.99
T	18.0 — 19.99
U	20.0 — 22.39
V	22.4 — and up

FIGURE 59-10 Reprinted with permission from NFPA 70–1987, National Electrical Code, copyright © 1993, National Fire Protection Association, Quincy, MA 02269. This reprinted material is not the complete and official position of the NFPA on the referenced subject, which is represented only by the standard in its entirety.

Table 430-152. Maximum Rating or Setting of Motor Branch-Circuit Short-Circuit and Ground-Fault Protective Devices

	Percent of Full-Load Current			
Type of Motor	**Nontime Delay Fuse**	**Dual Element (Time-Delay) Fuse**	**Instantaneous Trip Breaker**	*** Inverse Time Breaker**
Single-phase, all types				
No code letter	300	175	700	250
All ac single-phase and polyphase squirrel-cage and synchronous motors† with full-voltage, resistor or reactor starting:				
No code letter	300	175	700	250
Code letter F to V	300	175	700	250
Code letter B to E	250	175	700	200
Code letter A	150	150	700	150
All ac squirrel-cage and synchronous motors† with autotransformer starting:				
Not more than 30 amps				
No code letter	250	175	700	200
More than 30 amps				
No code letter	200	175	700	200
Code letter F to V	250	175	700	200
Code letter B to E	200	175	700	200
Code letter A	150	150	700	150
High-reactance squirrel-cage				
Not more than 30 amps				
No code letter	250	175	700	250
More than 30 amps				
No code letter	200	175	700	200
Wound-rotor —				
No code letter	150	150	700	150
Direct-current (constant voltage)				
No more than 50 hp				
No code letter	150	150	250	150
More than 50 hp				
No code letter	150	150	175	150

For explanation of Code Letter Marking, see Table 430-7(b).
For certain exceptions to the values specified, see Sections 430-52 through 430-54.
* The values given in the last column also cover the ratings of nonadjustable inverse time types of circuit breakers that may be modified as in Section 430-52.
† Synchronous motors of the low-torque, low-speed type (usually 450 rpm or lower), such as are used to drive reciprocating compressors, pumps, etc. that start unloaded, do not require a fuse rating or circuit-breaker setting in excess of 200 percent of full-load current.

FIGURE 59-11 Reprinted with permission from NFPA 70–1987, National Electrical Code, copyright © 1993, National Fire Protection Association, Quincy, MA 02269. This reprinted material is not the complete and official position of the NFPA on the referenced subject, which is represented only by the standard in its entirety.

EXAMPLE: A 100 hp, three-phase induction motor with a code letter J is to be connected to a 240-volt line. A dual-element time delay fuse is to be used as the short-circuit protective device. Determine the size fuse to be used.

Table 430-150 gives a current rating of 248 amperes for a 100 hp, three-phase induction motor

240-6. Standard Ampere Ratings. The standard ampere ratings for fuses and inverse time circuit breakers shall be considered 15, 20, 25, 30, 35, 40, 45, 50, 60, 70, 80, 90, 100, 110, 125, 150, 175, 200, 225, 250, 300, 350, 400, 450, 500, 600, 700, 800, 1000, 1200, 1600, 2000, 2500, 3000, 4000, 5000, and 6000.

Exception: Additional standard ratings for fuses shall be considered 1, 3, 6, 10, and 601.

FIGURE 59-12 Reprinted with permission from NFPA 70–1987, National Electrical Code, copyright © 1993, National Fire Protection Association, Quincy, MA 02269. This reprinted material is not the complete and official position of the NFPA on the referenced subject, which is represented only by the standard in its entirety.

connected to 240 volts. Table 430-152 gives a percent factor of 175 for a motor with code letters F to V.

$$248 \times 1.75 = 434 \text{ amperes}$$

The standard sizes of fuses and circuit breakers listed on 240-6 indicate that the nearest standard rating of fuse without going lower than 434 amps is 450 amperes. The fuses used as the short-circuit protective devices for this motor will be 450 amp dual element time delay fuses. If this fuse cannot withstand the starting current, 430-52 (exceptions 1 and 2) permit the rating of the branch-circuit short-circuit and ground-fault protective device to be increased. The value of a dual-element time delay fuse can be increased to a maximum of 225% of the full-load current.

Example Problems

Determine the conductor size, overload size, and short-circuit protection for the following motors:

1. A 40 hp, 240-volt dc motor has a nameplate current rating of 119 amperes. The conductors are to be copper with THWN insulation. The short-circuit protective device is to be an instantaneous trip circuit breaker. Refer to figure 59-13.

   The conductor size must be determined from the current listed in Table 430-147. This current is to be increased by 125%.

$$140 \times 1.25 = 175 \text{ amperes}$$

Table 310-16 is used to find the conductor size. According to the table, 2/0 conductors will be used.

This overload size is determined from 430-32. Since there is no service factor or temperature rise listed, the heading "All Other Motors" will be used. The motor

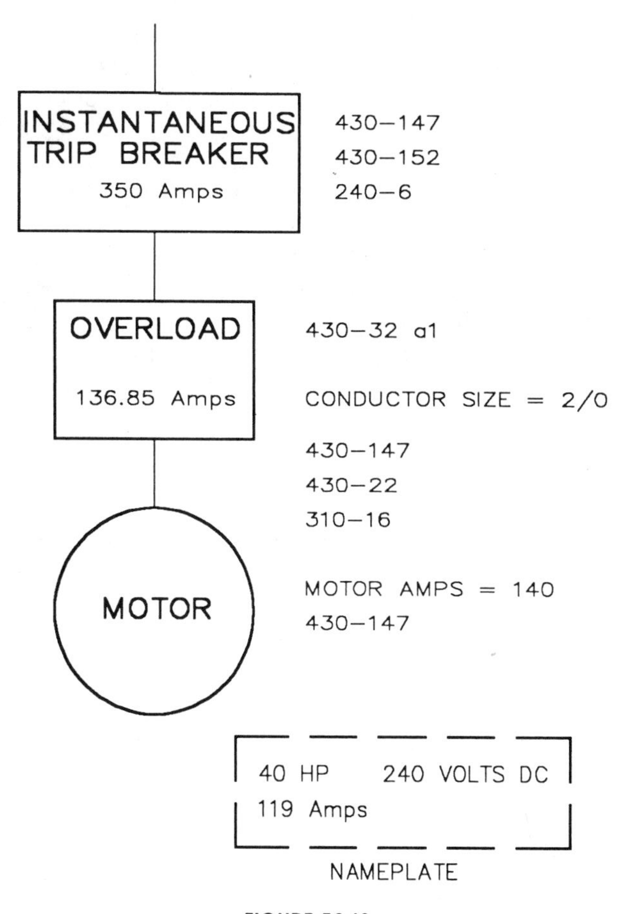

FIGURE 59-13

nameplate current is used instead of the value listed in the motor table when determining overload size.

$$119 \times 1.15 = 136.85 \text{ amperes}$$

The circuit breaker size is determined from Table 430-152. The current rating from Table 430-147 is used instead of the motor nameplate rating. Under direct-current motors not more than 50 hp, the instantaneous trip circuit breaker rating is given at 250%.

$$140 \times 2.50 = 350 \text{ amperes}$$

Since 350 amperes is one of the standard sizes of circuit breakers listed on 240-6, that size breaker will be used as the short-circuit protective device.

2. A 150 hp, three-phase squirrel-cage induction motor is connected to 440 volts. The motor nameplate lists the following information:
   Motor amps 175 SF 125 Code D
   The conductors are to be copper with type THHN insulation.
   The short-circuit protection device is to be an inverse time circuit breaker. Refer to figure 59-14.

The conductor size is determined from Table 430-150 and then increased 25%.

$$180 \times 1.25 = 225 \text{ amperes}$$

Table 310-16 is used to determine conductor size. Type THHN insulation is located in the fourth column. The conductor size will be 3/0 AWG.

The overload size is determined from the nameplate current and 430-32.

$$175 \times 1.25 = 218.75 \text{ amperes}$$

Table 430-152 is used to determine the breaker size. The table indicates a factor of 200% for motors with a code letter of B through E. The value of current from Table 430-150 is used in this computation.

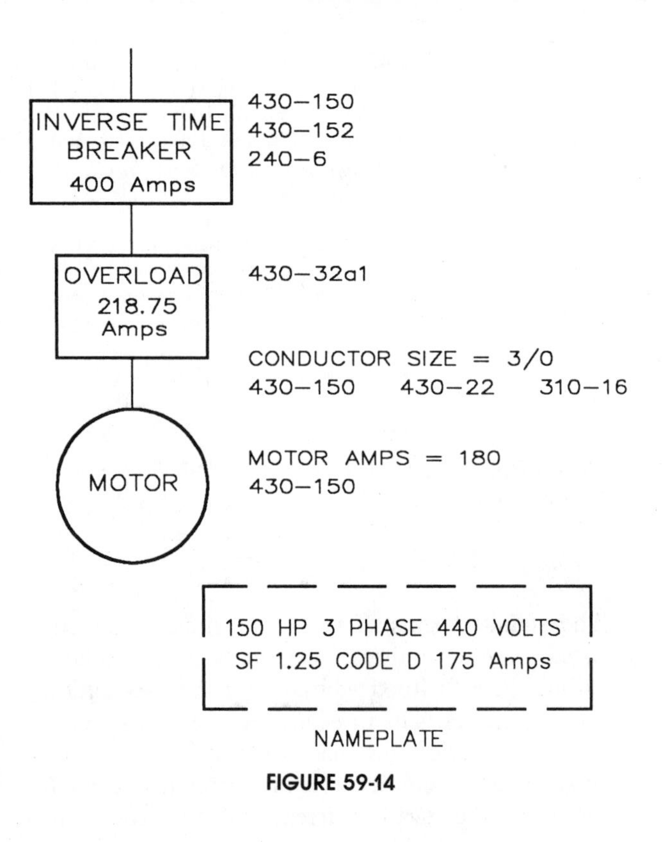

FIGURE 59-14

$$180 \times 2.00 = 360 \text{ amperes}$$

The nearest standard circuit breaker size listed on 240-6 without going under 360 amperes is 400 amps. A 400 ampere inverse time circuit breaker will be used as the short-circuit protective device.

Multiple Motor Calculations

The two primary codes for calculating the main feeder short-circuit protective device and conductor size for a multiple motor connection are 430-62a and 430-24. In the following example, three motors are connected to a common feeder. The feeder is 440-volts three-phase, and the conductors are to be copper with type THHN insulation. Each motor is to be protected with a non-time delay fuse and a separate overload device. The main feeder is also protected by a non-time delay fuse. The motor nameplate ratings are as follows:

	MOTOR #1	MOTOR #2	MOTOR #3
Phase	3	3	3
SF	1.25	—	—
Volts	440	440	440
Type	Induction	Induction	Synchronous
Hp	20	60	100
Code	C	J	A
Amps	23	72	96
Temp	—	40°C	—
PF	—	—	90%

Motor #1 Calculations

The first step is to calculate the values for amperage, conductor size, overload size, and short-circuit protective device size for each motor. Assume that each motor is supplied with copper conductors with type THHN insulation. The values for motor #1 are shown in figure 59-15. The motor amperage rating from Table 430-150 is used to determine the conductor and fuse size. The amperage rating must be increased by 25% for the conductor size.

$$27 \times 1.25 = 33.75 \text{ amps}$$

The conductor size is now chosen from 310-16. The table indicates a #10 AWG conductor can be used for this connection. The note at the bottom of the table, however, states that the overcurrent protective device cannot exceed a maximum rating of 30 amperes for a #10 AWG conductor. For this reason, a #8 AWG conductor will be used.

The overload size is computed for the nameplate current. The demand factors in 430-32a1 are used for the overload calculation.

$$23 \times 1.25 = 28.75 \text{ amps}$$

The fuse size is determined by using the motor current rating from Table 430-150 and the demand factor from Table 430-152.

$$27 \times 250\% = 67.5 \text{ amps}$$

The nearest standard fuse size listed in 240-6 is 70 amps. A 70 amp fuse will be used.

Motor #2 Calculation

Figure 59-16 illustrates an example for the calculation of motor #2. Table 430-150 lists a full-load current of 77 amps for this motor. This value of current is increased by 25% for the calculation of the conductor current.

$$77 \times 1.25 = 96.25 \text{ amps}$$

Table 310-16 indicates that a #3 AWG conductor should be used for this motor connection.

The overload size is determined from 430-32a1. The motor nameplate lists a temperature rise of 40°C for this motor. The nameplate current will be increased by 25%.

$$72 \times 1.25 = 90 \text{ amps}$$

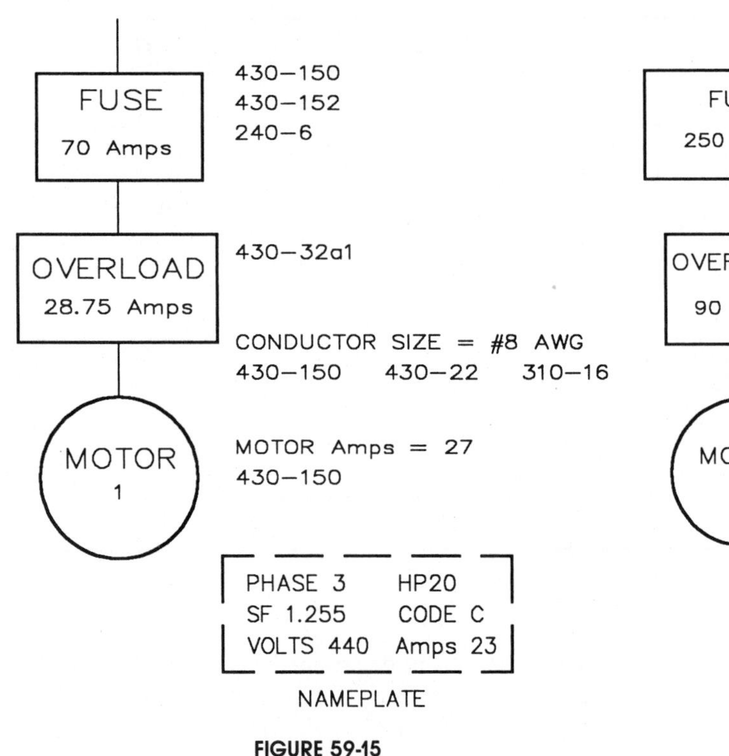

FIGURE 59-15

The fuse size will be determined from Table 430-152.

$$77 \times 300\% = 231 \text{ amps}$$

The nearest standard fuse size listed in 240-6 is 250 amps. 250 amp fuses will be used to protect this circuit.

Motor #3 Calculations

Motor #3 is a synchronous motor intended to operate with a 90% power factor. Figure 59-17 shows an example of this calculation. The current listed in 430-150 must be increased by 10% to find the full-load running current.

$$101 \times 1.10 = 111 \text{ amperes}$$

The conductor size is computed by using this current rating and increasing it by 25%.

$$111 \times 1.25 = 138.75 \text{ amperes}$$

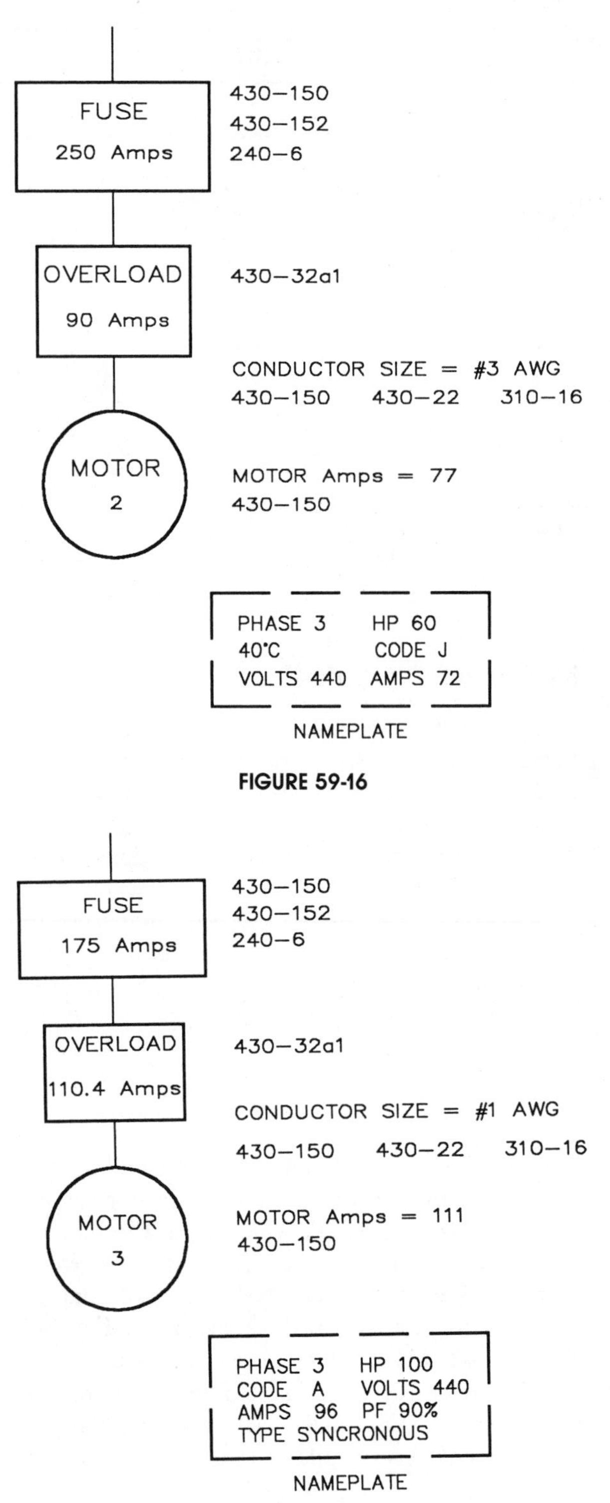

FIGURE 59-17

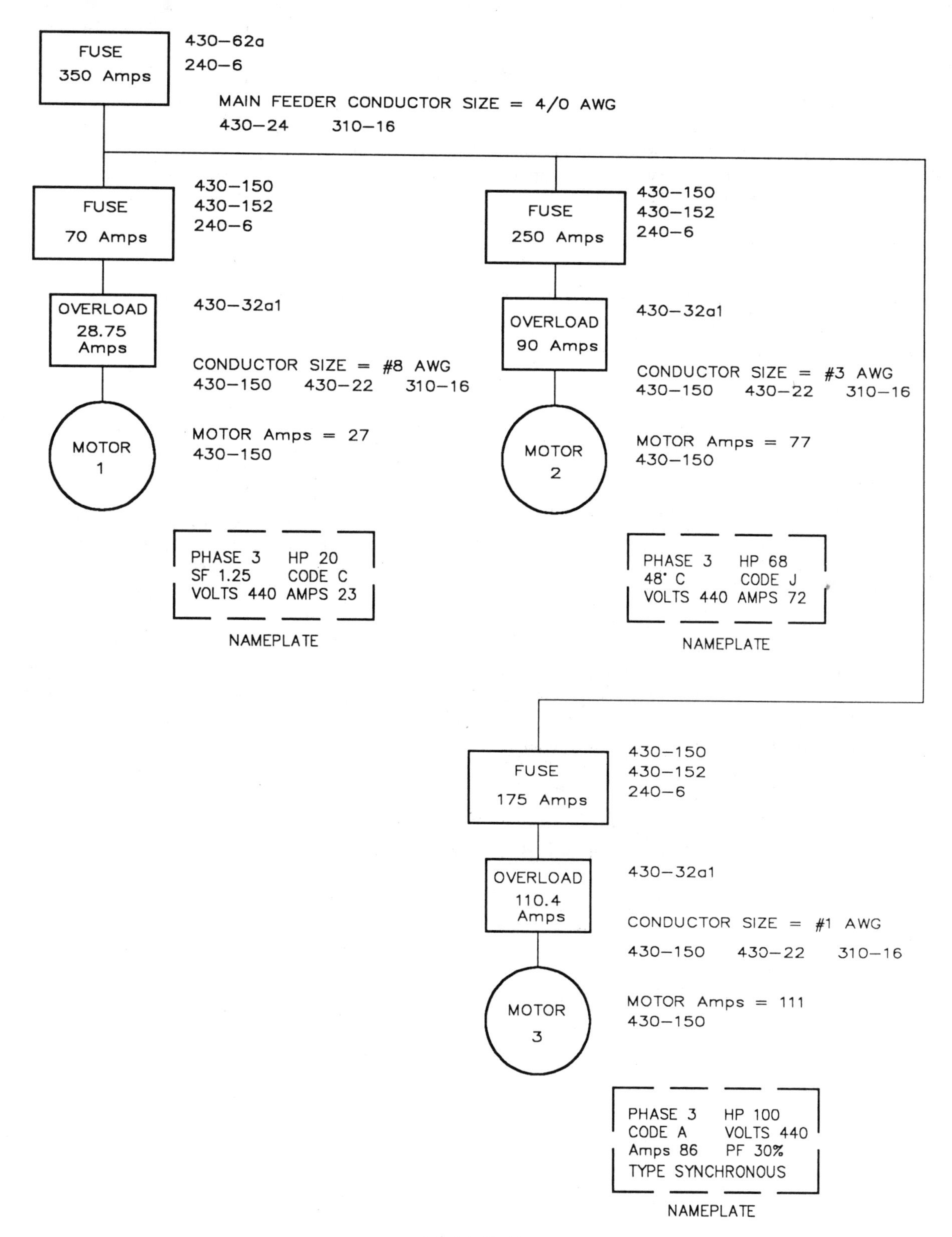
FUSE
350 Amps
430—62a
240—6
MAIN FEEDER CONDUCTOR SIZE = 4/0 AWG
430—24 310—16
FUSE
70 Amps
430—150
430—152
240—6
OVERLOAD
28.75
Amps
430—32a1
CONDUCTOR SIZE = #8 AWG
430—150 430—22 310—16
MOTOR
1
MOTOR Amps = 27
430—150
PHASE 3 HP 20
SF 1.25 CODE C
VOLTS 440 AMPS 23
NAMEPLATE
FUSE
250 Amps
430—150
430—152
240—6
OVERLOAD
90 Amps
430—32a1
CONDUCTOR SIZE = #3 AWG
430—150 430—22 310—16
MOTOR
2
MOTOR Amps = 77
430—150
PHASE 3 HP 68
48° C CODE J
VOLTS 440 AMPS 72
NAMEPLATE
FUSE
175 Amps
430—150
430—152
240—6
OVERLOAD
110.4
Amps
430—32a1
CONDUCTOR SIZE = #1 AWG
430—150 430—22 310—16
MOTOR
3
MOTOR Amps = 111
430—150
PHASE 3 HP 100
CODE A VOLTS 440
Amps 86 PF 30%
TYPE SYNCHRONOUS
NAMEPLATE

FIGURE 59-18

Table 310-16 indicates that a #1 AWG conductor will be used for this circuit.

This motor does not have a marked service factor or a marked temperature rise. The overload size will be calculated by increasing the nameplate current by 15% as indicated by 430-32a1.

$$96 \times 1.15 = 110.4 \text{ amperes}$$

The fuse size is determined from Table 430-152. The motor has a code letter of A which corresponds to an increase of 150%.

$$111 \times 1.50 = 166.5 \text{ amperes}$$

The nearest standard size fuse listed in 240-6 is 175 amps. 175 amp fuses will be used to protect this circuit.

Main Feeder Calculation

An example of the main feeder connection is shown in figure 59-18. The conductor size is computed by increasing the largest amperage rating of the motors connected to the feeder by 25% and then adding the amp rating of the other motors to this amount. In this example, the 100 hp synchronous motor has the largest running current. This current will be increased by 25%, and then the running currents, as determined from Table 430-150, will be added.

$$111 \times 1.25 = 138.75 \text{ amps}$$
$$138.75 + 77 + 27 = 242.75 \text{ amps}$$

Table 310-16 indicates that a #4/0 conductor is to be used as the main feeder conductor.

The size of the short-circuit protective device is determined by 430-62a. The code states that the rating or setting of the short-circuit protective device shall not be *greater* than the largest rating or setting of the branch-circuit short-circuit and ground-fault protective device, plus the sum of the full-load currents of the other motors. The largest fuse size was that of the 60 hp induction motor. The fuse calculation for this motor was 231 amperes. The running currents of the other two motors will be added to this value to determine the rating of the fuse for the main feeder.

$$231 + 111 + 27 = 369 \text{ amperes}$$

The closest standard fuse size listed in 240-6 is 350 amps. 350 amp fuses will be used as the short-circuit protective devices for this circuit.

REVIEW QUESTIONS

(NOTE: It will be necessary to use the *NEC* to answer some of the following questions.)

1. A 20 hp, dc motor is connected to a 500-volt dc line. What is the full-load running current of the motor?
2. What rating is used to determine the full-load current of a torque motor?
3. A 3/4 hp, single-phase motor is connected to a 208-volt ac line. What is the full-load current rating of the motor?
4. A 30 hp, two-phase motor is connected to a 230 volt ac line. What is the rated current of the phase conductors and the rated current of the neutral conductor?

   ____________ amp(s) phase ____________ amp(s) neutral
5. A 5 hp, three-phase adjustable voltage motor is connected to a maximum voltage of 230-volts three phase. No current is listed on the nameplate of the motor. What will be the full-load current used to determine the conductor size for this motor?

6. A 125 hp synchronous motor is connected to a 208-volt three-phase line. The motor is to be operated at a power factor of 80%. What will be the full-load current rating of this motor?
7. Must the terminal housing of a motor be made of metal?
8. A hermetic refrigerant compressor motor is started by connecting only part of its winding to the line to reduce starting current. Is this motor considered a part-winding motor?
9. How is the amount of full-load current used to determine conductor size for a hermetic refrigerant compressor determined?
10. What is the full-load current rating of a three-phase, 50 hp induction motor connected to a three-phase 600-volt line?
11. A 125 hp, three-phase squirrel-cage induction motor is connected to 560 volts. The nameplate current is 115 amperes. It has a marked temperature rise of 40°C and a code letter of J. The conductors are to be copper and run in conduit. The insulation of the conductors is to be type THHN. The short-circuit protective device is to be a dual-element time delay fuse. Find the conductor size, overload size, and fuse size for this motor.
    Conductor: ___________
    Overload: ___________
    Fuse: ___________
12. A 7.5 hp, single-phase motor is connected to 120 volts. The motor has a code letter of C. The nameplate current rating is 76 amperes. The conductors are to be copper with type TW insulation. The overcurrent device is to be a non-time delay fuse. Find the conductor, overload, and fuse size for this motor.
    Conductor: ___________
    Overload: ___________
    Fuse: ___________
13. The secondary current of a continuous duty wound rotor motor has a rated current draw of 63 amperes. The conductors connecting the conductors to the controller are copper and have type THWN insulation. What size must these conductors be?
14. A 1/3 hp, single-phase ac motor is connected to 120 volts. The motor has a thermal protector integral with it. What is the maximum setting (in amps) for this overload device?
15. A 75 hp, three-phase synchronous motor is connected to a 208-volt line. The motor is to be operated at 80% power factor. The motor nameplate lists a full-load current of 185 amperes, a temperature rise of 40° C, and a code letter A. The conductors are to be made of copper and have type THHN insulation. The short-circuit protective device is to be an inverse time circuit breaker. Determine the conductor, overload, and circuit breaker size for this motor.
    Conductor: ___________
    Overload: ___________
    Circuit breaker: ___________
16. How many thermal overload units must be used to protect a three-phase motor?
17. A 50 hp, three-phase squirrel-cage induction motor is connected to a 240-volt line. The motor has a code letter B. An inverse time circuit breaker is used

as the short-circuit protective device. The circuit breaker size is found to be sufficient to start the motor. What is the maximum size inverse time circuit breaker that can be used as the short-circuit protection for this motor?

18. Where the setting of an instantaneous trip circuit breaker is not sufficient for the starting current of the motor, it is permissible to increase the size of the circuit breaker. To what maximum size can this circuit breaker be increased?
19. In a large capacity installation, where heavy capacity feeders are installed to provide for future additions or changes, what determines the rating or setting of the feeder protective device?
20. Must a motor and its controller be supplied by the same disconnecting means?
21. What is the minimum torque of pressure terminals used with #14 AWG or smaller conductors when these conductors are used with control-circuit devices?
22. A 5 hp, three-phase motor is ten inches in diameter. What is the minimum volume of the terminal housing if the motor uses wire-to-wire connections?
23. Is it permissible to attach the equipment grounding connection outside the motor terminal housing?
24. When determining the highest rated motor of a multimotor connection, is the nameplate current rating used or the current ratings listed in Tables 430-147 through 430-150?
25. Three motors are connected to a single-branch circuit. The motors are connected to a 440-volt three-phase line. Motor #1 is a 50 hp induction motor with a code letter of B. Motor #2 is 40 hp with a code letter of H, and motor #3 is 50 hp with a code letter of J. Determine the size of the branch-circuit conductor supplying these three motors if the conductor is made of copper with type THHN insulation.
26. If the short-circuit protective device supplying the motors in question #5 is an inverse time circuit breaker, what size breaker should be used?
27. Is it permissible to use the service switch as the disconnecting means for a single motor?
28. For small motors not covered by Tables 430-147 through 430-150, how is the locked-rotor current determined?
29. Five 5 hp, three-phase induction motors, having a code letter J, are connected to a 240-volt line. The conductors are copper and have type THWN insulation. What size conductor should be used as the main feeder to supply all of these motors?
30. If a dual-element time delay fuse is used as the short-circuit protective device for the main feeder circuit supplying the motors in the preceding question, what size fuse should be used?

SECTION 7

Motor Drives

Direct Drives and Pulley Drives

Objectives *After studying this unit, the student will be able to:*

- State the advantages of direct and pulley drives
- Install directly coupled motor drives and pulley motor drives
- Check the alignment of the motor and machine shafts, both visually and with a dial indicator
- Install motors and machines in the proper positions for maximum efficiency
- Calculate pulley sizes using the equation:

$$\frac{\text{Drive revolutions per minute}}{\text{Driven revolutions per minute}} = \frac{\text{Driven Pulley Diameter}}{\text{Drive Pulley Diameter}}$$

DIRECTLY COUPLED DRIVE INSTALLATION

The most economical speed for an electrical motor is about 1800 revolutions per minute. Most electrically driven constant speed machines, however, operate at speeds below 1800 rpm. These machines must be provided with either a high-speed motor and some form of mechanical speed reducer, or a low-speed, directly coupled motor.

Synchronous motors can be adapted for direct coupling to machines operating at speeds from 3600 rpm to about 80 rpm, with horsepower ratings ranging from 20 to 5000 and above. It has been suggested that synchronous motors are less expensive to install than squirrel cage motors if the rating exceeds one hp. However, this recommendation considers only the first cost. It does not take into account the higher efficiency and better power factor of the synchronous motor. When the motor speed matches the machine input shaft speed, a simple mechanical coupling is used, preferably a flexible coupling.

Trouble-free operation can usually be obtained by following several basic recommendations for the installation of directly coupled drives, and pulley or chain drives. First, the motor and machine must be installed in a level position. When connecting the motor to its load, the alignment of the devices must be checked more than once from positions at right angles to each other. For example, when viewed from the side, two shafts may appear to be in line. When the same shafts are viewed from the top, as shown in figure 60-1, it is evident that the motor shaft is at an angle to the other shaft. A dial indicator should be used to check the alignment of the motor and the driven machinery, figure 60-2. If a dial indicator is not available, a feeler gauge may be used.

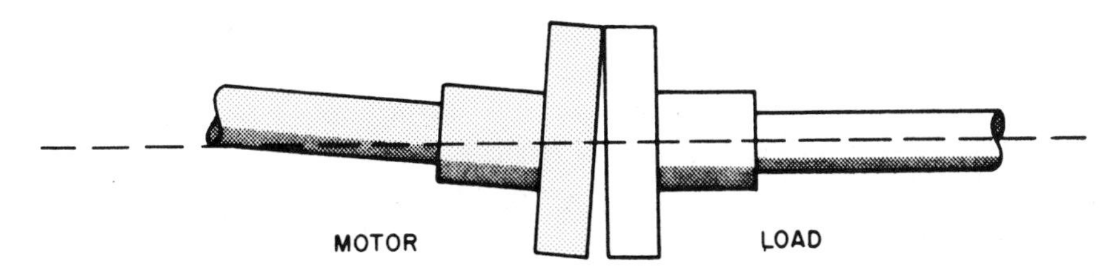

FIGURE 60-1 Every alignment check must be made from positions ninety degrees apart, or at right angles to each other

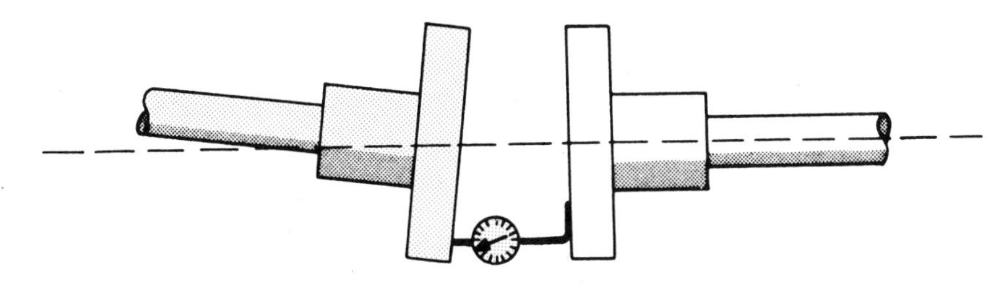

FIGURE 60-2 Angular check of direct motor couplings

During the installation, the shafts of the motor and the driven machine must be checked to insure that they are not bent. Both machine and motor should be rotated together, just as they rotate when the machine is running, and then rechecked for alignment. After the angle of the shafts is aligned, the shafts may appear to share the same axis. However, as shown in figure 60-3, the axes of the motor and the driven machine may really be off center. When viewed again from a position ninety degrees away from the original position, it can be seen that the shafts are not on the same axis.

To complete the alignment of the devices, the motor should be moved until rotation of both shafts shows that they share the same axis when viewed from four positions spaced ninety degrees apart around the shafts. The final test is to check the starting and running currents with the connected load to insure that they do not exceed specifications.

There are several disadvantages to the use of low-speed, directly coupled induction motors. They usually have a low power factor and low efficiency. Both of these characteristics increase electric power costs. Because of this, induction

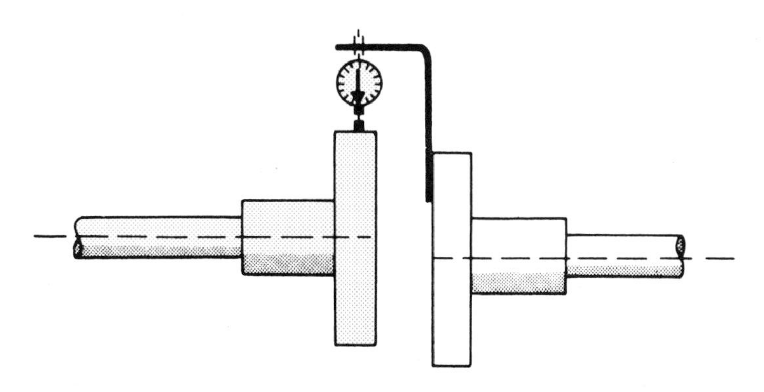

FIGURE 60-3 Axis alignment of direct motor couplings

motors are rarely used for operation at speeds below 500 rpm.

Constant speed motors are available with a variety of speed ratings. The highest possible speed is generally selected to reduce the size, weight, and cost of the motor. At five horsepower, a 1200 rpm motor is almost 50 percent larger than an 1800 rpm motor. At 600 rpm the motor is well over twice as large as the 1800 rpm motor. In the range from 1200 rpm to 900 rpm, the size and cost disadvantages may not be overwhelming factors. Where this is true, low-speed, directly coupled motors can be used. For example, this type of motor is used on most fans, pumps, and compressors.

PULLEY DRIVES

Installation

Flat belts, V-belts, chains, or gears are used on motors so that smooth speed changes at a constant rpm can be achieved. For speeds below 900 rpm, it is practical to use an 1800 rpm or 1200 rpm motor connected to the driven machine by a V-belt or a flat belt.

Machine shafts and bearings give long service when the power transmission devices are properly installed according to the manufacturer's instructions.

Offset drives, such as V-belts, gears, and chain drives, can be lined up more easily than direct drives. Both the motor and load shafts must be level. A straightedge can be used to insure that the motor is aligned on its axis and that it is at the proper angular position so that the pulley sheaves of the motor and the load are in line, figure 60-4. When belts are installed, they should be tightened just enough to assure nonslip engagement. The less cross tension there is, the less wear there will be on the bearings involved. Proper and firm positioning and alignment are necessary to control the forces that cause vibration and the forces that cause thrust.

The designer of a driven machine usually determines the motor mount and the type of drive to be used. This means the installer has little choice in the motor location. In many flat belt or V-belt applications, however, the construction or maintenance electrician may be called upon to make several choices. If a choice can be made, the motor should be placed where the force of gravity helps to increase the grip of the belts. A vertical drive can cause problems because gravity tends to pull the belts away from the lower sheave, figure 60-5. To counteract this action, the belts require far more tension than the bearings should have to withstand. The electrician should avoid this type of installation, if possible.

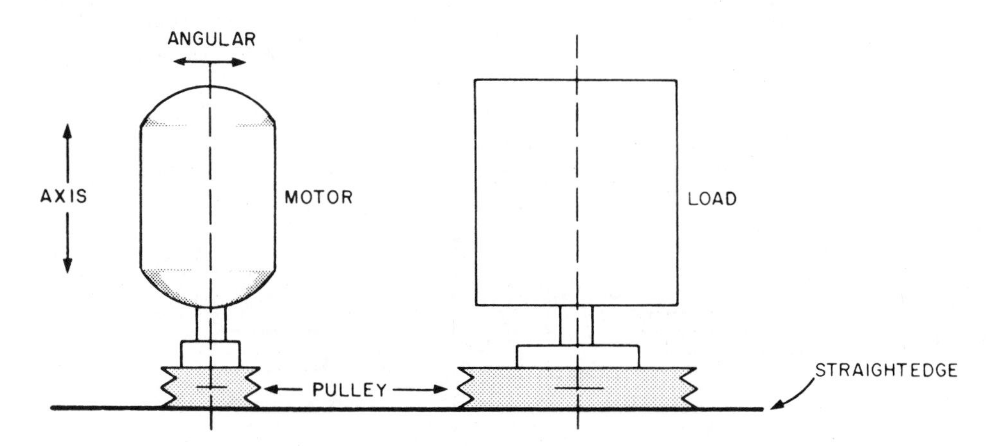

FIGURE 60-4 Using straightedge for angular and axis alignment

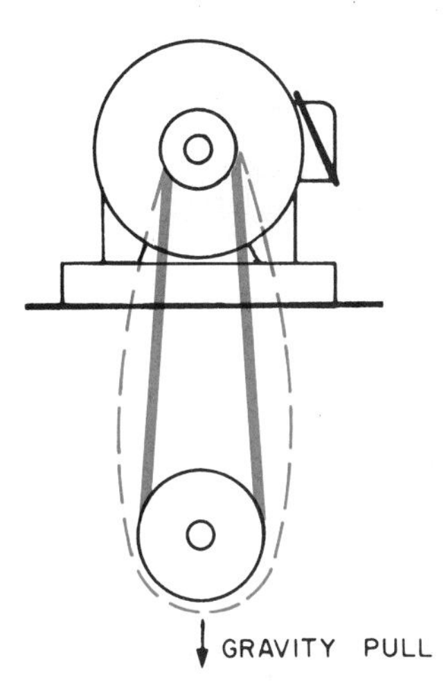

FIGURE 60-5 Greater belt tension is required in vertical drives where gravity opposes good belt traction. This greater tension generally exceeds that which the bearings should withstand.

There are correct and incorrect placements for horizontal drives. The location where the motor is to be installed can be determined from the direction of rotation of the motor shaft. It is recommended that the motor be placed with the direction of rotation so that the belt slack is on top. In this position, the belt tends to wrap around the sheaves. This problem is less acute with V-belts than it is with flat belts or chains. Therefore, if rotation is to take place in both directions, V-belts should be used. In addition, the motor should be placed on the side of the most frequent direction of rotation, figure 60-6.

Pulley Speeds

Motors and machines are frequently shipped without pulleys or with pulleys of incorrect sizes. The drive and driven speeds are given on the motor and machine nameplates, or in the descriptive literature accompanying the machines.

Four quantities must be known if the machinery is to be set up with the correct pulley sizes: the drive revolutions per minute, the driven revolutions per minute, the diameter of the drive pulley, and the diameter of the driven pulley. If three of these quantities are known, the fourth quantity can be determined. For example, if a motor runs at 3600 rpm, the driven speed is 400 rpm, and there is a four-inch pulley for the motor, the size of the pulley for the driven load can be determined from the following equation.

$$\frac{\text{Drive rpm}}{\text{Driven rpm}} = \frac{\text{Driven Pulley Diameter}}{\text{Drive Pulley Diameter}}$$

$$\frac{3600}{400} = \frac{X}{4} \text{ or } \frac{3600}{400} = \frac{X}{4}$$

By cross multiplying and then dividing, we arrive at the pulley size required:

$$4x = 144$$

$$x = \frac{144}{4} = 36\text{-inch diameter pulley}$$

If both the drive and driven pulleys are missing, the problem can be solved by estimating a reasonable pulley diameter for either pulley and then using the equation with this value to find the fourth quantity.

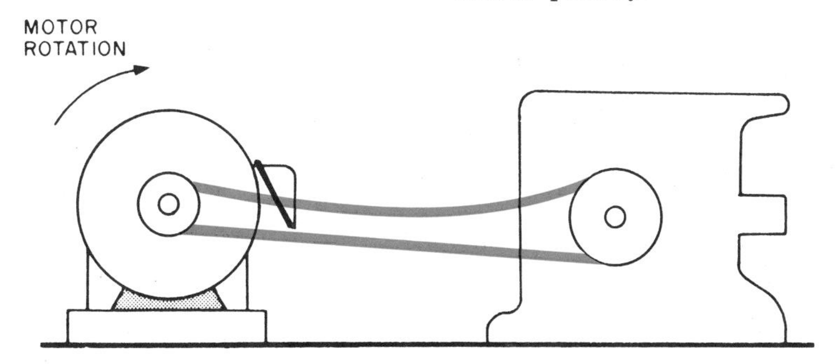

FIGURE 60-6 Direction of motor rotation can be used to advantage for good belt traction (with slack at top) if motor is placed in proper position.

REVIEW QUESTIONS

1. What are the disadvantages of low-speed, directly coupled induction motors?
2. What type of ac motor can be directly coupled at low rpm with larger horsepower ratings?
3. What three alignment checks should be made to insure satisfactory and long service for directly coupled and belt-coupled power transmissions?
4. What special tools, not ordinarily carried by an electrician, are required to align a directly coupled, motor-generator set?
5. Induction motors should not be used below a certain speed (rpm). What is this speed?
6. What is the primary reason for using a pulley drive?
7. How tightly should V-belts be adjusted?
8. Refer to figure 60-6 and assume the motor is to rotate in the opposite direction. How can the belt slack be maintained on top?
9. A machine delivered for installation has a two-inch pulley on the motor and a six-inch pulley on the load. The motor nameplate reads 180 rpm. At what speed in rpm will the driven machine rotate?

SECTION 8

Solid-State Motor Control

UNIT 61

Digital Logic

Objectives

After studying this unit, the student will be able to:

- **Discuss similarities between digital logic circuits and relay logic circuits**
- **Discuss different types of digital logic circuits**
- **Recognize gate symbols used for computer logic circuits**
- **Recognize gate symbols used for NEMA logic circuits**
- **Complete a truth table for the basic gates**

The electrician in today's industry must be familiar with solid-state digital logic circuits. Digital, of course, means a device that has only two states, on or off. Most electricians have been using digital logic for many years without realizing it. Magnetic relays, for instance, are digital devices. Relays are generally considered to be single-input, multi-output devices. The coil is the input and the contacts are the output. A relay has only one coil, but it may have a large number of contacts, figure 61-1.

Although relays are digital devices, the term "digital logic" has come to mean circuits that use solid-state control devices known as gates. There are five basic types of gates: the AND, OR, NOR, NAND, and INVERTER. Each of these gates will be covered later in this text.

There are also different types of logic. For instance, one of the earliest types of logic to appear was *RTL which stands for resistor-transistor logic.* This was followed by *DTL which stands for diode-transistor logic,* and *TTL which stands for transistor-transistor logic.* RTL and DTL are not used much anymore, but TTL is still used to a fairly large extent. TTL can be identified because it operates on 5 volts.

Another type of logic frequently used in industry is *HTL which stands for high-transit logic.* HTL is used because it does a better job of ignoring the voltage spikes and drops caused by the

FIGURE 61-1 Magnetic relay (Courtesy EATON Corp., Cutler-Hammer Products)

starting and stopping of inductive devices such as motors. HTL generally operates on 15 volts.

Another type of logic that has become very popular is CMOS, which has very high input impedance. *CMOS comes from COSMOS which means complementary-symmetry metal-oxide-semiconductor*. The advantage of CMOS logic is that it requires very little power to operate, but there are also some disadvantages. One disadvantage is that CMOS logic is so sensitive to voltage that the static charge of a person's body can sometimes destroy an IC just by touching it. People that work with CMOS logic often use a ground strap which straps around the wrist like a bracelet. This strap is used to prevent a static charge from building up on the body.

Another characteristic of CMOS logic is that unused inputs cannot be left in an indeterminate state. Unused inputs must be connected to either a high state or a low state.

THE AND GATE

While magnetic relays are single-input, multi-output devices, gate circuits are multi-input, single-output devices. For instance, an AND gate may have several inputs, but only one output. Figure 61-2 shows the USASI symbol for an AND gate with three inputs, labeled A, B, and C, and one output, labeled Y.

USASI symbols are more commonly referred to as computer logic symbols. Unfortunately for industrial electricians, there is another system known as NEMA logic which uses a completely different set of symbols. The NEMA symbol for a three-input AND gate is shown in figure 61-3.

Although both symbols mean the same thing, they are drawn differently. Electricians working in industry must learn both sets of symbols because both types of symbols are used. Regardless of which type of symbol is used, the AND gate operates the same way. An AND gate must have all of its inputs high in order to get an output. If it is assumed that TTL logic is being used, a high level is considered to be +5 volts and a low level is considered to be 0 volts. Figure 61-4 shows the truth table for a two-input AND gate.

The truth table is used to illustrate the state of a gate's output with different conditions of input. The number one represents a high state and zero represents a low state. Notice in figure 61-4 that the output of the AND gate is high only when

A	B	Y
0	0	0
0	1	0
1	0	0
1	1	1

FIGURE 61-4 Truth table for a two-input AND gate

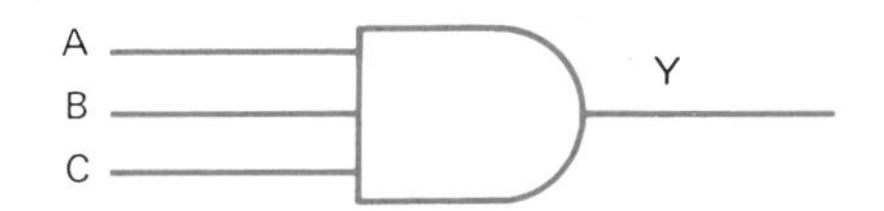

FIGURE 61-2 USASI symbol for a three-input AND gate

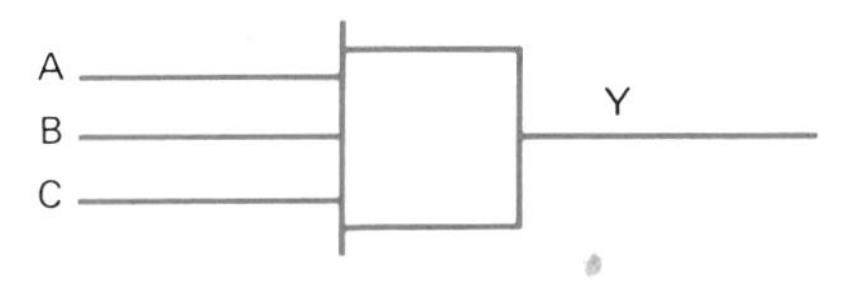

FIGURE 61-3 NEMA logic symbol for a three-input AND gate

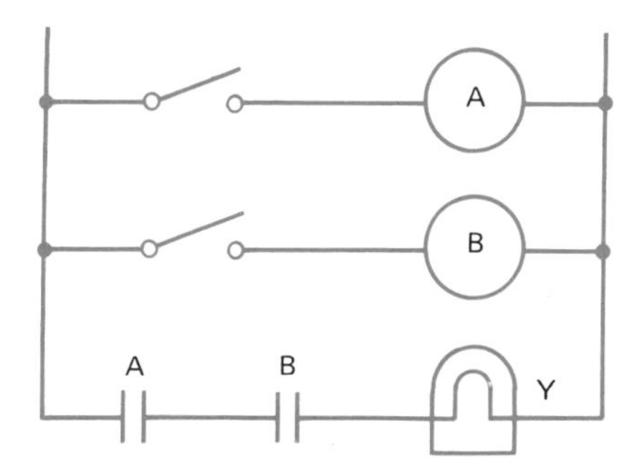

FIGURE 61-5 Relay equivalent circuit for a two-input AND gate

A	B	C	Y
0	0	0	0
0	0	1	0
0	1	0	0
0	1	1	0
1	0	0	0
1	0	1	0
1	1	0	0
1	1	1	1

FIGURE 61-6 Truth table for a three-input AND gate

both of its inputs are high. The operation of the AND gate is very similar to that of the simple relay circuit shown in figure 61-5.

If a lamp is used to indicate the output of the AND gate, both relay coils A and B must be energized before there can be an output. Figure 61-6 shows the truth table for a three-input AND gate. Notice that there is still only one condition that permits a high output for the gate, and that condition is when all inputs are high or at logic level one. *When using an AND gate, any zero input = a zero output.* An equivalent relay circuit for a three-input AND gate is shown in figure 61-7.

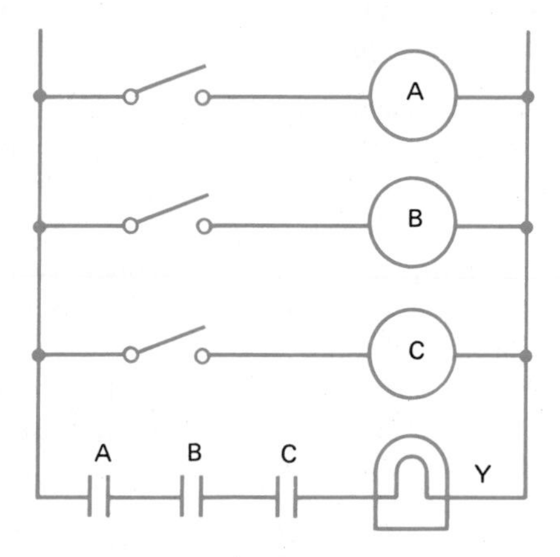

FIGURE 61-7 Relay equivalent circuit for a two-input AND gate

THE OR GATE

The computer logic symbol and the NEMA logic symbol for the OR gate are shown in figure 61-8. The OR gate has a high output when either

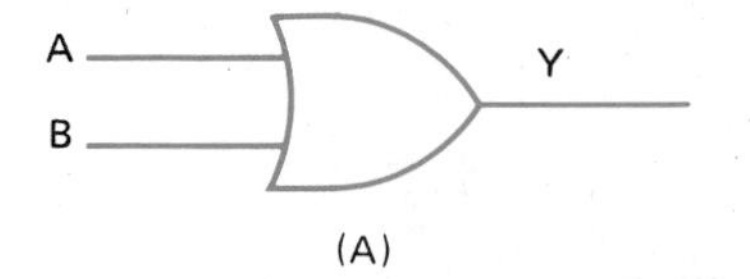

(A)

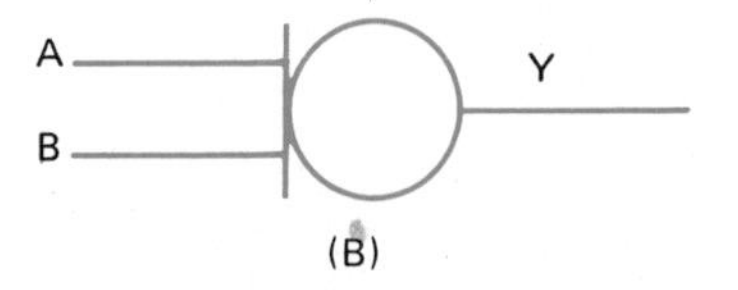

(B)

FIGURE 61-8 (A) Computer logic symbol for an OR gate (B) NEMA logic symbol for an OR gate

or both of its inputs are high. Refer to the truth table shown in figure 61-9. *An easy way to remember how an OR gate functions is to say that any one input = a one output.* An equivalent relay circuit for the OR gate is shown in figure 61-10. Notice in this circuit that if either or both of the relays are energized, there will be an output at Y.

Another gate which is very similar to the OR gate is known as an EXCLUSIVE OR gate. The symbol for an EXCLUSIVE OR gate is shown in figure 61-11. The EXCLUSIVE OR gate has a high output when either, but not both, of its inputs are high. Refer to the truth table shown in figure 61-12. An equivalent relay circuit for the EXCLUSIVE OR gate is shown in figure 61-13. Notice that if both relays are energized or de-energized at the same time, there is no output.

THE INVERTER

The simplest of all the gates is the INVERTER. The INVERTER has one input and one output. As its name implies, *the output is inverted, or the opposite of the input.* For example, if the input is high, the output is low, or if the input is low, the output is high. Figure 61-14 shows the computer logic and NEMA symbols for an INVERTER.

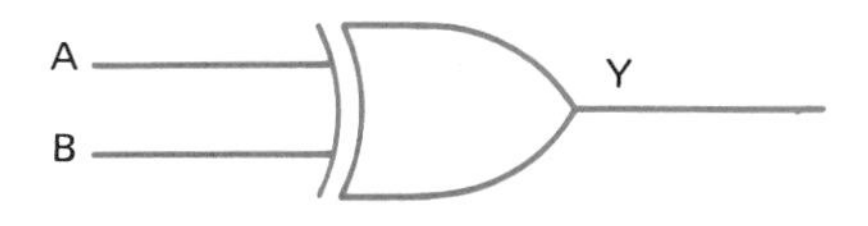

FIGURE 61-11

A	B	Y
0	0	0
0	1	1
1	0	1
1	1	1

FIGURE 61-9 Truth table for a two-input OR gate

A	B	Y
0	0	0
0	1	1
1	0	1
1	1	0

FIGURE 61-12 Truth table for an EXCLUSIVE OR gate

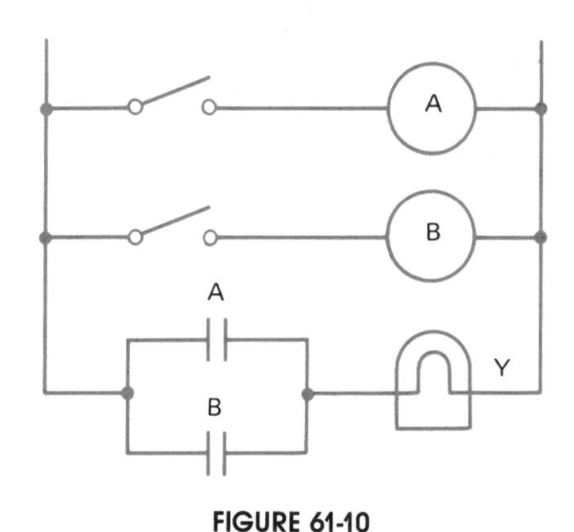

FIGURE 61-10

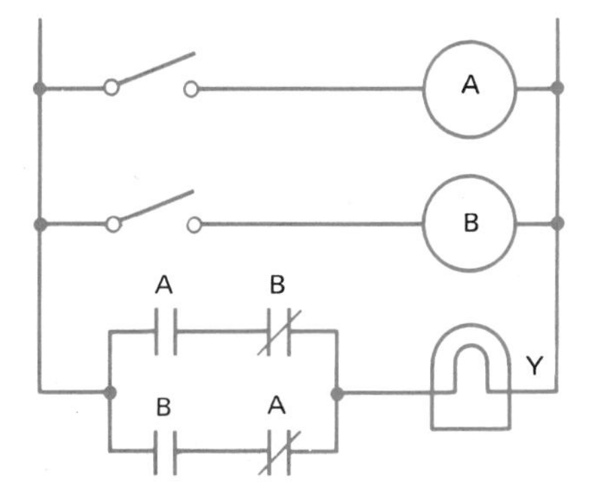

FIGURE 61-13 Equivalent relay circuit for an EXCLUSIVE OR gate

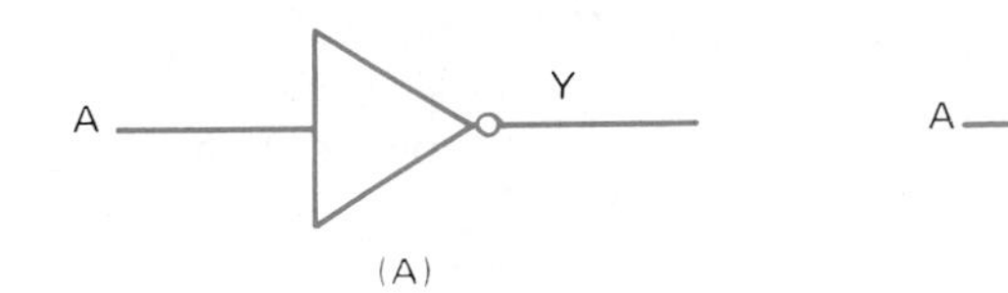

FIGURE 61-14 (A) Computer logic symbol for an INVERTER (B) NEMA logic symbol for an INVERTER

A	Y
0	1
1	0

FIGURE 61-15 Truth table for an INVERTER

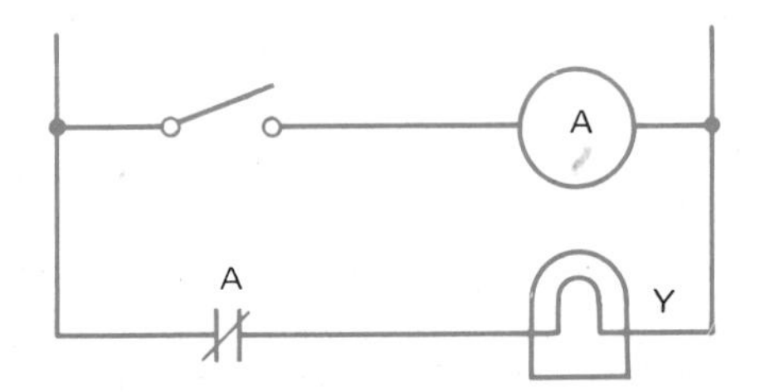

FIGURE 61-16 Equivalent relay circuit for an INVERTER

In computer logic, a circle drawn on a gate means to invert. Since the "O" appears on the output end of the gate, it means the output is inverted. In NEMA logic an X is used to show that a gate is inverted. The truth table for an INVERTER is shown in figure 61-15. The truth table clearly shows that the output of the INVERTER is the opposite of the input. Figure 61-16 shows an equivalent relay circuit for the INVERTER.

THE NOR GATE

The NOR gate is the "NOT OR" gate. Referring to the computer logic and NEMA logic symbols for a NOR gate in figure 61-17, notice that the symbol for the NOR gate is the same as the symbol for the OR gate with an inverted output. A NOR gate can be made by connecting an INVERTER to the output of an OR gate as shown in figure 61-18.

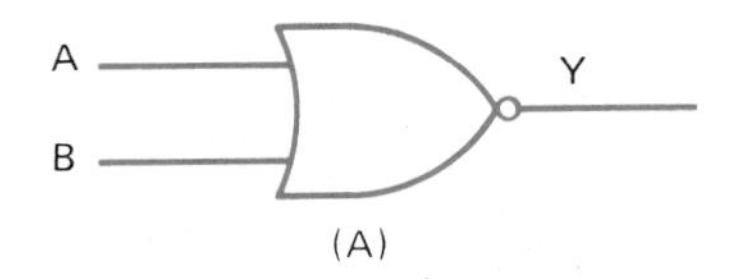

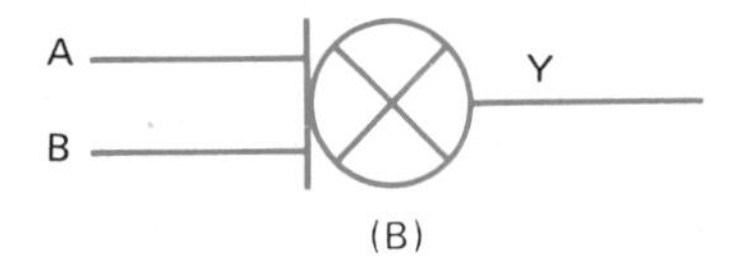

FIGURE 61-17 (A) Computer symbol for a two-input NOR gate (B) NEMA logic symbol for a two-input NOR gate

FIGURE 61-18 Equivalent NOR gate

A	B	Y
0	0	1
0	1	0
1	0	0
1	1	0

FIGURE 61-19 Truth table for a two-input NOR gate

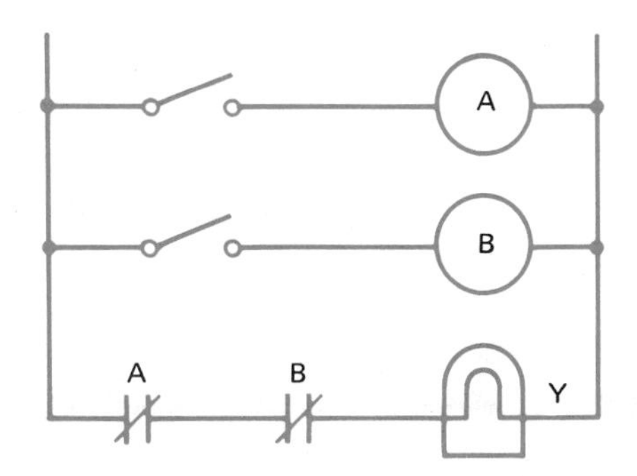

FIGURE 61-20 Equivalent relay circuit for a two-input NOR gate

The truth table shown in figure 61-19 shows that the output of a NOR gate is zero, or low, when any input is high. Therefore, it could be said that any *one input = a zero output for the NOR gate*. An equivalent relay circuit for the NOR gate is shown in figure 61-20. Notice in figure 61-20 that if either relay A or B is energized, there is no output at Y.

THE NAND GATE

The NAND gate is the "NOT AND" gate. Figure 61-21 shows the computer logic symbol and the NEMA logic symbol for the NAND gate. Notice that these symbols are the same as the symbols for the AND gate with inverted outputs. If any input of a NAND gate is low, the output is high. Refer to the truth table in figure 61-22. Notice that the truth table clearly indicates that *any zero input = a one output*. Figure 61-23 shows an equivalent relay circuit for the NAND gate. If eigher relay A or relay B is de-energized, there is an output at Y.

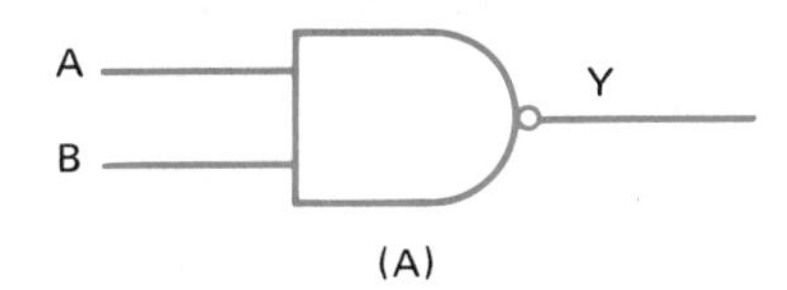

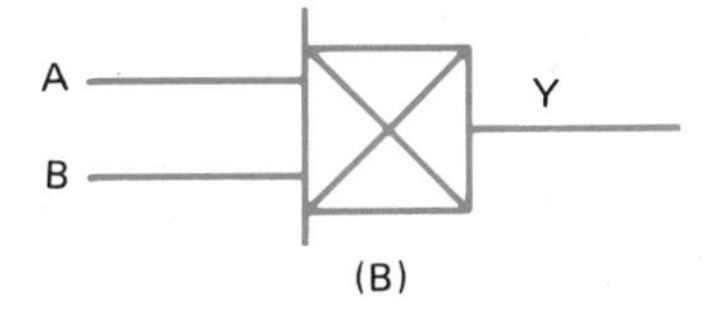

FIGURE 61-21 (A) Computer logic symbol for a two-input NAND gate (B) NEMA logic symbol for a two-input NAND gate

A	B	Y
0	0	1
0	1	1
1	0	1
1	1	0

FIGURE 61-22 Truth table for a two-input NAND gate

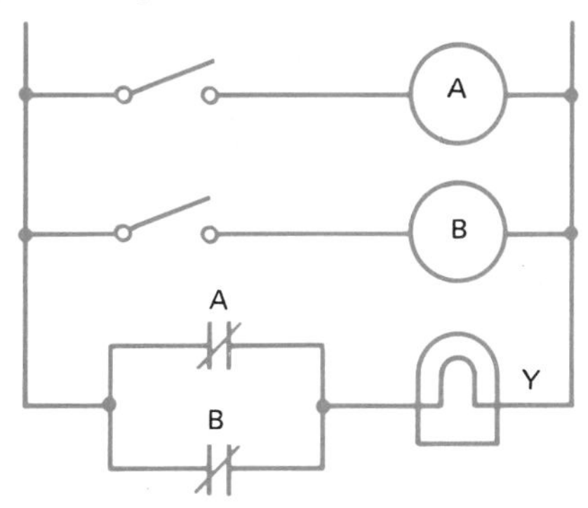

FIGURE 61-23 Equivalent relay circuit for a two-input NAND gate

FIGURE 61-24 NAND gate connected as an INVERTER

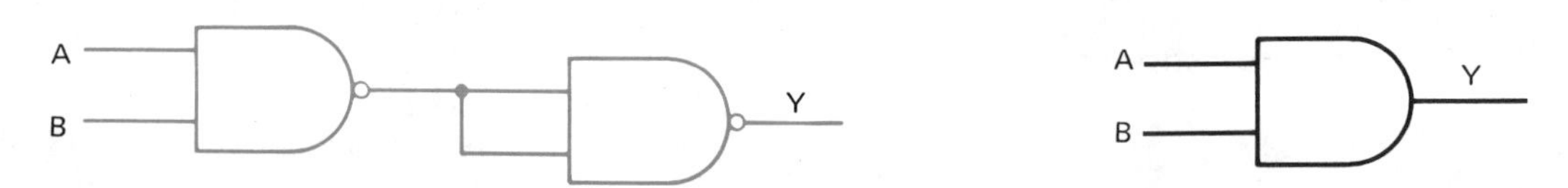

FIGURE 61-25 NAND gates connected as an AND gate

The NAND gate is often referred to as the basix gate because it can be used to make any of the other gates. For instance, figure 61-24 shows the NAND gate connected to make an INVERTER. If a NAND gate is used as an INVERTER and is connected to the output of another NAND gate, it will become an AND gate as shown in figure 61-25. When two NAND gates are connected as INVERTERS, and these INVERTERS are connected to the inputs of another NAND gate, an OR gate is formed, figure 61-26. If an INVERTER is added to the output of the OR gate shown in figure 61-26, a NOR gate is formed, figure 61-27.

INTEGRATED CIRCUITS

Digital logic gates are generally housed in fourteen-pin, IC packages. One of the old reliable types of TTL logic which is frequently used is the 7400 family of devices. For instance, a 7400 IC is a quad, two-input, positive NAND gate. The word quad means that there are four NAND gates

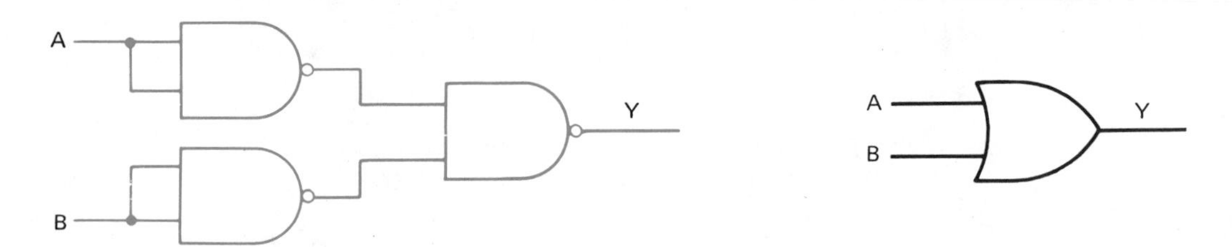

FIGURE 61-26 NAND gates connected as an OR gate

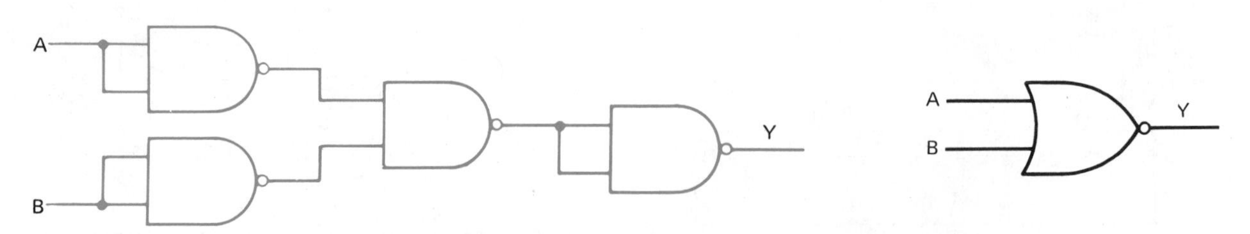

FIGURE 61-27 NAND gates connected as a NOR gate

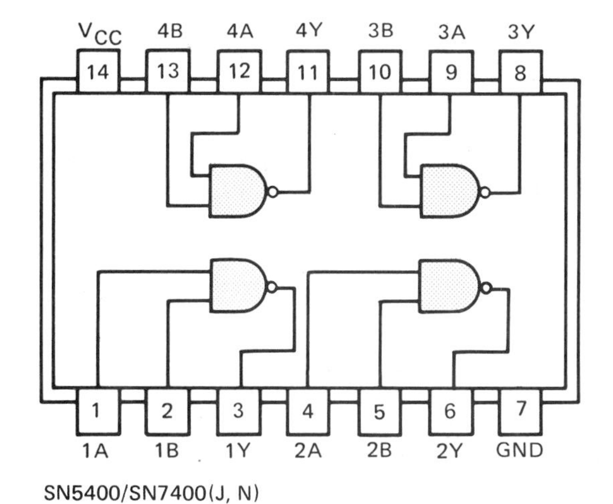

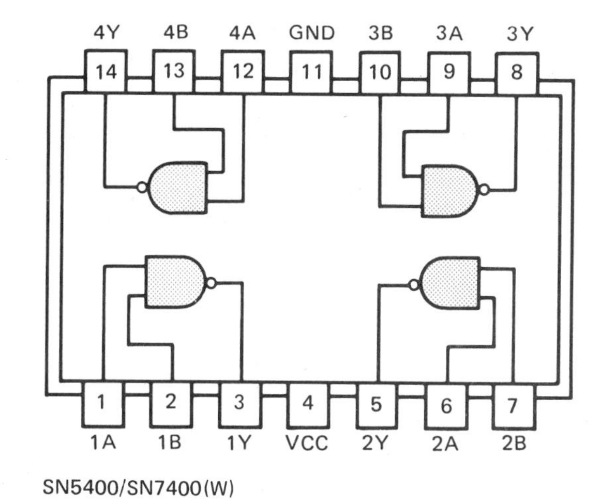

FIGURE 61-28 Integrated circuit connection of a quad, two-input NAND gate (Reproduced by permission of Tektronix, Inc., copyright © 1983)

contained in the package. Each NAND gate has two inputs, and positive means that a level one is considered to be a positive voltage.

There can, however, be a difference in the way ICs are connected. A 7400 (J or N) IC has a different pin connection than a 7400 (W) package. In figure 61-28, both ICs contain four two-input NAND gates, but the pin connections are different. For this reason, it is necessary to use a connection diagram when connecting or testing integrated circuits. A fourteen-pin IC is shown in figure 61-29.

FIGURE 61-29 Fourteen-pin, inline, integrated circuit used to house digital logic gates (Courtesy RCA Corp.)

TESTING INTEGRATED CIRCUITS

Integrated circuits cannot be tested with a volt-ohm-milliammeter. Most ICs must be tested by connecting power to them and then testing the inputs and outputs with special test equipment. Most industrial equipment is designed with different sections of the control system built in modular form. The electrician determines which section of the circuit is not operating and replaces that module. The defective module is then sent to the electronics department or to a company outside of the plant for repair.

REVIEW QUESTIONS

1. What type of digital logic operates on 5 volts?
2. What precautions must be taken when connecting CMOS logic?

3. What do the letters COSMOS stand for?
4. When using a two-input AND gate, what conditions of input must be met to have an output?
5. When using a two-input OR gate, what conditions of input must be met to have an output?
6. Explain the difference between an OR gate and an EXCLUSIVE OR gate.
7. When using a two-input NOR gate, what condition of input must be met to have an output?
8. When using a two-input NAND gate, what condition of input must be met to have an output?
9. If an INVERTER is connected to the output of a NAND gate, what logic gate is formed?
10. If an INVERTER is connected to the output of an OR gate, what gate is formed?
11. What symbol is used to represent "invert" when computer logic symbols are used?
12. What symbol is used to represent "invert" when NEMA logic symbols are used?

The Bounceless Switch

Objectives *After studying this unit, the student will be able to:*

- **Discuss why mechanical contacts should be spring loaded**
- **Discuss problems associated with contact bounce**
- **Describe methods of eliminating contact bounce**
- **Connect a bounceless switch circuit using digital logic gates**

When a control circuit is constructed, it must have sensing devices to tell it what to do. The number and type of sensing devices used are determined by the circuit. Sensing devices can range from a simple push button to float switches, limit switches, and pressure switches. Most of these sensing devices use some type of mechanical switch to indicate their condition. A float switch, for example, indicates its condition by opening or closing a set of contacts, figure 62-1. The float switch can "tell" the control circuit that a liquid is either at a certain level or not. Most of the other types of sensing devices use this same method to indicate some condition. A pressure switch indicates that a pressure is either at a certain level or not, and a limit switch indicates if some device has moved a certain distance or if a device is present or absent from some location.

Almost all of these devices employ a snap-action switch. When a mechanical switch is used, the snap action is generally obtained by spring loading the contacts. This snap action is necessary to insure good contact when the switch operates. Assume that a float switch is used to sense when water reaches a certain level in a tank. If the water rises at a slow rate, the contacts will come together at a slow rate, resulting in a poor connection. However, if the contacts are spring loaded, when the water reaches a certain level, the contacts will snap from one position to another.

Although most contacts have a snap action, they do not generally close with a single action. When the movable contacts meet the stationary contact, there is often a fast bouncing action. This means that the contacts may actually make and break contact three or four times in succession before the switch remains closed. When this type of switch is used to control a relay, contact bounce does not cause a problem because relays are relatively slow-acting devices, figure 62-2.

When this type of switch is used with an electronic control system, however, contact bounce can cause a great deal of trouble. Most digital logic circuits are very fast-acting and can count each pulse when a contact bounces. Depending on the specific circuit, each of these pulses may be interpreted as a command. Contact bounce can cause the control circuit to literally "loose its mind."

FIGURE 62-1 (A) Normally open float switch (B) Normally closed float switch

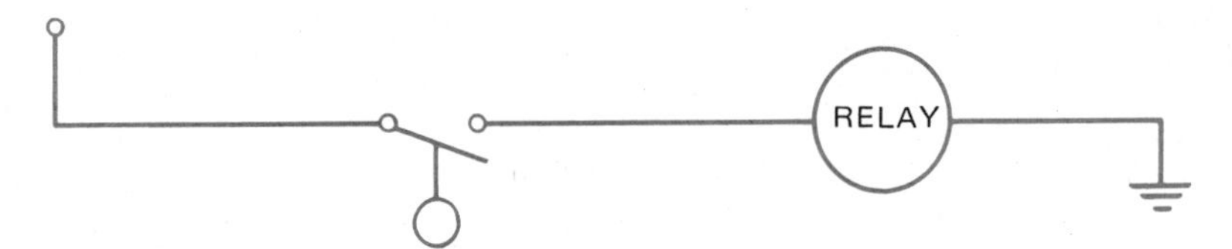

FIGURE 62-2 Contact bounce does not greatly affect relay circuits

Since contact bounce can cause trouble in an electronic control circuit, contacts are debounced before they are permitted to "talk" to the control system. When contacts must be debounced, a circuit called a *bounceless switch* is used. Several circuits can be used to construct a bounceless switch, but the most common construction method uses digital logic gates. Although any of the inverting gates can be used to construct a bounceless switch, in this example only two will be used.

Before construction of the circuit begins, the operation of a bounceless switch circuit should first be discussed. The idea is to construct a circuit that will lock its output either high or low when it detects the first pulse from the mechanical switch. If its output is locked in a position, it will ignore any other pulses it receives from the switch. The output of the bounceless switch is connected to the input of the digital control circuit. The control circuit will now receive only one pulse instead of a series of pulses.

The first gate used to construct a bounceless switch is the INVERTER. The computer symbol and the truth table for the INVERTER are shown in figure 62-3. The bounceless switch circuit using INVERTERS is shown in figure 62-4. The output of the circuit should be high with the switch in the position shown. The switch connects the input of INVERTER #1 directly to ground, or low. This causes the output of INVERTER #1 to be at a high state. The output of INVERTER #1 is connected to the input of INVERTER #2. Since the input of INVERTER #2 is high, its output is low. The output of INVERTER #2 is connected to the input of INVERTER #1. This causes a low condition to be maintained at the input of INVERTER #1.

A	Y
0	1
1	0

FIGURE 62-3 (A) Symbol for an INVERTER (B) Truth table for an INVERTER

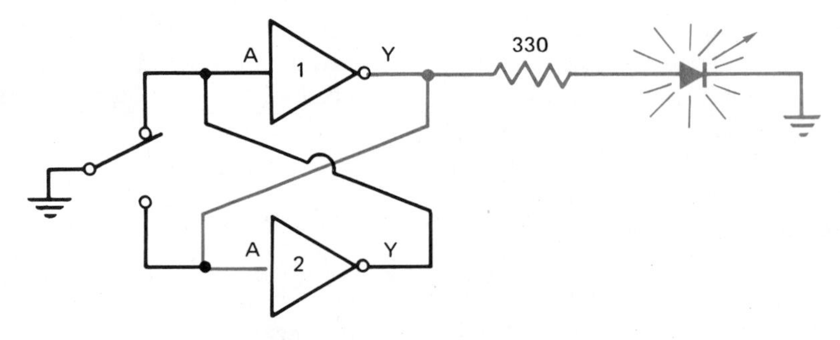

FIGURE 62-4 High output condition

If the position of the switch is changed as shown in figure 62-5, the output will change to low. The switch now connects the input of INVERTER #2 to ground, or low. The output of INVERTER #2 is, therefore, high. The high output of INVERTER #2 is connected to the input of INVERTER #1. Since the input connected to INVERTER #1 is now high, its output becomes low. The output of INVERTER #1 is connected to the input of INVERTER #2. This forces a low input to be maintained at INVERTER #2. Notice that the output of one INVERTER is used to lock the input of the other INVERTER.

The second logic gate used to construct a bounceless switch is the NAND gate. The computer symbol and the truth table for the NAND

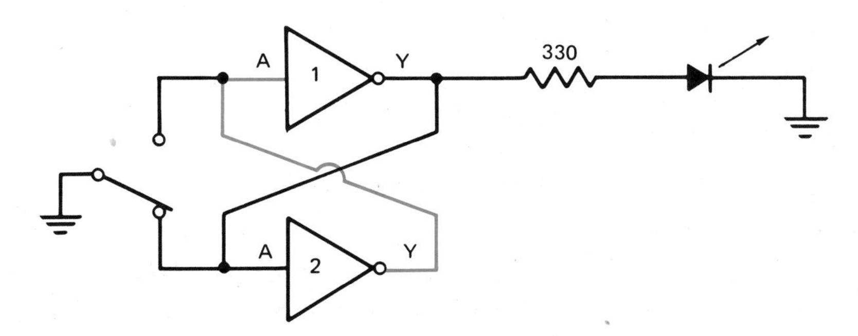

FIGURE 62-5 Low output condition

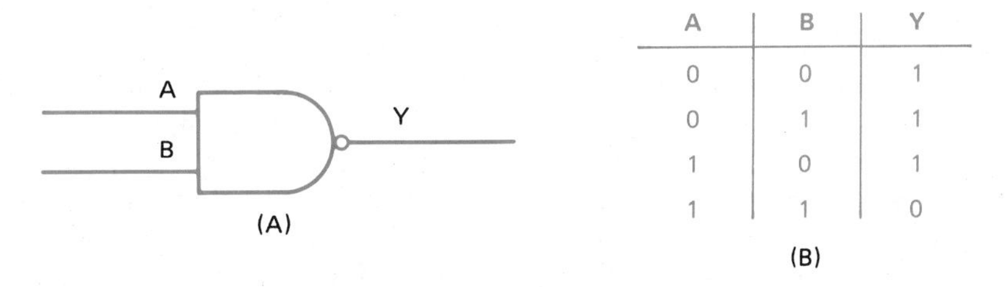

A	B	Y
0	0	1
0	1	1
1	0	1
1	1	0

(B)

FIGURE 62-6 (A) Symbol for a NAND gate (b) Truth table for a NAND gate

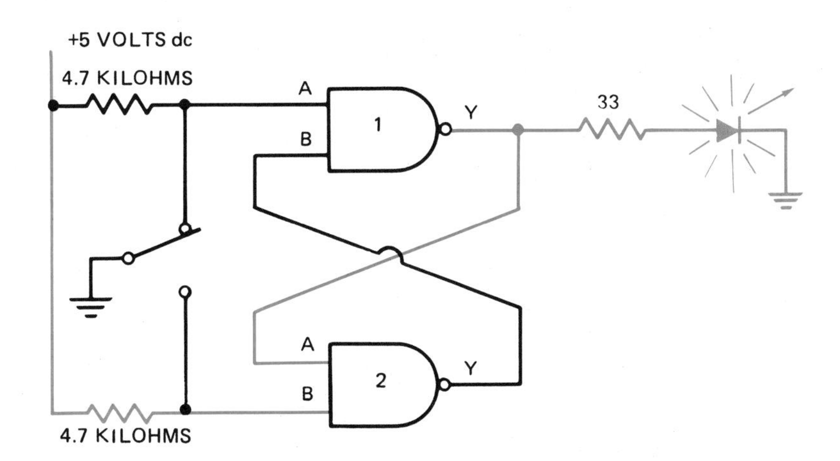

FIGURE 62-7 High output condition

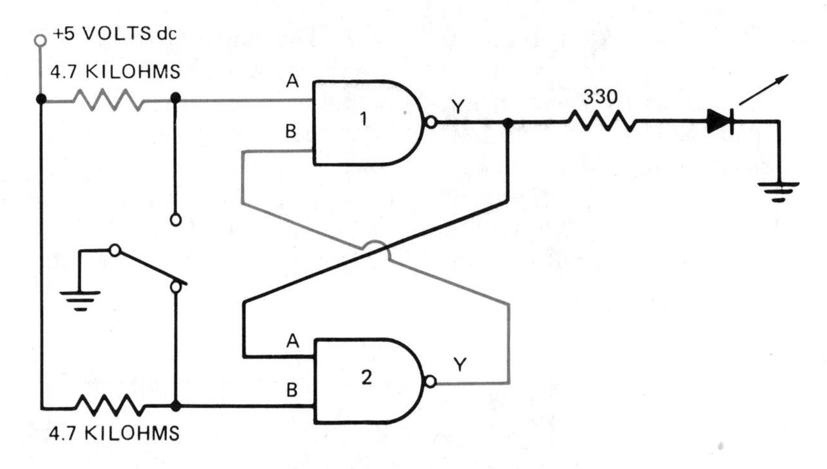

FIGURE 62-8 Low output condition

gate are shown in figure 62-6. The circuit in figure 62-7 shows the construction of a bounceless switch using NAND gates. In this circuit, the switch has input A of gate #1 connected to low, or ground. Since input A is low, the output is high. The output of gate #1 is connected to input A of gate #2. Input B of gate #2 is connected to a high through the 4.7 kilohm resistor. Since both inputs of gate #2 are high, its output is low. This low output is connected to input B of gate #1. Since gate #1 now has a low connected to input B, its output is forced to remain high even if contact bounce causes a momentary high at input A.

When the switch changes position as shown in figure 62-8, input B of gate #2 is connected to a low. This forces the output of gate #2 to become high. The high output of gate #2 is connected to input B of gate #1. Input A of gate #1 is connected to a high through a 4.7 kilohm resistor. Since both inputs of gate #1 are high, its output is low. This low is connected to input A of gate #2, which forces its output to remain high even if contact bounce causes a high to be momentarily connected to input B.

The output of this circuit will remain constant even if the switch contacts bounce. The switch has now been debounced and is ready to be connected to the input of an electronic control circuit.

REVIEW QUESTIONS

1. Why should mechanical contacts be spring loaded?
2. Name three examples of sensing devices.
3. Why must contacts be debounced before they are connected to electronic control circuits?
4. What function does a bounceless switch circuit perform?
5. Name two types of logic gates that can be used to construct a bounceless switch circuit.

UNIT 63

Start-Stop Push-button Control

Objectives

After studying this unit, the student will be able to:

- **Describe the operation of a start-stop relay control circuit**
- **Describe the operation of the basic gates used in this unit**
- **Describe the operation of the solid-state control circuit**
- **Discuss practical wiring techniques for connecting digital logic circuits**
- **Connect a start-stop, push-button control using logic gates**

In this unit a digital circuit will be designed to perform the same function as a common relay circuit. The relay circuit is a basic stop-start, push-button circuit with overload protection, figure 63-1.

Before beginning the design of an electronic circuit that will perform the same function as this relay circuit, the operation of the relay circuit should first be discussed. In the circuit shown in figure 63-1, no current can flow to relay coil M because the normally open start button and the normally open contact are controlled by relay coil M.

When the start button is pushed, current flows through the relay coil and normally closed overload contact to the power source, figure 63-2. When current flows through relay coil M, the contacts connected parallel to the start button close. These contacts maintain the circuit to coil M when the start button releases and returns to its open position, figure 63-3.

The circuit will continue to operate until the stop button is pushed and breaks the circuit to the coil, figure 63-4. When the current flow to the coil stops, the relay de-energizes and contact M re-opens. Since the start button is now open and con-

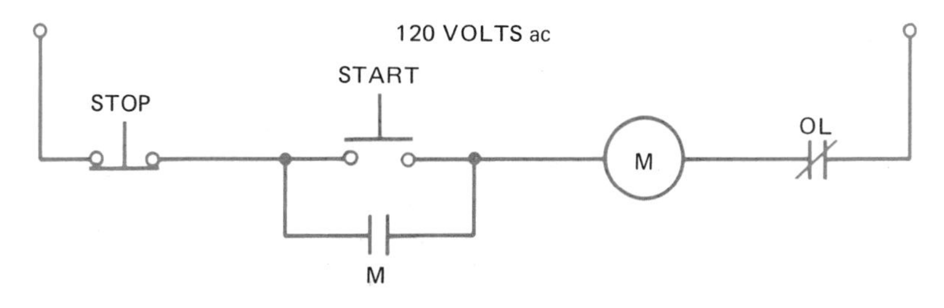

FIGURE 63-1 Start-stop, push-button circuit

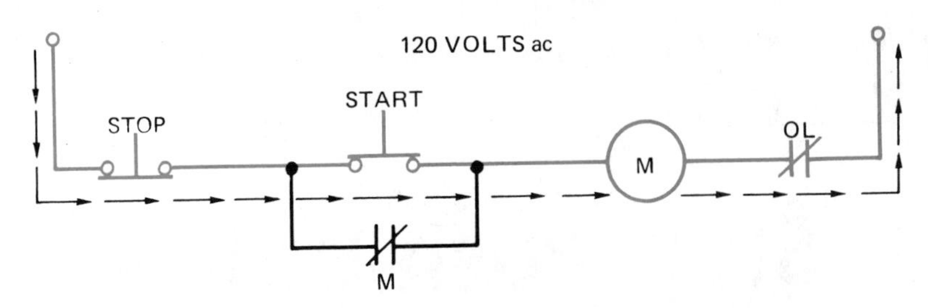

FIGURE 63-2 Start button energizes "M" relay coil

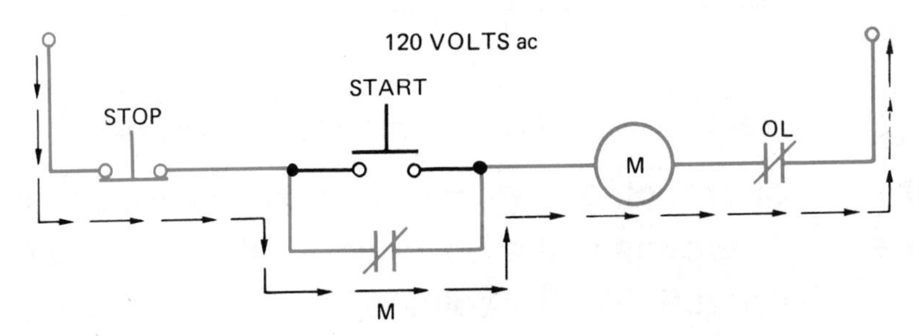

FIGURE 63-3 "M" contacts maintain the circuit

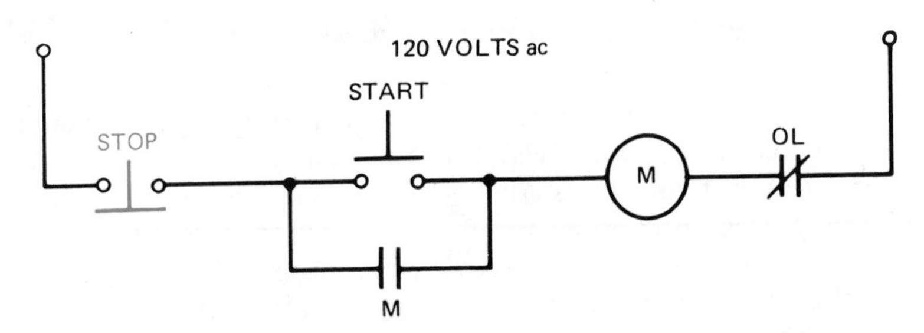

FIGURE 63-4 Stop button breaks the circuit

tact M is open, there is no complete circuit to the relay coil when the stop button is returned to its normally closed position. If the relay is to be restarted, the start button must be pushed again to provide a complete circuit to the relay coil.

The only other logic condition that can occur in this circuit is caused by the motor connected to the load contacts of relay M. The motor is connected in series with the heater of an overload relay, figure 63-5. When coil M energizes, it closes the load contact M as shown in figure 63-6. When the load contact closes, it connects the motor to the 120-volt, ac power line.

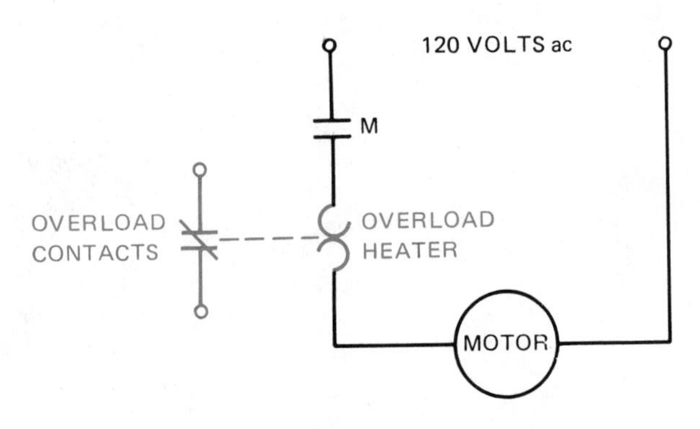

FIGURE 63-5

If the motor is overloaded, it will cause too much current to flow through the circuit. When a current greater than normal flows through the overload heater, the heater produces more heat than it does under normal conditions. If the current becomes high enough, it will cause the normally closed overload contact to open. Notice that the overload contact is electrically isolated from the heater. The contact, therefore, can be connected to a different voltage source than the motor.

If the overload contact opens, the control circuit is broken and the relay de-energizes as if the stop button had been pushed. After the overload contact has been reset to its normally closed position, the coil will remain de-energized until the start button is again pressed.

Now that the logic of the circuit is understood a digital logic circuit that will operate in this manner can be designed. The first problem is to find a circuit that can be turned on with one push button and turned off with another. The circuit shown in figure 63-7 can perform this function. This circuit consists of an OR gate and an AND gate. Input A of the OR gate is connected to a normally open push button which is connected to +5 volts dc. Input B of the OR gate is connected to the output of the AND gate. The output of the OR gate is connected to input A of the AND gate. Input B of the AND gate is connected through a normally closed push button to +5 volts dc. This normally closed push button is used as the stop button. The output of the AND gate is the output of the circuit.

To understand the logic of this circuit, assume that the output of the AND gate is low. This produces a low at input B of the OR gate. Since the push button connected to input A is open, a low is produced at this input also. When all inputs of an OR gate are low, its output is also low. The low output of the OR gate is connected to input A of the AND gate. Input B of the AND gate is connected to a high through the normally closed push-button switch. Since input A of the AND

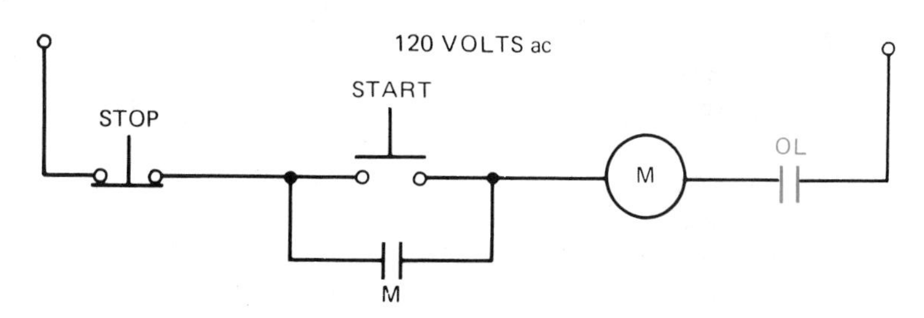

FIGURE 63-6 Overload contacts break the circuit

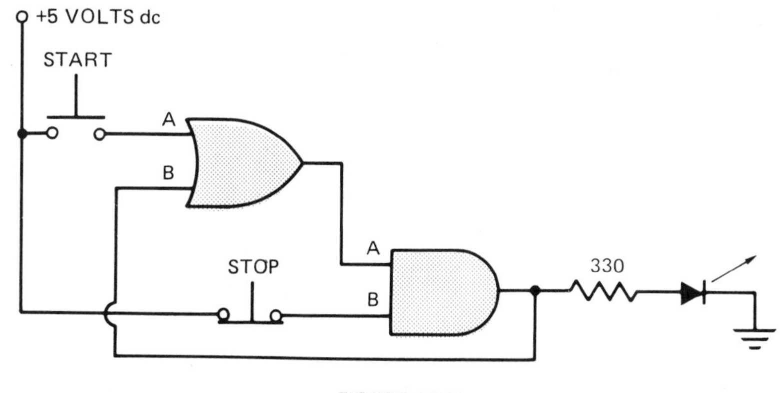

FIGURE 63-7

gate is low, the output of the AND gate is forced to remain in a low state.

When the start button is pushed, a high is connected to input A of the OR gate. This causes the output of the OR gate to change to high. This high output is connected to input A of the AND gate. The AND gate now has both of its inputs high, so its output changes from a low to a high state. When the ouput of the AND gate changes to a high state, input B of the OR gate becomes high also. Since the OR gate now has a high connected to its B input, its output will remain high when the push button is returned to its open condition and input A becomes low. Notice that this circuit operates the same as the relay circuit when the start button is pushed. The output changes from a low state to a high state and the circuit locks in this condition so the start button can be reopened.

When the normally closed stop button is pushed, input B of the AND gate changes from high to low. When input B changes to a low state, the output of the AND gate changes to a low state also. This causes a low to appear at input B of the OR gate. The OR gate now has both of its inputs low, so its output changes from a high state to a low state. Since input A of the AND gate is now low, the output is forced to remain low when the stop button returns to its closed position and input B becomes high. The circuit designed here can be turned on with the start button and turned off with the stop button.

The next design task is to connect the overload contact to the circuit. The overload contact must be connected in such a manner that it will cause the output of the circuit to turn off if it opens. One's first impulse might be to connect the overload contact to the circuit as shown in figure 63-8. In this circuit, the output of AND gate #1 has been connected to input A of AND gate #2. Input B of AND gate #2 has been connected to a high through the normally closed overload contact. If the overload contact remains closed, input B will remain high. The output of AND gate #2 is, therefore, controlled by input A. If the output of AND gate #1 changes to a high state, the output of AND gate #2 will also change to a high state. If the output of AND gate #1 becomes low, the output of AND gate #2 will become low also.

If the output of AND gate #2 is high and the overload contact opens, input B will become low and the output will change from a high to a low state. This circuit appears to operate with the same logic as the relay circuit until the logic is examined closely. Assume that the overload contacts are closed and the output of AND gate #1 is high. Since both inputs of AND gate #2 are high, the output is also high. Now assume that the overload contact opens and causes input B to change to a low condition. This forces the output of AND

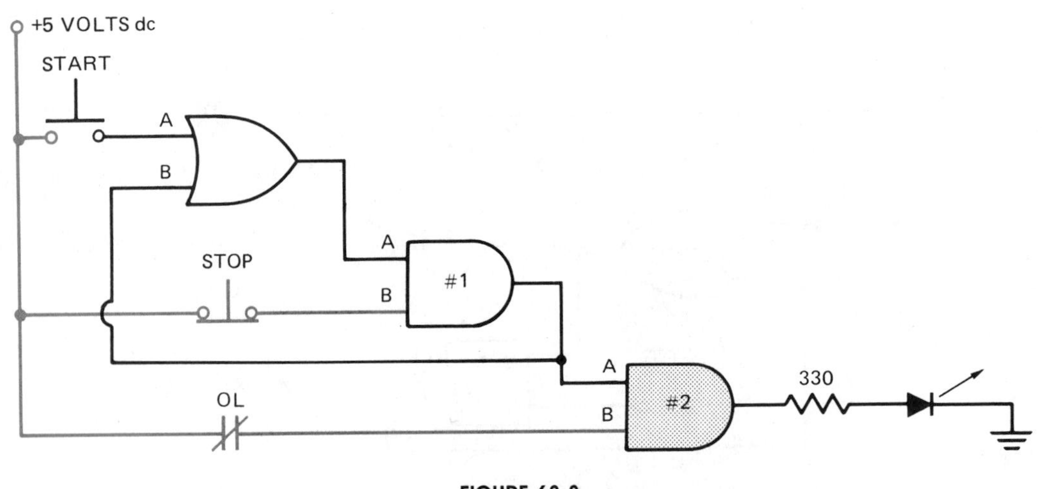

FIGURE 63-8

gate #2 to change to a low state also. Input A of AND gate #2 is still high, however. If the overload contact is reset, the output will immediately change back to a high state. If the overload contact opens and is then reset in the relay circuit, the relay will not restart itself. The start button must be pushed to restart the circuit. Although this is a small difference in circuit logic, it could become a safety hazard in some cases.

This fault can be corrected with a slight design change. Refer to figure 63-9. In this circuit, the normally closed stop button has been connected to input A of AND gate #2, and the normally closed overload switch has been connected to input B. As long as both of these inputs are high, the output of AND gate #2 will provide a high to input B of AND gate #1. If either the stop button or the overload contact opens, the output of AND gate #2 will change to a low state. When input B of AND gate #2 changes to a low state, it will cause the output of AND gate #1 to change to a low state and unlock the circuit, just as pushing the stop button did in the circuit shown in figure 63-8. The logic of this digital circuit is now the same as the relay circuit.

Although the logic of this circuit is now correct, there are still some problems that must be corrected. When gates are used, their inputs must be connected to a definite high or low. When the start button is in its normal position, input A of the OR gate is not connected to anything. When an input is left in this condition, the gate may not be able to determine if the input should be high or low. The gate could, therefore, assume either condition. To prevent this, inputs must always be connected to a definite high or low.

When using TTL logic, inputs are always pulled high with a resistor as opposed to being pulled low. If a resistor is used to pull an input low as shown in figure 63-10, it will cause the gate to have a voltage drop at its output. This means that in the high state, the output of the gate may

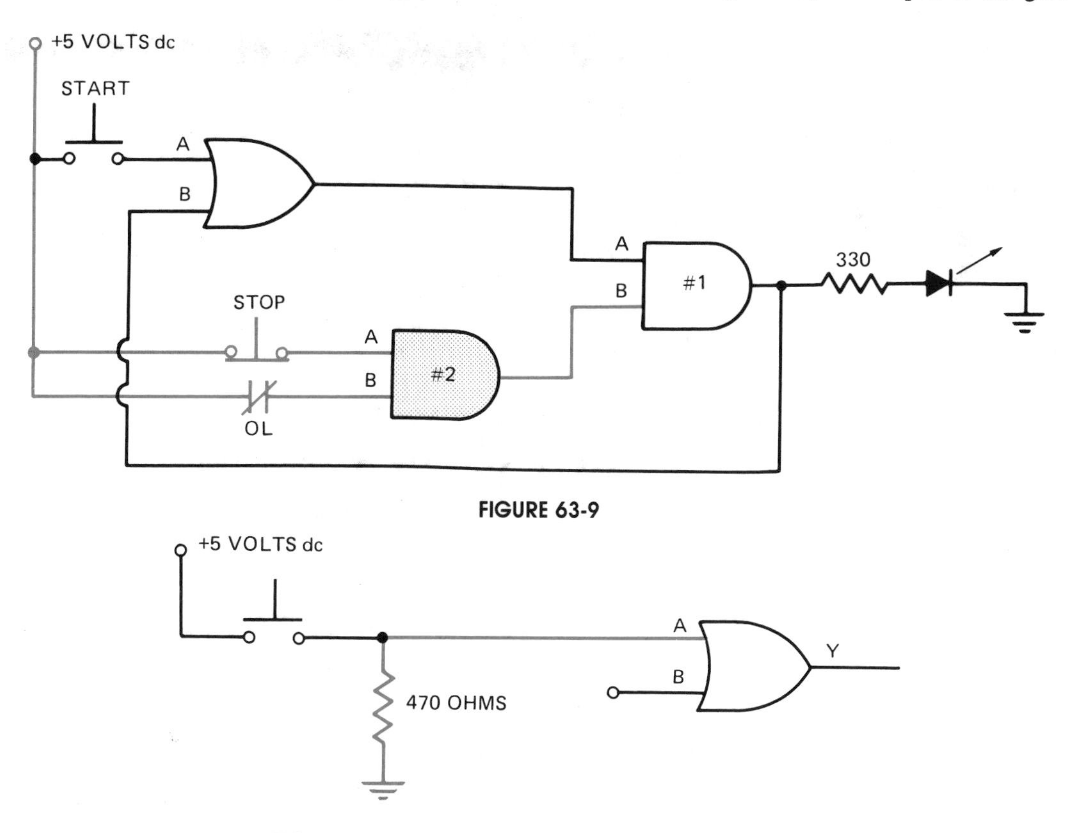

FIGURE 63-9

FIGURE 63-10 Resistor used to lower the input of a gate

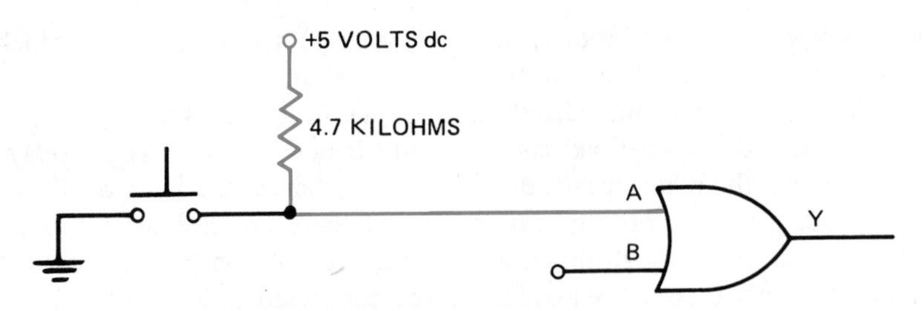

FIGURE 63-11 Resistor used to raise the input of a gate

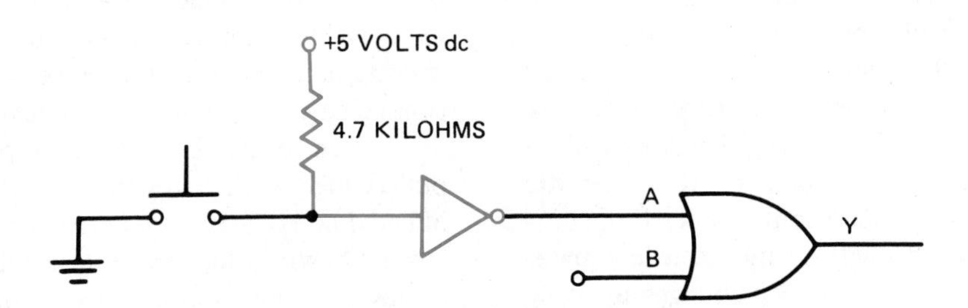

FIGURE 63-12 Push button produces a high at the input

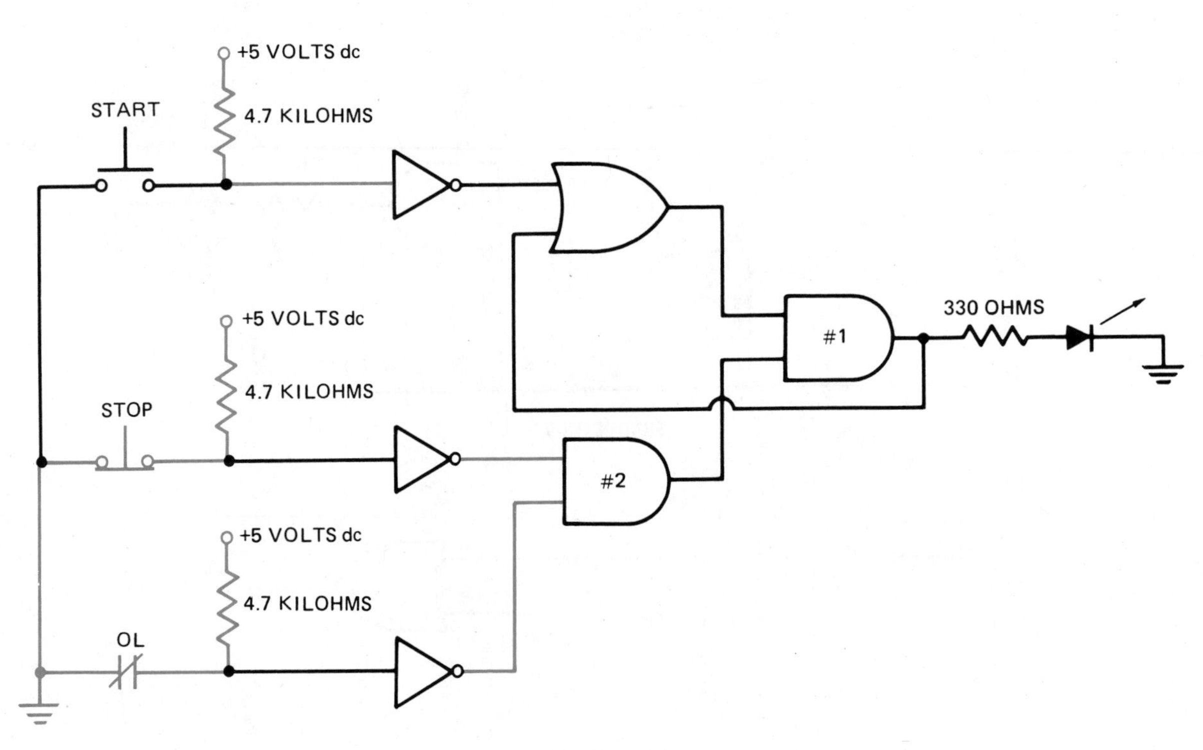

FIGURE 63-13

be only 3 or 4 volts instead of 5 volts. If this output is used as the input of another gate, and the other gate has been pulled low with a resistor, the output of the second gate may be only 2 or 3 volts. Notice that each time a gate is pulled low through a resistor, its output voltage becomes low. If this were done through several steps, the output voltage would soon become so low it could not be used to drive the input of another gate.

Figure 63-11 shows a resistor used to pull the input of a gate high. In this circuit, the push button is used to connect the input of the gate to ground, or low.

The push button can be adapted to produce a high at the input instead of a low by adding an INVERTER as shown in figure 63-12. In this circuit, a pull-up resistor is connected to the input of an INVERTER. Since the input of the INVERTER is high, its output will produce a low at input A of the OR gate. When the normally open push button is pressed, a low will be produced at the input of the INVERTER. When the input of the INVERTER becomes low, its output becomes high. Notice that the push button will now produce a high at input A of the OR gate when it is pushed.

Since both of the push buttons and the normally closed overload contact are used to provide high inputs, the circuit is changed as shown in figure 63-13. Notice that the normally closed push button and the normally closed overload switch connected to the inputs of AND gate #2 are connected to ground instead of Vcc. When the switches are connected to ground, a low is provided to the input of the INVERTERS to which they are connected. The INVERTERS, therefore, produce a high at the input of the AND gate. If one of these normally closed switches opens, a high will be provided to the input of the INVERTER. This will cause the output of the INVERTER to become low. If the logic of the circuit shown in figure 63-13 is checked, it can be seen that it is the same as the logic of circuit 63-9.

The final design problem for this circuit concerns the output. So far, a light-emitting diode has been used as the load. The LED is used to indi-

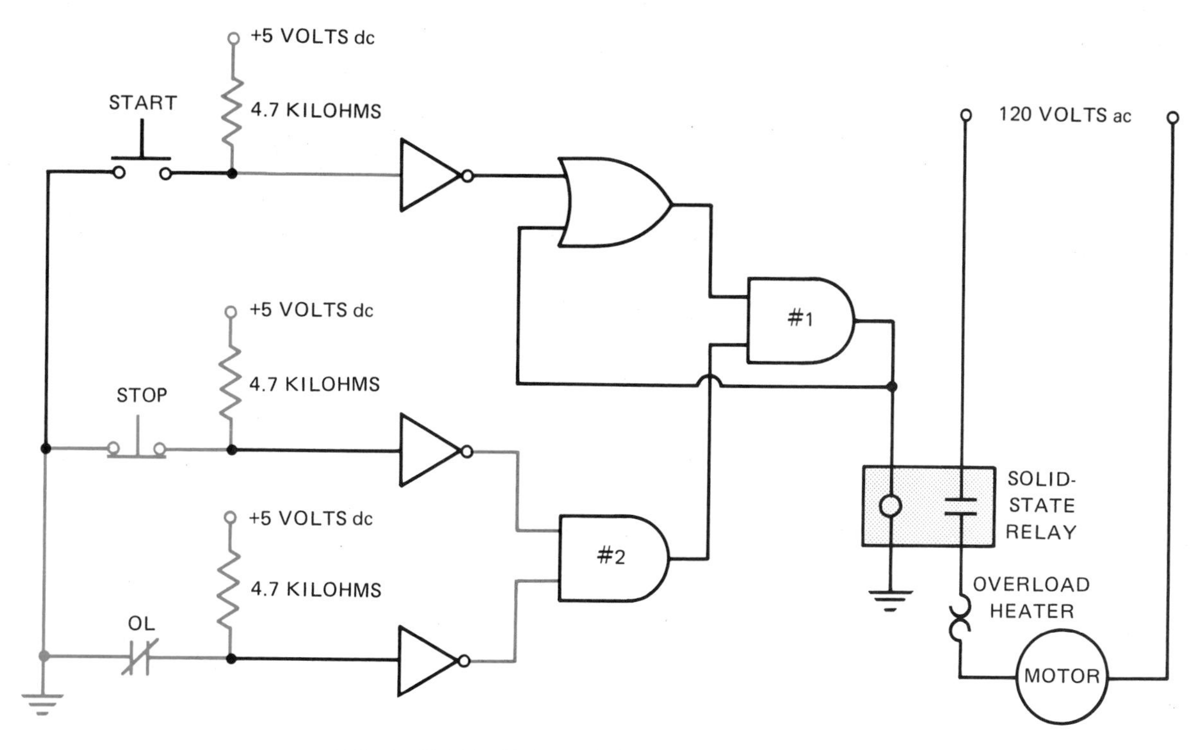

FIGURE 63-14 Solid-state, start-stop, push-button control

cate when the output is high and when it is low. The original circuit, however, was used to control a 120-volt ac motor. This control can be accomplished by connecting a solid-state relay to the output in place of the LED, figure 63-14. In this circuit, the output of AND gate #1 is connected to the input of an opto-isolated, solid-state relay. When the output of the AND gate changes to a high condition, the solid-state relay turns on and connects the 120-volt ac load to the line.

REVIEW QUESTIONS

1. In a relay circuit, what function is served by the holding contacts?
2. What is the function of the overload relay in a motor control circuit?
3. What conditions of input must exist if an OR gate is to produce a high output?
4. What conditions of input must exist if an AND gate is to produce a high output?
5. When connecting TTL logic, why are inputs pulled high instead of low?
6. Referring to figure 63-9, how would this circuit operate if input B of the OR gate was reconnected to input A of AND gate #1 instead of its output?
7. Referring to figure 63-12, what function does the INVERTER serve in this circuit?

Programmable Controllers

Objectives ***After studying this unit, the student will be able to:***

- **List the principal parts of a programmable controller**
- **Describe the differences between programmable controllers and other types of computers**
- **Discuss the operation of the programming terminal, the central processor, and the I/O track**
- **Draw basic diagrams of how the input and output modules function**

Programmable controllers were first used by the automobile industry in the late 1960s. Each time a change was made in design, it was necessary to change the control systems operating the machinery. This consisted of physically rewiring the control system to make it perform the new operation. Rewiring the system, of course, was extremely time consuming and costly. What the industry needed was some type of control system that could be changed without the extensive rewiring required to change relay control systems.

One of the first questions that is generally asked is, "Is a programmable controller a computer?" The answer to that question is "yes." The programmable controller, or PC, is a special computer designed to perform a special function.

DIFFERENCES BETWEEN THE PC AND THE COMMON COMPUTER

Some differences between a PC and a home and business computer are:

1. The PC is designed to be operated in an industrial environment. Any computer used in industry must be able to operate in extremes of temperature, ignore voltage spikes and drops on the incoming power line, survive in an atmosphere that often contains corrosive vapors, oil, and dirt, and withstand shock and vibration.
2. Most programmable controllers are designed to be programmed with relay schematic or ladder diagrams instead of the common computer languages such as Basic or Fortran. An electrician who is familiar with relay logic diagrams can generally be trained to program a PC in a few hours, whereas it generally takes several months to train someone to program a standard computer.

BASIC COMPONENTS

Programmable controllers can be divided into four basic parts:

A. The power supply

B. The central processing unit

C. The program loader or terminal

D. The I/O (pronounced eye-oh) track

The Power Supply

The power supply is used to lower the incoming ac voltage to the desired level, rectify it to direct current, and then filter and regulate it. The internal logic circuits of programmable controllers operate on 5 to 15 volts dc depending on the type of controller. This voltage must be free of voltage spikes and other electrical noise. It must also be regulated to within 5% of the required voltage value. Some manufacturers of PCs use a separate power supply, and others build the power supply into the central processor.

The CPU

The central processing unit, or CPU, is the brain of the programmable controller. It contains the microprocessor chip and related integrated circuits to perform all the logic functions. The microprocessor chip used in most PCs is the same as the common computer chip used in many home and business machines, figure 64-1.

The central processing unit generally has a key switch located on the front panel. This switch must be turned on before the CPU can be programmed. This is done to prevent the circuit from being changed accidently. Plug connections mounted on the central processor are used to provide connections for the programming terminal and the I/O tracks, figures 64-2A and 64-2B. Most CPUs are designed so that once the program has been tested, it can be stored on tape or disc. In this way if a central processing unit fails and has to be replaced, the new unit can be reprogrammed from the tape or disc. This eliminates the time consuming process of having to reprogram by hand.

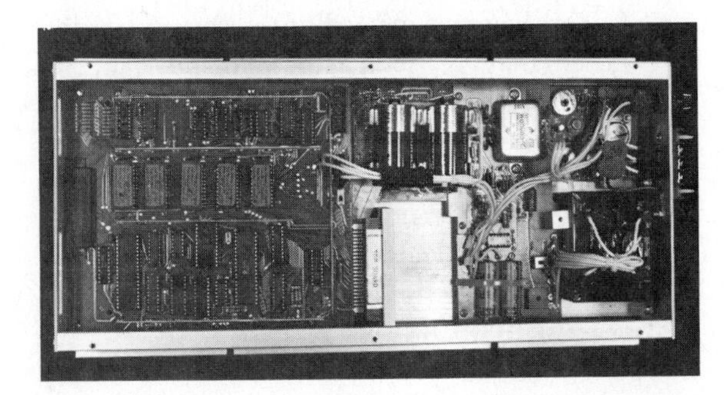

FIGURE 64-1 Inside the processor (Courtesy Struthers-Dunn, Inc.)

FIGURE 64-2A The central processor unit (Courtesy Struthers-Dunn, Inc.)

FIGURE 64-2B The central processor unit (Courtesy Allen-Bradley Co., Systems Division)

The Programming Terminal

The programming terminal or loading terminal is used to program the CPU. Most terminals are one of two types. One type is a small hand-held device that uses a liquid crystal display to show the program, figure 64-3. This terminal, however, will display only one line of the program at a time.

The other type of terminal uses a cathode ray tube, or CRT, to show the program. This terminal looks similar to a portable television set with a keyboard attached, figures 64-4A, 64-4B, and 64-5. It will display from four to six lines of the program at a time, depending on the manufacturer.

FIGURE 64-3 Small programmable controller and hand-held programming terminal (Courtesy EATON Corp., Cutler-Hammer Products)

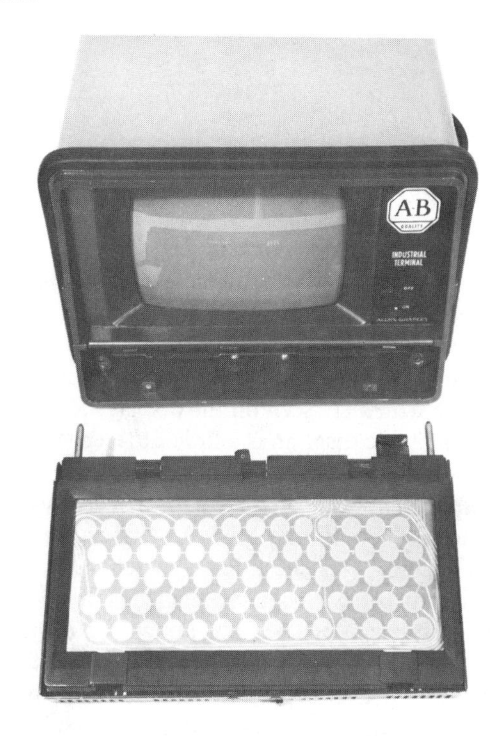

FIGURE 64-4B Programming terminal with key pad (Courtesy Allen-Bradley Co., Systems Division)

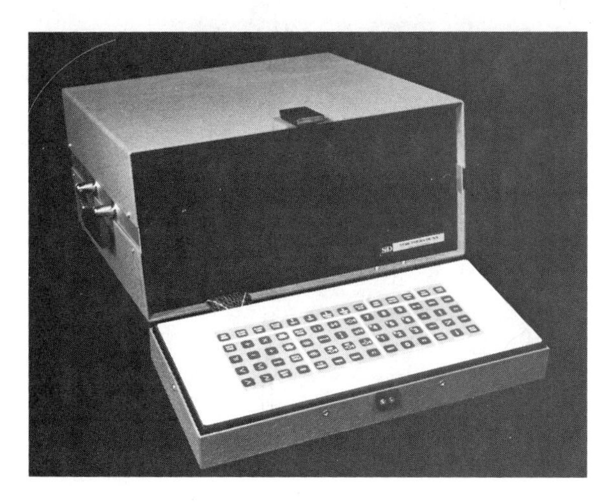

FIGURE 64-4A Programming terminal (Courtesy Struthers-Dunn, Inc.)

FIGURE 64-5 A central processor unit, I/O track, and programming terminal (Courtesy General Electric Co.)

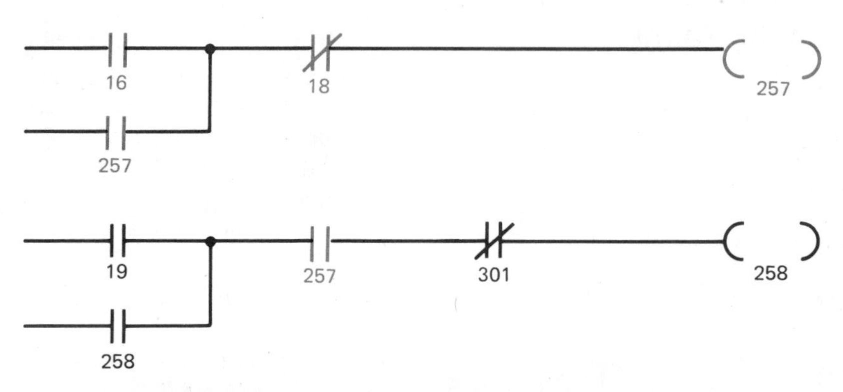

FIGURE 64-6 Analyzing circuit operation with the terminal

The terminal is not only used to program the controller, but it is also used to troubleshoot the circuit. When the terminal is connected to the CPU, the circuit can be examined while it is in operation. Figure 64-6 illustrates a circuit typical of those which are seen on the display. Notice that this schematic diagram is a little different from the typical ladder diagram. All of the line components are shown as normally open or normally closed contacts. There are no NEMA symbols for push-buttons, float switches, limit switches, etc. The programmable controller recognizes only open or closed contacts. It does not know if a contact is controlled by a push button, a limit switch, or a float switch. Each contact, however, does have a number. The number is used to distinguish one contact from another.

The coil symbols look like a set of parentheses instead of a circle as shown on most ladder diagrams. Each line ends with a coil, and each coil has a number. When a contact symbol has the same number as a coil, it means the contact is controlled by that coil. Figure 64-6 shows a coil numbered 257, and two contacts numbered 257. When relay coil 257 is energized, the controller interprets both of these contacts to be closed.

Notice that the 257 contacts, contacts 16 and 18, and coil 257 are drawn with dark heavy lines. When a contact has a complete circuit through it, or a coil is energized, the terminal will illuminate that contact or coil. Contact 16 is illuminated, which means that it is closed and providing a current path. Contact 18 is closed, providing a current path to coil 257. Since coil 257 is energized,

FIGURE 64-7A I/O track with input and output modules (Courtesy Struthers-Dunn, Inc.)

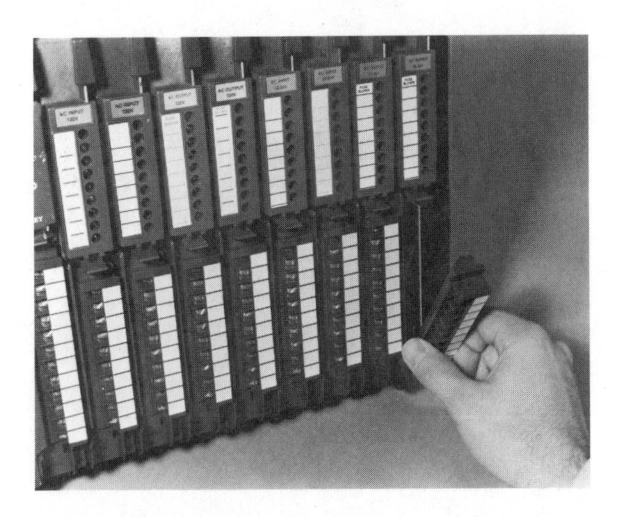

FIGURE 64-7B I/O track with input and output modules (Courtesy Allen-Bradley Co., Systems Division)

both 257 contacts are closed and providing current paths.

Contacts 19, 258, and 301 are not illuminated. This means that contacts 19 and 258 are de-energized and open. Contact 301, however, has been energized. This contact is shown as normally closed. Since it is not illuminated, it is open and no current path exists through it. Notice that the illumination of a contact does not mean that the contact has energized or changed position; it means that there is a complete path for current flow.

When the terminal is used to load a program into the central processing unit, contact and coil symbols on the keyboard are used. These symbol keys are used to load a ladder diagram similar to the one shown in figure 64-6 into the CPU. Programming will be discussed in Unit 65.

FIGURE 64-8 I/O track with a capacity for 80 inputs and outputs (Courtesy General Electric Co.)

The I/O Track

The I/O track is used to connect the central processing unit to the outside world. It contains input modules which carry information to the CPU, and output modules which carry information from the CPU. An I/O track with input and output modules is shown in figures 64-7A and B. Most modules contain more than one input or output. Any number from two to eight is common depending on the manufacturer. The modules shown in figure 64-7A can each handle four connections. This means that each input module can handle four different inputs from pilot devices such as push buttons, float switches, or limit switches. Each output module can control four external devices such as pilot lights, solenoids, or motor starter coils. The operating voltage of modules can be alternating current or direct current and are generally either 120 or 24 volts. The I/O track in figure 64-7A can handle eight modules. Since each module can accommodate four devices, this I/O track can control 32 inputs or outputs.

I/O Capacity One factor which determines the size and cost of a programmable controller is its I/O capacity. Many small units are designed to handle only 32 inputs or outputs. Large units can handle several hundred. The controller shown in

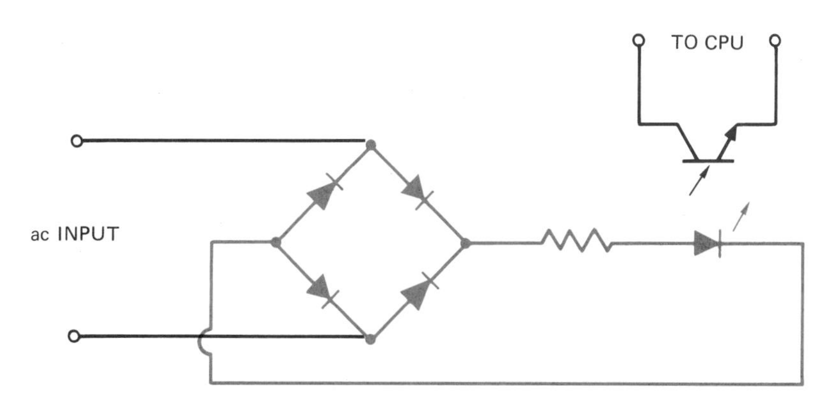

FIGURE 64-9 Input circuit

figure 64-2A is designed to handle eight I/O tracks. Since each I/O track has 32 inputs or outputs, the controller has an I/O capacity of 256. The I/O track in figure 64-8 has a capacity for 80 inputs and outputs.

The Input Module The central processing unit of a programmable controller is extremely sensitive to voltage spikes and electrical noise. For this reason the input I/O uses opto-isolation to electrically separate the incoming signal from the CPU. Another job performed by the input I/O is debouncing any switch contacts connected to it.

Figure 64-9 shows a typical circuit used for the input. The bridge rectifier changes the ac voltage into dc voltage. A resistor is used to limit current to the light-emitting diode. When the LED

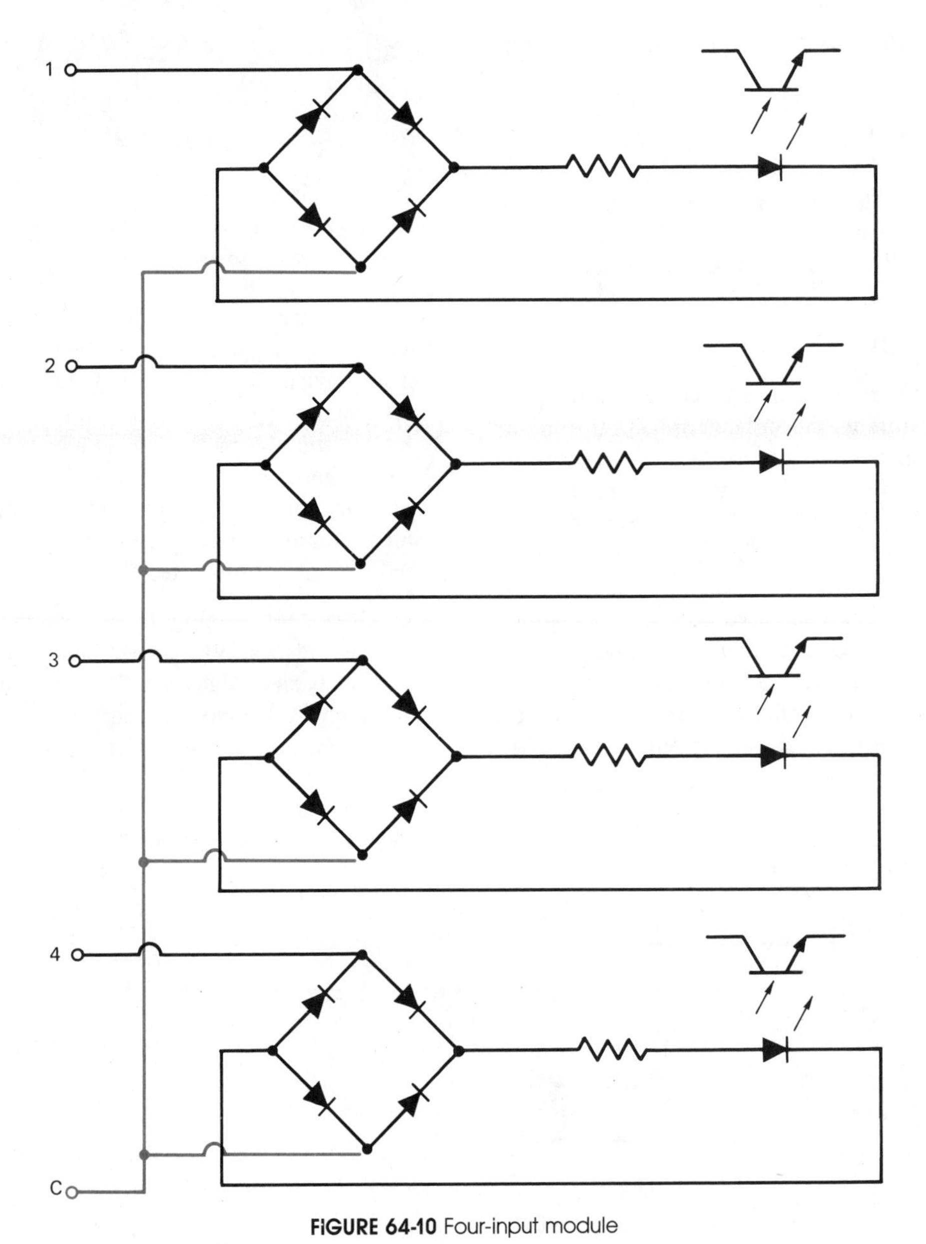

FIGURE 64-10 Four-input module

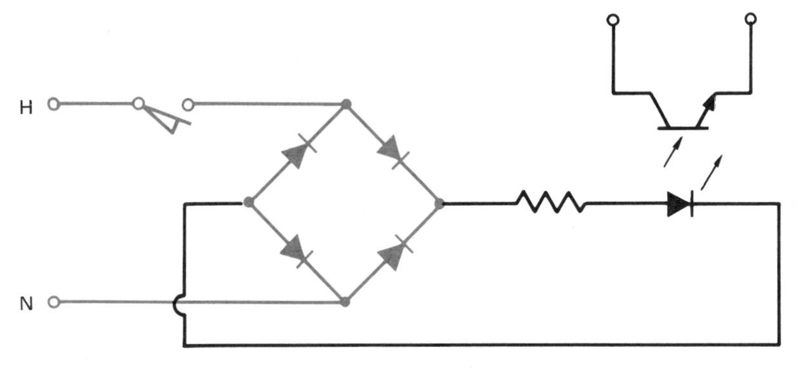

FIGURE 64-11 Limit switch completes circuit to rectifier

turns on, the light is detected by the phototransistor which signals the CPU that there is a voltage present at the input terminal.

When the module has more than one input, the bridge rectifiers are connected together on one side to form a common terminal. On the other side the rectifiers are labeled 1, 2, 3, and 4. Figure 64-10 shows four bridge rectifiers connected together to form a common terminal.

Figure 64-11 shows a limit switch connected to the input. Notice that the limit switch completes a circuit from the ac line to the bridge rectifier. When the limit switch closes, 120-volts ac is applied to the rectifier causing the LED to turn on.

The Output Module The output module is used to connect the central processing unit to the load. The output is an opto-isolated, solid-state relay. The current rating can range from .5 to 3 amps depending on the manufacturer. Voltage ratings are generally 24 or 120 volts and can be ac or dc.

If the output is designed to control a dc voltage, a power transistor is used to control the load, figure 64-12. The transistor is a phototransistor which is operated by a light-emitting diode. The LED is operated by the CPU.

If the output is designed to control an ac load, a triac, rather than a power transistor, is used as the control device, figure 64-13. A photodetector connected to the gate of the triac is used to control the output. When the LED is turned on by the CPU, the photodetector permits current to flow through the gate of the triac and turn it on.

If more than one output is contained in a module, the control devices are connected together on one side to form a common terminal. Figure 64-14 shows an output module that contains four outputs. Notice that one side of each triac has been connected to form a common terminal. On the other side the triacs are labeled 1, 2, 3, and 4. If power transistors are used as the control devices, the emitters or the collectors can be connected to form a common terminal.

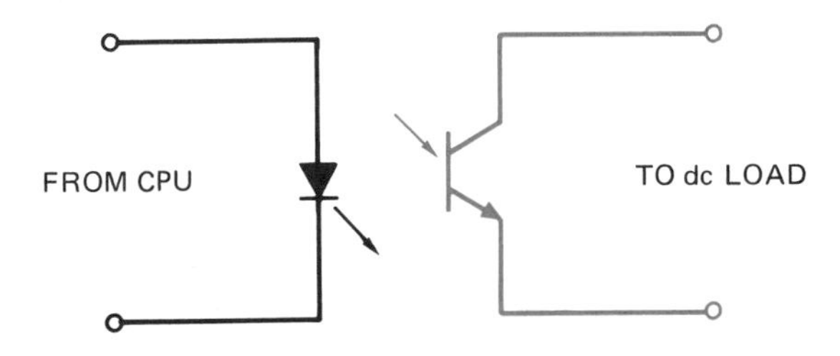

FIGURE 64-12 Output module used to control a dc module

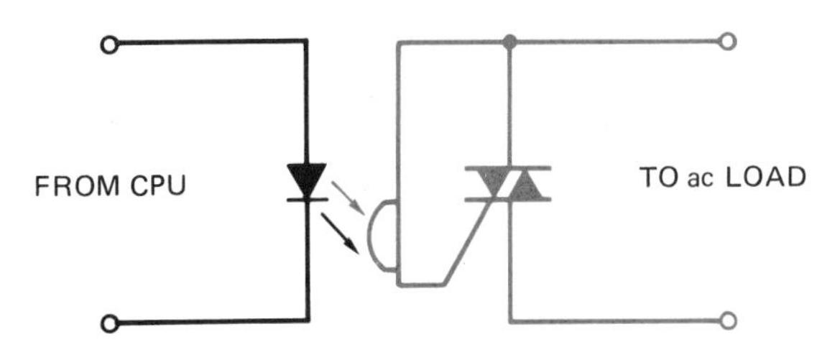

FIGURE 64-13 Output module used to control an ac module

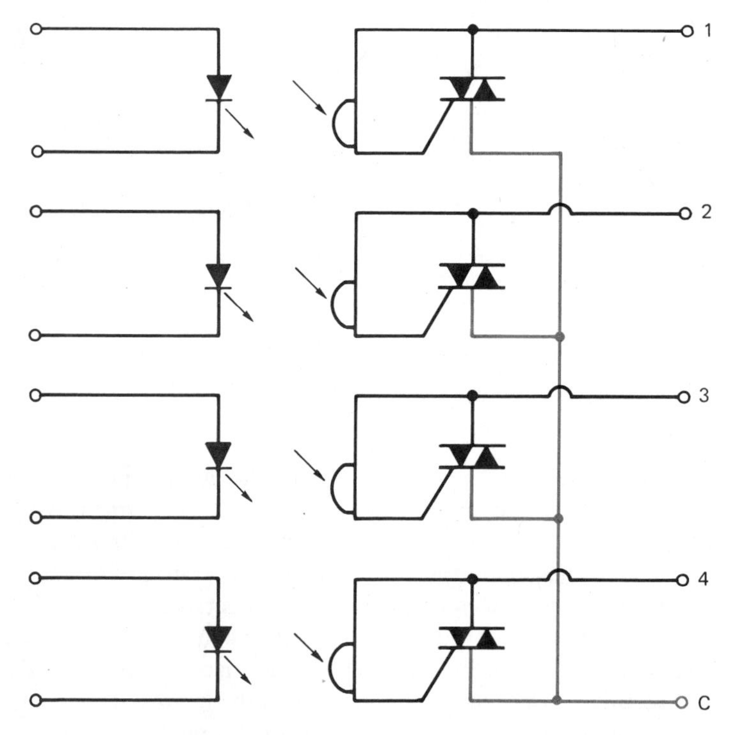

FIGURE 64-14 Four ac outputs in one module

Figure 64-15 shows a solenoid coil connected to an ac output module. Notice that the triac is used as a switch to complete a circuit so that current can flow through the coil. The output module does not provide power to operate the load. The power must be provided by an external power source. The amount of current an output can control is limited. Small current loads, such as solenoid coils and pilot lights, can be controlled directly by the I/O output, but large current loads, such as motors, can not. When a large amount of current must be controlled, the output is used to operate the coil of a motor starter or contactor which can be used to control almost anything.

INTERNAL RELAYS

The actual logic of the control circuit is performed by *internal relays*. An internal relay is an imaginary device that exists only in the logic of the computer. It can have any number of contacts,

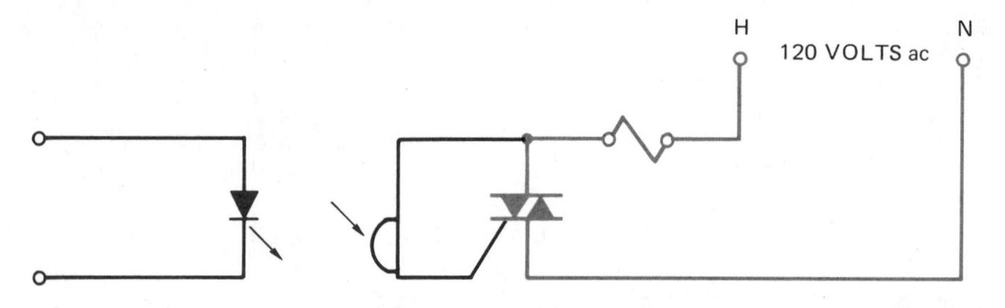

FIGURE 64-15 An output controls a solenoid

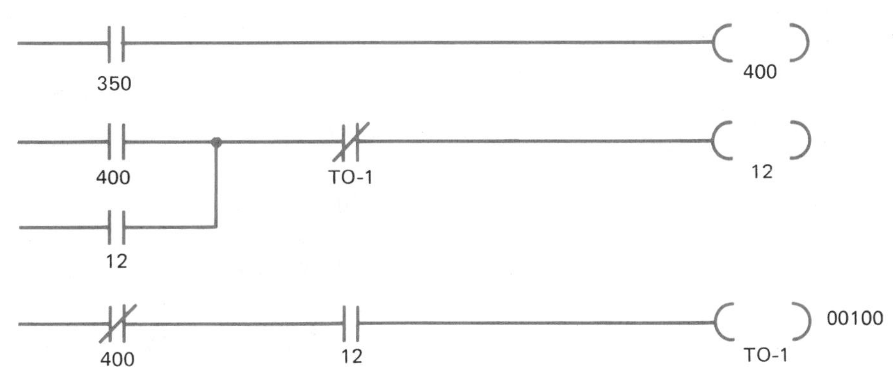

FIGURE 64-16 Off-delay circuit

from one to several hundred, and the contacts can be normally open or normally closed. Internal relays can be programmed into the computer by assigning a coil some number greater than the I/O capacity. For example, assume that the programmable controller has an I/O capacity of 256. If a coil is programmed into the computer and assigned a number greater than 256, 257 for instance, it is an internal relay. Any number of contacts can be controlled by relay 257 by inserting a contact symbol in the program and numbering it 257. If a coil is numbered 256 or less, it can turn on an output when energized.

Inputs are programmed in a similar manner. If a contact is inserted in the program and assigned the number 256 or less, the contact will be changed when a voltage is sensed at that input point. For example, assume that a normally open contact has been programmed into the circuit and assigned number 22. When voltage is applied to input number 22, the contact will close. Since 22 is used as an input in this circuit, care must be taken not to assign number 22 to a coil. Terminal number 22 cannot be used as both an input and an output at the same time.

Counters and Timers

The internal relays of a programmable controller can be used as counters and timers. When timers are used, most of them are programmed in .1 second time intervals. For example, assume that a timer is to be used to provide a delay of 10 seconds. When the delay time is assigned to the timer, the number 00100 is used. This means the timer has been set for 100 tenths of a second which is 10 seconds.

Off-Delay Circuit

The internal timers of a programmable controller function as on-delay relays. A simple circuit can be used, however, to change the sense of the on-delay timer to make it perform as an off-delay timer. Figure 64-16 is this type of circuit. The desired operation of the circuit is as follows. When contact 350 closes, relay coil 12 energizes immediately and turns on a solenoid valve. When contact 350 opens, coil 12 remains energized for 10 seconds before it de-energizes and turns off the solenoid.

This logic is accomplished as follows:

A. When contact 350 closes, internal relay 400 energizes.

B. When coil 400 energizes, normally open contact 400 closes and completes a circuit to coil 12, and normally closed contact 400 connected in series with timer TO-l opens.

C. When relay coil 12 energizes, both normally open 12 contacts close, and the I/O output at terminal 12 connects the solenoid coil to the power line.

D. When contact 350 opens, internal relay 400 de-energizes.

E. This causes both 400 contacts to change back to their original positions.

F. When normally open contact 400 returns to its open state, a continued current path to coil 12 is maintained by the now closed contact 12 connected parallel to it.

G. When normally closed contact 400 returns to its closed position, a circuit is completed through the now closed contact 12 to coil TO-1.

H. When coil TO-1 energizes, a 10-second timer starts. At the end of this time period, contact TO-1 opens and de-energizes coil 12.

I. When coil 12 de-energizes, both 12 contacts return to the open position and the I/O output turns the solenoid off.

J. Timer TO-1 de-energizes when contact 12 opens and the circuit is back in its original start condition.

The number of internal relays and timers contained in a programmable controller is determined by the memory capacity of the computer. As a general rule, PCs that have a large I/O capacity will have a large memory, and machines that have less I/O capacity will have less memory.

The use of programmable controllers has steadily increased since their invention in the late 1960s. A PC can replace hundreds of relays and occupy only a fraction of the space. The circuit logic can be changed easily and quickly without requiring extensive hand rewiring. They have no moving parts or contacts to wear out, and their down time is less than an equivalent relay circuit. A programmable controller used to control a dc drive unit is shown in figure 64-17.

FIGURE 64-17 Dc drive unit controlled by a programmable controller (Courtesy Allen-Bradley Co., Drives Division)

The programming methods presented in this text are for one type of programmable controller. Although there are about as many different methods of programming a programmable controller as there are manufacturers, the concepts presented here are basic to all controllers. It will be necessary, however, to consult the instruction manual when using a particular brand of controller.

REVIEW QUESTIONS

1. What industry first started using programmable controllers?
2. Name two differences between PCs and home or business computers.
3. Name the four basic sections of a programmable controller.
4. In what section of the programmable controller is the actual circuit logic performed?
5. What device is used to program the PC?
6. What device separates the programmable controller from the outside circuits?
7. What two functions are performed by an input I/O?
8. If an output I/O controls dc voltage, what electronic device controls the circuit?
9. If an output I/O controls ac voltage, what electronic device controls the circuit?
10. What is an internal relay?

UNIT 65

Programming a PC

Objectives ***After studying this unit, the student will be able to:***

- **Convert a relay schematic to a schematic used for programming a PC**
- **Enter a program into a programmable controller**

In this unit a relay schematic will be converted into a diagram used to program a programmable controller. The process to be controlled is shown in figure 65-1. A tank is used to mix two liquids. The control circuit operates as follows:

A. When the start button is pressed, solenoids A and B energize. This permits the two liquids to begin filling the tank.

B. When the tank is filled, the float switch trips. This de-energizes solenoids A and B and starts the motor used to mix the liquids together.

C. The motor is permitted to run for one minute. After one minute has elapsed, the motor turns off and solenoid C energizes to drain the tank.

D. When the tank is empty, the float switch de-energizes solenoid C.

E. A stop button can be used to stop the process at any point.

F. If the motor becomes overloaded, the action of the entire circuit will stop.

G. Once the circuit has been energized it will continue to operate until it is manually stopped.

CIRCUIT OPERATION

A relay schematic that will perform the logic of this circuit is shown in figure 65-2. The logic of this circuit is as follows:

A. When the start button is pushed, relay coil CR is energized. This causes all CR contacts to close. Contact CR-1 is a holding contact used to maintain the circuit to coil CR when the start button is released.

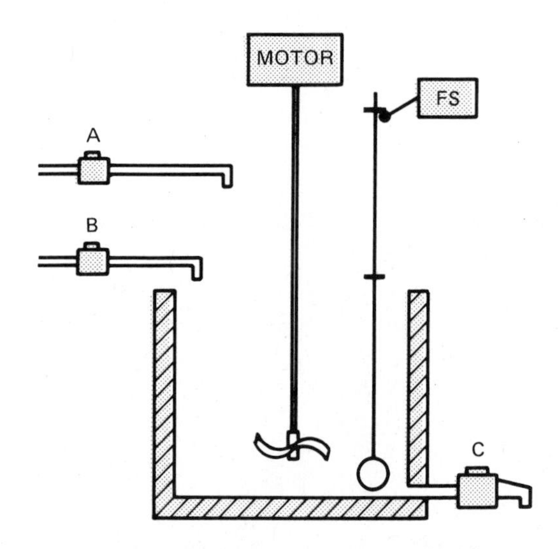

FIGURE 65-1 Tank used to mix two liquids

B. When contact CR-2 closes, a circuit is completed to solenoid coils A and B. This permits the two liquids that are to be mixed together to begin filling the tank.

C. As the tank fills, the float rises until the float switch is tripped. This causes the normally closed float switch contact to open and the normally open contact to close.

D. When the normally closed float switch opens, solenoid coils A and B de-energize and stop the flow of the two liquids into the tank.

E. When the normally open contact closes, a circuit is completed to the coil of a motor starter and the coil of an on-delay timer. The motor is used to mix the two liquids together.

F. At the end of the one minute time period, all of the TR contacts change position. The normally closed TR-2 contact connected in series with the motor starter coil opens and stops the operation of the motor. The normally open TR-3 contact closes and energizes solenoid coil C which permits liquid to begin draining from the tank. The normally closed TR-1 contact is used to assure that valves A and B cannot be re-energized until solenoid C de-energizes.

G. As liquid drains from the tank, the float drops. When the float drops far enough, the float switch trips and its contacts return to their normal positions. When the normally open float switch contact reopens and de-energizes coil TR, all TR contacts return to their normal positions.

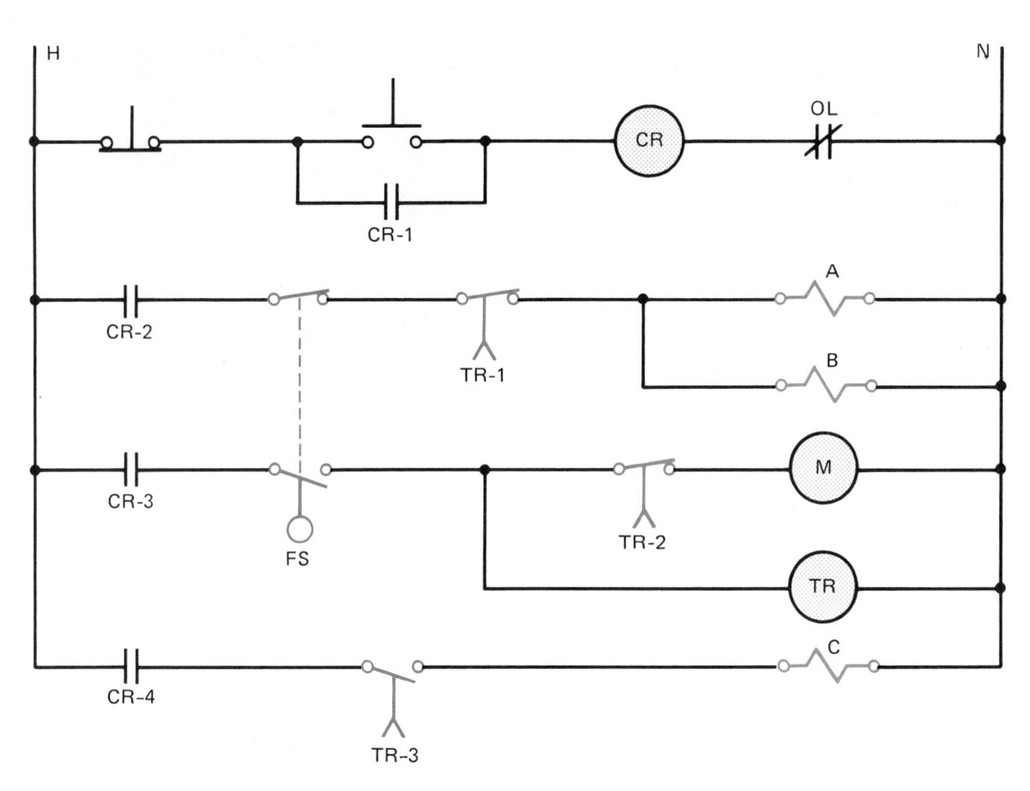

FIGURE 65-2 Relay schematic

H. When the normally open TR-3 contact reopens, solenoid C de-energizes and closes the drain valve. Contact TR-2 recloses, but the motor cannot restart because of the normally open float switch contact. When contact TR-1 recloses, a circuit is completed to solenoids A and B. This permits the tank to begin refilling, and the process starts over again.

I. If the stop button or overload contact opens, coil CR de-energizes and all CR contacts open. This de-energizes the entire circuit.

DEVELOPING A PROGRAM

This circuit will now be developed into a program which can be loaded into the programmable controller. Figure 65-3 shows a program being developed on a programming terminal. Assume that the controller has an I/O capacity of 32, that I/O terminals 1 through 16 are used as inputs, and that terminals 17 through 32 are used as outputs.

The first line of the relay schematic in figure 65-2 shows a normally closed stop button connected in series with a normally open start button. A normally open CR contact is connected parallel to the start button, and coil CR and the normally closed overload contact are connected in series with the start button.

Figure 65-4 shows lines 1 and 2 of the program which will be used to program the PC. Contact #1 is controlled by the normally closed stop push button connected to terminal #1 of the I/O track. Notice that contact #1 is programmed normally open instead of normally closed. Since the stop button is a normally closed push button, power will be applied to terminal #1 of the I/O track during the normal operation of the circuit. When power is applied to an input terminal of the I/O track, the central processing unit interprets it as an instruction to change the position of the contact that corresponds to that terminal number. Therefore, the CPU will interpret contact #1 to be closed during the normal operation of the circuit.

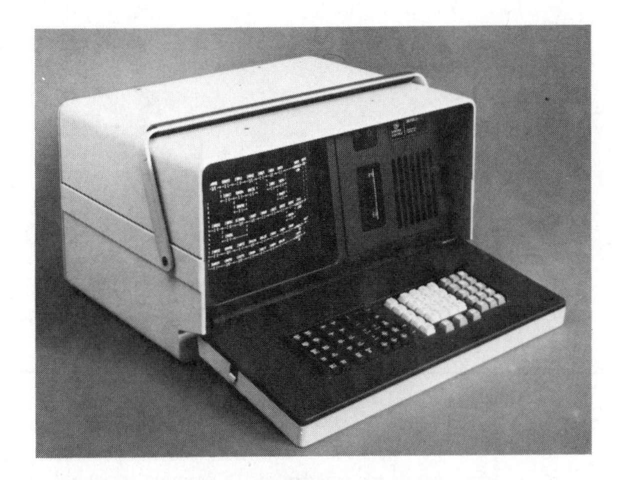

FIGURE 65-3 A program being developed on a programming terminal (Courtesy General Electric Co.)

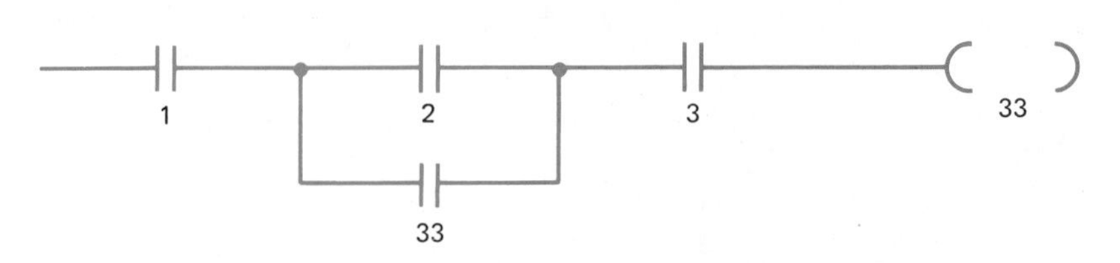

FIGURE 65-4 Lines 1 and 2 of the program

Contact #2 is connected to the normally open start button. Contact #3 is connected to a normally closed overload contact. Notice that this contact is programmed normally open for the same reason that contact #1 is programmed normally open. Notice also that the position of the overload contact has been changed from the right side of the coil to the left side. This is done because a coil ends the line of a program. All contacts used to control a coil must be placed ahead of it. The coil is assigned #33. Since this programmable controller has an I/O capacity of 32, coil 33 is an internal relay.

The normally open contact 33 is connected parallel to contact #2. Although this contact is the only component on the line, it counts as one full line of the program.

In figure 65-5 two more lines have been added to the program. Contact #4 is connected to the float switch. The float switch controls the operation of internal relay 34. Internal relay 35 is controlled by contacts 33, 34, and TO-1.

Figure 65-6 shows the addition of two more lines of the program. When internal relay 35 is energized, outputs 17 and 18 turn on. Solenoid coil A is connected to terminal 17 of the I/O track and solenoid coil B is connected to terminal 18.

Figure 65-7 shows the complete program. Coil 19 controls output 19 which is connected to the coil of a motor starter. Coil TO-1 is an internal timer that has been programmed for 600 tenths of a second, or 60 seconds. Coil 20 controls output 20 which is connected to solenoid coil C.

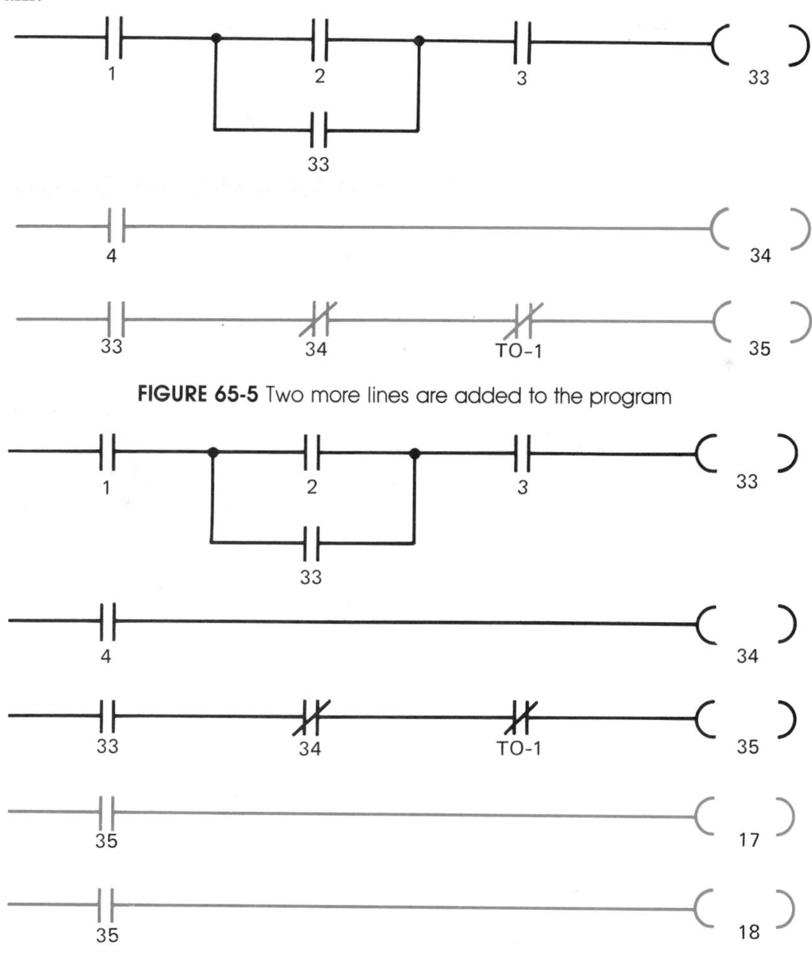

FIGURE 65-5 Two more lines are added to the program

FIGURE 65-6 Solenoids A and B are connected to outputs 17 and 18

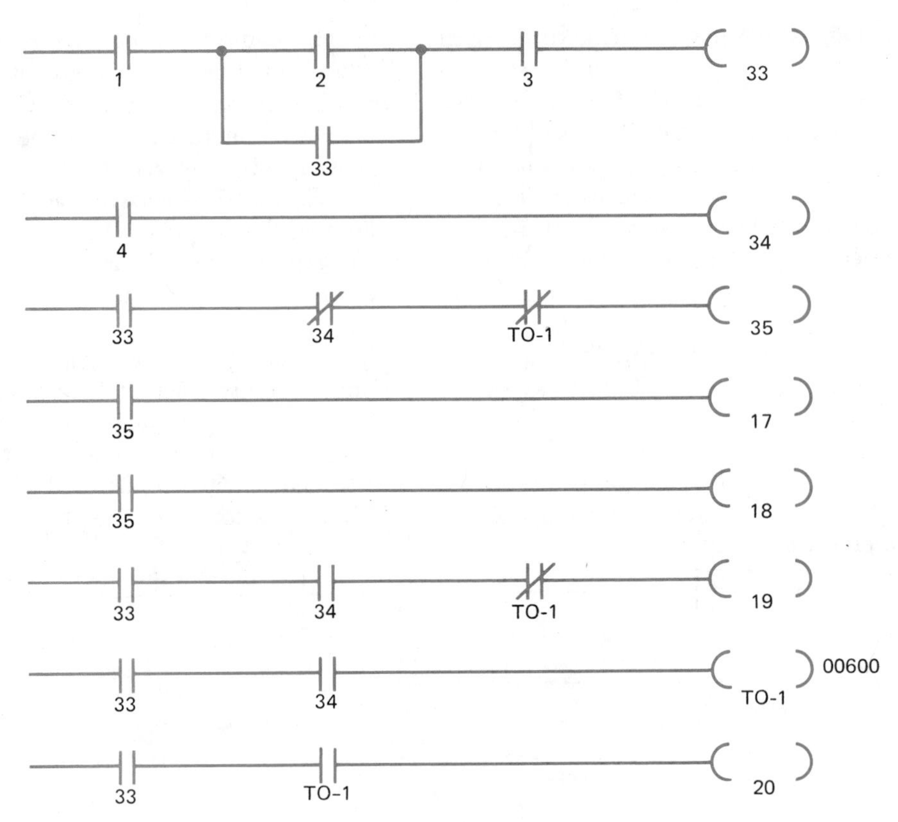

FIGURE 65-7 Complete program

OPERATION OF THE PC CIRCUIT

This circuit operates as follows:

A. Since the stop button and the overload contact are normally closed, inputs #1 and #3 have continuous voltage applied to them. The central processing unit, therefore, interprets contacts #1 and #3 to be closed.

B. When the start button is pushed, contact #2 closes, completing a circuit to coil 33.

C. When coil 33 energizes, all 33 contacts close. The start button can now be released and the circuit will be maintained by the 33 contact connected parallel to contact #2. Internal relay coil 35 energizes when a 33 contact closes.

D. When contacts 35 close, coils 17 and 18 energize. This permits solenoids A and B to energize and begin filling the tank with liquid.

E. When the float rises high enough, the float switch activates and supplies power to terminal #4 of the I/O track. The CPU interprets contact #4 to be closed, which causes internal relay coil 34 to energize.

F. The normally closed 34 contact opens and de-energizes coil 35. This causes both of the 35 contacts to open and de-energize coils 17 and 18. When output terminals 17 and 18 turn off, solenoids A and B de-energize.

When the normally open 34 contact closes, coil 19 energizes, which connects power to the motor starter coil, and timer coil TO-1 energizes and starts the 60-second timer.

G. At the end of the 60-second time period, all TO-1 contacts change position. One of the normally closed contacts opens and prevents coil 35 from energizing until the timer has been reset. The other normally closed contact opens and de-energizes output 19 which de-energizes the motor starter coil to stop the mixing motor. The normally open TO-1 contact closes and energizes output 20. This output energizes solenoid C which begins draining the tank. As liquid drains from the tank, the float drops. When the float drops far enough, the float switch opens and breaks the circuit to input terminal #4. This causes contact #4 to open and de-energize coil 34.

I. When coil 34 de-energizes, all 34 contacts return to their normal positions. This causes coil TO-1 to de-energize and all TO-1 contacts to return to their normal positions.

J. This de-energizes coil 20 which turns off solenoid C, and completes the circuit to coil 35. When coil 35 energizes, outputs 17 and 18 turn on and energize solenoids A and B to start the process over again.

K. If the stop push button is pushed or if the overload contact opens, coil 33 will de-energize and stop the operation of the circuit.

Notice that this circuit operates with the same logic as the relay circuit. The circuit in figure 65-7 is now ready to be programmed into the controller. Push buttons located on the programming terminal which represent open and closed con-

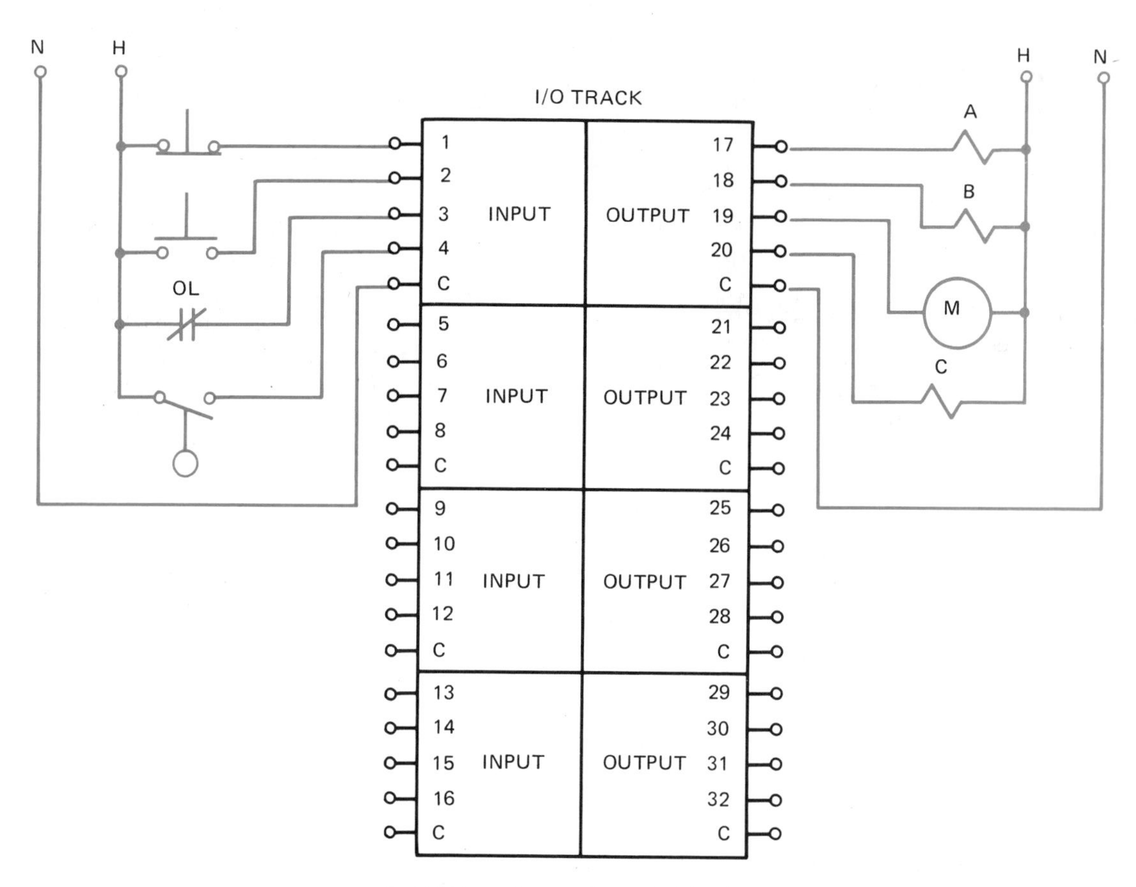

FIGURE 65-8 Connection to I/O track

tacts, coils and timers are used to insert the program into the central processing unit.

Figure 65-8 shows the connection of external components to the I/O track. The push buttons, the overload contact, and the float switch are connected to the input terminals. The solenoid coils and the motor starter coil are connected to the output terminals.

PROGRAMMING IN BOOLEAN

The preceding example circuit was developed for one specific type of programmable controller. It was intended as an example of how to develop and enter a program into the logic of the CPU using a programming terminal similar to the one shown in figure 64-4A. There may be times when it is necessary to use a small programming device which is hand held or which attaches directly to the CPU when entering a program. A unit of this type is shown in figure 65-9. This programming unit can be used with the SERIES ONE group of programmable controllers manufactured by GE Fanuc Automation. The following program will be developed for entry into the SERIES ONE using the hand-held programmer.

DEVELOPING THE PROGRAM

The following program will be used as a trouble annunciator: A pressure switch is to be connected to the input of a programmable controller. When the pressure rises to a preset point, an audible alarm will be sounded and a warning light will flash on and off. When the operator acknowledges the trouble, the audible alarm will be silenced, but the warning light will continue to flash on and off until the pressure returns to a safe level.

PARAMETERS OF THE PROGRAMMABLE CONTROLLER

Before the program can be developed, the parameters of the programmable controller being used must be known. Because the SERIES ONE programmable controller is being used in this example, its parameters will be discussed. An operations and programming guide for the SERIES ONE is shown in figure 65-10. All coil and I/O references must be entered in *OCTAL*. *OCTAL* is a number system which contains only eight digits, 0 through 7. The numbers 8 and 9 are not used because they do not exist as far as the computer is concerned. This does not mean that the numbers 8 and 9 cannot be used when entering times for a timer. This applies only to the way inputs, outputs, and internal relays are identified. For example, any programmable controller that is octal base will not use the numbers 8 or 9. The I/O points for this unit are 000 through 157. Assume the first I/O module used with this controller contains eight units, and these eight units are inputs. The inputs will number from 0 to 7. Now, assume the next set of I/O's is an output module. Numbers 10 through 17 can be used as an output. Notice that numbers 8 and 9 are omitted. The programming guide indicates that a total of 144 internal coils exists. Coils 160 through 337 are nonretentive and coils 340 through 373 are retentive. There are a total of 64 timers and counters which begin with 600 and go through 677. Remember that there are no 8s or 9s. After timer number 607 is used, the next timer will be 610.

The circuit shown in figure 65-11 will be programmed into the controller using the small programming unit. The contacts labeled 0 and 1 are inputs. Contact 0 is connected to the normally open pressure switch which is used to sense the high pressure condition. Contact 1 is connected to the normally open push button used to acknowledge the fault and to turn off the audible alarm. Coils 10 and 11 are outputs. Coil 10 is connected to the warning light and coil 11 is connected to the audible alarm. Coils T600 and T601 are timers used to produce the flashing action of the warning light. The time setting of these timers determines the flash sequence of the warning light. In this circuit, the warning light will be on for 0.5 second and off for 0.5 second. Coil 160 is an internal relay.

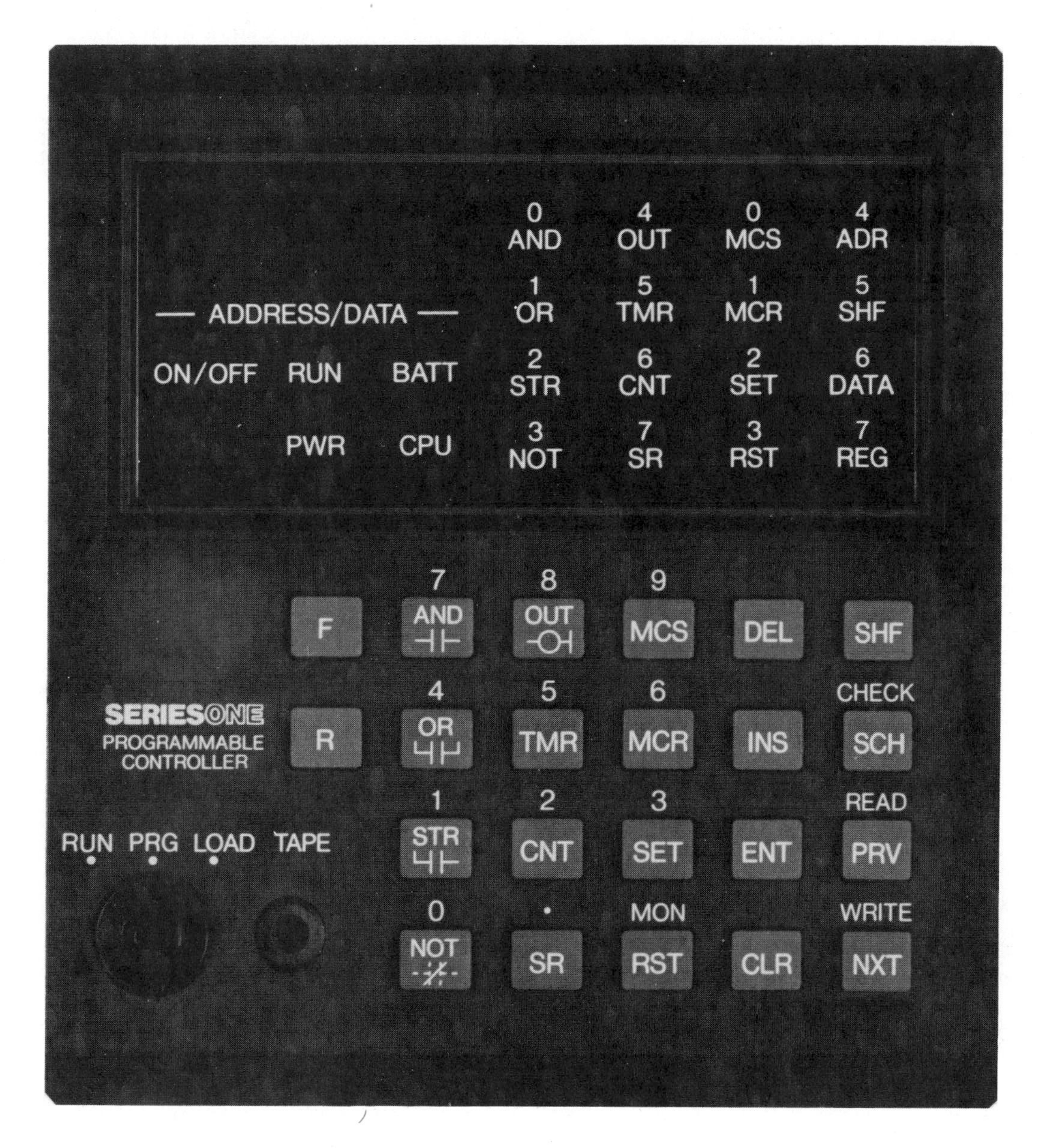

FIGURE 65-9 Small programming unit attaches directly to the programming unit (Courtesy GE Fanuc Automation North America, Inc.)

OPERATION OF THE CIRCUIT

The circuit operates in the following manner: When the pressure switch closes, all 0 inputs change position. This provides a current path to timer T600 which begins timing. A current path is provided to output 10 which turns on the warning light, and a current path is provided to the audible alarm turning it on. The normally open 0 contact connected in series with coil 160 closes. At the end of a half second, timer T600 times out and

MEMORY TYPE	VALID REFERENCES (OCTAL)	QUANTITY (DECIMAL)
SERIES ONE		
I/O Points	000- 157	112 total
Internal Coils		144 total
Non-Retentive	160 - 337	112
Retentive Coils	340 - 373	28
Initial Reset	374	1
0.1 Second Clock	375	1
Disable All Outputs	376	1
Back-Up Battery Status	377	1
Shift Registers	400-577	128 steps
Timer/Counters	600-677	64 (1)
Sequencers	600-677	64 (1000 steps)
SERIES ONE PLUS		
I/O Points	000-157 700-767	168 total
Internal Coils		144 total
Non-Retentive	160 - 337	112
Retentive Coils	340 - 373	28
Initial Reset	374	1
0.1 Second Clock	375	1
Disable All Outputs	376	1
Back-Up Battery Status	377	1
Shift Registers	400-577 (2)	128 steps
Timer/Counters	600-677	64 (1)
Sequencers	600-677	64 (1000 steps)
Data Registers	400-577 (2)	64 (16-bit)

(1) Total maximum number of Timers and/or Counters
(2) Shift register and data register references are identical, however, shift registers operate on bits, while data registers (located in a totally different area of memory) operate on bytes

MODE SWITCH	
POSITION	FUNCTION
RUN	CPU scans logic, outputs enabled
PROG	Enter/Edit logic, no scanning
LOAD	Controls transfer to and from external device (recorder, printer, PROM writer)

STATUS INDICATORS	
ON OFF	Operating state of I/O, internal coils or shift registers
RUN	ON=CPU in RUN mode
BATT	ON=Lithium battery voltage low
PWR	ON=Power supply DC voltage normal
CPU	ON=CPU internal fault. Watchdog timer timed out. Low DC voltage

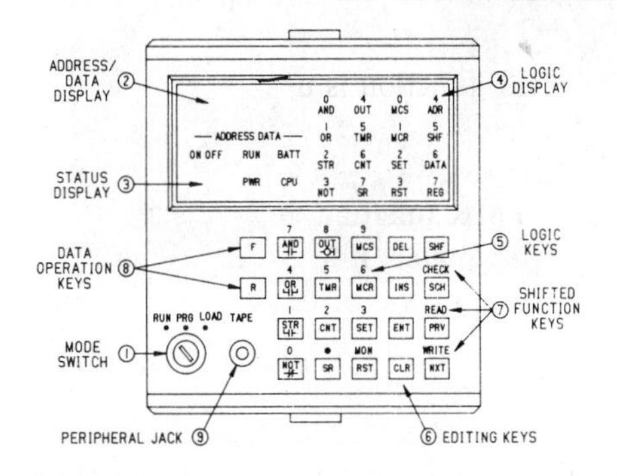

LOGIC AND EDITING KEYS

KEY	DESCRIPTION
F	(Series One Plus Only) Entered before a 2-digit number to select a data operation.
R	(Series One Plus Only) Entered before a 3-digit data register or 2-digit group reference when programming data operations.
AND	Places logic in series with previous logic.
OR	Places logic in parallel with previous logic.
STR	Starts a new line or group of logic.
NOT	Specifies a normally closed contact when used with AND/OR.
OUT	Ends line of logic with a coil, can be an output.
TMR	Specifies a timer function.
CNT	Specifies a counter function.
SR	Specifies a shift register function.
MCS	Begins a master control relay function.
MCR	Ends a master control relay function.
SET	Specifies a latched coil or used to force an I/O reference on.
RST	Turns off a latched coil or forces an I/O reference off.
DEL	Included in sequence for removing (deleting) an instruction from program memory.
INS	Included in sequence for adding (inserting) an instruction in program memory.
ENT	Causes logic to be placed in program memory.
CLR	Removes (clears) previous logic entry, acknowledges error codes, causes memory address to be displayed when monitoring a program.
SHF	Selects shifted functions (upper label above keys).
SCH	Used when initiating a search function.
PRV	Selects previous logic or function, and when monitoring, selects the previous group of 8 references.
NXT	Selects the next logic function. When monitoring selects the next group of 8 references.
0 to 9	SHIFTED FUNCTION. Selects numerical values.
•	SHIFTED FUNCTION. Selects decimal point when entering numerical values, (timers using XXX.X seconds).
MON	SHIFTED FUNCTION. Selects monitor operation.
CHECK	SHIFTED FUNCTION. Initiates verify operation with peripheral.
READ	SHIFTED FUNCTION. Initiates loading of CPU memory from a peripheral.
WRITE	SHIFTED FUNCTION. Initiates writing (recording) program in CPU memory to a peripheral.

FIGURE 65-10 Programming guide for a SERIES ONE programmable controller (Courtesy of GE Fanuc Automation North America Inc.)

changes the position of all T600 contacts. The normally closed contact connected in series with the warning light opens and turns off output 10. The normally open T600 contact closes and permits timer T601 to begin timing. At the end of a half second, timer T601 opens its normally closed contact connected in series with timer T600. This causes timer T600 to reset and return all of its

PROGRAMMER OPERATION

OPERATION	KEYSTROKES	MODE* R	P	L
Clear all memory.	[CLR] [SHF] [3] [4] [8] [DEL] [NXT]		X	
Display present address.	[CLR]	X	X	
Display present function.	[NXT]	X	X	
Next function.	[NXT]	X	X	
Previous function.	[PRV]	X	X	
Go to first function in program memory (address 0000).	[SHF] [NXT]	X	X	
Go to a specific memory address.	[SHF] (Address) [NXT]	X	X	
Search for a specific function.	(Function) [SHF] (Ref No.) [SCH] [NXT]	X	X	
Search for a specific reference number.	[SHF] (Ref. No.) [SCH] [NXT]	X	X	
Insert function before the displayed function (or address)	(Function) [SHF] (Ref No.) [INS] [NXT]		X	
Delete function	(Address) [DEL] [PRV]		X	
Edit a program	(Address) (Function) [SHF] (Ref No.) [ENT]		X	
Check program for errors. If none, next empty address is displayed	[CLR] [SCH]	X	X	
Change T/C preset	(Address) [SHF] (preset) [ENT]	X		
Mon. ON/OFF state of contact or coil	Observe On/OFF LED when coil or contact is selected	X		
Monitor group of 8 consecutive references (I/O, internal coils, SR coils)	[SHF] (Beginning Ref. No.) [MON]	X		
Monitor timer or counter accumulate register	[SHF] (T/C No.) [MON]	X		
Force a reference ON (will be overridden by user logic)	[SET] [SHF] (Ref. No.) [ENT]	X		
Force a reference OFF (will be overridden by user logic)	[RST] [SHF] (Ref No.) [ENT]	X		
Enter a function into program memory	(Function) [SHF] (Ref No.) [ENT]		X	
Write to tape, printer, or PROM writer	(Optional Program ID) [WRITE]			X
Load program memory from tape.	(Optional Program ID) [READ]			X
Verify data on tape or in PROM writer RAM against program memory.	(Optional Program ID) [CHECK]			X

*R=RUN, P=PROGRAM, L=LOAD

FIGURE 65-10 *(continued)*

contacts to their normal position. The normally closed T600 contact permits output 10 to turn on again, and the normally open T600 contact resets timer T601. When timer T601 resets, its contact returns to its normal position, and timer T600 begins timing again.

This condition continues until the operator presses the acknowledge button causing input contact 1 to close. Contact 1 completes a current path to internal relay 160. When internal relay 160 energizes, the normally open 160 contact closes and seals the circuit around contact 1. The normally closed 160 contact opens and turns off the audible alarm. At this time in the circuit, the audible alarm has been turned off, but the warning light is flashing on and off at half-second intervals. This will continue until the pressure drops to a safe level and input 0 reopens all of its contacts causing the circuit to reset to its normal position.

ENTERING THE PROGRAM

Now that the circuit has been developed, it must be entered into the memory of the CPU. When using a small programming terminal as shown in figure 65-9, the program must be entered in a language called *Boolean*. When programming in Boolean, to connect one contact in series with another, the AND function must be used. To connect a contact in parallel with another, the OR function is used. To change a contact from open to closed, the NOT function is used. To start a line of the program, the STR function must be used. To end a line of the program, the OUT function is used except when programming a special function such as a timer or counter. When ending a line of the program with a timer, the TMR function is used; when ending the line with a counter, the CNT function is used. Each component of the program must be entered into memory using the ENT key. Some of the keys on this programming unit serve two functions. The AND key, for example, is also used to enter the number 7 into the program. The NOT key is also used to enter the number 0 into the program. The second function keys are very similar to the dual purpose keys on a typewriter where the shift key is used to access the second function of a key. The same is true for this unit. The SHF key is used to cause the keys to perform their second function. Once the SHF key has been

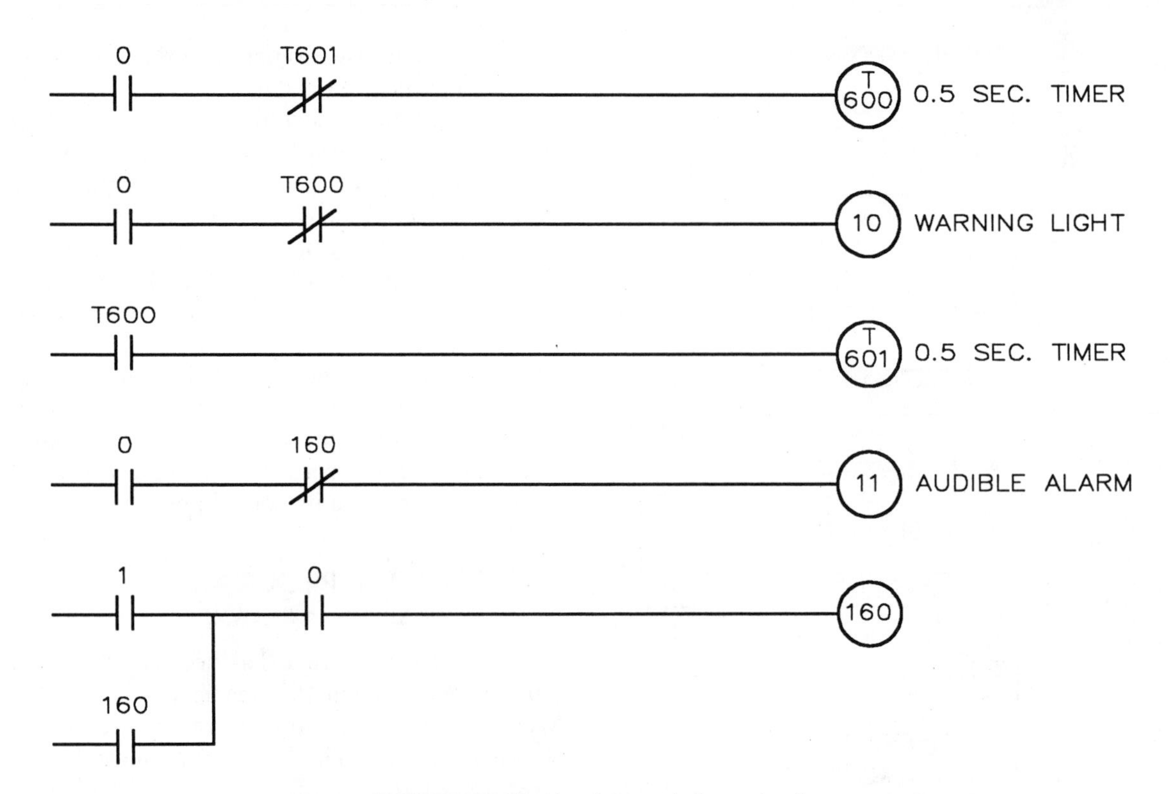

FIGURE 65-11 Warning light and alarm circuit

pressed, it will remain in effect until the ENT key is pressed. There is no need to hold the SHF key down when entering more than one digit into the program.

The first line of logic will be entered as follows:

STR SHF 0 ENT
AND NOT TMR SHF 601 ENT
TMR SHF 600 ENT
SHF .5 ENT

Notice that the STR command is used to start the line of logic. The SHF key must be pressed in order to permit the number 0 to be entered. The ENT command causes that instruction to be entered into the logic of the CPU. The AND function causes the next contact entered to be connected in series with the first contact, and the NOT command instructs the CPU that the contact is to be normally closed instead of normally open. The TMR command instructs the programmable controller that the contact is to be controlled by a timer. Since this line of logic is ended with a timer instead of a normal output or internal relay, the TMR command is used again to instruct the CPU that the last coil is a timer and not an internal relay or output. The CPU can interpret this last timer command to be a coil instead of a contact because directly following this command, the time of the timer had been entered instead of a tie command such as AND or OR. The time is entered with the use of a decimal point in this controller instead of assuming each time interval to be 0.1 second. Different programmable controllers use different methods to enter the time.

The second line of logic is entered as follows:

STR SHF 0 ENT
AND NOT TMR SHF 600 ENT
OUT SHF 10 ENT

The third line of logic is entered as follows:

STR TMR SHF 600 ENT
TMR SHF 601 ENT
SHF .5 ENT

The fourth line of logic is entered as follows:

STR SHF 0 ENT
AND NOT SHF 160 ENT
OUT SHF 11 ENT

The fifth and sixth lines of logic will be entered together because the sixth line of logic is connected in parallel with the fifth:

STR SHF 1 ENT
OR SHF 160 ENT
AND SHF 0 ENT
OUT SHF 160 ENT

This completes the programming of the circuit into the CPU.

REVIEW QUESTIONS

1. Why are NEMA symbols representing such components as push buttons, limit switches, and float switches not used in a programmable controller schematic?
2. Explain how to program an internal relay into the controller.
3. Why are the contacts used to represent stop buttons and overload contacts programmed normally open?
4. Why is the output I/O used to energize a motor starter instead of energizing the motor directly?
5. A timer is to be programmed for a delay of 3 minutes. What number is used to set this timer?
6. When programming in Boolean, what command is used to connect two circuit components together is series?
7. When programming in Boolean, what command is used to connect two circuit components together in parallel?
8. When programming in Boolean, what command is used to change a contact from normally open to normally closed?
9. Why are the numbers 8 and 9 not used in an *OCTAL* based system?

UNIT 66

Analog Sensing for Programmable Controllers

Objectives *After studying this unit, the student will be able to:*

- **Describe the differences between analog and digital inputs**
- **Discuss precautions that should be taken when using analog inputs**
- **Describe the operation of a differential amplifier**

Many of the programmable controllers found in industry are designed to accept analog as well as digital inputs. Analog means continuously varying. These inputs are designed to sense voltage, current, speed, pressure, temperature, humidity, etc. When an analog input is used, a special module mounts on the I/O track of the PC. An analog sensor may be designed to operate between a range of settings, such as 50 to 300 C°, or 0 to 100 psi. These sensors are used to indicate between a range of values instead of merely operating in an on or off mode. An analog pressure sensor designed to indicate pressures between 0 and 100 psi would have to indicate when the pressure was 30 psi, 50 psi, or 80 psi. It would not just indicate whether the pressure had or had not reached 100 psi. A pressure sensor of this type can be constructed in several ways. One of the most common methods is to let the pressure sensor operate a current generator which produces currents between 4 and 20 milliamperes. It is desirable for the sensor to produce a certain amount of current instead of a certain amount of voltage because it eliminates the problem of voltage drop on lines. For example, assume a pressure sensor is designed to sense pressures between 0 and 100 psi. Also assume that the sensor produces a voltage output of 1 volt when the pressure is 0 psi and a voltage of 5 volts when the pressure is 100 psi. Since this is an analog sensor, when the pressure is 50 psi, the sensor should produce a voltage of 3 volts. This sensor is connected to the analog input of a programmable controller, figure 66-1. The analog input has a sense resistance of 250Ω. If the wires between the sensor and the input of the programmable controller are short enough (so that there is almost no wire resistance), the circuit will operate without a problem. Because the sense resistor in the input of the programmable controller is the only resistance in the circuit, all of the output voltage of the pres-

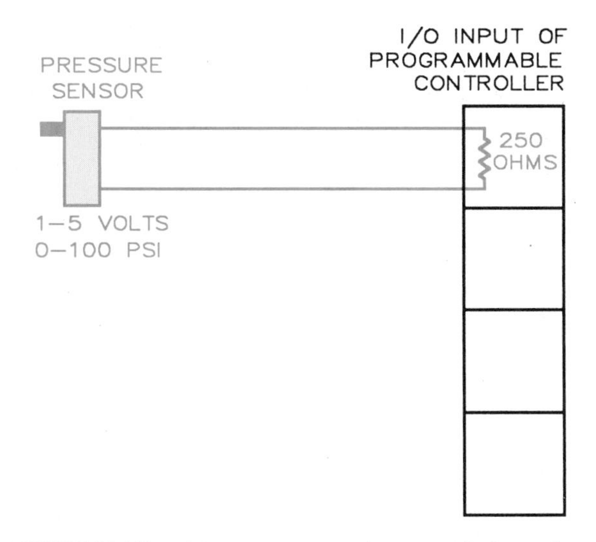

FIGURE 66-1 The pressure sensor produces one to five volts

sure sensor will appear across it. If the pressure sensor produces a 3-volt output, 3 volts will appear across the sense resistor.

If the pressure sensor is located some distance away from the programmable controller, however, the resistance of the two wires running between the pressure sensor and the sense resistor can cause inaccurate readings. Assume that the pressure sensor is located far enough from the programmable controller so that the two conductors have a total resistance of 50Ω, figure 66-2. This means that the total resistance of the circuit is now 300Ω (250 + 50 = 300). If the pressure sensor produces an output voltage of 3 volts when the pressure reaches 50 psi, the current flow in the circuit will be 0.010 amp (3/300 = 0.010). Since there is a current flow of 0.010 through the 250Ω sense resistor, a voltage of 2.5 volts will appear across it. This is substantially less than the 3 volts being produced by the pressure sensor.

If the pressure sensor is designed to operate a current generator with an output of 4 to 20 ma., the resistance of the wires will not cause an inaccurate reading at the sense resistor. Since the sense resistor and the resistance of the wire between the pressure sensor and the programmable controller form a series circuit, the current must be the same at the point in the circuit. If the pressure sensor produces an output current of 4 ma. when the pressure is 0 psi and a current of 20 ma. when the pressure is 100 psi, at 50 psi it will produce a current of 12 ma. When a current of 12 ma. flows through the 250 sense resistor, a voltage of 3 volts will be dropped across it, figure 66-3. Because the pressure sensor produces a certain amount of current instead of a certain amount of voltage with a change in pressure, the amount of wire resistance between the pressure sensor and programmable controller is of no concern.

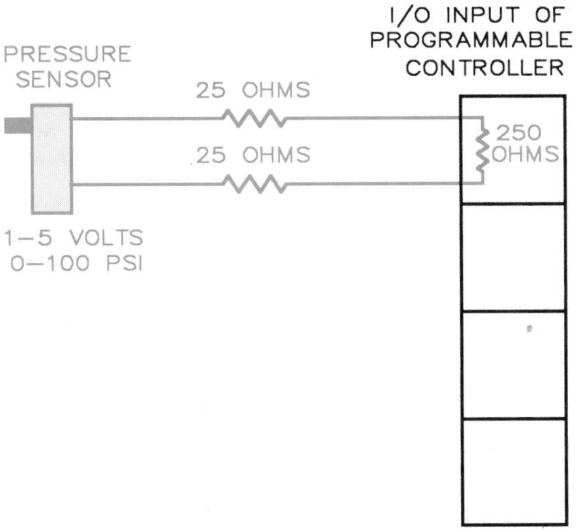

FIGURE 66-2 Resistance in the lines can cause problems

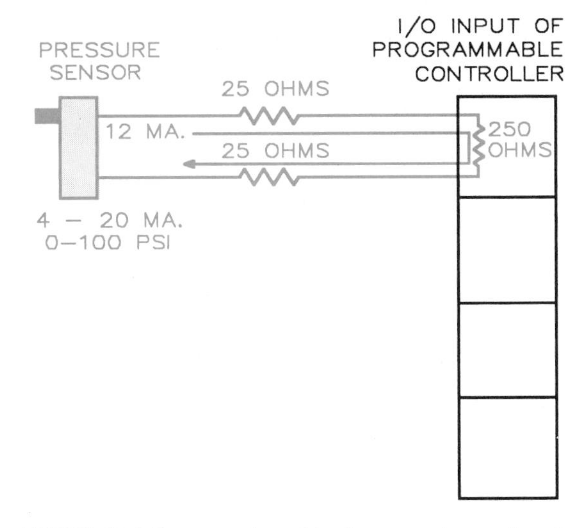

FIGURE 66-3 The current must be the same in a series circuit

INSTALLATION

Most analog sensors can produce only very weak signals—0 to 10 volts or 4 to 20 milliamps are common. In an industrial environment where intense magnetic fields and large voltage spikes abound, it is easy to loose the input signal in the electrical noise. For this reason, special precautions should be taken when installing the signal wiring between the sensor and the input module. These precautions are particularly important when using analog inputs, but they should also be followed when using digital inputs.

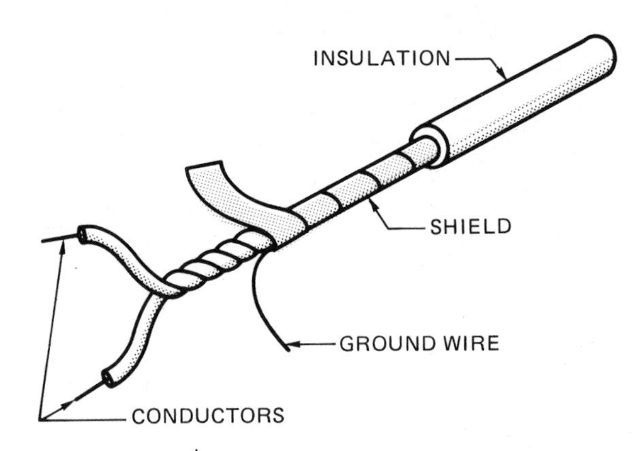

FIGURE 66-5 Shielded cable

Keep Wire Runs Short

Try to keep wire runs as short as possible. A long wire run has more surface area of wire to pick up stray electrical noise.

Plan the Route of the Signal Cable

Before starting, plan how the signal cable should be installed. *Never run signal wire in the same conduit with power wiring.* Try to run signal wiring as far away from power wiring as possible. When it is necessary to cross power wiring, install the signal cable so that it crosses at a right angle as shown in figure 66-4.

Use Shielded Cable

Shielded cable is used for the installation of signal wiring. One of the most common types, shown in figure 66-5, uses twisted wires with a Mylar foil shield. The ground wire must be grounded if the shielding is to operate properly. This type of shielded cable can provide a noise reduction ratio of about 30,000:1.

Another type of signal cable uses a twisted pair of signal wires surrounded by a braided shield. This type of cable provides a noise reduction of about 300:1.

Common coaxial cable should be avoided. This cable consists of a single conductor surrounded by a braided shield. This type of cable offers very poor noise reduction.

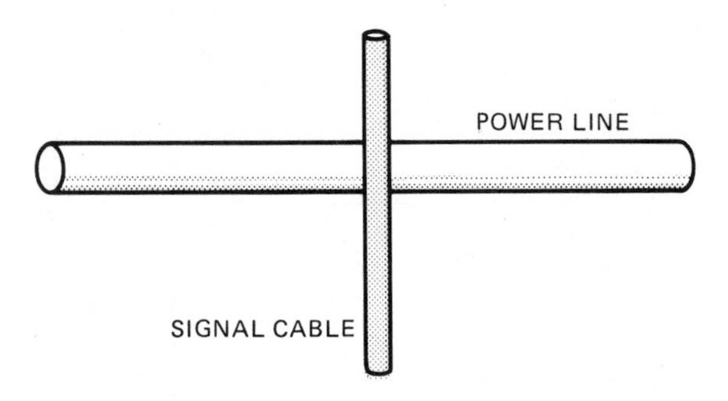

FIGURE 66-4 Signal cable crosses power line at right angle

Grounding

Ground is generally thought of as being electrically neutral, or zero at all points. However, this may not be the case in practical application. It is not uncommon to find that different pieces of equipment have ground levels that are several volts apart, figure 66-6.

To overcome this problem large cable is sometimes used to tie the two pieces of equipment together. This forces them to exist at the same potential. This method is sometimes referred to as the brute-force method.

Where the brute-force method is not practical, the shield of the signal cable is grounded at only one end. The preferred method is generally to ground the shield at the sensor.

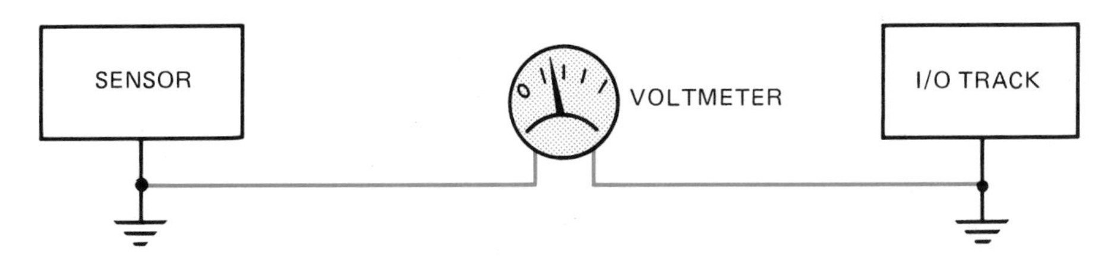

FIGURE 66-6 All grounds are not equal

THE DIFFERENTIAL AMPLIFIER

An electronic device that is often used to help overcome the problem of induced noise is the differential amplifier, figure 66-7. This device detects the voltage difference between the pair of signal wires and amplifies this difference. Since the induced noise level should be the same in both conductors, the amplifier will ignore the noise. For example, assume an analog sensor is producing a 50-millivolt signal. This signal is applied to the input module, but induced noise is at a level of 5 volts. In this case the noise level is 100 times greater than the signal level. The induced noise level, however, is the same for both of the input conductors. Therefore, the differential amplifier ignores the 5-volt noise and amplifies only the voltage difference which is 50 millivolts.

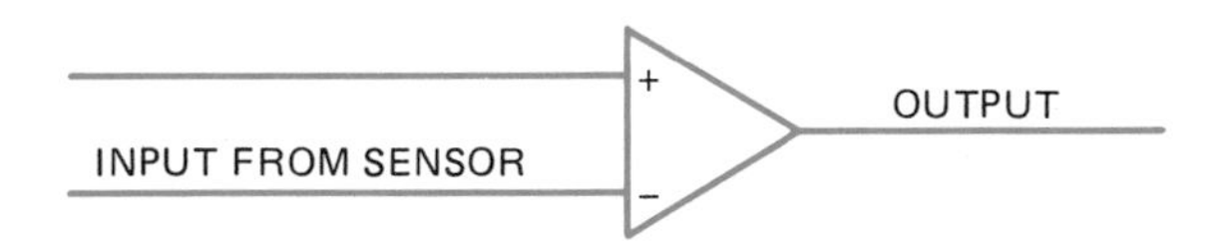

FIGURE 66-7 Differential amplifier detects difference in signal level

REVIEW QUESTIONS

1. Explain the difference between digital inputs and analog inputs.
2. Why should signal wire runs be kept as short as possible?
3. When signal wiring must cross power wiring, how should the wires be crossed?
4. Why is shielded wire used for signal runs?
5. What is the brute-force method of grounding?
6. Explain the operation of the differential amplifier.

APPENDIX

TESTING SOLID-STATE COMPONENTS

1. Testing a Diode

1. Connect the ohmmeter leads to the diode. Notice if the meter indicates continuity through the diode or not.
2. Reverse the diode connection to the ohmmeter. Notice if the meter indicates continuity through the diode or not. The ohmmeter should indicate continuity through the diode in only one direction. NOTE: If continuity is not indicated in either direction, the diode is open. If continuity is indicated in both directions, the diode is shorted.

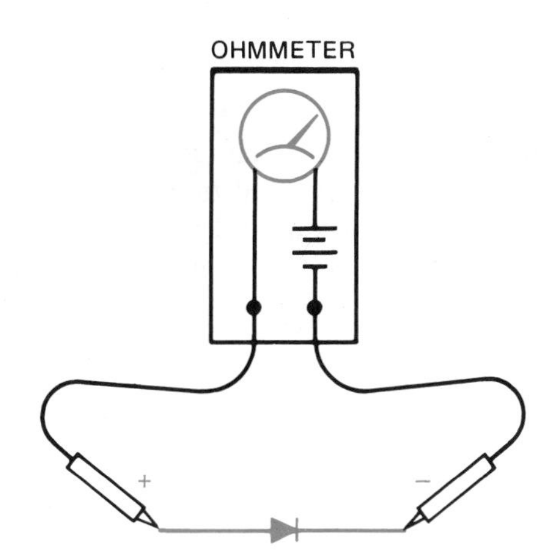

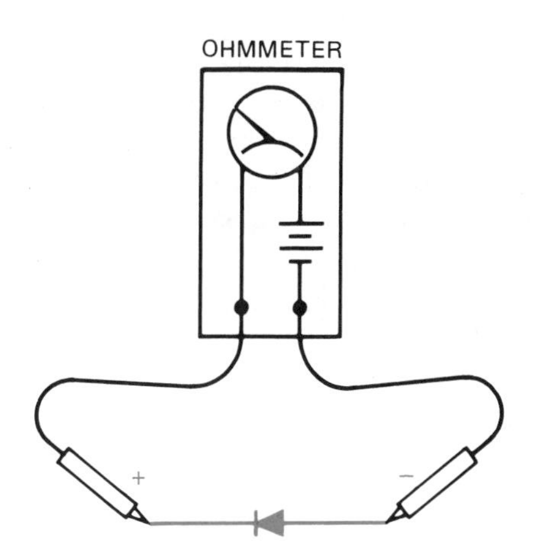

2. Testing a Transistor

1. Using a diode, determine which ohmmeter lead is positive and which is negative. The ohmmeter will indicate continuity through the diode only when the positive lead is connected to the anode and the negative lead is connected to the cathode.

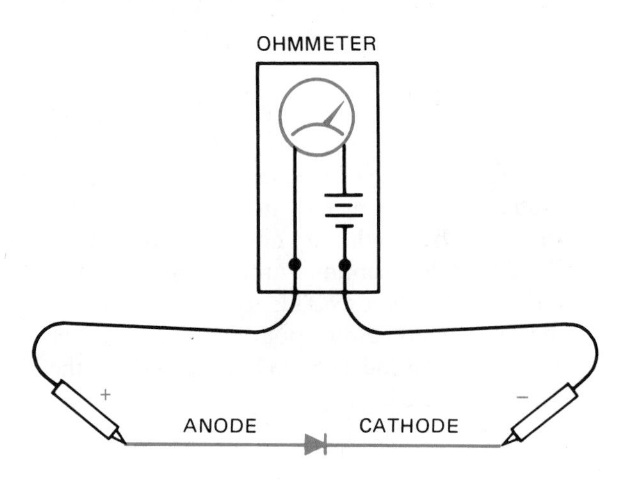

2. If the transistor is an NPN, connect the positive ohmmeter lead to the base and the negative lead to the collector. The ohmmeter should indicate continuity. The reading should be about the same as the reading obtained when the diode was tested.

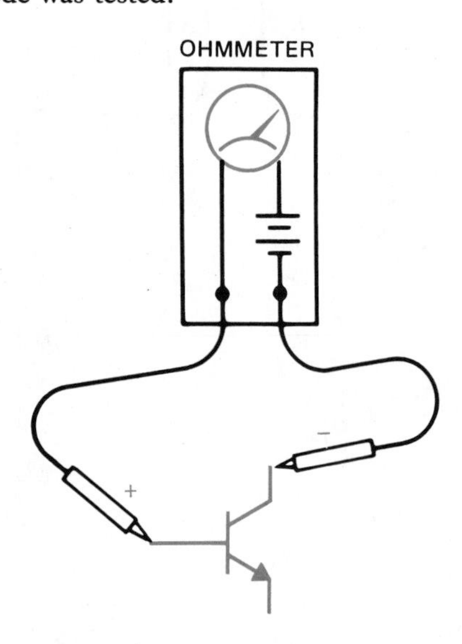

3. With the positive ohmmeter lead still connected to the base of the transistor, connect the negative lead to the emitter. The ohmmeter should again indicate a forward diode junction. NOTE: If the ohmmeter does not indicate continuity between the base-collector or the base-emitter, the transistor is open.

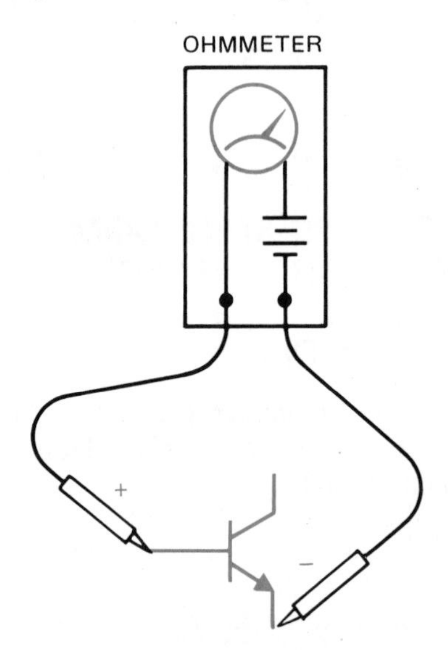

4. Connect the negative ohmmeter lead to the base and the positive lead to the collector. The ohmmeter should indicate infinity or no continuity.

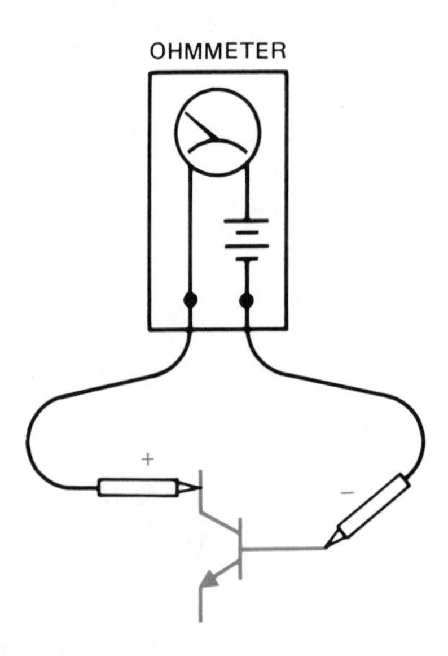

5. With the negative ohmmeter lead connected to the base, reconnect the positive lead to the emitter. There should, again, be no indication of continuity. NOTE: If a very high resistance is indicated by the ohmmeter, the transistor is "leaky" but it may still operate in the circuit. If a very low resistance is seen, the transistor is shorted.

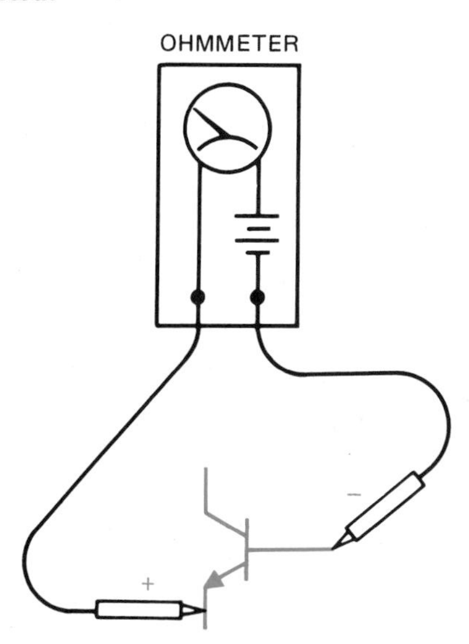

6. To test a PNP transistor, reverse the polarity of the ohmmeter leads and repeat the test. When the negative ohmmeter lead is connected to the base, a forward diode junction should be indicated when the positive lead is connected to the collector or emitter.

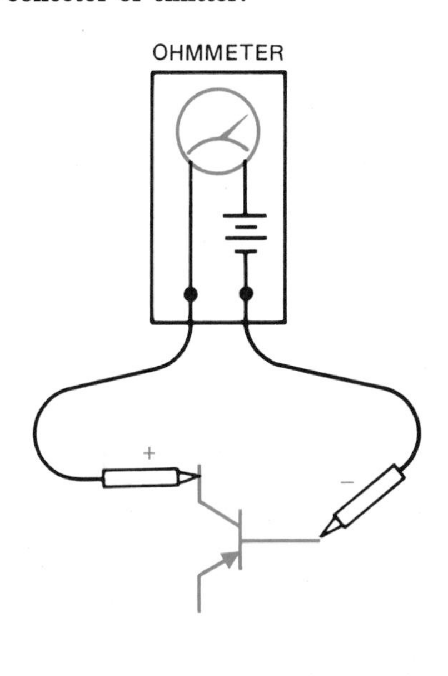

7. If the positive ohmmeter lead is connected to the base of a PNP transistor, no continuity should be indicated when the negative lead is connected to the collector or the emitter.

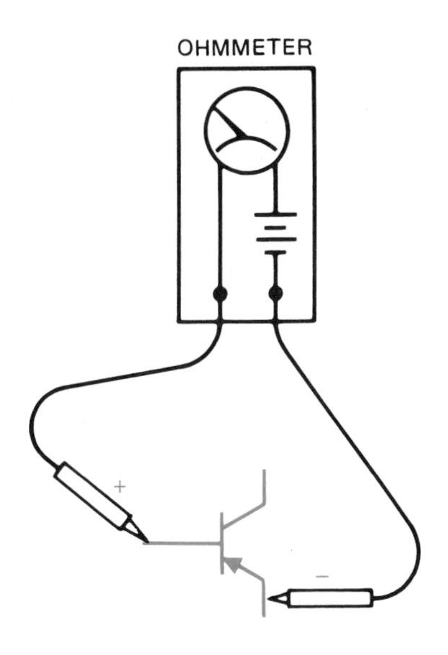

3. Testing a Unijunction Transistor

1. Using a junction diode, determine which ohmmeter lead is positive and which is negative. The ohmmeter will indicate continuity when the positive lead is connected to the anode and the negative lead is connected to the cathode.

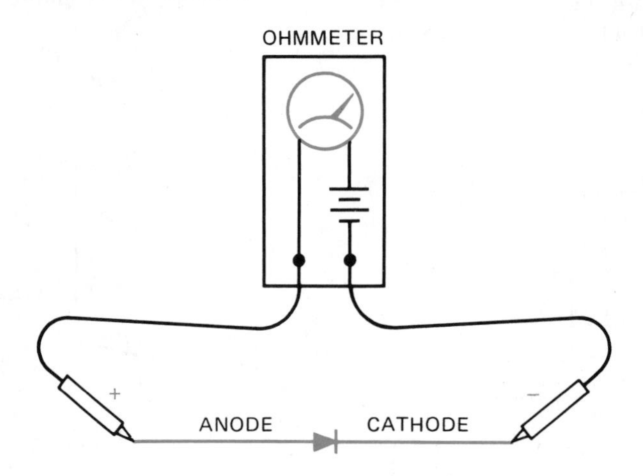

2. Connect the positive ohmmeter lead to the emitter lead and the negative lead to base #1. The ohmmeter should indicate a forward diode junction.

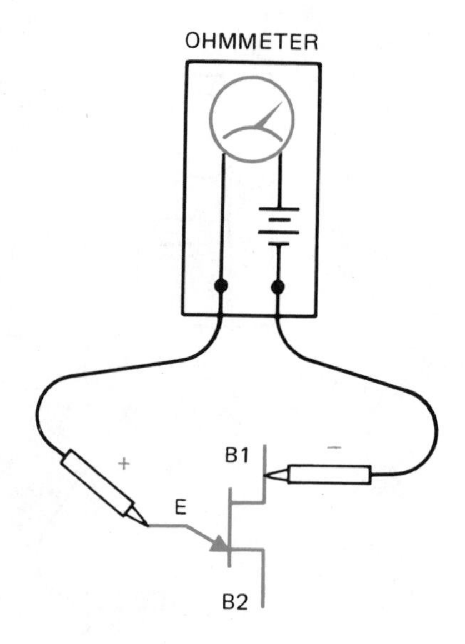

3. With the positive ohmmeter lead connected to the emitter, reconnect the negative lead to base #2. The ohmmeter should again indicate a forward diode junction.

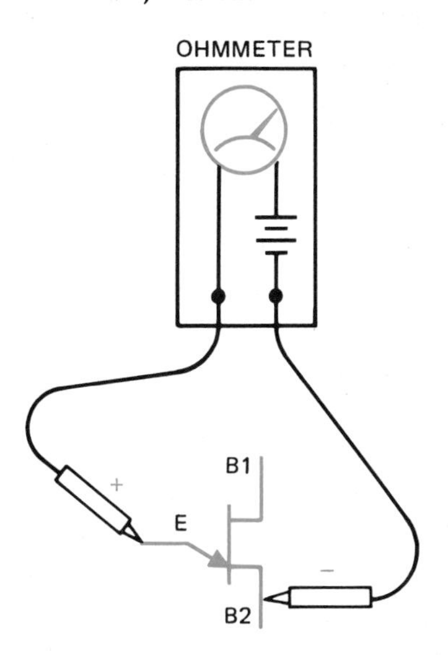

4. If the negative ohmmeter lead is connected to the emitter, no continuity should be indicated when the positive lead is connected to base #1 or base #2.

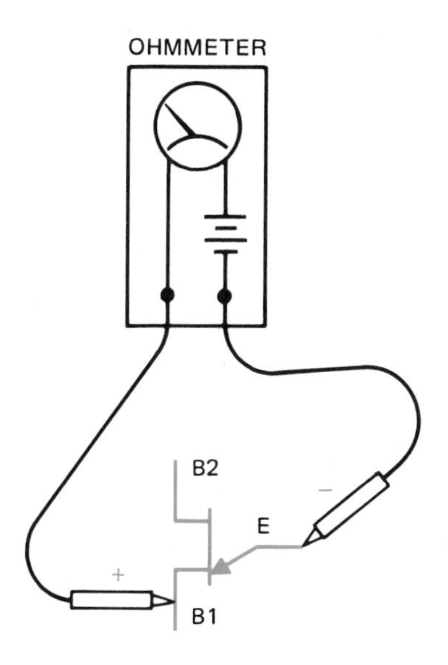

4. Testing an SCR

1. Using a junction diode, determine which ohmmeter lead is positive and which is negative. The ohmmeter will indicate continuity only when the positive lead is connected to the anode of the diode and the negative lead is connected to the cathode.

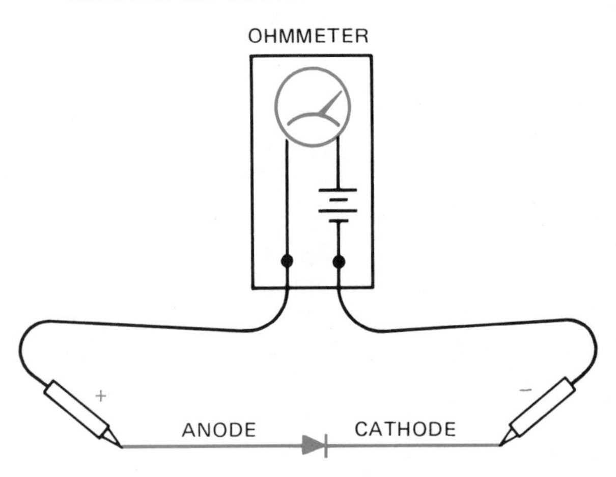

2. Connect the positive ohmmeter lead to the anode of the SCR and the negative lead to the cathode. The ohmmeter should indicate no continuity.

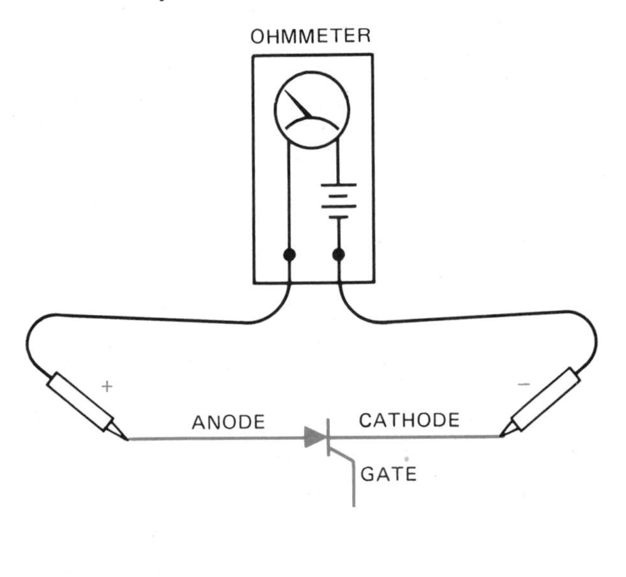

3. Using a jumper lead, connect the gate of the SCR to the anode. The ohmmeter should indicate a forward diode junction when the connection is made. NOTE: If the jumper is removed, the SCR may continue to conduct or it may turn off. This will be determined by whether or not the ohmmeter can supply enough current to keep the SCR above its holding current level.

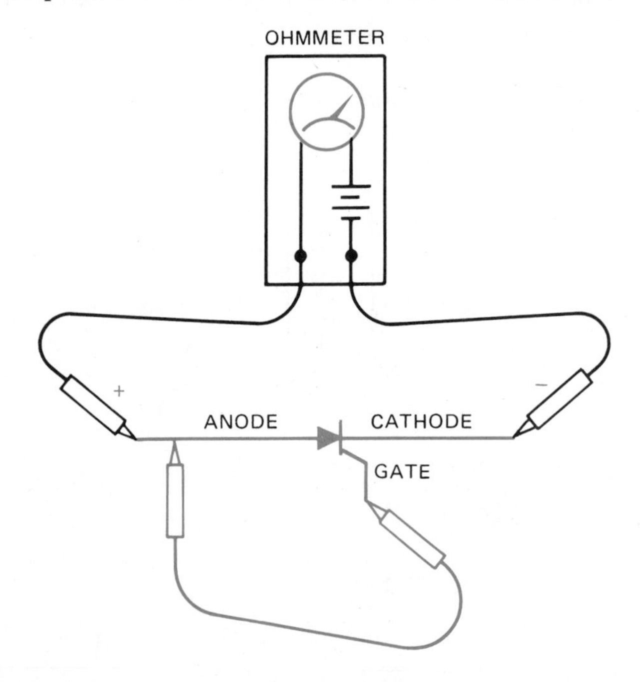

4. Reconnect the SCR so that the cathode is connected to the positive ohmmeter lead and the anode is connected to the negative lead. The ohmmeter should indicate no continuity.

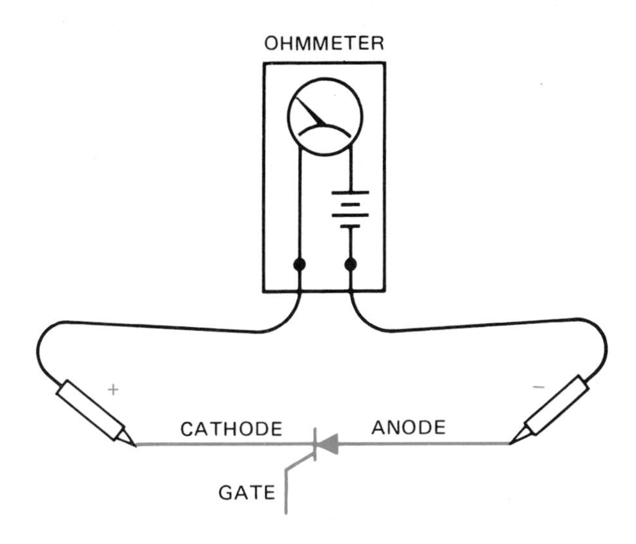

5. If a jumper lead is used to connect the gate to the anode, the ohmmeter should indicate no continuity. NOTE: SCRs designed to switch large currents (50 amperes or more) may indicate some leakage current with this test. This is normal for some devices.

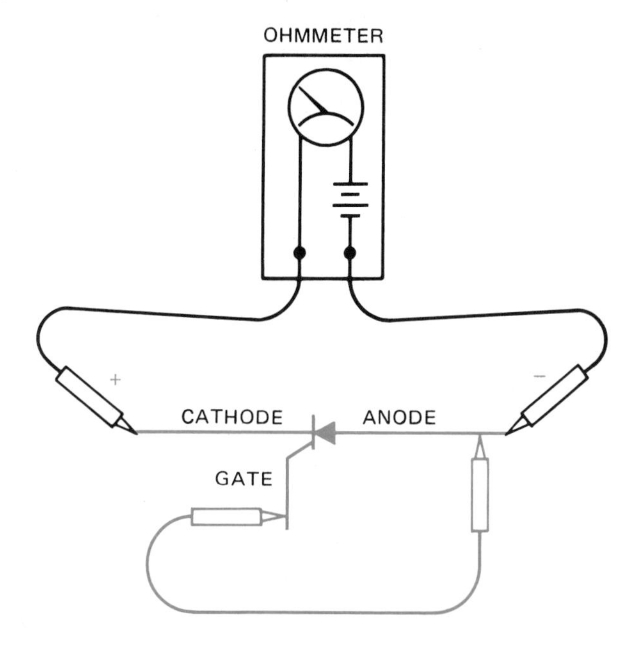

5. Testing a Triac

1. Using a junction diode determine which ohmmeter lead is positive and which is negative. The ohmmeter will indicate continuity only when the positive lead is connected to the anode and the negative lead is connected to the cathode.

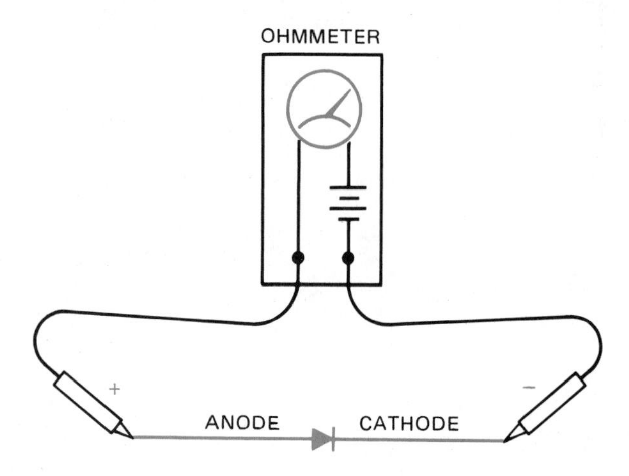

2. Connect the positive ohmmeter lead to MT2 and the negative lead to MT1. The ohmmeter should indicate no continuity through the triac.

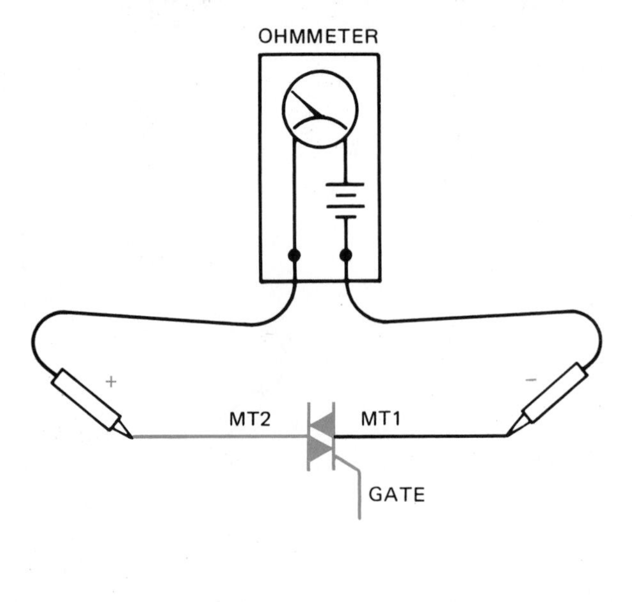

3. Using a jumper lead, connect the gate of the triac to MT2. The ohmmeter should indicate a forward diode junction.

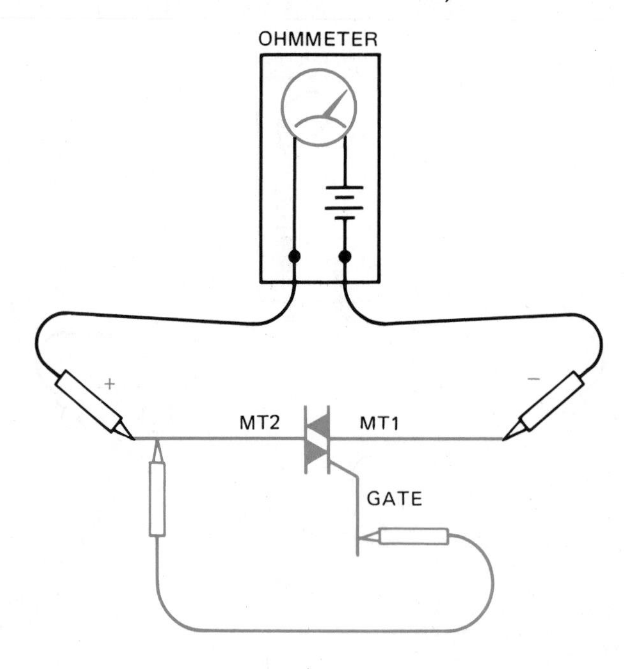

4. Reconnect the triac so that MT1 is connected to the positive ohmmeter lead and MT2 is connected to the negative lead. The ohmmeter should indicate no continuity through the triac.

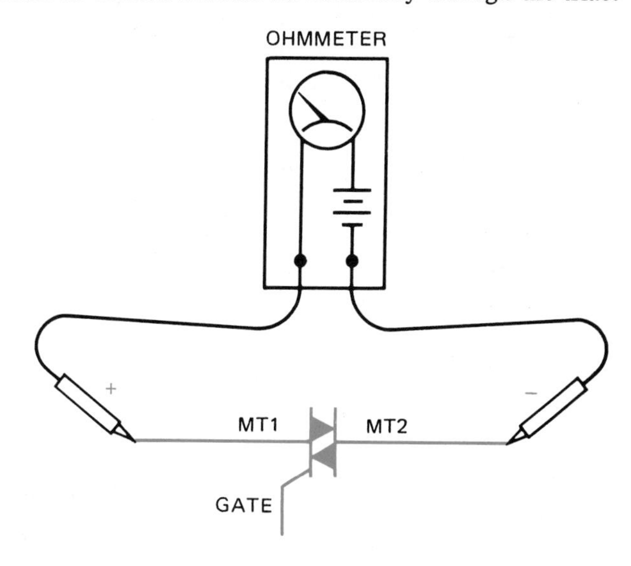

5. Using a jumper lead, again connect the gate to MT2. The ohmmeter should indicate a forward diode junction.

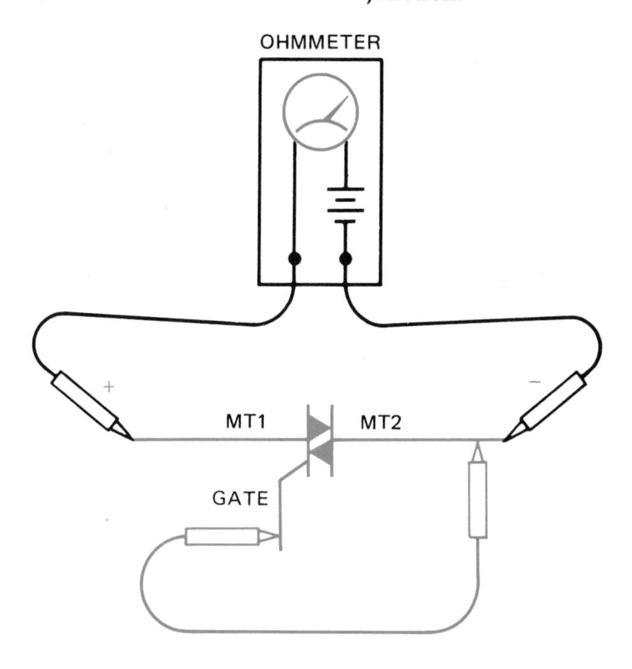

IDENTIFYING THE LEADS OF A THREE-PHASE, WYE-CONNECTED, DUAL-VOLTAGE MOTOR

The terminal markings of a three-phase motor are standardized and used to connect the motor for operation on 240 or 480 volts. Figure #1 shows these terminal markings and their relationship to the other motor windings. If the motor is to be connected to a 240-volt line, the motor windings are connected parallel to each other as shown in figure #2. If the motor is to be operated on a 480-volt line, the motor windings are connected in series is shown in figure #3.

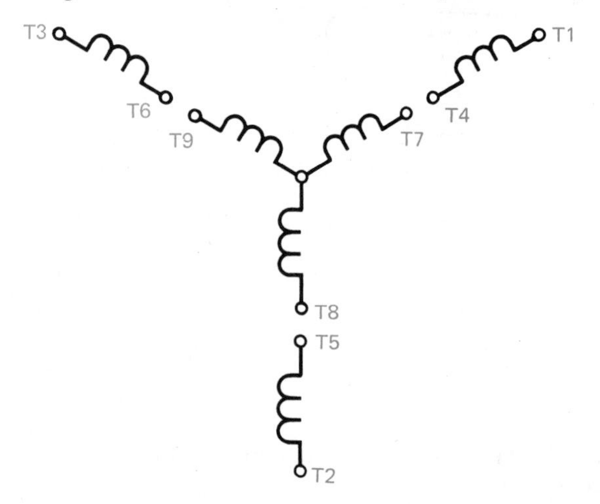

FIGURE 1 Standard terminal markings for a three-phase motor

As long as these motor windings remain marked with the proper numbers, connecting the motor for operation on a 240- or 480-volt power line is relatively simple. If these numbers are removed or damaged, however, the leads must be reidentified before the motor can be connected. The following procedure can be used to identify the proper relationship of the motor windings.

1. Using an ohmmeter, divide the motor windings into four separate circuits. One circuit will have continuity to three leads, and the other three circuits will have continuity between only two leads (see figure #1).

   Caution: the circuits that exhibit continuity between two leads must be identified as pairs, but do not let the ends of the leads touch anything.

2. Mark the three leads that have continuity with each other as T7, T8, and T9. Connect these three leads to a 240-volt, three-phase power source, figure #4. (Note: Since these windings are rated at 240 volts each, the motor can be safely operated on one set of windings as long as it is not connected to a load.)

3. With the power turned off, connect one end of one of the paired leads to the terminal marked T7. Turn the power on, and using

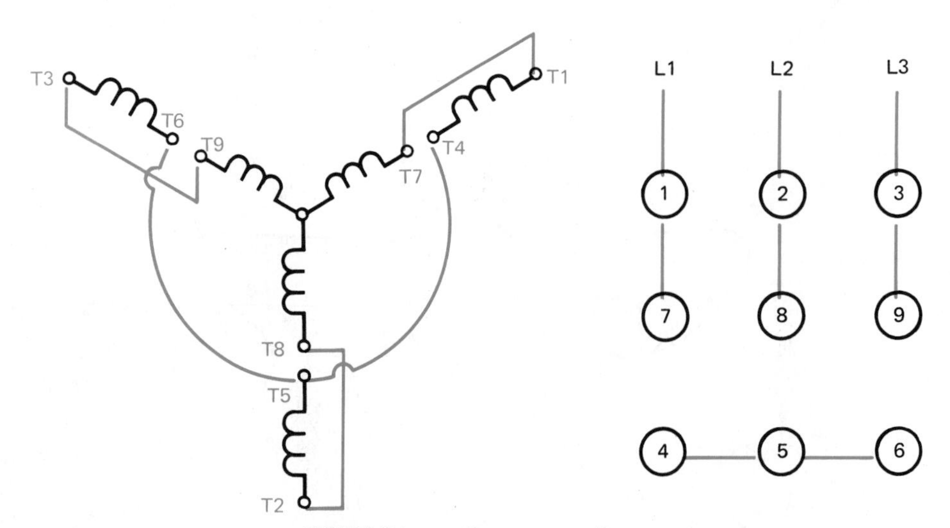

FIGURE 2 Low voltage connection

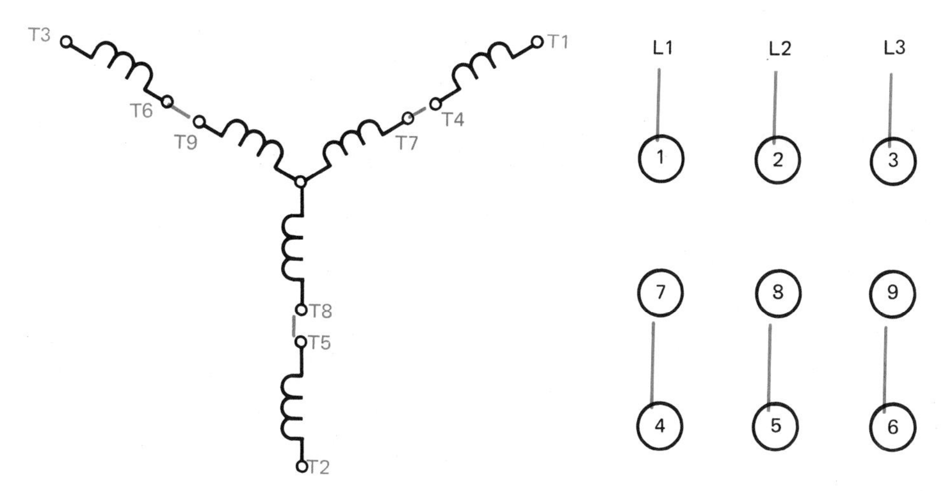

FIGURE 3 High voltage connection

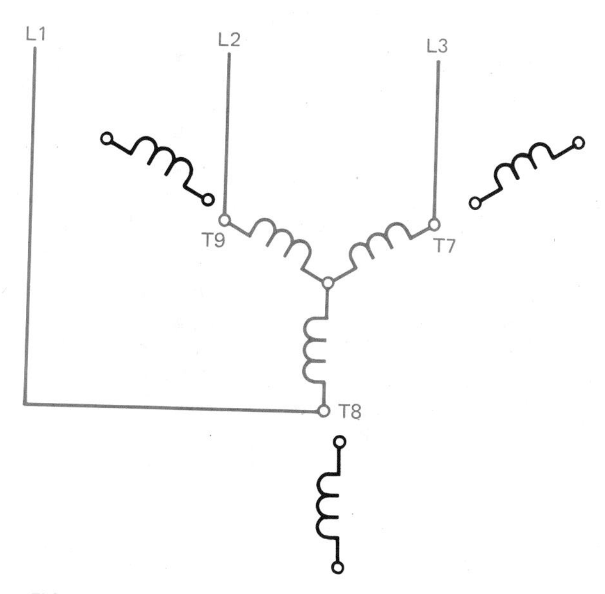

FIGURE 4 T7, T8, and T9 connected to a three-phase, 240-volt line

an ac voltmeter set for a range not less than 480 volts, measure the voltage from the unconnected end of the paired lead to terminals T8 and T9, figure #5. If the measured voltages are unequal, the wrong paired lead is connected to terminal T7. Turn the power off, and connect another paired lead to T7. When the correct set of paired leads is connected to T7, the voltage readings to T8 and T9 will be equal.

4. After finding the correct pair of leads, a decision must be made as to which lead should be labeled T4 and which should be labeled T1. Since an induction motor is basically a transformer, the phase windings act very similar to a multiwinding autotransformer. If terminal T1 is connected to terminal T7, it will operate similar to a transformer with its windings connected to form subtractive polarity. If an ac voltmeter is connected to T4,

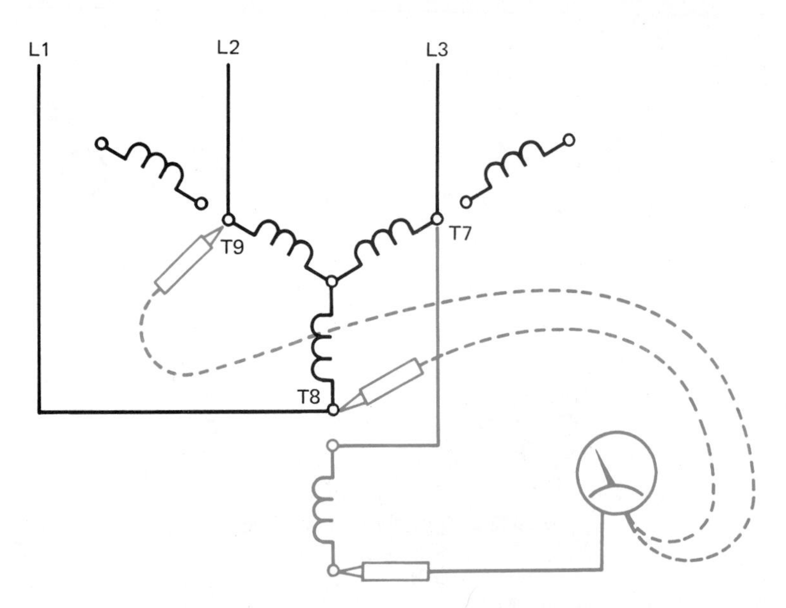

FIGURE 5 Measure voltage from unconnected paired lead to T8 and T9

a voltage of about 140 volts should be seen between T4 and T8 or T4 and T9, figure #6.

If terminal T4 is connected to T7, the winding will operate similar to a transformer with its windings connected for additive polarity. If an ac voltmeter is connected to T1, a voltage of about 360 volts will be indicated when the other lead of the voltmeter is connected to T8 or T9, figure #7.

Label leads T1 and T4 using the preceding procedure to determine which lead is correct. Then disconnect and separate T1 and T4.

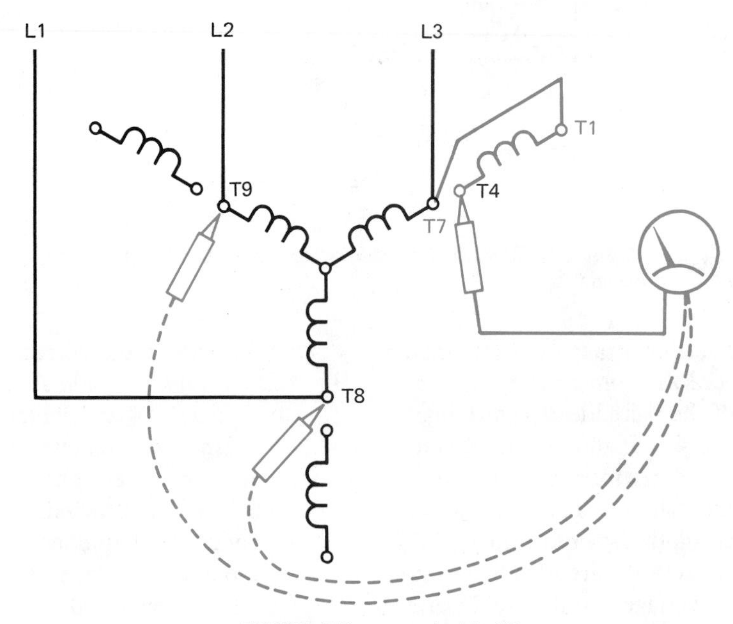

FIGURE 6 T1 connected to T7

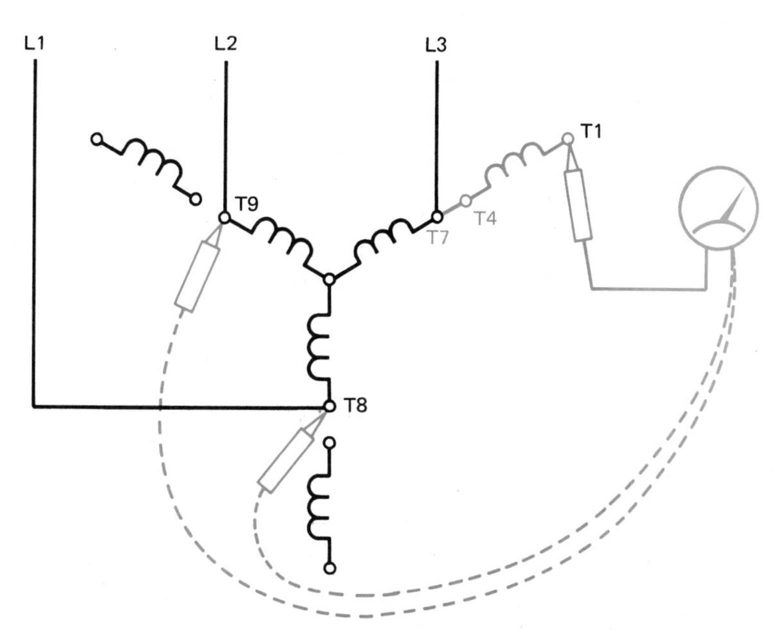

FIGURE 7 T4 connected to T7

5. To identify the other leads, follow the same basic procedure. Connect one end of one of the remaining pairs to T8. Measure the voltage between the unconnected lead and T7 and T9 to determine if it is the correct lead pair for terminal T8. When the correct lead pair is connected to T8, the voltage between the unconnected terminal and T7 or T9 will be equal. Then determine which is T5 or T2 by measuring for a high or low voltage. When T5 is connected to T8, about 360 volts can be measured between T2 and T7 or T2 and T9.

6. The remaining pair can be identified as T3 or T6. When T6 is connected to T9, a voltage of about 360 volts can be measured between T3 and T7 or T3 and T8.

OHM'S LAW FORMULAS

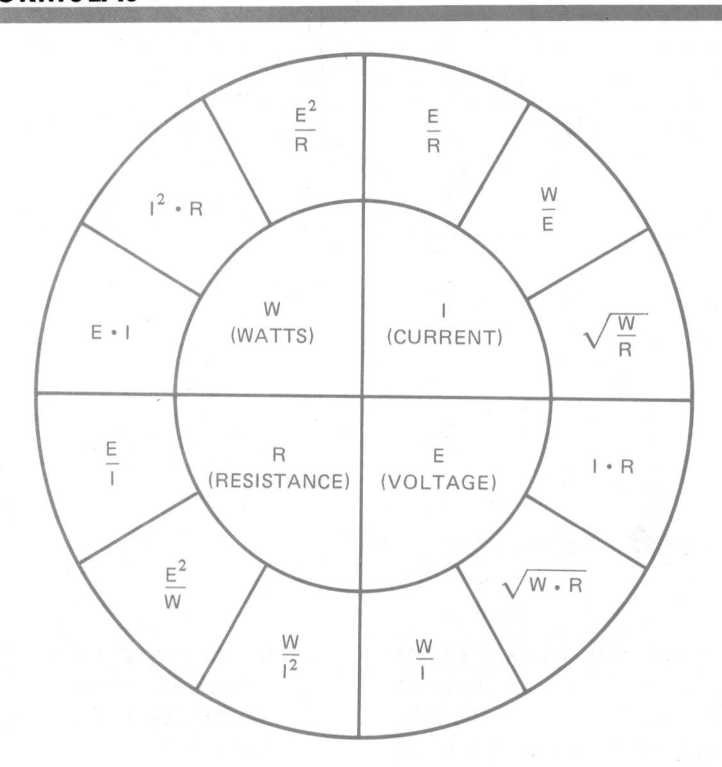

STANDARD WIRING DIAGRAM SYMBOLS

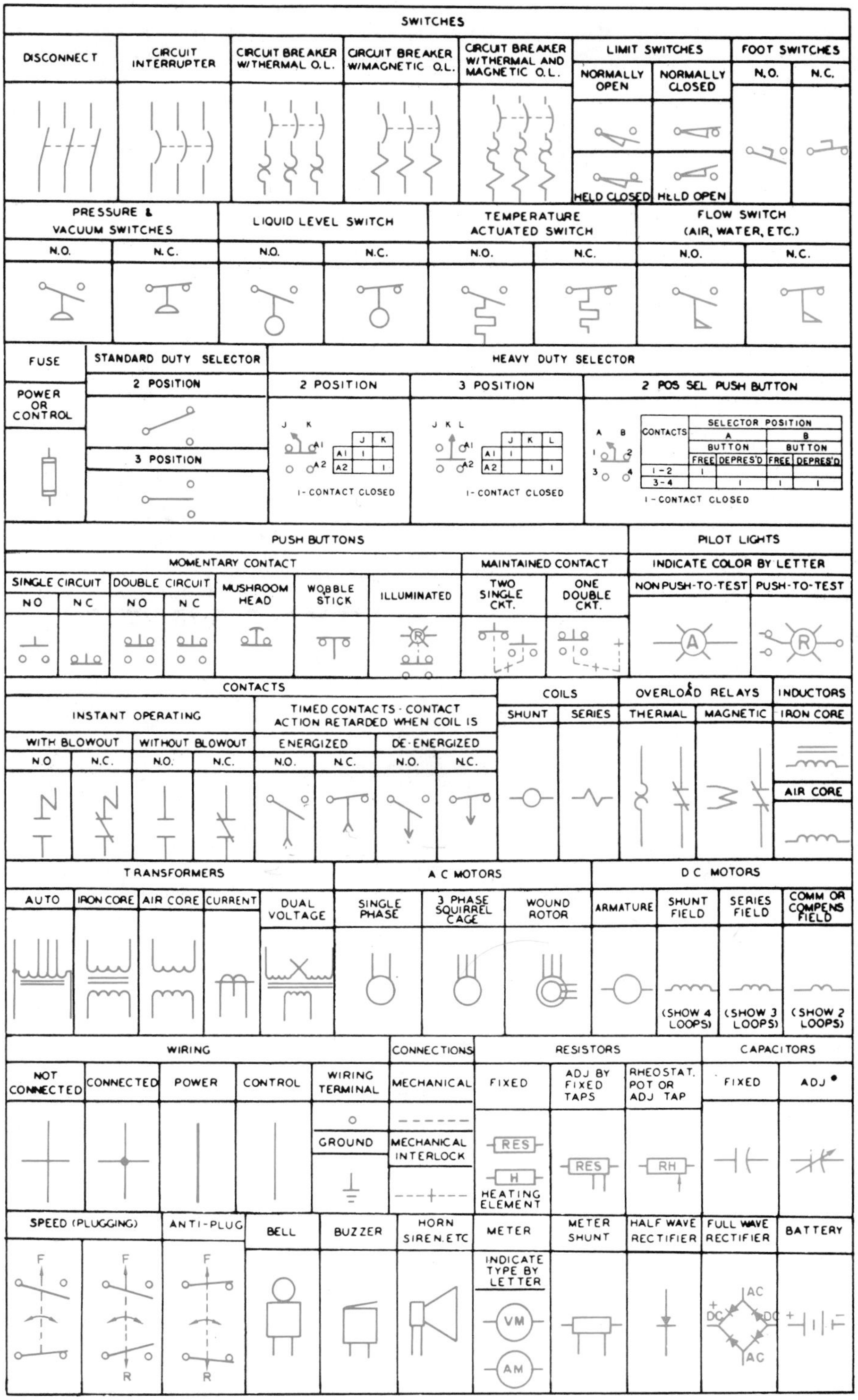

SWITCHES

DISCONNECT | CIRCUIT INTERRUPTER | CIRCUIT BREAKER W/THERMAL O.L. | CIRCUIT BREAKER W/MAGNETIC O.L. | CIRCUIT BREAKER W/THERMAL AND MAGNETIC O.L.

LIMIT SWITCHES: NORMALLY OPEN, NORMALLY CLOSED, HELD CLOSED, HELD OPEN

FOOT SWITCHES: N.O., N.C.

PRESSURE & VACUUM SWITCHES: N.O., N.C.

LIQUID LEVEL SWITCH: N.O., N.C.

TEMPERATURE ACTUATED SWITCH: N.O., N.C.

FLOW SWITCH (AIR, WATER, ETC.): N.O., N.C.

FUSE

POWER OR CONTROL

STANDARD DUTY SELECTOR: 2 POSITION, 3 POSITION

HEAVY DUTY SELECTOR

2 POSITION

	J	K
A1	1	
A2		1

1 - CONTACT CLOSED

3 POSITION

	J	K	L
A1	1		
A2			1

1 - CONTACT CLOSED

2 POS SEL PUSH BUTTON

CONTACTS	SELECTOR POSITION			
	A BUTTON		B BUTTON	
	FREE	DEPRES'D	FREE	DEPRES'D
1-2	1			
3-4		1	1	1

1 - CONTACT CLOSED

PUSH BUTTONS

MOMENTARY CONTACT: SINGLE CIRCUIT (NO, NC), DOUBLE CIRCUIT (NO, NC), MUSHROOM HEAD, WOBBLE STICK, ILLUMINATED

MAINTAINED CONTACT: TWO SINGLE CKT., ONE DOUBLE CKT.

PILOT LIGHTS

INDICATE COLOR BY LETTER: NON PUSH-TO-TEST, PUSH-TO-TEST

CONTACTS

INSTANT OPERATING: WITH BLOWOUT (N.O., N.C.), WITHOUT BLOWOUT (N.O., N.C.)

TIMED CONTACTS - CONTACT ACTION RETARDED WHEN COIL IS: ENERGIZED (N.O., N.C.), DE-ENERGIZED (N.O., N.C.)

COILS

SHUNT, SERIES

OVERLOAD RELAYS

THERMAL, MAGNETIC

INDUCTORS

IRON CORE, AIR CORE

TRANSFORMERS

AUTO, IRON CORE, AIR CORE, CURRENT, DUAL VOLTAGE

A C MOTORS

SINGLE PHASE, 3 PHASE SQUIRREL CAGE, WOUND ROTOR

D C MOTORS

ARMATURE, SHUNT FIELD (SHOW 4 LOOPS), SERIES FIELD (SHOW 3 LOOPS), COMM OR COMPENS FIELD (SHOW 2 LOOPS)

WIRING

NOT CONNECTED, CONNECTED, POWER, CONTROL, WIRING TERMINAL, GROUND

CONNECTIONS

MECHANICAL, MECHANICAL INTERLOCK

RESISTORS

FIXED (RES; H HEATING ELEMENT), ADJ BY FIXED TAPS (RES), RHEOSTAT, POT OR ADJ TAP (RH)

CAPACITORS

FIXED, ADJ

SPEED (PLUGGING) (F, R), ANTI-PLUG (F, R), BELL, BUZZER, HORN SIREN, ETC, METER (INDICATE TYPE BY LETTER: VM, AM), METER SHUNT, HALF WAVE RECTIFIER, FULL WAVE RECTIFIER (AC, DC, +), BATTERY

ELECTRONIC SYMBOLS

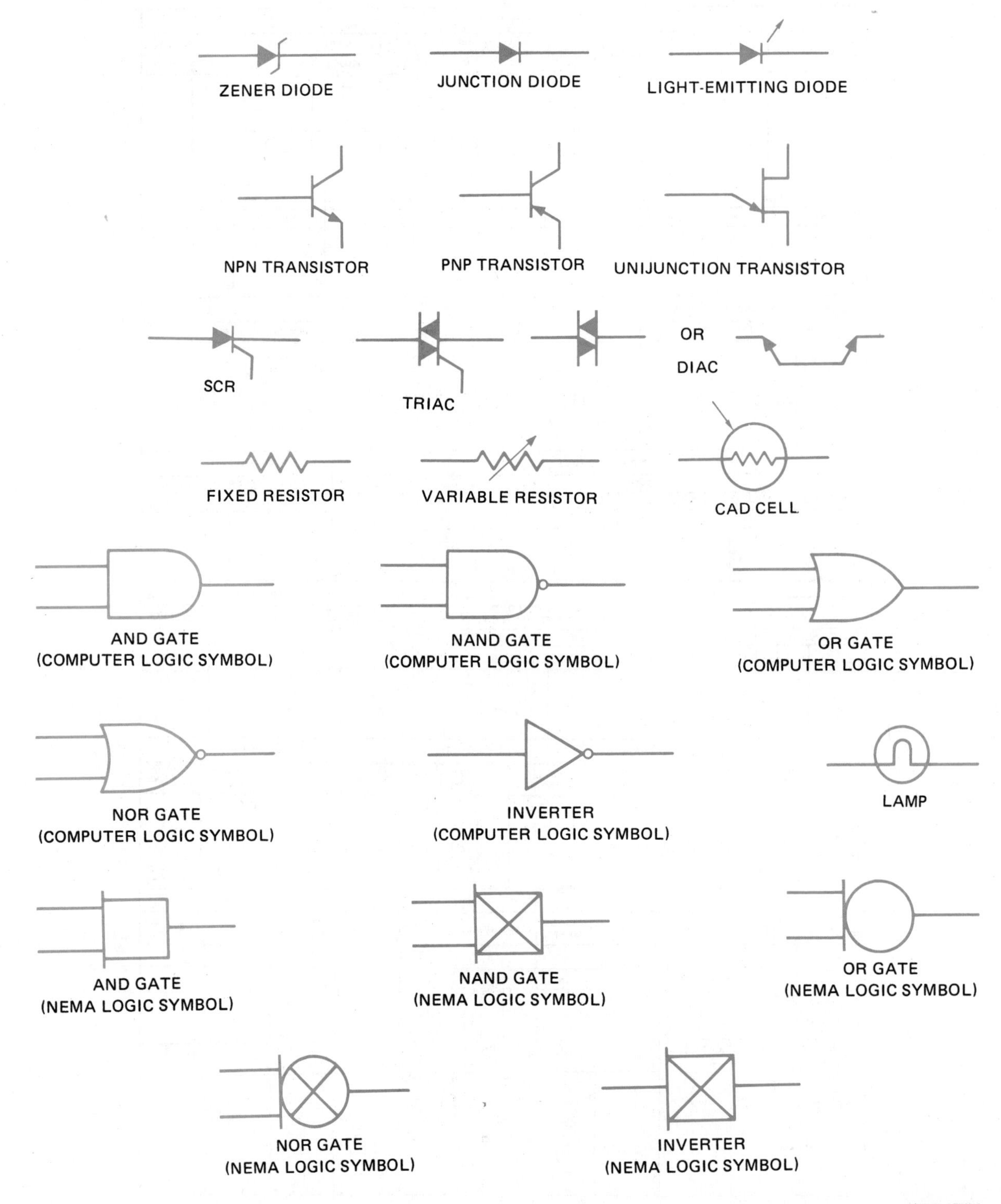

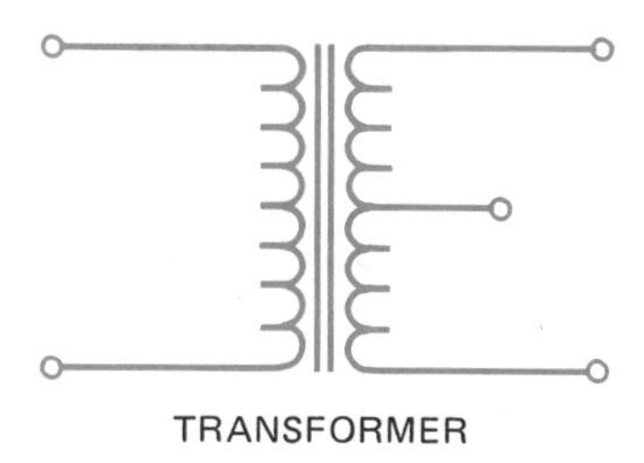

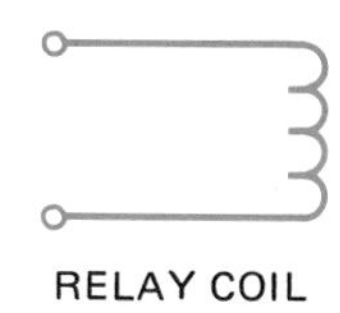

MOTOR TYPES AND LINE DIAGRAMS

Dc Shunt Motor. Main field winding is designed for parallel connection to the armature; stationary field; rotating armature with commutator; has a no-load speed; full speed at full load is less than no-load speed; torque increases directly with load.

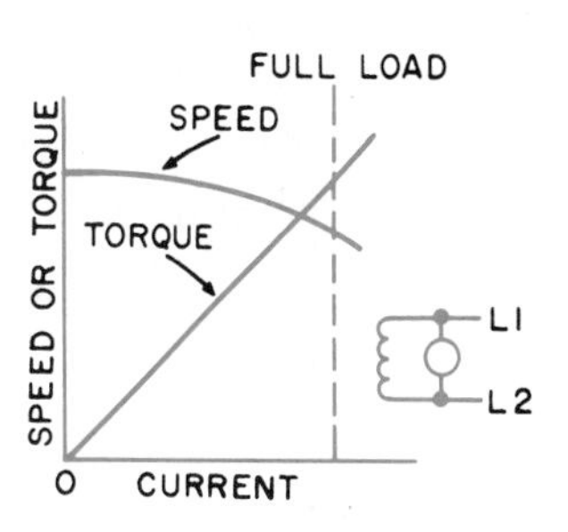

Dc Series Motor. Main field winding is designed for series connection to the armature; stationary field; rotating armature with commutator; does not have a no-load speed; requires solid direct connection to the load to prevent runaway at no-load; speed decreases rapidly with increase in load; torque increases as square of armature current; main motor for crane hoists; excellent starting torque.

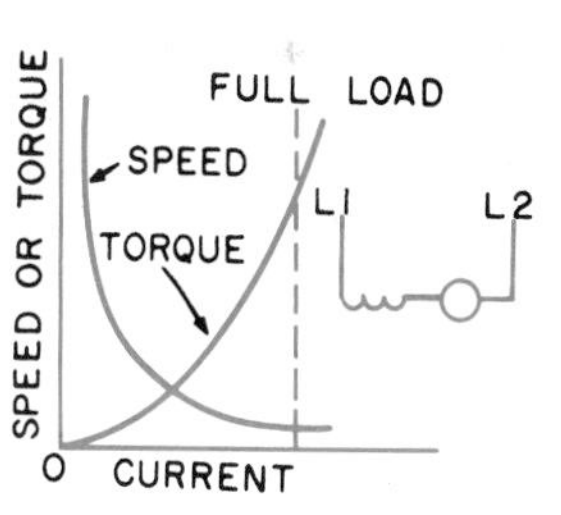

Dc Compound Motor. Main field both shunt (parallel) and series; stationary fields; rotating armature with commutator; combination shunt and series fields produce characteristics between straight shunt or series dc motor; good starting torque; main motor for dc driven machinery (mills or presses).

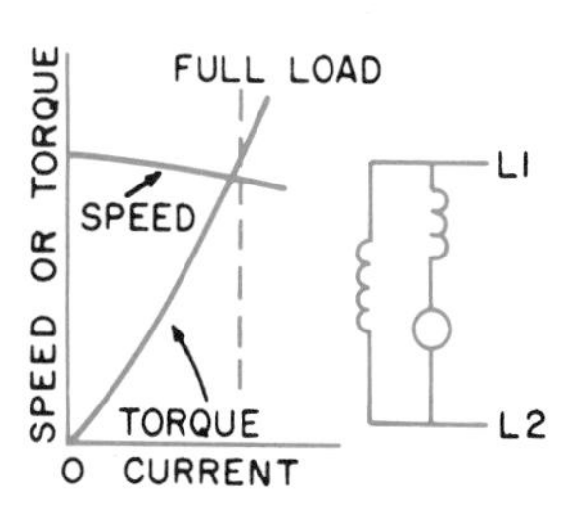

Ac Squirrel Cage Motor. Single or three phase; single phase requires a starting winding; three phase, self starting; stationary stator winding; no electrical connection to short-circuited rotor; torque produced from magnetic reaction of stator and rotor fields; speed a function of supply frequency and number of electrical poles wound on stator; considered as constant speed even though speed decreases slightly with increased load; good starting torque; high inrush currents during starting on full voltage; rugged construction; easily serviced and maintained; high efficiency; good running power factor when delivering full load; requires motor control for stator windings only.

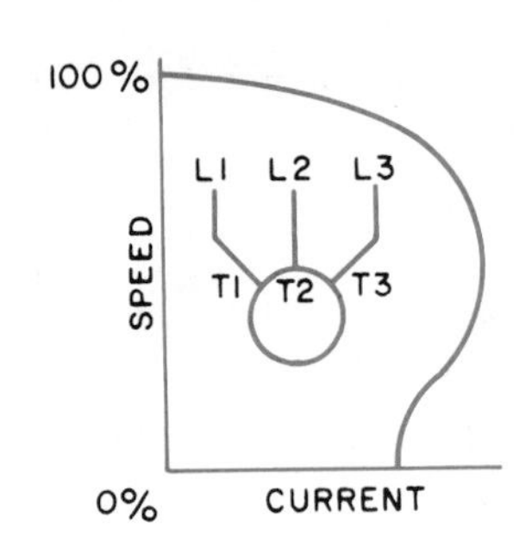

Ac Wound-Rotor Induction Motor. Characteristics similar to squirrel cage motor; stationary stator winding; rotor windings terminate on slip rings; external addition of resistance to rotor circuit for speed control; good starting torque; high inrush current during starting on full voltage; low efficiency when resistor is inserted in rotor windings; good running power factor; requires motor controls for stator and rotor circuits.

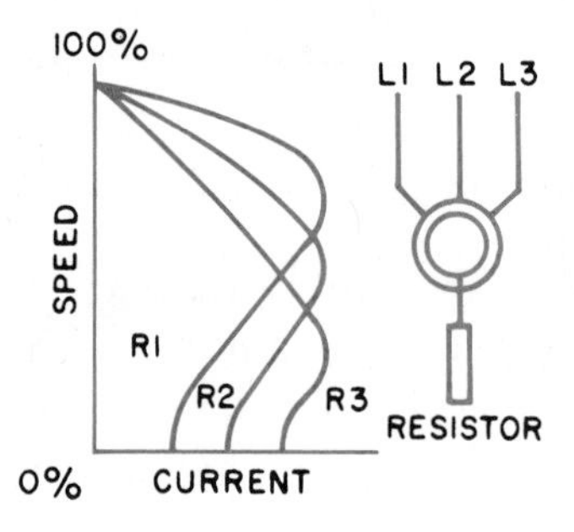

Ac Synchronous Motor. Stationary ac stator windings; rotating dc field winding; no starting torque unless motor has starting winding; generally poor starting torque; constant speed when motor up to speed and dc field winding energized; can provide power factor correction with proper dc field excitation; requires special motor control for both ac and dc windings to prevent the dc field winding from being energized until a specified percent of running speed has been obtained.

POWER SUPPLIES

All electrical power supplied as ac or dc; primarily ac; generation, transmission, and some distribution of power at high voltage (above 5000 volts) or medium voltage (600 to 5000 volts); most power distribution of voltage for industrial and residential use is 600 volts and under; ac power generally at 60-Hertz frequency; ac distribution at use location single or three phase.

Single-phase. Two wire, 120 volts, one line grounded; 120/240 volts, three-wire center-line grounded; residential distribution, lighting, heat, fractional horsepower motors and business machines.

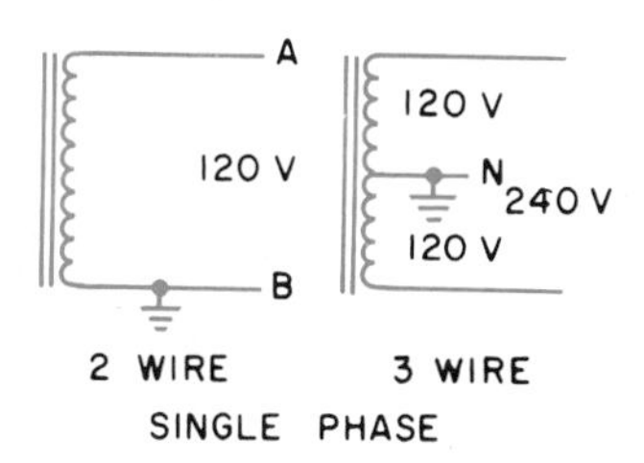

Three-phase. Three-wire delta 230/460 volts, 580 volts; four-wire wye 208/440, 277/480 volts neutral line grounded; primary industrial power distribution; main motor drives, integral horsepower motors, lighting, heating, fractional horsepower motors, and business machines; used as three-phase or single-phase power supply.

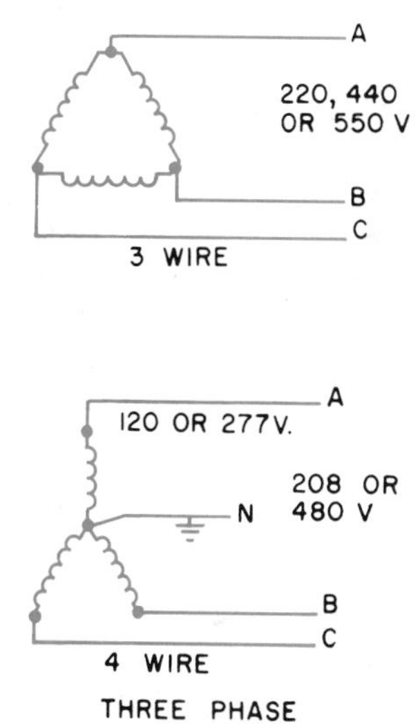

MOTOR CIRCUIT ELEMENTS

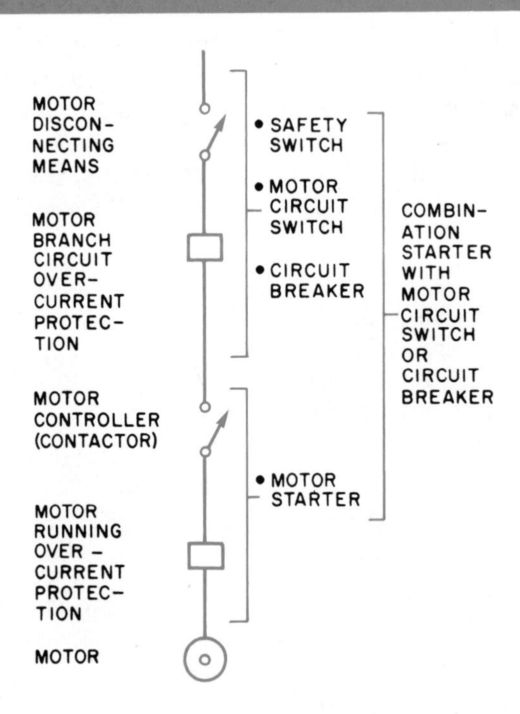

DEVELOPING CONTROL CIRCUITS

Circuit #1: Two Pump Motors

The water for a housing development is supplied by a central tank. The tank is pressurized by the water as it is filled. Two separate wells supply water to the tank, and each well has a separate pump. It is desirable that water be taken from each well equally, but it is undesirable that both pumps operate at the same time. A circuit is to be constructed that will let the pumps work alternately. Also, a separate switch must be installed that will override the automatic control and let either pump operate independently of the other in the event one pump fails. The requirements of the circuit are as follows:

1. The pump motors are operated by a 480-volt three-phase system, but the control circuit must operate on a 120-volt supply.
2. Each pump motor contains a separate overload protector. If one pump overloads, it will not prevent operation of the second pump.
3. A manual ON-OFF switch can be used to control power to the circuit.
4. A pressure switch mounted on the tank controls the operation of the pump motors. When the pressure of the tank drops to a certain level, one of the pumps will be started. When the tank has been filled with water, the pressure switch will turn the pump off. When the pressure of the tank drops low enough again, the other pump will be started and run until the pressure switch is satisfied. Each time the pressure drops to a low enough level, the alternate pump motor will be used.
5. An override switch can be used to select the operation of a particular pump, or to permit the circuit to operate automatically.

When developing a control citcuit, the logic of the circuit is developed one stage at a time until the circuit operates as desired. The first stage of the circuit is shown in figure 1. In this stage, a control transformer has been used to step the 480-volt supply line voltage down to 120 volts for use by the control circuit. A fuse is used as short-circuit protection for the control wiring. A manually operated ON-OFF switch permits the control circuit to be disconnected from the power source. The pressure switch must close when the pressure drops. For this reason, it will be connected as normally closed. This is a normally closed held-open switch. A set of normally closed overload contacts are connected in series with coil 1M, which will operate the motor starter of pump motor #1.

To understand the operation of this part of the circuit, assume that the manual power switch has been set to the ON position. When the tank pressure drops sufficiently, pressure switch PS will close and energize coil 1M, starting pump #1. As water fills the tank, the pressure increases. When the pressure has increased sufficiently, the pressure switch opens and disconnects coil 1M, stopping the operation of pump #1.

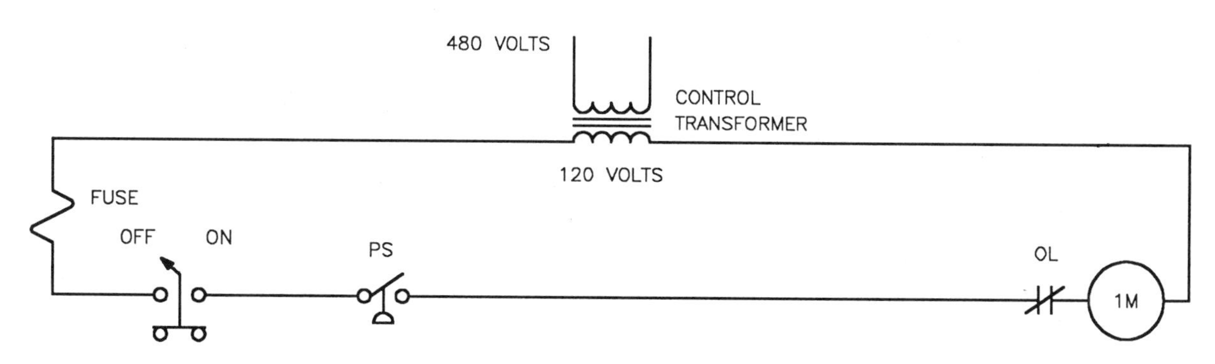

FIGURE 1 The pressure switch starts pump #1.

If pump #1 is to operate alternately with pump #2, some method must be devised to remember which pump operated last. This function will be performed by control relay CR. Since relay CR is to be used as a memory device, it must be permitted to remain energized when either or both of the motor starters are not energized. For this reason, this section of the circuit is connected to the input side of pressure switch PS. This addition to the circuit is shown in figure 2.

The next stage of circuit development can be seen in figure 3. In this stage of the circuit, motor starter 2M has been added. When pressure switch PS closes and energizes motor starter coil 1M, all 1M contacts change position. Contacts $1M_1$ and $1M_2$ close at the same time. When $1M_1$ contact closes, coil CR is energized, changing the position of all CR contacts. Contact CR_1 opens, but the current path to coil 1M is maintained by contact $1M_1$. Contact CR_2 is used as a holding contact around contact $1M_2$. Notice that each motor starter coil is protected by a separate overload contact. This fulfills the requirement that an overload on either motor will not prevent the operations of the other motor. Also notice that this section of the circuit has been connected to the output side of pressure switch PS. This permits the the pressure switch to control the operation of both pumps.

To understand the operation of the circuit, assume pressure switch PS closes. This provides a current path to motor starter coil 1M. When coil 1M energizes, all 1M contacts change position and pump #1 starts. Contact $1M_1$ closes and energizes coil CR. Contact $1M_2$ closes to maintain a current path to coil 1M. Contact $1M_3$ opens to provide interlock with coil 2M, which prevents it from energizing whenever coil 1M is energized.

When coil CR energizes, all CR contacts change position. Contact CR_1 opens to break the circuit to coil 1M. Contact CR_2 closes to maintain a current path around contact $1M_1$, and contact CR_3 closes to provide a current path to motor starter coil 2M. Coil 2M cannot be energized, however, because of the now open $1M_3$ contact.

When the pressure switch opens, coil 1M will deenergize, permitting all 1M contacts to return to their normal positions, and the circuit will be left as shown in figure 4. Note that this diagram is intended to show the condition of the circuit when the pressure switch is opened—it is not intended to show the contacts in their normal deenergized position. At this point in time, a current path is maintained to control relay CR.

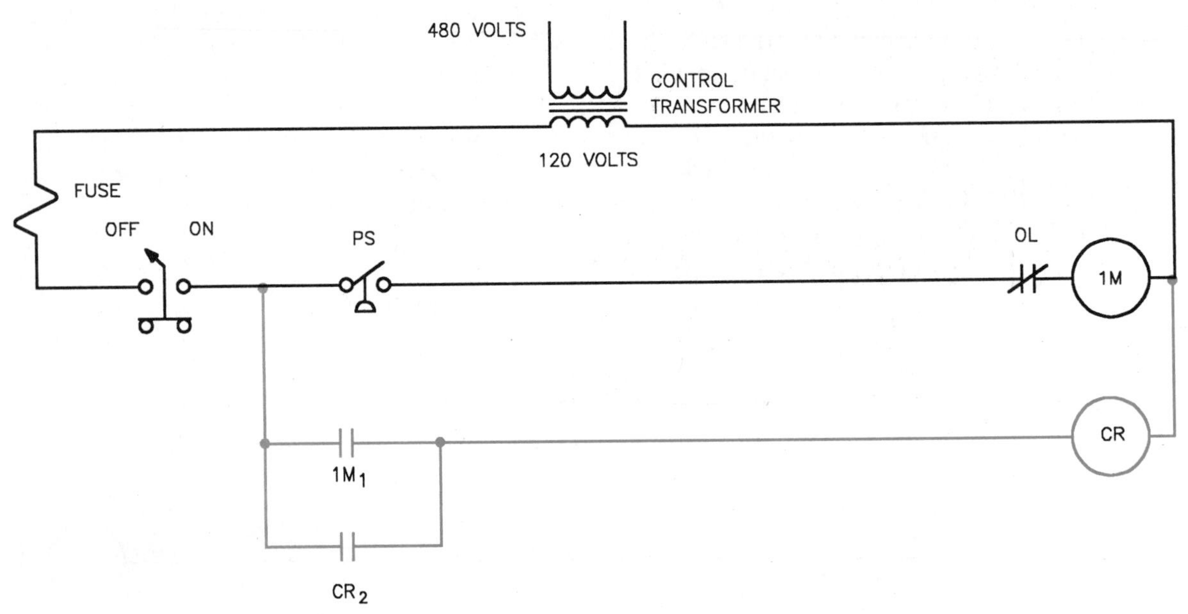

FIGURE 2 The control relay is used as a memory device.

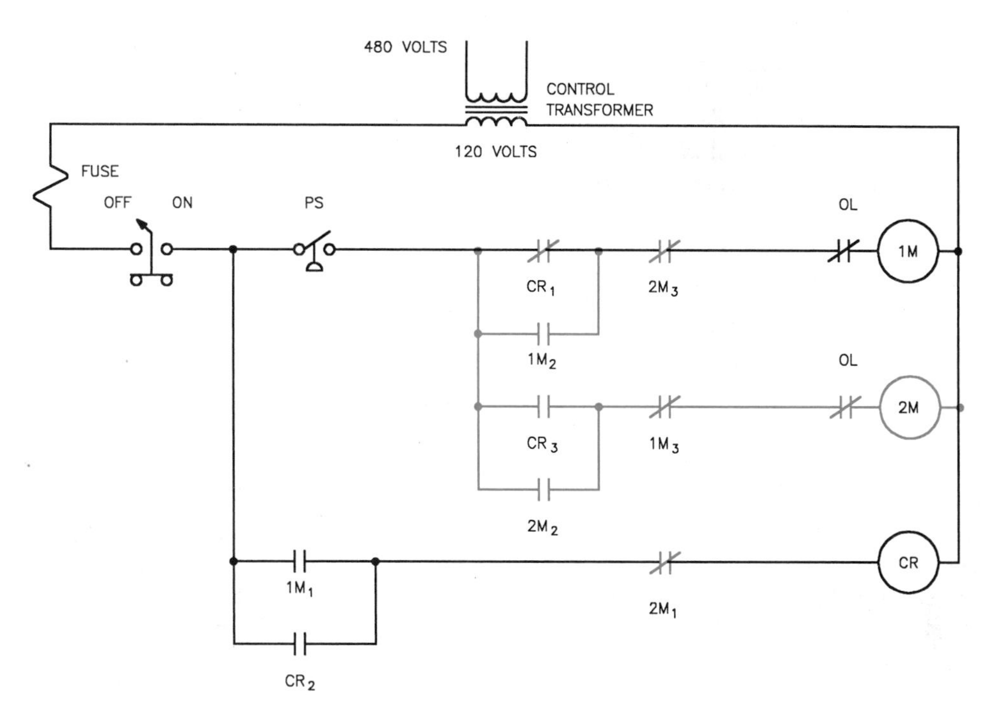

FIGURE 3 The addition of the second motor starter

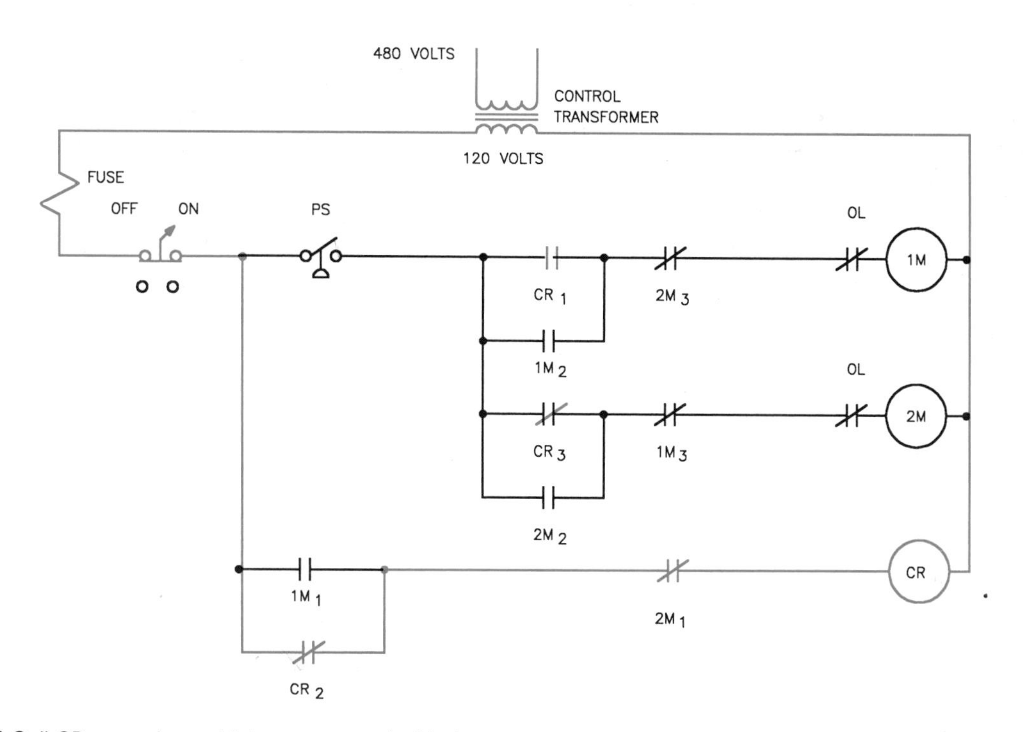

FIGURE 4 Coil CR remembers which pump operated last.

When pressure switch PS closes again, contact CR_1 prevents a current path from being established to coil 1M, but contact CR_3 permits a current path to be established to coil 2M. When coil 2M energizes, pump #2 starts and all 2M contacts change position.

Contact $2M_1$ opens and causes coil CR to deenergize. Contact $2M_2$ closes to maintain a circuit to coil 2M when contact CR_3 returns to its normally open position, and contact $2M_3$ opens to prevent coil 1M from being energized when contact CR_1 returns to its normally closed position. The circuit will continue to operate in this manner until pressure switch PS opens and disconnects coil 2M from the line. When this happens, all 2M contacts will return to their normal positions as shown in figure 3.

The only requirement not fulfilled is a switch that permits either pump to operate independently if one pump fails. This addition to the circuit is shown in figure 5. A three-position selector switch is connected to the output of the pressure switch. The selector switch will permit the circuit to alternate operation of the two pumps, or permit the operation of one pump only.

Although the logic of the circuit is now correct, there is a potential problem. After pump #1 has completed a cycle and the circuit is set as shown in figure 4, there is a possibility that contact CR_3 will reopen before contact $2M_2$ closes to seal the circuit. If this happens, coil 2M will deenergize and coil 1M will be energized. This is often referred to as a contact race. To prevent this problem, an OFF DELAY timer will be added as shown in figure 6. In this circuit, coil CR has been replaced by coil TR of the timer. When coil TR energizes, contact TR will close immediately, energizing coil CR. When coil TR deenergizes, con-

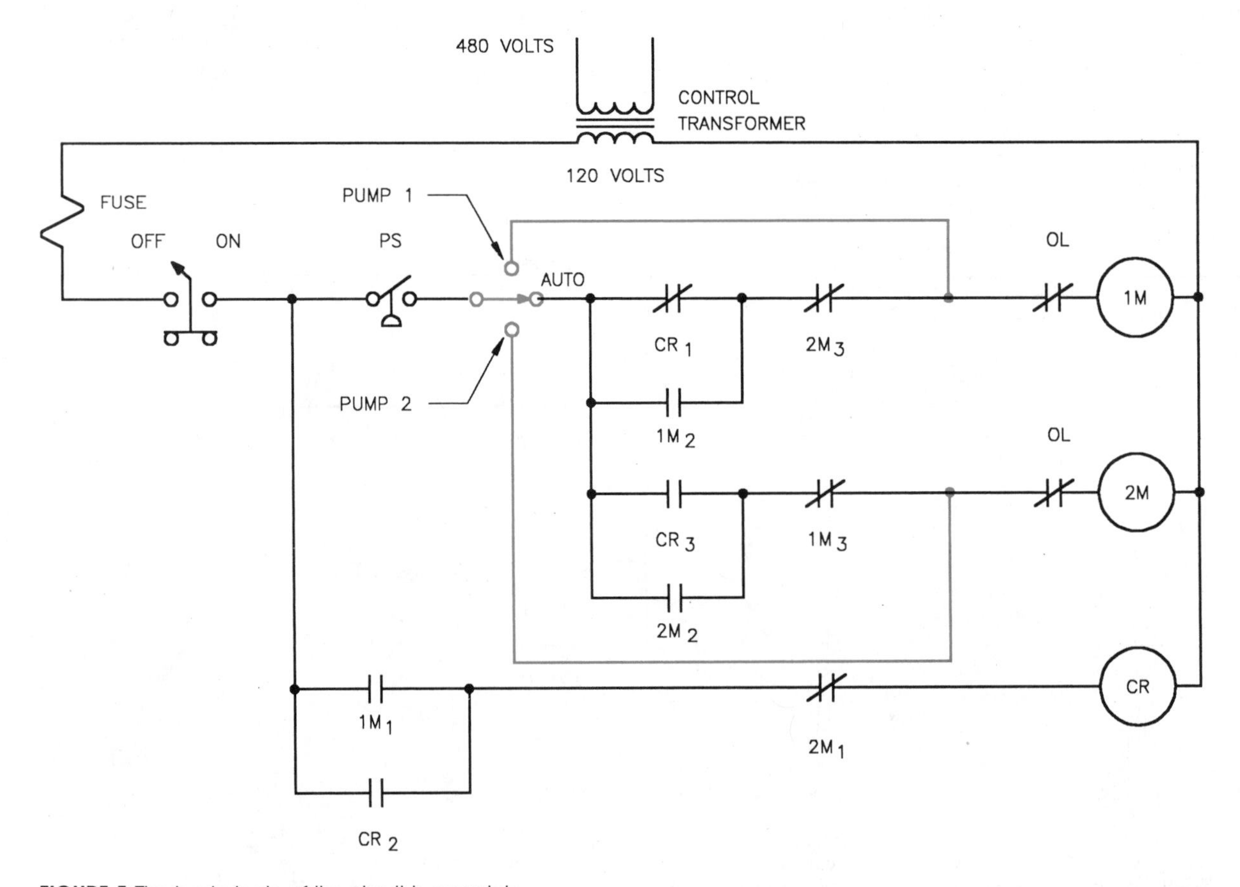

FIGURE 5 The basic logic of the circuit is complete.

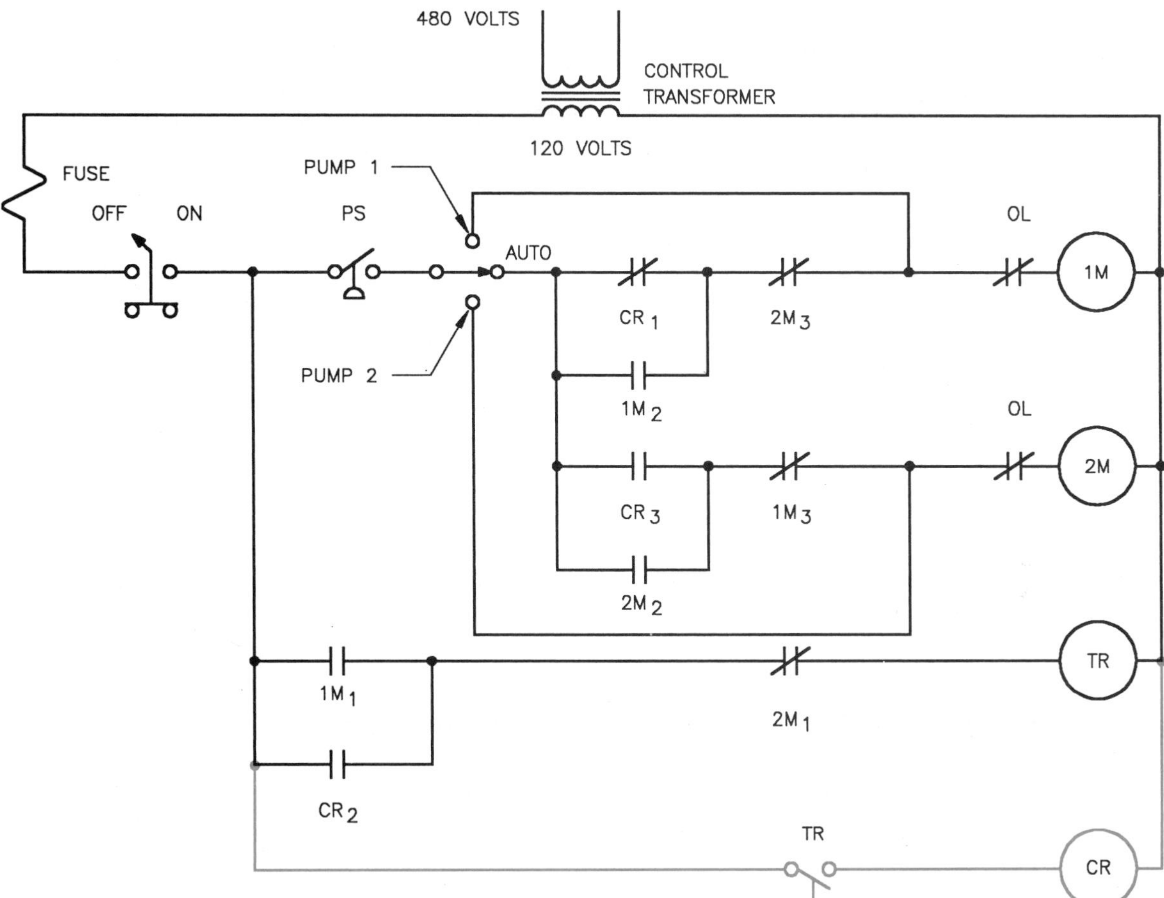

FIGURE 6 A timer is added to ensure proper operation.

tact TR will remain closed for one second before reopening and permitting coil CR to deenergize. This short delay time will ensure proper operation of the circuit.

Circuit #2: Speed Control of a Wound Rotor Induction Motor

The second circuit to be developed will control the speed of a wound rotor induction motor. The motor will have three steps of speed. Separate push buttons are used to select the speed of operation. The motor will accelerate automatically to the speed selected. For example, if second speed is selected, the motor must start in the first or lowest speed and then accelerate to second speed. If third speed is selected, the motor must start in first speed, accelerate to second speed, and then accelerate to third speed. The requirements of the circuit are as follows:

1. The motor is to operate on a 480-volt three-phase power system, but the control system is to operate on 120 volts.
2. One stop button can stop the motor regardless of which speed has been selected.
3. The motor will have overload protection.
4. Three separate push buttons will select first, second, or third speed.
5. There will be a three-second time delay between accelerating from one speed to another.

6. If the motor is in operation and a higher speed is desired, it can be obtained by pushing the proper button. If the motor is operating and a lower speed is desired, the stop button must be pressed first.

Recall that speed control for a wound rotor motor is obtained by placing resistance in the secondary or rotor circuit as shown in figure 7. In this circuit, load contacts 1M are used to connect the stator or primary of the motor to the power line. Two banks of three-phase resistors have been connected to the rotor. When power is applied to the stator, all resistance is connected in the rotor circuit and the motor will operate in its lowest or first speed. Second speed is obtained by closing contacts 1S and shorting out the first three-phase resistor bank. Third speed is obtained by closing contacts 2S. This shorts the rotor winding and the motor operates as a squirrel cage motor. A control transformer is connected to two of the three-phase lines to provide power for the control system.

The first speed can be obtained by connecting the circuit shown in figure 8. When the first speed button is pressed, motor starter coil 1M will close and connect the stator of the motor to the power line. Because all the resistance is in the rotor circuit, the motor will operate in its lowest speed.

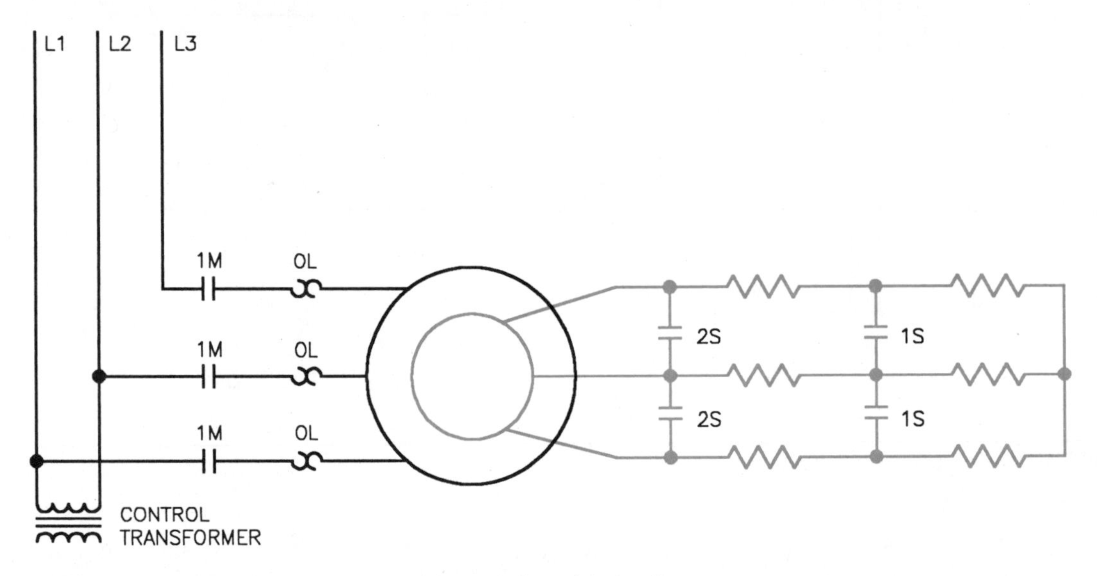

FIGURE 7 Speed is controlled by connecting resistance in the rotor circuit.

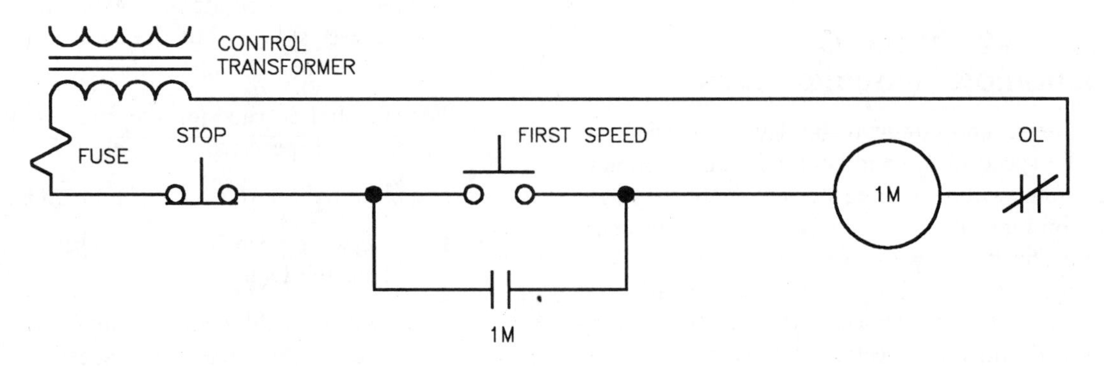

FIGURE 8 First speed

Auxiliary contact $1M_1$ is used as a holding contact. A normally closed overload contact is connected in series with coil 1M to provide overload protection. Notice that only one overload contact is shown, indicating the use of a three-phase overload relay.

The second stage of the circuit can be seen in figure 9. When the second speed button is pressed, the coil of ON delay timer 1TR is energized. Since the motor must be started in the first speed position, instantaneous timer contact $1TR_1$ closes to energize coil 1M and connect the stator of the motor to the line. Contact $1TR_2$ is used as a holding contact to keep coil 1TR energized when the second speed button is released. Contact $1TR_3$ is a timed contact. At the end of three seconds, it will close and energize contactor coil 1S, causing all 1S contacts to close and shunt the first set of resistors. The motor now operates in second speed.

The final stage of the circuit is shown in figure 10. The third speed button is used to energize the coil of control relay 1CR. When coil 1CR is energized, all 1CR contacts change position. Contact $1CR_1$ closes to provide a current path to motor starter coil 1M, causing the motor to start in its lowest speed. Contact $1CR_2$ closes to provide a current path to timer 1TR. This permits timer 1TR to begin its timer operation. Contact $1CR_3$ maintains a current path to coil 1CR after the third speed button is opened, and contact $1CR_4$ permits a current path to be established to timer 2TR. This contact is also used to prevent a current path to coil 2TR when the motor is to be operated in the second speed.

After timer 1TR has been energized for a period of three seconds, contact $1TR_3$ closes and energizes coil 1S. This permits the motor to accelerate to the second speed. Coil 1S also closes auxiliary contact $1S_1$ and completes a circuit to timer 2TR.

After a delay of three seconds, contact 2TR closes and energizes coil 2S. This causes contacts

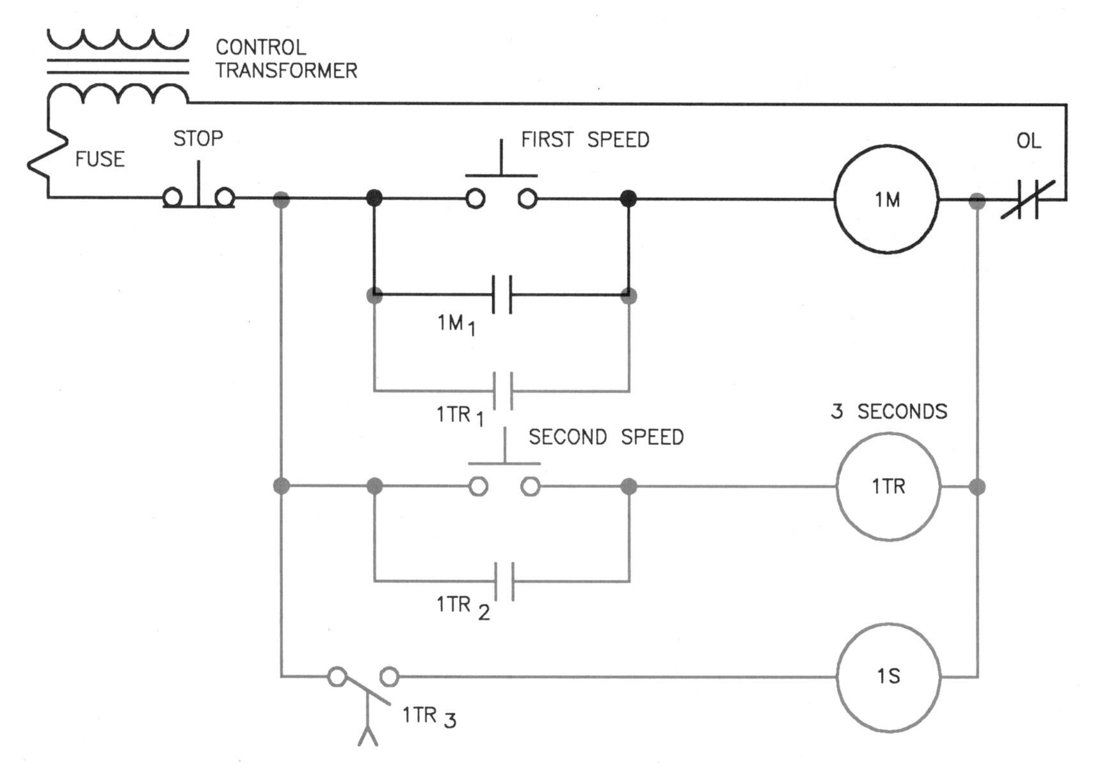

FIGURE 9 Second speed

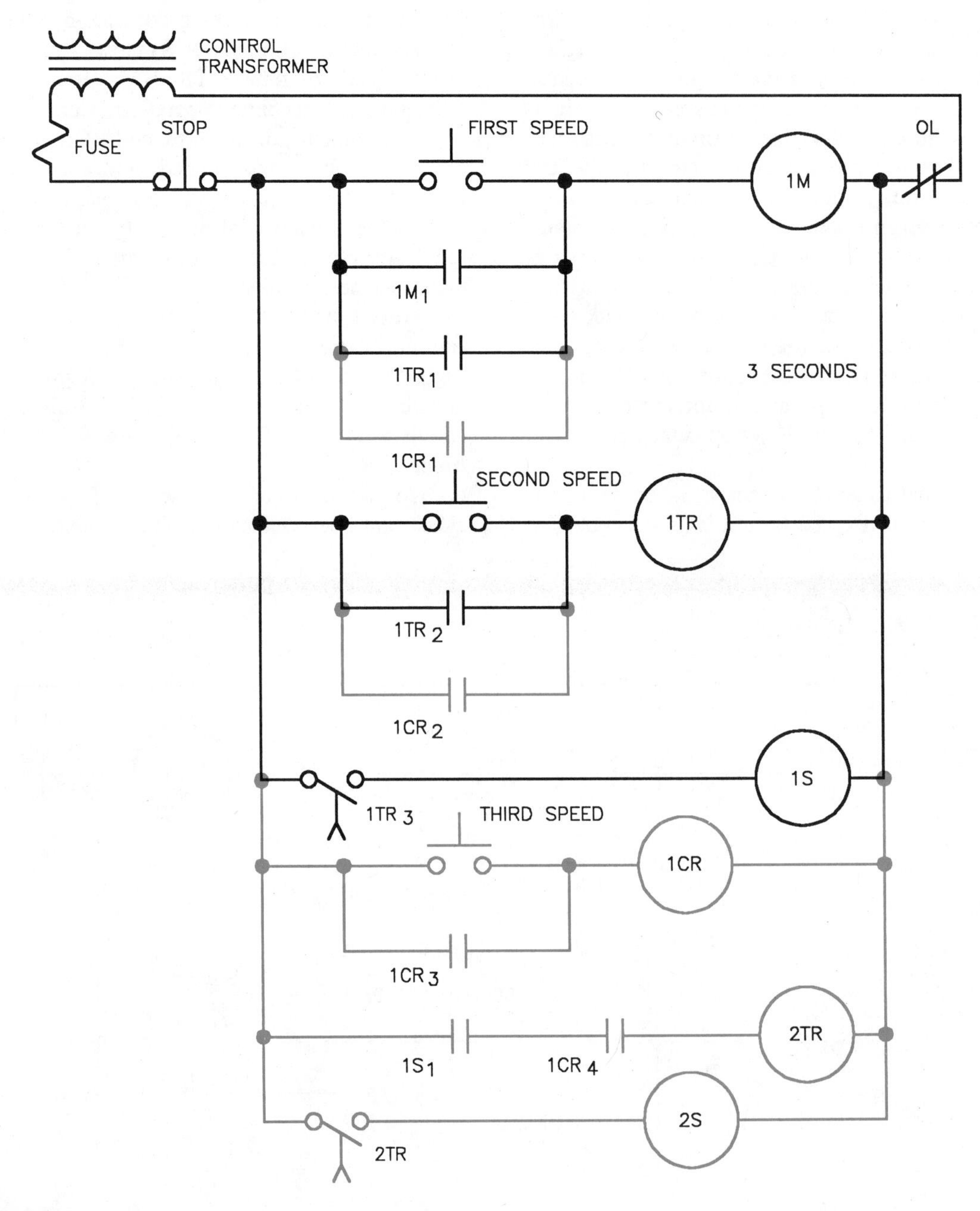

FIGURE 10 Third speed

2S to close and the motor operates in its highest speed.

Circuit #3: An Oil Heating Unit

In the circuit shown in figure 11, motor starter 1M controls a motor that operates a high-pressure pump. The pump is used to inject fuel oil into a combustion chamber where it is burned. Motor starter 2M operates an air induction blower that forces air into the combustion chamber when the oil is being burned. Motor starter 3M controls a squirrel cage blower which circulates air across a heat exchanger to heat a building. A control transformer is used to change the incoming voltage from 240 volts to 120 volts, and a separate OFF-ON switch can be used to disconnect power from the circuit. Thermostat TS1 senses temperature inside the building and thermostat TS2 is used to sense the temperature of the heat exchanger.

To understand the operation of the circuit, assume the manual OFF-ON switch is set in the ON position. When the temperature inside the building drops to a low enough level, thermostat TS1 closes and provides power to starters 1M and 2M. This permits the pump motor and air induction blower to start. When the temperature of the heat exchanger rises to a high enough level, thermostat TS2 closes and energizes starter 3M. The blower circulates the air inside the building across the heat exchanger and raises the temperature inside the building. When the building temperature rises to a high enough level, thermostat TS1 opens and disconnects the pump motor and air induction motor. The blower will continue to operate until the heat exchanger has been cooled to a low enough temperature to permit thermostat TS2 to open its contact.

After some period of operation, it is discovered that the design of this circuit can lead to some serious safety hazards. If the overload contact connected to starter 2M should open, the high-pressure pump motor will continue to operate without sufficient air being injected into the combustion chamber. Also, there is no safety switch to turn the pump motor off if the blower motor fails to provide cooling air across the heat exchanger. It is recommended that the following changes be made to the circuit:

1. If an overload occurs to the air induction motor, it will stop operation of both the high-pressure pump motor and the air induction motor.
2. An overload of the high pressure pump motor will stop only that motor and permit the air induction motor to continue operation.
3. The air induction motor will continue operating for one minute after the high-pressure

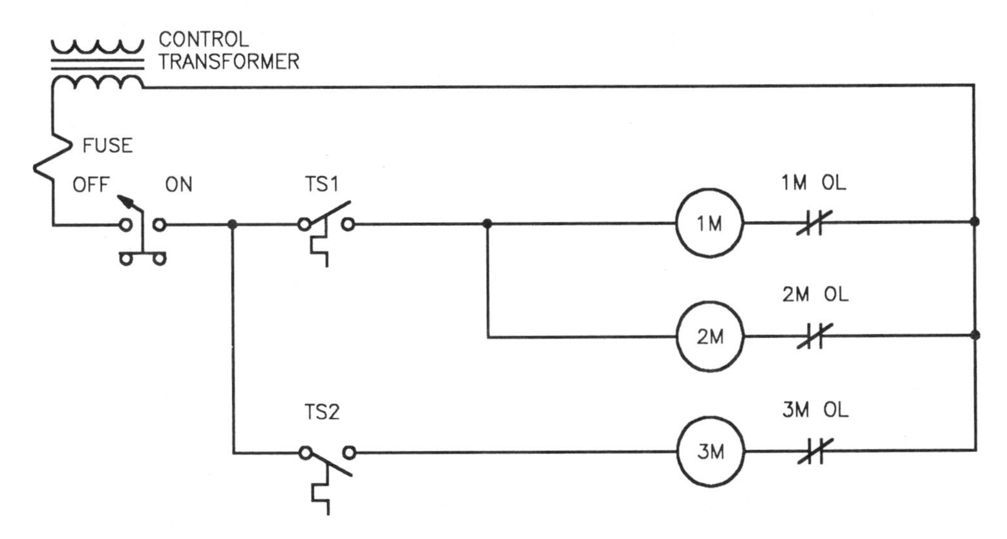

FIGURE 11 Heating system control

pump motor has been turned off. This will clear the combustion chamber of excessive smoke and fumes.

4. A high-limit thermostat is added to the heat exchanger to turn the pump motor off if the temperature of the heat exchanger should become excessive.

These circuit changes can be seen in figure 12. Thermostat TS3 is the high limit thermostat. Since it is to be used to perform the function of stop, it is normally closed and connected in series with motor starter 1M. An off delay timer is used to control starter 2M, and the overload contact of starter 2M has been connected in such a manner that it can stop the operation of both the air induction blower and the high-pressure pump. Notice, however, that if 1M overload contact opens, it will not stop the operation of the air induction blower motor. The air induction blower motor would continue to operate for a period of one minute before stopping.

The logic of the circuit is as follows: When thermostat TS1 closes its contact, coils 1M and TR are energized. Because timer TR is an off delay timer, contact TR closes immediately, permitting motor starter 2M to energize. When thermostat TS1 is satisfied and reopens its contact, or if thermostat TS3 opens its contact, coils 1M and TR will deenergize. Contact TR will remain closed for a period of one minute before opening and disconnecting starter 2M from the power line.

Although the circuit in figure 12 satisfies the basic circuit requirement, there is still a potential problem. If the air induction blower fails for some reason other than the overload contact opening, the high-pressure pump motor will continue to inject oil into the combustion chamber. To prevent this situation, an airflow switch, FL1, is added to the circuit as shown in figure 13. This flow switch is mounted in such a position that it can sense the movement of air produced by the air induction blower.

When thermostat contact TS1 closes, coil TR

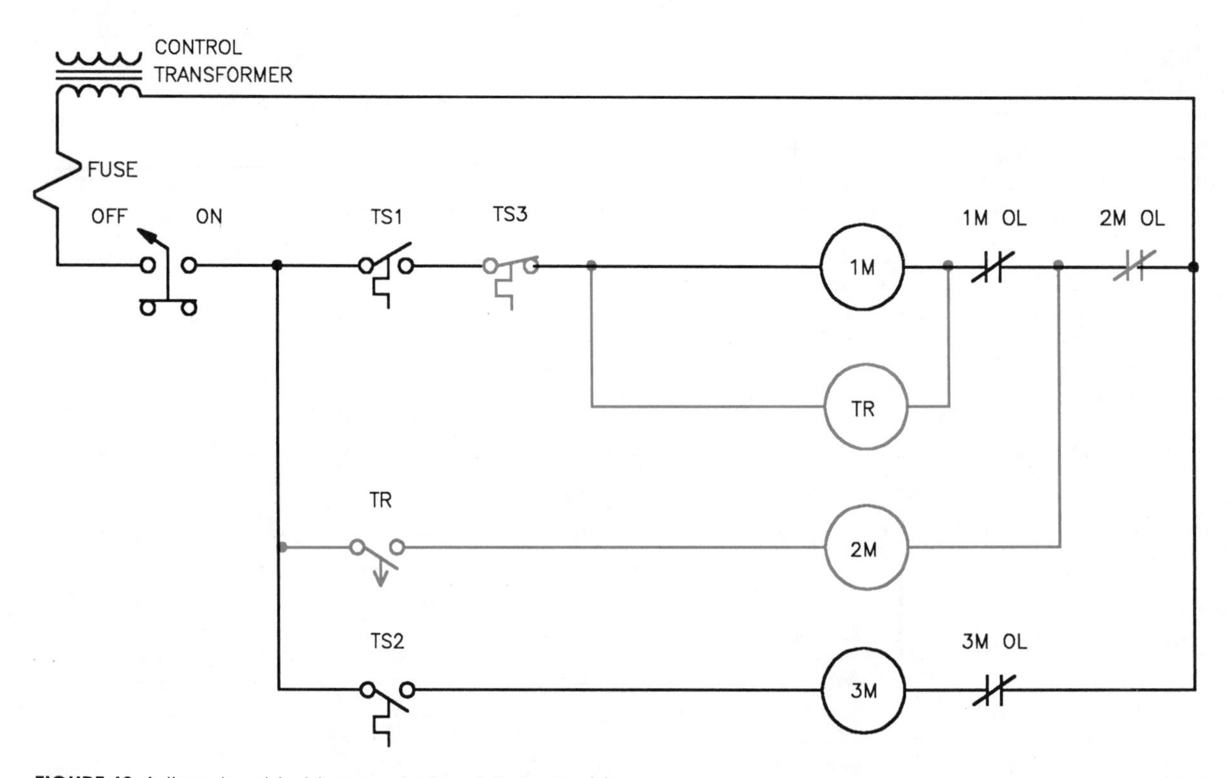

FIGURE 12 A timer is added to operate the air induction blower.

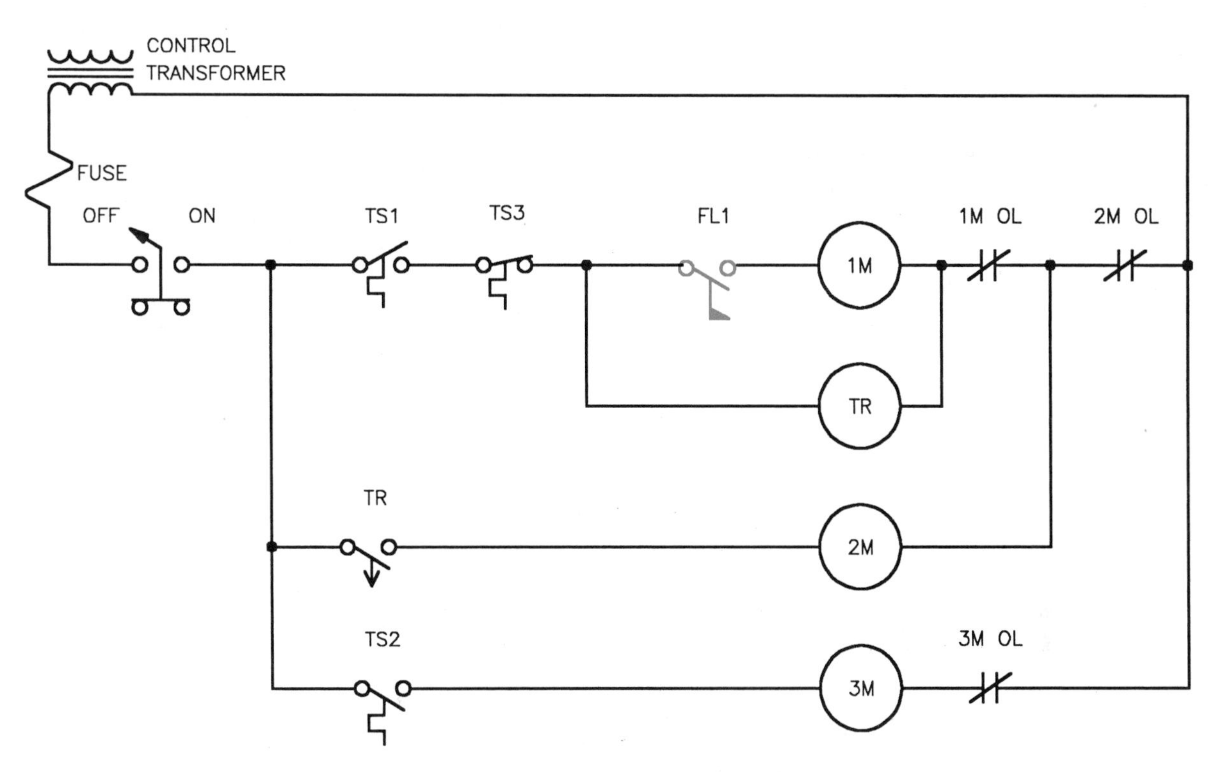

FIGURE 13 An air flow switch controls operation of the high-pressure burner motor.

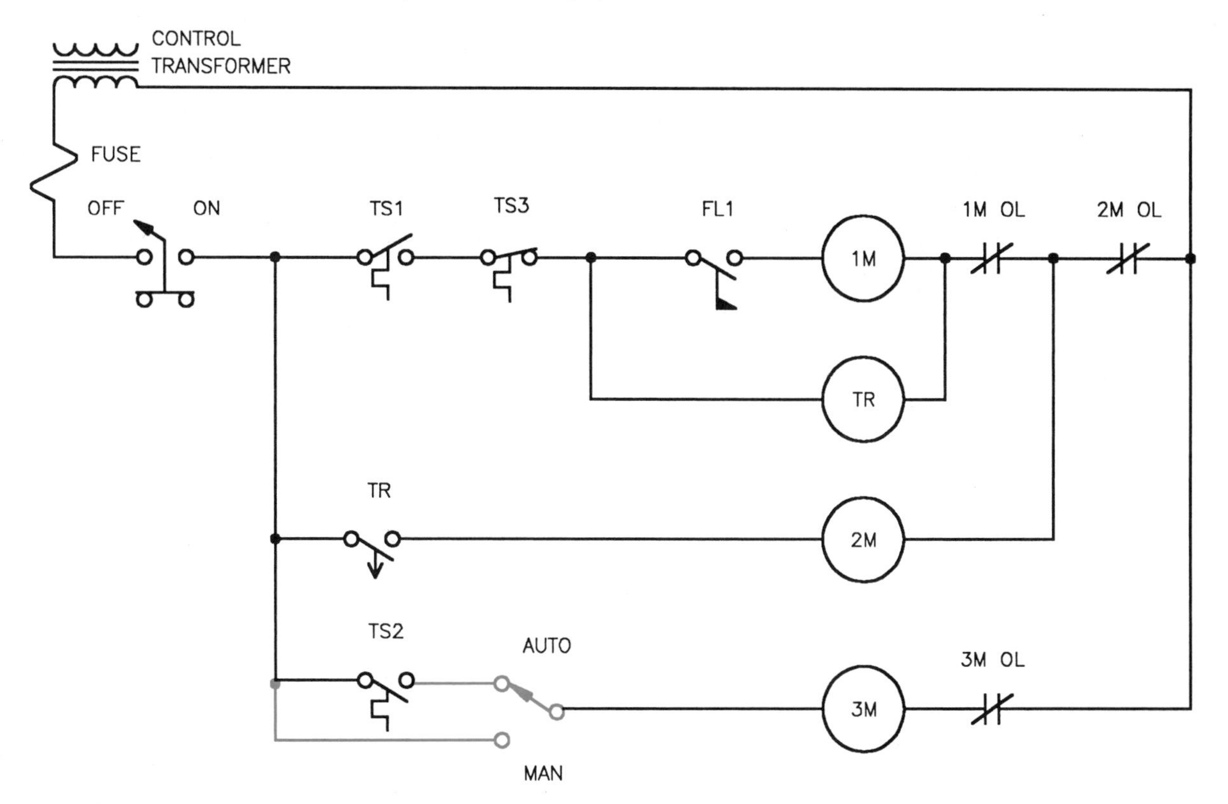

FIGURE 14 An AUTO-MANUAL switch is added to the blower motor.

energizes and closes contact TR. This provides a circuit to motor starter 2M. When the air injection blower starts, flow switch FL1 closes its contact and permits the high-pressure pump motor to start. If the air injection blower motor stops for any reason, flow switch FL1 will disconnect motor starter 1M from the power line and stop operation of the high-pressure pump.

Although the circuit now operates as desired, the owner of the building later decides the blower should circulate air inside the building when the heating system is not in use. To satisfy this request, an AUTO-MANUAL switch is added as shown in figure 14. When the switch is set in the AUTO position, it permits the blower motor to be controlled by the thermostat TS2. When the switch is set in the MANUAL position, it connects the coil of starter 3M directly to the power line and permits the blower motor to operate independently of the heating system.

GLOSSARY

Accelerating Relay Any type of relay used to aid in starting a motor or to accelerate a motor from one speed to another. Accelerating relays may function by: motor armature current (current limit acceleration); armature voltage (counter emf acceleration); or definite time (definite time acceleration).

Accessory (control use) A device that controls the operation of magnetic motor control. (Also see Master Switch, Pilot Device, and Push Button.)

Across-the-line Method of motor starting which connects the motor directly to the supply line on starting or running. (Also called Full Voltage Control.)

Alternating Current (Ac) Current changing both in magnitude and direction; most commonly used current.

Alternator A machine used to generate alternating current by rotating conductors through a magnetic field.

Ambient Temperature The temperature surrounding a device.

Ampacity The maximum current rating of a wire or cable.

Ampere Unit of electrical current.

Amplifier A device used to increase a signal.

Amplitude The highest value reached by a signal, voltage, or current.

AND Gate A digital logic gate that must have all of its inputs high to produce an output.

Anode The positive terminal of an electronic device.

Applied Voltage The amount of voltage connected to a circuit or device.

ASA American Standards Association.

Astable Mode The state in which an oscillator can continually turn itself on and off, or continually change from positive to negative output.

Atom The smallest part of an element that contains all the properties of that element.

Attenuator A device that decreases the amount of signal, voltage, or current.

Automatic Self-acting, operating by its own mechanism when actuated by some triggering signal such as a change in current strength, pressure, temperature, or mechanical configuration.

Automatic Starter A self-acting starter which is completely controlled by master or pilot switches or other sensing devices; designed to control automatically the acceleration of a motor during the acceleration period.

Auxiliary Contacts Contacts of a switching device in addition to the main circuit contacts; auxiliary contacts operate with the movement of the main contacts.

Barrier Charge The potential developed across a semiconductor junction.

Base The semiconductor region between the collector and emitter of a transistor. The base controls the current flow through the collector-emitter circuit.

Base Current The amount of current that flows through the base-emitter section of a transistor.

Bias A dc voltage applied to the base of a transistor to preset its operating point.

Bimetal Strip A strip made by bonding two unlike metals together that, when heated, expand at different rates. This causes a bending or warping action.

Blowout Coil Electromagnetic coil used in contactors and starters to deflect an arc when a circuit is interrupted.

Bounceless Switch A circuit used to eliminate contact bounce in mechanical contacts.

Branch Circuit That portion of a wiring system that extends beyond the final overcurrent device protecting the circuit.

Brake An electromechanical friction device to stop and hold a load. Generally electric release spring applied—coupled to motor shaft.

Breakdown Torque (of a motor) The maximum torque that will develop with the rated voltage applied at the rated frequency, without an abrupt drop in speed. (ASA)

Bridge Circuit A circuit that consists of four sections connected in series to form a closed loop.

Bridge Rectifier A device constructed with four diodes, which converts both positive and negative cycles of ac voltage into dc voltage.

Busway A system of enclosed power transmission that is current and voltage rated.

Cad Cell A device that changes its resistance with a change of light intensity.

Capacitance The electrical size of a capacitor.

Capacitive Any circuit or device having characteristics similar to those of a capacitor.

Capacitor A device made with two conductive plates separated by an insulator or dielectric.

A single-phase induction motor with a main winding arranged for direct connection to the power source and an auxiliary winding connected in series with a capacitor. The capacitor phase is in the circuit only during starting. (NEMA)

Cathode The negative terminal of a device.

Cathode-Ray Tube (CRT) An electron beam tube in which the beam of electrons can be focused to any point on the face of the tube. The electron beam causes the face of the tube to produce light when it is struck by the beam.

Center-Tapped A transformer that has a wire connected to the electrical midpoint of its winding. Generally the secondary is tapped.

Charge Time The amount of time necessary to charge a capacitor.

Choke An inductor designed to present an impedance to ac current, or to be used as the current filter of a dc power supply.

Circuit Breaker Automatic device that opens under abnormal current in carrying circuit; circuit breaker is not damaged on current interruption; device is ampere, volt, and horsepower rated.

Clock Timer A time-delay device that uses an electric clock to measure the delay period.

Collapse (of a magnetic field) When a magnetic field suddenly changes from its maximum value to a zero value.

Collector The semiconductor region of a transistor which must be connected to the same polarity as the base.

Comparator A device or circuit that compares two like quantities such as voltage levels.

Conduction Level The point at which an amount of voltage or current will cause a device to conduct.

Conductor A device or material that permits current to flow through it easily.

Contact A conducting part of a relay which acts with another conducting part to complete or to interrupt a circuit.

Contactor A device that repeatedly establishes or interrupts an electric power circuit.

Continuity A complete path for current flow.

Controller A device or group of devices that governs, in a predetermined manner, the delivery of electric power to apparatus connected to it.

Controller Function Regulate, accelerate, decelerate, start, stop, reverse, or protect devices connected to an electric controller.

Controller Service Specific application of controller. General Purpose: standard or usual service. Definite Purpose: service condition for specific application other than usual.

Current The rate of flow of electrons. Measured in amperes.

Current Flow The flow of electrons.

Current Rating The amount of current flow a device is designed to withstand.

Current Relay A relay that functions at a predetermined value of current. A current relay may be either an overcurrent relay or an undercurrent relay.

Dashpot Consists of a piston moving inside a cylinder filled with air, oil, mercury, silicon, or other fluid. Time delay is caused by allowing the air or fluid to escape through a small orifice in the piston. Moving contacts actuated by the piston close the electrical circuit.

Definite Time (or Time Limit) Definite time is a qualifying term indicating that a delay in action is purposely introduced. This delay remains substantially constant regardless of the magnitude of the quantity that causes the action.

Definite-Purpose Motor Any motor designed, listed, and offered in standard ratings with standard operating characteristics or mechanical construction for use under service conditions other than usual or for use on a particular type of application. (NEMA)

Delta Connection A circuit formed by connecting three electrical devices in series to form a closed loop. Most often used in three-phase connections.

Device A unit of an electrical system that is intended to carry but not utilize electrical energy.

Diac A bidirectional diode.

Dielectric An electrical insulator.

Digital Device A device that has only two states of operation.

Digital Logic Circuit elements connected in such a manner as to solve problems using components that have only two states of operation.

Digital Voltmeter A voltmeter that uses a direct-reading, numerical display as opposed to a meter movement.

Diode A two-element device that permits current to flow through it in only one direction.

Direct Current (Dc) Current that does not reverse its direction of flow. A continuous nonvarying current in one direction.

Disconnecting Means (Disconnect) A device, or group of devices, or other means whereby the conductors of a circuit can be disconnected from their source of supply.

Drum Controller Electrical contacts made on the surface of a rotating cylinder or section; contacts made also by operation of a rotating cam.

Drum Switch A switch having electrical connecting parts in the form of fingers held by spring pressure against contact segments or surfaces on the periphery of a rotating cylinder or sector.

Duty Specific controller functions. Continuous (time) Duty: constant load, indefinite long time period. Short Time Duty: constant load, short or specified time period. Intermittent Duty: varying load, alternate intervals, specified time periods. Periodic Duty: intermittent duty with recurring load conditions. Varying duty: varying loads, varying time intervals, wide variations.

Dynamic Braking Using a dc motor as a generator, taking it off the line and applying an energy dissipating resistor to the armature. Dynamic braking for an ac motor is accomplished by disconnecting the motor from the line and connecting dc power to the stator windings.

Eddy Currents Circular induced currents contrary to the main currents; a loss of energy that shows up in the form of heat.

Electrical Interlocking Accomplished by control circuits in which the contacts in one circuit control another circuit.

Electric Controller A device, or group of devices, which governs, in some predetermined manner, the electric power delivered to the apparatus to which it is connected.

Electron One of the three major subatomic parts of an atom. The electron carries a negative charge.

Electronic Control Control system using gas and/or vacuum tubes, or solid-state devices.

Emitter The semiconductor region of a transistor which must be connected to a polarity different than the base.

Enclosure Mechanical, electrical, and environmental protection for control devices.

Eutectic Alloy Metal with low and sharp melting point; used in thermal overload relays; converts from a solid to a liquid state at a specific temperature; commonly called solder pot.

Exclusive OR Gate A digital logic gate that will produce an output when its inputs have opposite states of logic level.

Feeder The circuit conductor between the service equipment, or the generator switchboard of an isolated plant and the branch circuit overcurrent device.

Feeler Gauge A precision instrument with blades in thicknesses of thousandths of an inch for measuring clearances.

Filter A device used to remove the ripple produced by a rectifier.

Frequency Number of complete variations made by an alternating current per second; expressed in Hertz. (See Hertz)

Full Load Torque (of a motor) The torque necessary to produce the rated horsepower of a motor at full load speed.

Full Voltage Control *(Across-the-line)* Connects equipment directly to the line supply on starting.

Fuse An overcurrent protective device with a fusible member, which is heated directly and destroyed by the current passing through it to open a circuit.

Gain The increase in signal power produced by an amplifier.

Gate A device that has multiple inputs and a single output; or one terminal of some solid-state devices such as SCRs or triacs.

General-Purpose Motor Any open motor that has a continuous 40C rating and is designed, listed, and offered in standard ratings with standard operating characteristics and mechanical construction for use under usual service conditions without restrictions to a particular application or type of application. (NEMA)

Heat Sink A metallic device designed to increase the surface area of an electronic component to remove heat at a faster rate.

Hertz International unit of frequency, equal to one cycle per second of alternating current.

High Voltage Control Formerly, all control above 600 volts. Now, all control above 5,000 volts. See Medium Voltage for 600- to 5,000-volt equipment.

Holding Contacts Contacts used for the purpose of maintaining current flow to the coil of a relay.

Holding Current The amount of current needed to keep an SCR or a triac turned on.

Horsepower Measure of the time rate of doing work (working rate).

Hysteresis Loop A graphic curve that shows the value of magnetizing force for a particular type of material.

Impedance The total opposition to current flow in an electrical circuit.

Induced Current produced in a conductor by the cutting action of a magnetic field.

Inductor A coil used to introduce inductance into an electrical circuit.

Input Power delivered to an electrical device.

Input Voltage The amount of voltage connected to a device or circuit.

Instantaneous A qualifying term indicating that no delay is purposely introduced in the action of a device.

Insulator A material used to electrically isolate two conductive surfaces.

Integral Whole or complete; not fractional.

Interlock To interrelate with other controllers; an auxiliary contact. A device is connected in such a way that the motion of one part is held back by another part.

Internal Relay Digital logic circuits in a programmable controller that can be programmed to operate in the same manner as control relays.

Inverse Time A qualifying term indicating that a delayed action is introduced purposely. This delay decreases as the operating force increases.

Inverter (Gate) A digital logic gate that has an output opposite its input.

Isolation Transformer A transformer whose secondary winding is electrically isolated from its primary winding.

Jogging (Inching) Momentary operations; the quickly repeated closure of the circuit to start a motor from rest for the purpose of accomplishing small movements of the driven machine.

Jumper A short length of conductor used to make a connection between terminals or around a break in a circuit.

Junction Diode A diode that is made by joining two pieces of semiconductor material.

Kick-Back Diode A diode used to eliminate the voltage spike induced in a coil by the collapse of a magnetic field.

Lattice Structure An orderly arrangement of atoms in a crystalline material.

Led (Light-Emitting Diode) A diode that will produce light when current flows through it.

Limit Switch A mechanically operated device which stops a motor from revolving or reverses it when certain limits have been reached.

Load Center Service entrance; controls distribution; provides protection of power; generally of the circuit breaker type.

Local Control Control function, initiation, or change accomplished at the same location as the electric controller.

Locked Rotor Current (of a motor) The steady-state current taken from the line with the rotor locked (stopped) and with the rated voltage and frequency applied to the motor.

Locked Rotor Torque (of a motor) The minimum torque that a motor will develop at rest for all angular positions of the rotor with the rated voltage applied at a rated frequency. (ASA)

Lockout A mechanical device that may be set to prevent the operation of a push button.

Logic A means of solving complex problems through the repeated use of simple functions which define basic concepts. Three basic logic functions are: and, or, and not.

Low Voltage Protection (LVP) Magnetic control only; nonautomatic restarting; three-wire control; power failure disconnects service; power restored by manual restart.

Low Voltage Release (LVR) Manual and magnetic control; automatic restarting; two-wire control; power failure disconnects service; when power is restored, the controller automatically restarts the motor.

Magnet Brake Friction brake controlled by electromagnetic means.

Magnetic Contactor A contactor that is operated electromechanically.

Magnetic Field The space in which a magnetic force exists.

Magnetic Controller An electric controller; device functions operated by electromagnets.

Maintaining Contact A small control contact used to keep a coil energized; usually actuated by the same coil. Holding contact; Pallet switch.

Manual Controller An electric controller; device functions operated by mechanical means or manually.

Master Switch A main switch to operate contactors, relays, or other remotely-controlled electrical devices.

Medium Voltage Control Formerly known as High Voltage; includes 600- to 5000-volt apparatus; air break or oil-immersed main contactors; high interrupting capacity fuses; 150,000 kVa at 2,300 volts; 250,000 kVA at 4,000–5,000 volts.

Microprocessor A small computer. The central processing unit is generally made from a single integrated circuit.

Mode A state or condition.

Monostable (Mode) The state in which an oscillator or timer will operate through only one sequence of events.

Motor Device for converting electrical energy to mechanical work through rotary motion; rated in horsepower.

Motor Circuit Switch Motor branch circuit switch rated in horsepower; capable of interrupting overload motor current.

Motor Controller A device used to control the operation of a motor.

Motor-Driven Timer A device in which a small pilot motor causes contacts to close after a predetermined time.

Multispeed Motor A motor that can be operated at more than one speed.

Multispeed Starter An electric controller with two or more speeds; reversing or nonreversing; full or reduced voltage starting.

NAND Gate A digital logic gate that will produce a high output only when all of its inputs are in a low state.

Negative One polarity of voltage, current, or a charge.

Negative Resistance The property of a device in which an increase of current flow causes an increase of conductance. The increase of conductance causes a decrease in the voltage drop across the device.

NEMA National Electrical Manufacturers Association.

NEMA Size Electric controller device rating; specific standards for horsepower, voltage, current, and interrupting characteristics.

Neutron One of the principal parts of an atom. The neutron has no charge and is part of the nucleus.

Nonautomatic Controller Requires direct operation to perform function; not necessarily a manual controller.

Noninductive Load An electrical load that does not have induced voltages caused by a coil. Noninductive loads are generally resistive, but can be capacitive.

Nonreversing Operation in one direction only.

NOR Gate A digital logic gate that will produce a high output when any of its inputs are low.

Normally Open and Normally Closed When applied to a magnetically-operated switching device, such as a contactor or relay, or to the contacts of these devices, these terms signify the position taken when the operating magnet is de-energized. The terms apply only to nonlatching types of devices.

Off-Delay Timers A timer in which the contacts change position immediately when the coil or circuit is energized, but delay returning to their normal positions when the coil or circuit is de-energized.

Ohmmeter A meter used to measure resistance.

On-Delay Timer A timer in which the contacts delay changing position when the coil or circuit is energized, but change back immediately to their normal positions when the coil or circuit is de-energized.

Operational Amplifier (OP-AMP) An integrated circuit used as an amplifier.

Optoisolator A device used to connect sections of a circuit by means of a light beam.

Oscillator A device or circuit used to change dc voltage into ac voltage.

Oscilloscope An instrument that measures the amplitude of voltage with respect to time.

Out-of-phase Voltage A voltage that is not in phase when compared to some other voltage or current.

Output Devices Elements such as solenoids, motor starters, and contactors that receive input.

Output Pulse A short duration voltage or current which can be negative or positive, produced at the output of a device or circuit.

Overload Protection Overload protection is the result of a device that operates on excessive current, but not necessarily on short circuit, to cause and maintain the interruption of current flow to the device governed. NOTE: Operating overload means a current that is not in excess of six times the rated current for alternating-current motors, and not in excess of four times the rated current for direct-current motors.

Overload Relay Running overcurrent protection; operates on excessive current; not necessarily protection for short circuit; causes and maintains interruption of device from power supply. Overload Relay Heater Coil: Coil used in thermal overload relays; provides heat to melt eutectic alloy. Overload Relay Reset: Push button used to reset thermal overload relay after relay has operated.

Panelboard Panel, group of panels, or units; an assembly that mounts in a single panel; includes buses, with or without switches and/or automatic overcurrent protective devices; provides control of light, heat, power circuits; placed in or against wall or partition; accessible from front only.

Parallel Circuit A circuit that has more than one path for current flow.

Peak Inverse/Peak Reverse Voltage The rating of a semiconductor device which indicates the maximum amount of voltage that can be applied to the device in the reverse direction.

Peak-To-Peak Voltage The amplitude of voltage measured from the negative peak of an ac waveform to the positive peak.

Peak Voltage The amount of voltage of a waveform measured from the zero voltage point to the positive or negative peak.

Permanent-split Capacitor Motor A single-phase induction motor similar to the capacitor start motor except that it uses the same capacitance which remains in the circuit for both starting and running. (NEMA)

Permeability The ease with which a material will conduct magnetic lines of force.

Phase Relation of current to voltage at a particular time in an ac circuit. Single Phase: A single voltage and current in the supply. Three Phase: Three electrically-related (120-degree electrical separation) single-phase supplies.

Phase-Failure Protection Phase-failure protection is provided by a device that operates when the power fails in one wire of a polyphase circuit to cause and maintain the interruption of power in all the wires of the circuit.

Phase-Reversal Protection Phase-reversal protection is provided by a device that operates when the phase rotation in a polyphase circuit reverses to cause and maintain the interruption of power in all the wires of the circuit.

Phase Rotation Relay A relay that functions in accordance with the direction of phase rotation.

Phase Shift A change in the phase relationship between two quantities of voltage or current.

Photodetector A device that responds to change in light intensity.

Photodiode A diode that conducts in the presence of light, but not in darkness.

Pilot Device Directs operation of another device. Float Switch: A pilot device that responds to liquid levels. Foot Switch: A pilot device operated by the foot of an operator. Limit Switch: A pilot device operated by the motion of a power-driven machine; alters the electrical circuit with the machine or equipment.

Plugging Braking by reversing the line voltage or phase sequence; motor develops retarding force.

Pneumatic Timer A device that uses the displacement of air in a bellows or diaphragm to produce a time delay.

Polarity The characteristic of a device that exhibits opposite quantities, such as positive and negative, within itself.

Pole The north or south magnetic end of a magnet; a terminal of a switch; one set of contacts for one circuit of main power.

Potentiometer A variable resistor with a sliding contact, which is used as a voltage divider.

Power Factor A comparison of the true power (WATTS) to the apparent power (VOLT AMPS) in an ac circuit.

Power Rating The rating of a device that indicates the amount of current flow and voltage drop that can be permitted.

Pressure Switch A device that senses the presence or absence of pressure and causes a set of contacts to open or close.

Printed Circuit A board on which a predetermined pattern of printed connections has been formed.

Proton One of the three major parts of an atom. The proton carries a positive charge.

Pull-up Torque (of alternating-current motor) The minimum torque developed by the motor during the period of acceleration from rest to the speed at which breakdown occurs. (ASA)

Push Button A master switch; manually-operable plunger or button for an actuating device; assembled into push-button stations.

RC Time Constant The time constant of a resistor and capacitor connected in series. The time in seconds is equal to the resistance in ohms multiplied by the capacitance in farads.

Reactance The opposition to current flow in an ac circuit offered by pure inductance or pure capacitance.

Rectifier A device that converts alternating current into direct current.

Regulator A device that maintains a quantity at a predetermined level.

Relay Operated by a change in one electrical circuit to control a device in the same circuit or another circuit; rated in amperes; used in control circuits.

Remote Control Controls the function initiation or change of an electrical device from some remote point, or location.

Remote Control Circuit Any electrical circuit that controls any other circuit through a relay or an equivalent device.

Residual Magnetism The retained or small amount of remaining magnetism in the magnetic material of an electromagnet after the current flow has stopped.

Resistance The opposition offered by a substance or body to the passage through it of an electric current; resistance converts electrical energy into heat; resistance is the reciprocal of conductance.

Resistance Start Induction Run Motor One type of split-phase motor that uses the resistance of the start winding to produce a phase shift between the current in the start winding and the current in the run winding.

Resistor A device used primarily because it possesses the property of electrical resistance. A resistor is used in electrical circuits for purposes of operation, protection, or control; commonly consists of an aggregation of units.

- *Starting Resistors* Used to accelerate a motor from rest to its normal running speed without damage to the motor and connected load from excessive currents and torques, or without drawing undesirable inrush current from the power system.
- *Armature Regulating Resistors* Used to regulate the speed of torque of a loaded motor by resistance in the armature or power circuit.
- *Dynamic Braking Resistors* Used to control the current and dissipate the energy when a motor is decelerated by making it act as a generator

to convert its mechanical energy to electrical energy and then to heat in the resistor.

- *Field Discharge Resistors* Used to limit the value of voltage that appears at the terminals of a motor field (or any highly inductive circuit) when the circuit is opened.
- *Plugging Resistors* Used to control the current and torque of a motor when deceleration is forced by electrically reversing the motor while it is still running in the forward direction.

Rheostat A resistor that can be adjusted to vary its resistance without opening the circuit in which it may be connected.

Ripple An ac component in the output of a dc power supply caused by improper filtering.

RMS Value The value of ac voltage that will produce as much power when connected across a resistor as a like amount of dc voltage.

Safety Switch Enclosed manually-operated disconnecting switch; horsepower and current rated; disconnects all power lines.

Saturation The maximum amount of magnetic flux a material can hold.

Schematic An electrical diagram which shows components in their electrical sequence without regard for physical location.

Selector Switch A master switch that is manually operated; rotating motion for actuating device; assembled into push-button master stations.

Semiautomatic Starter Part of the operation of this type of starter is nonautomatic while selected portions are automatically controlled.

Semiconductor A material that contains four valence electrons and is used in the production of solid-state devices. The most common types are silicon and germanium.

Semimagnetic Control An electric controller in which functions are partly controlled by electromagnets.

Sensing Device A pilot device that measures, compares, or recognizes a change or variation in the system which it is monitoring; provides a controlled signal to operate or control other devices.

Series-Aiding Two or more voltage producing devices connected in series in such a manner that their voltages add to produce a higher total voltage.

Series Circuit An electric circuit formed by the connection of one or more components in such a manner that there is only one path for current flow.

Service The conductors and equipment necessary to deliver energy from the electrical supply system to the premises served.

Service Equipment Necessary equipment, circuit breakers, or switches and fuses with accessories mounted near the entry of the electrical supply; constitutes the main control or cutoff for supply.

Service Factor (of a general-purpose motor) An allowable overload; the amount of allowable overload is indicated by a multiplier which, when applied to a normal horsepower rating, indicates the permissible loading.

Shaded-Pole Motor A single-phase induction motor provided with an auxiliary short-circuited winding or windings displaced in magnetic position from the main winding. (NEMA)

Shading Loop A large copper wire or band connected around part of a magnetic pole piece to oppose a change of magnetic flux.

Short Circuit An electrical circuit that contains no resistance to limit the flow of current.

Signal The event, phenomenon, or electrical quantity that conveys information from one point to another.

Signal Generator A text instrument used to produce a low-value, ac voltage for the purpose of testing or calibrating electronic equipment.

Silicon-Controlled Rectifier (SCR) A four-layer semiconductor device that is a rectifier and must be triggered by a pulse applied to the gate before it will conduct.

Sine-Wave Voltage A voltage waveform; its value at any point is proportional to the trigonometric sine of the angle of the generator producing it.

Slip Difference between the rotor rpm and the rotating magnetic field of an ac motor.

Snap Action The quick opening and closing action of a spring-loaded contact.

Solenoid A magnetic device used to convert electrical energy into linear motion. A tubular, current-carrying coil that provides magnetic action to perform various work functions.

Solenoid-and-Plunger A solenoid-and-plunger is a solenoid provided with a bar of soft iron or steel called a plunger.

Solenoid Valve A valve operated by an electric solenoid.

Solder Pot See Eutectic Alloy.

Solid-State Devices Electronic components that control electron flow through solid materials such as crystals; e.g., transistors, diodes, integrated circuits.

Special-Purpose Motor A motor with special operating characteristics or special mechanical construction, or both, designed for a particular application and not falling within the definition of a general-purpose or definite-purpose motor. (NEMA)

Split-Phase A single-phase induction motor with auxiliary winding, displaced in magnetic position from, and connected parallel to, the main winding. (NEMA)

Starter A starter is a controller designed for accelerating a motor to normal speed in one direction of rotation. NOTE: A device designed for starting a motor in either direction of rotation includes the additional function of reversing and should be designated as a controller.

Startup The time between equipment installation and the full operation of the system.

Static Control Control system in which solid-state devices perform the functions. Refers to no moving parts or without motion.

Stealer Transistor A transistor used in such a manner as to force some other component to remain in the off state by shunting its current to electrical ground.

Step-Down Transformer A transformer that produces a lower voltage at its secondary winding than is applied to its primary winding.

Step-Up Transformer A transformer that produces a higher voltage at its secondary winding than is applied to its primary winding.

Surge A transient variation in the current and/or potential at a point in the circuit; unwanted, temporary.

Switch A device for making, breaking, or changing the connections in an electric circuit.

Switchboard A large, single panel with a frame or assembly of panels; devices may be mounted on the face of the panels, on the back, or both; contains switches, overcurrent, or protective devices; instruments accessible from the rear and front; not installed in wall-type cabinets. (See Panelboard)

Synchronous Speed The speed of the rotating magnetic field of an ac induction motor.

Tachometer Generator Used for counting revolutions per minute. Electrical magnitude or impulses are calibrated with a dial-gauge reading in rpm.

Temperature Relay A relay that functions at a predetermined temperature in the apparatus protected. This relay is intended to protect some other apparatus such as a motor or controller and does not necessarily protect itself.

Terminal A fitting attached to a circuit or device for convenience in making electrical connections.

Thermal Compound A grease-like substance used to thermally bond two surfaces together for the purpose of increasing the rate of heat transfer from one object to another.

Thermal Protector (as applied to motors) An inherent overheating protective device that is responsive to motor current and temperature. When properly applied to a motor, this device protects the motor against dangerous overheating due to overload or failure to start.

Thermistor A resistor that changes its resistance with a change of temperature.

Thyristor An electronic component that has only two states of operation, on and off.

Time Limit See Definite Time.

Timer A pilot device that is also considered a timing relay; provides adjustable time period to perform function; motor driven; solenoid-actuated; electronic.

Torque The torque of a motor is the twisting or turning force which tends to produce rotation.

Transducer A device that transforms power from one system to power of a second system: for example, heat to electrical.

Transformer An electromagnetic device that converts voltages for use in power transmission and operation of control devices.

Transient See Surge.

Transistor A solid-state device made by combining three layers of semiconductor material. A small amount of current flow through the base-emitter can control a larger amount of current flow through the collector-emitter.

Triac A bidirectional, thyristor device used to control ac voltage.

Trigger Pulses A voltage or current of short duration used to activate the gate, base, or input of some electronic device.

Trip Free Refers to a circuit breaker that cannot be held in the on position by the handle on a sustained overload.

Troubleshoot To locate and eliminate the source of trouble in any flow of work.

Truth Table A chart used to show the output condition of a logic gate or circuit as compared to different conditions of input.

Undervoltage Protection The result when a device operates on the reduction or failure of voltage to cause and maintain the interruption of power to the main circuit.

Undervoltage Release Occurs when a device operates on the reduction or failure of voltage to cause the interruption of power to the main circuit, but does not prevent the reestablishment of the main circuit on the return of voltage.

Unijunction Transistor (UJT) A special transistor that is a member of the thyristor family of devices and operates like a voltage-controlled switch.

Valence Electron The electron in the outermost shell or orbit of an atom.

Variable Resistor A resistor in which the resistance value can be adjusted between the limits of its minimum and maximum value.

Varistor A resistor that changes its resistance value with a change of voltage.

Volt/Voltage An electrical measure of potential difference, electromotive force, or electrical pressure.

Voltage Divider A series connection of resistors used to produce different values of voltage drop across them.

Voltage Drop The amount of voltage required to cause an amount of current to flow through a certain value of resistance or reactance.

Voltage Rating A rating that indicates the amount of voltage that can safely be connected to a device.

Voltage Regulator A device or circuit that maintains a constant value of voltage.

Voltage Relay A relay that functions at a predetermined value of voltage. A voltage relay may be either an overvoltage or an undervoltage relay.

Voltmeter An instrument used to measure a level of voltage.

Volt-Ohm-Milliammeter (VOM) A test instrument so designed that it can be used to measure voltage, resistance, or milliamps.

Watt A measure of true power.

Waveform The shape of a wave as obtained by plotting a graph with respect to voltage and time.

Wye Connection A connection of three components made in such a manner that one end of each component is connected. This connection generally connects devices to a three-phase power system.

Zener Diode A special diode that exhibits a constant voltage drop when connected in such a manner that current flows through it in the reverse direction.

Zener Region The region current enters into when it flows through a diode in the reverse direction.

Zero Switching A feature of some solid-state relays that causes current to continue flowing through the device until the ac waveform returns to zero.

INDEX